Green Energy and Technology

Climate change, environmental impact and the limited natural resources urge scientific research and novel technical solutions. The monograph series Green Energy and Technology serves as a publishing platform for scientific and technological approaches to "green"—i.e. environmentally friendly and sustainable—technologies. While a focus lies on energy and power supply, it also covers "green" solutions in industrial engineering and engineering design. Green Energy and Technology addresses researchers, advanced students, technical consultants as well as decision makers in industries and politics. Hence, the level of presentation spans from instructional to highly technical.

Indexed in Scopus.

Indexed in Ei Compendex.

Guangxue Wu
Editor

Anaerobic Digestion

Fundamentals, Modelling, and Applications

Springer

Editor
Guangxue Wu
Civil Engineering, School of Engineering
University of Galway
Galway, Ireland

ISSN 1865-3529 ISSN 1865-3537 (electronic)
Green Energy and Technology
ISBN 978-3-031-69377-9 ISBN 978-3-031-69378-6 (eBook)
https://doi.org/10.1007/978-3-031-69378-6

This Springer imprint is published by the registered company Springer Nature Switzerland AG
The registered company address is: Gewerbestrasse 11, 6330 Cham, Switzerland

Contents

Anaerobic Digestion and Its Understanding by Introducing the Design-Build-Test-Learn Cycle

Guangxue Wu

Abstract Anaerobic digestion is crucial for tackling environmental pollution and energy crisis. Its development is mainly affected by societal requirements, relevant regulations, and technological innovation. Through anaerobic digestion, the products of biogas, water, and digestates can be recovered as resources for further applications. The design-build-test-learn cycle can be adopted for better understanding and application of anaerobic digestion. Especially, the establishment of a comprehensive platform with pertinent information can significantly promote the widespread adoption of anaerobic digestion.

Keywords Anaerobic digestion · Bibliometric analysis · Design-build-test-learn · Material flow

1 Introduction

A cleaner environment is now a global societal demand. This requires cleaner sources of energy and more efficient waste management. According to the International Energy Agency, the total world primary energy supply increased by approximately 2.6 times between 1971 and 2018 (IEA 2018a). To achieve the Sustainable Development Goals (SDGs), it is necessary to increase the share of renewable energy from one quarter in 2018 to two-thirds in 2040 (IEA 2018b). In parallel, it will be necessary, by 2030, to halve the 80% of wastewater that currently goes into waterways without adequate treatment (UNDP 2015). Anaerobic digestion is a powerful sustainable mechanism to address these challenges. It has the potential to simultaneously generate renewable bioenergy and reduce waste pollution, thereby mitigating greenhouse gas emissions. Nowadays, 50 million microdigesters and 132,000 full-scale digesters are operating globally for waste management and energy recovery (Feedback 2020), and the global energy generation potential from major feedstocks

G. Wu (✉)
Civil Engineering, School of Engineering, College of Science and Engineering, University of Galway, Galway H91 TK33, Ireland
e-mail: guangxue.wu@universityofgalway.ie

G. Wu (ed.), *Anaerobic Digestion*, Green Energy and Technology,
https://doi.org/10.1007/978-3-031-69378-6_1

(including liquid wastewater and solid waste) by anaerobic digestion can meet 23–32% of the world's coal consumption or 16–22% of the electricity consumed (WBA 2019). The International Energy Agency reports (IEA 2020) showed that for maintaining the global temperature rise below 2 °C, the natural gas equivalent of global biogas will be around 370 billion m^3 by 2040.

During anaerobic digestion, organic materials are broken down to produce biogas in the absence of oxygen, through syntrophic activities of various microorganisms from two entirely different kingdoms, i.e. the bacteria and the archaea, in different interactive processes, i.e. hydrolysis, acidification, acetogenesis, and methanogenesis (McCarty 2001). First, complex organic wastes, such as proteins, lipids, and carbohydrates, are hydrolysed into smaller molecular organic substrates of amino acids, saturated fatty acids, and monosaccharides, respectively, by fermenting bacteria. These are then converted to volatile fatty acids (VFAs). Second, the hydrogen gas (H_2)-producing acetogenic bacteria use VFAs to further produce acetate and H_2. Finally, methanogens catabolize acetate, H_2, and carbon dioxide (CO_2) to produce the end-product of methane (CH_4). Usually, for solid waste disposal, the hydrolysis step is the reaction rate limiting process, while for wastewater treatment, methanogenesis is the reaction rate limiting process (Gijzen 2002). Furthermore, anaerobic digestion is also capable of destroying and detoxifying recalcitrant organic carbons, such as chlorinated hazardous compounds of pesticides, chlorinated solvents, and polychlorinated biphenyls (McCarty 2001).

2 Development of Anaerobic Digestion

Despite as a green technology, the development of anaerobic digestion is driven mainly by the requirement for pollution control and energy security (or energy crisis) (Náthia-Neves et al. 2018). The rapid development periods include the 1970s energy crisis and the transition to renewable energy from the coal-based fuel. Furthermore, the relevant policies from the government, such as the feed-in-tariffs for renewable energy implemented in Germany, the Renewable Source Directive in European Union (EU), and the CO_2-emission abatement for targeting greenhouse gas mitigation, also contribute significantly to its development. For example, the adoption of the Kyoto Protocol has advanced the promotion of green technologies for biofuel production (Rodríguez-Jiménez et al. 2022). In EU, regulations such as the European Green Deal for transforming the EU into a modern, resource-efficient, and competitive economy, and the Renewable Energy Directive for developing clean energy across all sectors of the EU economy, have a significant impact on the application of anaerobic digestion in the diverse sectors of renewable energy, agriculture, and waste management.

The development of novel technologies promotes the application of anaerobic digestion, such as the wide adoption of high-rate up-flow anaerobic sludge blanket technology. For the commercialization of anaerobic digestion, other factors such as the cost of construction and feedstocks (especially for energy crops), the availability and transportation of feedstock, and the market of the produced biogas, etc., need to be considered. Therefore, it is necessary to adopt the systems thinking concept to assess all relevant aspects of anaerobic digestion, which can drive its wide application. Among all of these, the training of professionals and qualified specialists is fundamental to support the business chain of anaerobic digestion.

All the above-mentioned factors drive the interest of researchers working on anaerobic digestion. The bibliometric method can be applied to examine the development trends based on the publication information. It can provide insights into research gaps and future development, proposing valuable perspectives and suggestions for anaerobic digestion. Therefore, the Scopus database (Elsevier) was adopted for the bibliometric analysis of publications relating to anaerobic digestion. The search was conducted in December 2023 with the keywords "anaerobic treatment" or "anaerobic digestion" and "waste" or "wastewater treatment". The distribution of 16,778 results was analysed (Fig. 1), and articles came out as the most predominant document type comprising 78.5% of the total publications. All the original articles were utilized for further analysis as relevant citable items. Reviews were identified to account for almost 9.1%, and the other document types were conference papers (1061), book chapters (794), conference reviews (45), books (43), notes (17), short surveys (11), and so on.

The VOSviewer software (https://www.vosviewer.com) was used to create networks and clusters, and a Bibliometrix R-tool (https://www.bibliometrix.org) was employed to draw thematic map from the previous several years data, with results shown in Fig. 2.

Most of the top 30 keywords are bioenergy and gas production (bioenergy, biogas, biogas production, methane, energy recovery, and renewable energy) (4114, 24.5%), waste streams and feedstocks (digestate, food waste, municipal solid waste, organic

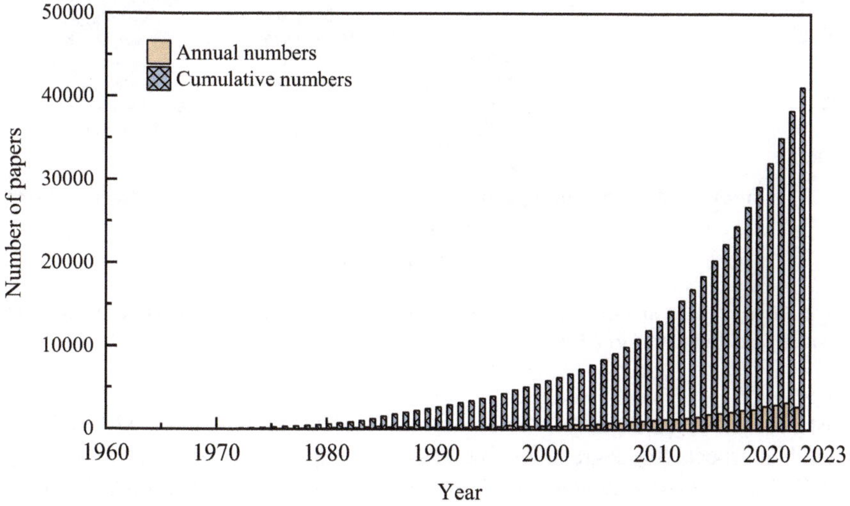

Fig. 1 Publication of papers on anaerobic digestion indexed in Scopus from 1960 to 2023

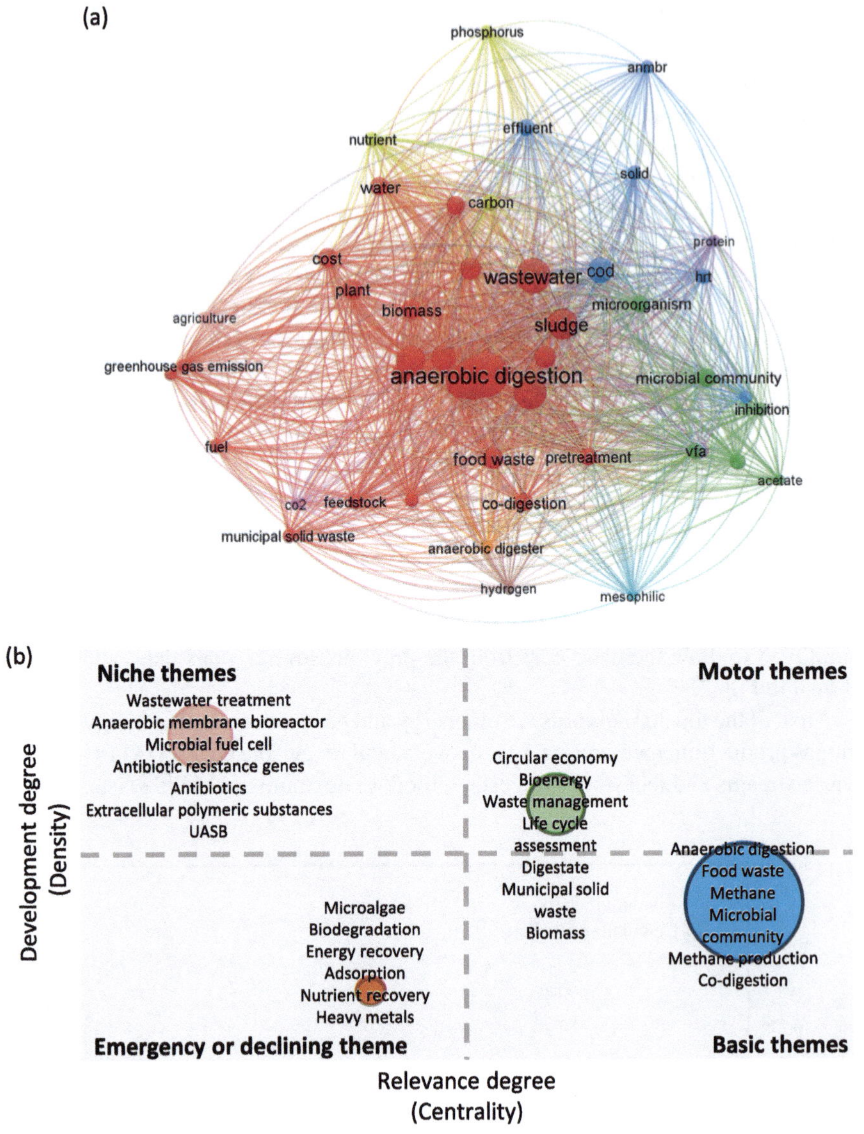

Fig. 2 Term map (**a**) and thematic map (**b**) of the clustered 100 most frequent keywords in the previous years from 2018 to 2023

waste, waste activated sludge, wastewater, and wastewater treatment) (3727, 22.2%), microbial aspects (biodegradation, biomass, microalgae, and microbial community) (1317, 7.8%), environmental impact and assessment (circular economy, life cycle assessment, and waste management) (946, 5.6%), and technological components and intermediates (biochar and volatile fatty acids) (581, 3.5%). This indicates that the research on bioenergy and gas production has attracted intensive attention.

The thematic map includes (i) motor-themes, (ii) basic themes, (iii) emerging or declining themes, and (iv) very specialized/niche themes. For example, from the analysis, the keywords "bioenergy", "circular economy", "digestate", and "life cycle assessment" are considered motor-themes, meaning that they are well-developed and important themes, while "anaerobic digestion", "methane", "methane production", "food waste", and "co-digestion" are considered basic themes, which are fundamental to anaerobic digestion.

3 Material Flow During Anaerobic Digestion

The overall material flow chain of anaerobic digestion is shown in Fig. 3, covering the feedstock, the anaerobic digestion process, and end products of biogas, water, and digestates. Also, the further application of the end products is summarized, demonstrating how they can contribute to the circular bioeconomy through recycling and reuse of all end products.

The feedstock for anaerobic digestion mainly refers to the liquid wastewater and solid waste with high organic contents, and it can also be extended to gaseous feedstock such as H_2 and CO_2 (Nikolausz and Kretzschmar 2020). Wastes such as biowastes, and agricultural and industrial organic wastes can be used as either the mono-feedstock or co-feedstock. Usually, mixed organic waste rather than a single type of organic waste is used in anaerobic digestion. The organic carbon compositions in the feedstock can be regulated to improve system performance, such as adjusting

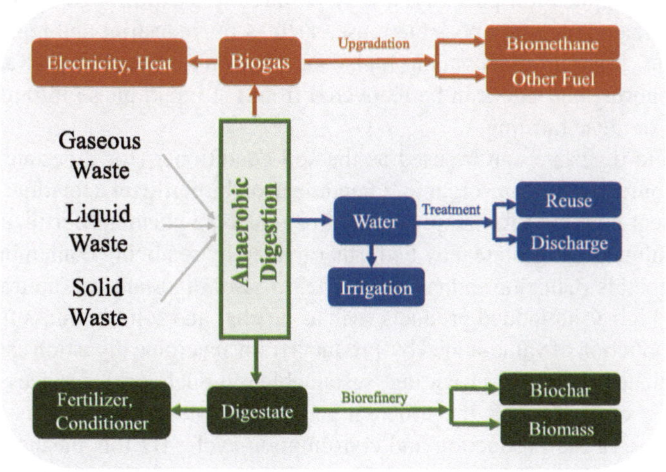

Fig. 3 Anaerobic digestion for the conversion of waste to value-added products, contributing to circular bioeconomy and sustainability

the carbon to nitrogen ratio. According to the Global Intelligence Alliance (GIA 2010), the potential sources of biogas globally are 75% in agricultural crops, by-products, and manure; 17% in municipal and industrial organic waste; and 8% in sewage wastewater treatment facilities.

For solid waste, anaerobic digestion can be classified as the large-scale centralized plant and the farm-scale plant (Vasco-Correa et al. 2018). Centralized plants usually treat waste from several farms or large-scale sources, while farm-scale plants are mainly applied for single large farms or dairy farms. For wastewater treatment, this can also be classified as centralized and decentralized plants, where the decentralized plants are applied for one industry while centralized plants can be applied for industrial parks with wastewater produced from lots of industries.

The value-added products from anaerobic digestion include gaseous green biogas, liquid treated wastewater, and solid digestate. Biogas is a mixture of diverse types of gases, mainly CH_4 and CO_2, accompanying with traces of other gases such as H_2, hydrogen sulphide (H_2S), ammonia, siloxanes, hydrocarbons, and water (Wilkinson 2011). The trace gases produced during anaerobic digestion are depending on the feedstock composition. The biogas can be used for generating electricity and heat in a combined heat and power unit. Sometimes, the produced biogas may be flared without being used due to the low benefit. After upgrading with the removal of CO_2 (such as applying high-pressure reactors) and other nuisance components such as the removal of corrosive H_2S through the micro-aeration process (Jeison 2015), the biogas can be converted to biomethane, which is then injected into natural gas grid for long-distance distribution or compressed and used as a transportation fuel for vehicles. When applying anaerobic digestion for wastewater treatment, the produced biogas is mainly self-used within the treatment plant such as for heating and lighting, and the excess is flared (Richard et al. 2019).

The treated liquid wastewater is also a resource rather than waste. Depending on the potential application pathway, it may be directly used through land spreading or further treated for other possible reuse, such as for industrial and replenishment applications. If wastewater contains high concentrations of nutrients such as nitrogen and phosphorus, nutrients can be recovered from the liquid phase through struvite formation or algae farming.

The solid digestate can be used as the soil conditioner (for digestate with low nutrient contents) for soil organic amendment or biofertilizer (for digestate with high nutrient contents) for supplementing or replacing chemical fertilizers. Under certain conditions, digestate may be further treated by removing contaminants such as heavy metals (Chozhavendhan et al. 2023). Also, digestate can be transformed into other high-value-added products such as biochar and solid biomass fuel.

The production of value-added by-products from anaerobic digestion can improve circular bioeconomy by providing sustainable products and services in many economic sectors through the efficient reuse of waste resources (materials and energy), closing the production and consumption cycle. By this means, anaerobic digestion has a potential to be a zero-waste technology. Thereby, anaerobic digestion can support the achievement of most SDGs directly and indirectly (Piadeh et al. 2023). As a viable solution and environmentally friendly technology, anaerobic digestion can be applied for the efficient management of wastewater and solid waste, curbing pollutants, and hygiene management (contributing to SDGs 3, 6, 12, 14, and 15), and at the same time, renewable energy (contributing to SDG 7) and fertilizer can be produced (contributing to SDG 2). All these can benefit the tackling of climate change (SDG 13).

4 Understanding of Anaerobic Digestion from the Design-Build-Test-Learn Cycle

4.1 The Design-Build-Test-Learn (DBTL) Cycle

The DBTL cycle has been applied widely for the discovery and optimization of microbial synthetic pathways for better product generation. The iterative application of the DBTL cycle is an engineering approach (Carbonell et al. 2018). Lawson et al. (2019) proposed that structuring research and technology developments based on the DBTL cycle can advance microbiome engineering and spur new discoveries such as the basic scientific principles governing microbiome function. The DBTL can also be applied to complex anaerobic digestion ecosystems (Fig. 4), where microbial interactions are highly nonlinear and dynamic within this microbial factory for resource recovery.

With problems identified and goals determined in the design stage, the metabolic pathways, functional microorganisms, and preliminary model systems will be integrated to develop multiple design candidates. Microbial ecosystems will be constructed or acclimated based on the designed process and the underlying principles. Following, system performance and microbial function will be tested according to a set of specified metrics, and further determine whether the design–build solution(s) would achieve the desired objective. Finally, the test results will be analysed

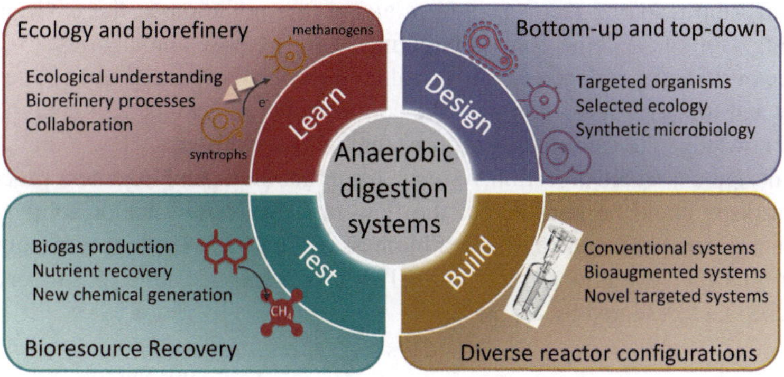

Fig. 4 DBTL cycle and its application to anaerobic digestion

to assess what has worked and what has not worked (and why), and all these insights will inform the subsequent DBTL cycle. Through these iterative cycles, improved design principles for assembling anaerobic digestion ecosystems with predictable function can be achieved, leading to better system design and operation to recover more resources from waste sources through anaerobic digestion.

4.2 Design Principles

Anaerobic digestion employs complex microbial communities to facilitate the desired biological conversions, where diverse microbial interactions occur to accomplish reactions that a single community cannot perform independently and respond effectively to external inputs for maintaining stable microbial environment and community structure (Babson et al. 2014). Within anaerobic digestion ecosystems, functional specialization or division of labour occurs through syntrophic metabolism of organic matter, benefiting all microbial communities (Giri et al. 2019). The division of labour can (i) reduce the metabolic burden imposed on each species, (ii) promote separate engineering or optimization of metabolic pathways that normally would not efficiently operate simultaneously, (iii) achieve recombinant assembly of long multistep pathways divided over two or more engineered species, and (iv) allow the unique metabolic features and cellular properties present in different species to be exploited (Babson et al. 2014).

Therefore, when designing anaerobic digestion systems, the fundamentals of ecological theory should be introduced to guide system design and improve the prediction of process behaviour and system performance, while the deliberately designed engineering ecosystem can be valuable new platforms for ecological research providing specific ecosystem services (Graham and Smith 2004).

Due to the complexity, variability, and uncertainty of anaerobic digestion systems, some design principles could be adopted from the ecological engineering discipline (Bergen et al. 2001), such as (1) design guided by ecological principles to maximize the resource recovery and minimize the pollution to the environment, (2) design for site-specific context targeting fit-for-purpose applications, (3) maintenance of the independence of functional requirements, (4) design for efficiency in energy and material utilization with the consideration of cost–benefit analysis, and (5) acknowledge the values and purposes that motivate design.

To design anaerobic digestion systems, like other biological treatment processes, three aspects could be considered: the flow rate (Q, wastewater or solid waste), concentrations of pollutant, nutrient or biomass (C), and time (t; the reaction duration and/or microbial evolution period, etc.). This was proposed as the QCt theory (Yin et al. 2021), represented by Eq. 1:

$$M = f\left(Q, X, \frac{C}{C + K}, t\right) \quad (1)$$

where M means the microbial function, Q indicates the flow rate, X means the microbial population, C means the substrate concentration, t means the time, and K means the half-saturation coefficient for substrate when the Monod equation is applied for substrate metabolism. This equation can also be revised for the inhibition process in anaerobic digestion systems, where t can represent the exposure duration of inhibitors, and parameters of C_I (the inhibitor concentration) and I (the inhibition constant) could be adopted. The multiplication of all three parameters will obtain the loading rate, which can be the influent loading rate for the anaerobic digestion system. The incorporation of this equation when designing anaerobic digestion processes would be quite helpful.

4.3 *Build and Test of Anaerobic Digestion Systems*

Guided by the ecological principles, mathematical modelling and experimental systems can be adopted for building and testing of anaerobic digestion systems (Fig. 5).

Mathematical modelling can be applied to the anaerobic digestion ecosystem to test the proposed hypothesis. Mathematical modelling can also direct experimentation by validating theoretical principles of microbial ecology (Martinez-Rabert et al. 2023), while its validation should be based on experimental data. The interlink of mathematical modelling and laboratory experiment helps to improve the capability of each other for describing and predicting anaerobic digestion performance. The prediction of anaerobic digestion is highly dependent on the integrative development of both mathematical models and experimental methodology (Widder et al. 2016).

Mathematical modelling usually depends on the underlying kinetic growth model and its parameters, which are obtained from both lab- and large-scale tests (Widder et al. 2016). The conventional Anaerobic Digestion Model 1 is mainly based on kinetic and stoichiometric coefficients, which is deterministic. Nowadays, novel

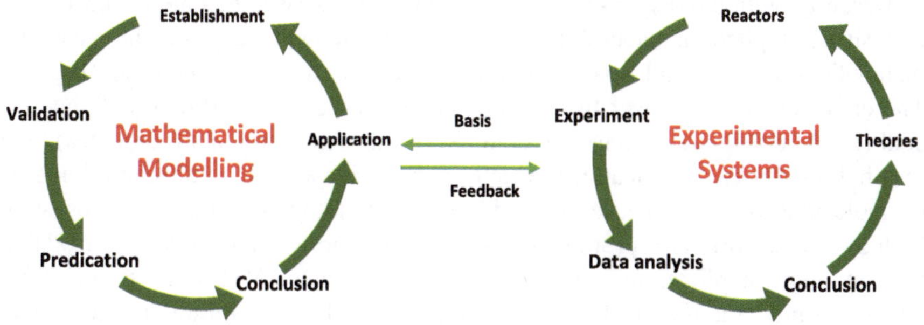

Fig. 5 Integrated modelling and experimental methodologies for building and testing of anaerobic digestion ecosystems

modelling, such as modelling with the thermodynamic method, modelling with the genome-based method, and machine learning based methods, has been developed. This is driven by a better understanding of microbial principles and new methodologies to calculate kinetic coefficients, such as thermodynamics of microbial metabolism (i.e. Desmond-Le Quéméner and Bouchez 2014). A collision frequency theory for microbial growth was also proposed (Button 1998) and coupled to thermodynamic considerations (Liu 2006), which was the first attempt to conceptualize the growth phenomena on a physical basis (Delattre et al. 2019). Furthermore, using these models, microbial interactions based on metabolic pathways (considering both ecology and evolution, and other possible theories such as the resource allocation theory) can be simulated to deeply understand of microbial principles and enhance the prediction accuracy of anaerobic digestion systems. In the anaerobic digestion field, biofilm-based technologies (either with or without carriers) are the typical process component, which requires the modelling of spatial distribution of microbial communities and further their potential function. The newly discovered principles, such as the direct interspecies electron transfer between syntrophic bacteria and methanogens, also require the integration of spatial distribution of microbial communities for the potential electron transfer through microbial contact.

Both system performance and theory validation require adequate replication and reproducibility in the experimental test. For anaerobic digestion, organic carbon degradation and methane production are usually tested from both substrate concentrations and their reaction rates. Nowadays, advanced molecular techniques have been developed rapidly and applied in anaerobic digestion systems to clarify microbial ecology, including diversity, richness, abundance, adaptability, and others. Usually, complementary experimental approaches can be applied to better understand system performance. The application of key taxa, genes, transcripts or metabolites identification and overall microbial community fingerprinting at different levels is essential for the transition to a new level of "microbial community monitoring" in anaerobic digestion (De Vrieze 2020). Specifically, quantitative polymerase chain reaction, flow cytometry, species-specific fluorescence in situ hybridization or novel combinations of single-cell and functional-targeting methods with genomics do yield information on absolute abundances (Berry et al. 2015; Widder et al. 2016).

Recently, with the decreased analytical cost, omics-based techniques have been intensively applied in anaerobic digestion ecosystems to explore the diversity, metabolic potential, and spatiotemporal dynamics of microorganisms (Fig. 6). The omics techniques cover the analysis based on the extracted deoxyribonucleic acid (DNA, potential), ribonucleic acid (RNA, activity), proteins (outcomes), and metabolites (activity verification), which can be corresponding to the central dogma of molecular biology. These omics-based methodologies within an ecosystem biology framework are promising for creating more predictable, controllable biotechnology applications for anaerobic microbiomes (McDaniel et al. 2021; Narayanasamy et al. 2015). For better comparing and a deep mining of these data, a standardization approach like the earth microbiome project could be developed and adopted in anaerobic digestion and other engineered microbial ecosystems (De Vrieze 2020). Although these techniques generate a plethora of new and

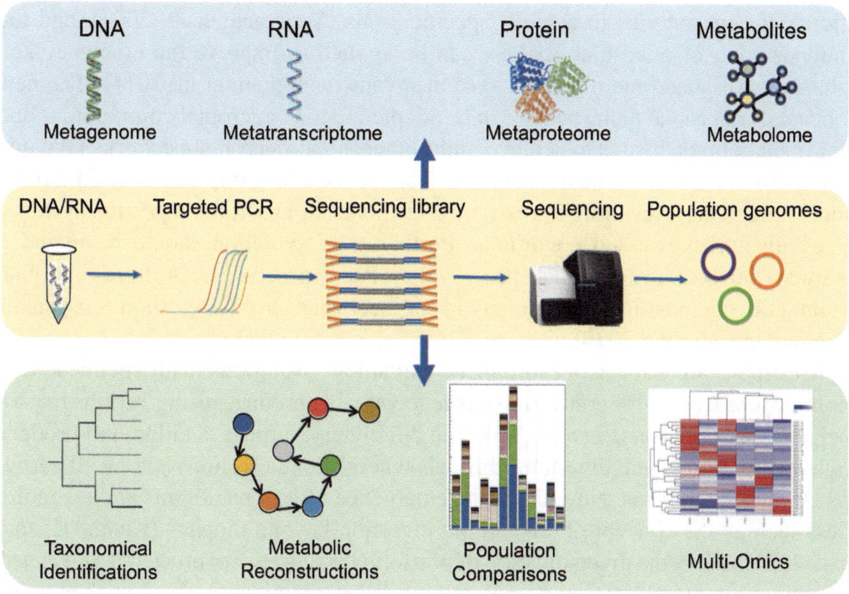

Fig. 6 Microbial community and function analysis through omics-based techniques

more comprehensive observations of microorganisms, the value of these advanced techniques is to address important questions and test important hypotheses, and theoretical tools are also required to detect underlying principles and mechanisms (Prosser et al. 2007, 2022). With the help of novel sequencing technologies, the emergence of new reactor systems or technology concepts based on ecological knowledge would be possible (Nikolausz and Kretzschmar 2020).

4.4 *Learn from Academic and Industrial Experience*

As a mature biotechnology, anaerobic digestion is still receiving intensive attention (Nikolausz and Kretzschmar 2020). However, key challenges, such as low system performance efficiency, long start-up period, and high capital cost, still exist for anaerobic digestion technologies (Fu et al. 2021). Especially, microbial structure and interactions in anaerobic digestion ecosystems could be easily inhibited by environmental stressors such as free ammonia, VFAs accumulation, and antibiotics. Therefore, mechanistic understanding of anaerobic digestion is still necessary for advancing its application.

Anaerobic digestion is the microbial ecology of anaerobes (Nikolausz and Kretzschmar 2020), offering an excellent platform for studying complex microbial communities and testing ecological theories (Vanwonterghem et al. 2014). Within this ecosystem, microbial resource management can be applied to manage or alter microbial communities to achieve specific goals (Verstraete et al. 2007), and the understanding of microbial ecology can be applied to improve the efficiency and robustness of anaerobic digestion systems (Vanwonterghem et al. 2014). The new concept of microbial niche nexus can be adopted to tune microbial communities and their metabolic pathways to achieve multidimensional microbial networks (Wu and Yin 2020). Such multidimensional microbial networks will improve microbial functional redundancy, resulting in resistance/resilience of anaerobic digestion systems, especially under stressed conditions. Furthermore, evolution should be included in microbial ecology, and such "eco-evolutionary" processes can drive microbial community composition and diversity in engineered anaerobic digestion ecosystems (Prosser and Martiny 2020).

Anaerobic digestion is a biorefinery technology using physical, chemical, and biological processes for converting waste to valuable products using various reactor technologies (Verstraete et al. 2005). The biorefinery purpose is tailored to produce high quality biogas and digestates through synergistic interactions among all participated microorganisms. Nowadays, the emergence of new paradigms and technologies, such as the concept of "anaerobic digestion beyond biogas" (Kumar Khanal et al. 2021), drives the diversification of portfolios of biorefinery processes and anaerobic digestion products. Other types of high-value chemicals such as methanol can be produced with biogas as feedstock, or through new biological or thermochemical reactions such as chain elongation rather than conventional anaerobic digestion reactions. The high diversity in both products and technologies improves the potential of anaerobic digestion towards a better circular bioeconomy and more sustainable development.

Anaerobic digestion combines both services for the management of waste and wastewater, and products of biogas for energy and digestates for fertilizer (Archana et al. 2024). The installation of anaerobic digestion requires the collaboration among different sectors, such as environmental engineering for technology innovation, government for policy support, the public and private entities for financial investment, and the public for awareness and promotion. For example, the feed-in-tariff for renewable energy is one excellent evidence showing the policy advancing the application of anaerobic digestion (Wilkinson 2011). Only through all these commitments, sustainable industrial chain of anaerobic digestion will be formed, covering raw feedstock collection, biorefinery, and product marketing. Due to the complexity and promise of anaerobic digestion systems, platforms can be established to share all experience, both from research and practical applications. The DBTL cycle could be adopted for building these platforms, incorporating all components of the DBTL cycle and cooperation from different sectors/disciplines. These platforms would benefit the anaerobic digestion field significantly by sharing knowledge through open access. Some examples of diverse knowledge and case studies of anaerobic digestion include the official information portal on anaerobic digestion from UK (https://www.biogas-info.co.uk) and from US EPA (https://www.epa.gov/anaerobic-digestion). Some useful databases relating to anaerobic digestion are PhyMet^2 (Michał et al. 2018), MCycDB (Qian et al. 2022), and MiDAS (Dueholm et al. 2022).

5 Take-Home Messages

(1) Anaerobic digestion is a cost-effective and environmentally friendly technology for renewable energy production from waste, contributing to the achievement of a circular bioeconomy and SDGs.
(2) The development of anaerobic digestion has been driven and intensified by requirements for pollution management, energy crisis, and regulation by government policies.
(3) Anaerobic digestion can recover biogas, water, and digestates, which closes the loop of material flow for circular bio-economy and sustainable development.
(4) Anaerobic digestion is a dynamic and complex ecosystem, and its mechanistic understanding and system operation require the guidance by both ecological and evolutionary theories.
(5) The DBTL cycle can be adopted to anaerobic digestion for better unification regarding its design, build, test, and learn from both academia and industries. A platform with detailed DBTL results or experience can be established for better communication and for promoting its application.

References

Archana K, Visckram AS, Kumar PS, Manikandan S, Saravanan A, Natrayan L (2024) A review on recent technological breakthroughs in anaerobic digestion of organic biowaste for biogas generation: challenges towards sustainable development goals. Fuel 358:130298

Babson DM, Held M, Schmidt-Dannert C (2014) Designer microbial ecosystems–toward biosynthesis with engineered microbial consortia. In: Osbourn A, Goss RJ, Carter GT (eds) Natural products: discourse, diversity, and design. John Wiley & Sons Inc., pp 23–38

Bergen SD, Bolton SM, Fridley JL (2001) Design principles for ecological engineering. Ecol Eng 18(2):201–210

Berry D, Mader E, Lee TK, Woebken D, Wang Y, Zhu D, Palatinszky M, Schintlmeister A, Schmid MC, Hanson BT, Shterzer N, Mizrahi I, Rauch I, Decker T, Bocklitz T, Popp J, Gibson CM, Fowler PW, Huang WE, Wagner M (2015) Tracking heavy water (D_2O) incorporation for identifying and sorting active microbial cells. PNAS 112(2):E194–E203

Button DK (1998) Nutrient uptake by microorganisms according to kinetic parameters from theory as related to cytoarchitecture. Microbiol Mol Biol Rev 62(3):636–645

Carbonell P, Jervis AJ, Robinson CJ, Yan C, Dunstan M, Swainston N, Vinaixa M, Hollywood KA, Currin A, Rattray NJW, Taylor S, Spiess R, Sung R, Williams AR, Fellows D, Stanford NJ, Mulherin P, Le Feuvre R, Barran P, Goodacre R, Turner NJ, Goble C, Chen GG, Kell DB, Micklefield J, Breitling R, Takano E, Faulon JL, Scrutton NS (2018) An automated design-build-test-learn pipeline for enhanced microbial production of fine chemicals. Commun Biol 1:66

Chozhavendhan S, Karthigadevi G, Bharathiraja B, Praveen Kumar R, Abo LD, Venkatesa Prabhu S, Balachandar R, Jayakumar M (2023) Current and prognostic overview on the strategic exploitation of anaerobic digestion and digestate: a review. Environ Res 216:114526

De Vrieze J (2020) The next frontier of the anaerobic digestion microbiome: from ecology to process control. Environ Sci Ecotechnol 3:100032

Delattre H, Desmond-Le Quéméner E, Duquennoi C, Filali A, Bouchez T (2019) Consistent microbial dynamics and functional community patterns derived from first principles. ISME J 13:263–276

Desmond-Le Quéméner E, Bouchez T (2014) A thermodynamic theory of microbial growth. ISME J 8:1747–1751

Dueholm MK, Nierychlo M, Andersen KS, Rudkjøbing V, Knutsson S, Albertsen M, Nielsen PH (2022) MiDAS 4: a global catalogue of full-length 16S rRNA gene sequences and taxonomy for studies of bacterial communities in wastewater treatment plants. Nat Commun 13(1):1908

Feedback (2020) Bad energy: defining the true role of biogas in a net zero future. London. https://feedbackglobal.org/wp-content/uploads/2020/09/Feedback-2020-Bad-Energy-report.pdf

Fu S, Angelidaki I, Cetecioglu Z, Kong Q, Zheng Y, Tsapekos P (2021) Editorial: biological strategies to enhance the anaerobic digestion performance: fundamentals and process development. Front Microbiol 12:762875

GIA (2010) How to profit from biogas market developments GIA industries white paper. Glob Intell Alliance, United States

Gijzen HJ (2002) Anaerobic digestion for sustainable development: a natural approach. Water Sci Technol 45(10):321–328

Giri S, Waschina S, Kaleta C, Kost C (2019) Defining division of labor in microbial communities. J Mol Biol 431(23):4712–4731

Graham DW, Smith VH (2004) Designed ecosystem services: application of ecological principles in wastewater treatment engineering. Front Ecol Environ 2(4):199–206

IEA (2018a) World energy balance 2018. International Energy Agency. https://www.iea.org/weo2018/

IEA (2018b) World energy outlook 2018. International Energy Agency. https://www.iea.org/statistics/balances/

IEA (2020) Outlook for biogas and biomethane: prospects for organic growth, IEA, Paris. Available from https://www.iea.org/reports/outlook-for-biogasand-biomethane-prospects-for-organicgrowth/the-outlook-for-biogas-and-biomethaneto-2040

Jeison D (2015) Anaerobic digestion: closing cycles for sustainability. Rev Environ Sci Bio/technol 14:535–536

Kumar Khanal S, Lü F, Wong JWC, Wu D, Oechsner H (2021) Anaerobic digestion beyond biogas. Biores Technol 337:125378

Lawson CE, Harcombe WR, Hatzenpichler R, Lindemann SR, Löffler FE, O'Malley MA, García Martín H, Pfleger BF, Raskin L, Venturelli OS, Weissbrodt DG, Noguera DR, McMahon KD (2019) Common principles and best practices for engineering microbiomes. Nat Rev Microbiol 17:725–741

Liu Y (2006) A simple thermodynamic approach for derivation of a general Monod equation for microbial growth. Biochem Eng J 31(1):102–105

Martinez-Rabert E, Sloan WT, Gonzalez-Cabaleiro R (2023) Multiscale models driving hypothesis and theory-based research in microbial ecology. Interface Focus 13:20230008

McCarty PL (2001) The development of anaerobic treatment and its future. Water Sci Technol 44(8):149–156

McDaniel EA, Wahl SA, Ishii S, Pinto A, Ziels R, Nielsen PH, McMahon KD, Williams RBH (2021) Prospects for multi-omics in the microbial ecology of water engineering. Water Res 205:117608

Michał B, Gagat P, Jabłoński S, Chilimoniuk J, Gaworski M, Mackiewicz P, Marcin Ł (2018) PhyMet^2: a database and toolkit for phylogenetic and metabolic analyses of methanogens. Environ Microbiol Rep 10(3):378–382

Narayanasamy S, Muller EE, Sheik AR, Wilmes P (2015) Integrated omics for the identification of key functionalities in biological wastewater treatment microbial communities. Microb Biotechnol 8(3):363–368

Náthia-Neves G, Berni M, Dragone G, Mussatto SI, Forster-Carneiro T (2018) Anaerobic digestion process: technological aspects and recent developments. Int J Environ Sci Technol 15:2033–2046

Nikolausz M, Kretzschmar J (2020) Anaerobic digestion in the 21st century. Bioengineering 7(4):157

Piadeh F, Offie I, Behzadian K, Rizzuto JP, Bywater A, Córdoba-Pachón JR, Walker M (2023) A critical review for the impact of anaerobic digestion on the sustainable development goals. J Environ Manage 349:119458

Prosser JI (2022) How and why in microbial ecology: an appeal for scientific aims, questions, hypotheses and theories. Environ Microbiol 24(11):4973–4980

Prosser JI, Martiny JB (2020) Conceptual challenges in microbial community ecology. Philos Trans R Soc B 375:20190241

Prosser JI, Bohannan BJ, Curtis TP, Ellis RJ, Firestone MK, Freckleton RP, Green JL, Green LE, Killham K, Lennon JJ, Osborn AM, Solan M, van der Gast CJ, Young JP (2007) The role of ecological theory in microbial ecology. Nat Rev Microbiol 5(5):384–392

Qian L, Yu X, Zhou J, Gu H, Ding J, Peng Y, He Q, Tian Y, Liu J, Wang S, Wang C (2022) MCycDB: a curated database for comprehensively profiling methane cycling processes of environmental microbiomes. Mol Ecol Resour 22(5):1803–1823

Richard EN, Hilonga A, Machunda RL, Njau KN (2019) A review on strategies to optimize metabolic stages of anaerobic digestion of municipal solid wastes towards enhanced resources recovery. Sustain Environ Res 29:36

Rodríguez-Jiménez LM, Pérez-Vidal A, Torres-Lozada P (2022) Research trends and strategies for the improvement of anaerobic digestion of food waste in psychrophilic temperatures conditions. Heliyon 8(10):e11174

UNDP (2015) United Nations development programme. https://www.undp.org/sustainable-development-goals/clean-water-and-sanitation

Vanwonterghem I, Jensen PD, Dennis PG, Hugenholtz P, Rabaey K, Tyson GW (2014) Deterministic processes guide long-term synchronised population dynamics in replicate anaerobic digesters. ISME J 8:2015–2028

Vasco-Correa J, Khanal S, Manandhar A, Shah A (2018) Anaerobic digestion for bioenergy production: global status, environmental and techno-economic implications, and government policies. Biores Technol 247:1015–1026

Verstraete W, Morgan-Sagastume F, Aiyuk S, Waweru M, Rabaey K, Lissens G (2005) Anaerobic digestion as a core technology in sustainable management of organic matter. Water Sci Technol 52(1–2):59–66

Verstraete W, Wittebolle L, Heylen K, Vanparys B, De Vos P, Van de Wiele T, Boon N (2007) Microbial resource management: the road to go for environmental biotechnology. Eng Life Sci 7(2):117–126

WBA (2019) Global potential of biogas. World Biogas Association. www.worldbiogasassociation.org

Widder S, Allen RJ, Pfeiffer T, Curtis TP, Wiuf C, Sloan WT, Cordero OX, Brown SP, Momeni B, Shou W, Kettle H (2016) Challenges in microbial ecology: building predictive understanding of community function and dynamics. ISME J 10:2557–2568

Wilkinson KG (2011) A comparison of the drivers influencing adoption of on-farm anaerobic digestion in Germany and Australia. Biomass Bioenerg 35(5):1613–1622

Wu G, Yin Q (2020) Microbial niche nexus sustaining biological wastewater treatment. NPJ Clean Water 3(1):33

Yin Q, Feng Z, Hu Y, Zhan X, Wu G (2021) Microbial interactions in pollution control ecosystems. Curr Pollut Rep 7:104–114

A Holistic Metabolic Pathway of Anaerobic Digestion Integrating Substrate Degradation, Electron Transfer, Energy Conservation, and Information Flow

Qidong Yin and Guangxue Wu

Abstract Main stages of anaerobic digestion include the sequential breakdown of organic matter through hydrolysis, acidogenesis, acetogenesis, and methanogenesis. Substrate degradation, electron transfer, energy conservation, and information flow in these stages determine the performance, stability, and resilience of anaerobic digestion. Direct interspecies electron transfer and quorum sensing are novel mechanisms potentially governing microbial interactions yet need deeper exploration. A comprehensive understanding of these mechanisms is crucial for developing new anaerobic digestion technologies.

Keywords Electron transfer · Energy conservation · Information flow · Metabolic pathway · Substrate degradation

1 Introduction

Anaerobic digestion is a complex biotechnological process underpinned by the collaborative activity of diverse microbial consortia, converting organic waste into primary methane (CH_4) and carbon dioxide (CO_2), in the absence of oxygen. Four central metabolic pathways enable the sequential breakdown and transformation of organic substrates. Anaerobic metabolisms in these pathways (i.e., substrate degradation, electron transfer, energy conservation, and information flow) largely determine the performance, stability, and resilience of microbial communities in these pollution control and energy recovery ecosystems.

Q. Yin (✉)
School of Civil Engineering, Sun Yat-Sen University, Guangzhou 51000, Guangdong, China
e-mail: yinqd3@mail.sysu.edu.cn

G. Wu
Civil Engineering, School of Engineering, College of Science and Engineering, University of Galway, Galway H91 TK33, Ireland
e-mail: guangxue.wu@universityofgalway.ie

G. Wu (ed.), *Anaerobic Digestion*, Green Energy and Technology,
https://doi.org/10.1007/978-3-031-69378-6_2

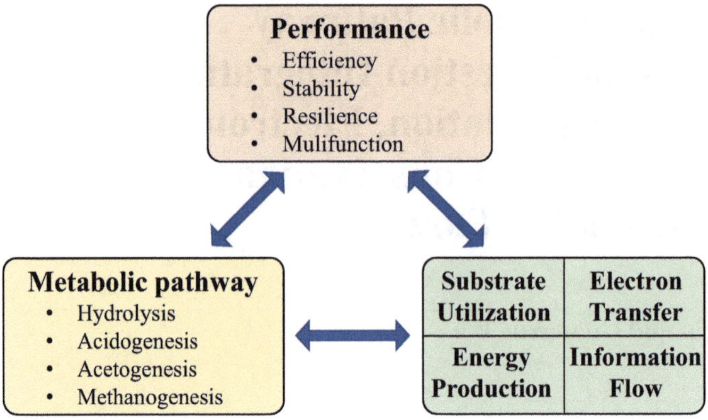

Fig. 1 Metabolic pathway and metabolisms determine the performance and development of anaerobic digestion

Exploring the fundamental metabolic pathways in the degradation of organic matter is the basis for understanding anaerobic digestion principles. As diverse types of organic carbon compounds can be used as the feedstock, dynamics of substrate utilization can be highlighted through understanding basic processing and transforming principles within the anaerobic digestion system. The electron transfer mechanisms and energy production are essential processes that drive microbial activity and biogas generation, especially for those in an equilibrium state. The complex information flow within microbial communities governs microbial ecology, and efficient microbial communication and coordination contribute to the stability of anaerobic digestion processes. A holistic view of the anaerobic digestion landscape can shed light on the scientific principles and microbial interactions that make this process a cornerstone of environmental sustainability and energy production (Fig. 1).

2 Key Metabolic Pathways

Anaerobic digestion involves a series of interconnected metabolic pathways that are essential for the breakdown of organic matter and the production of biogas (Lettinga 1995). These pathways, executed by various microorganisms, transform complex organic compounds into CH_4 and CO_2, the primary components of biogas. Metabolic pathways in anaerobic digestion can be broadly categorized into four main stages: hydrolysis, acidogenesis (or fermentation), acetogenesis, and methanogenesis (Fig. 2) (Adekunle and Okolie 2015). Each stage is characterized by distinct biochemical reactions facilitated by diverse microbial consortia. In the hydrolysis stage, complex organic compounds such as lipids, proteins, and polysaccharides are broken down into simpler molecules. This is followed by acidogenesis, where these simpler compounds are converted into volatile fatty acids (VFAs), alcohols,

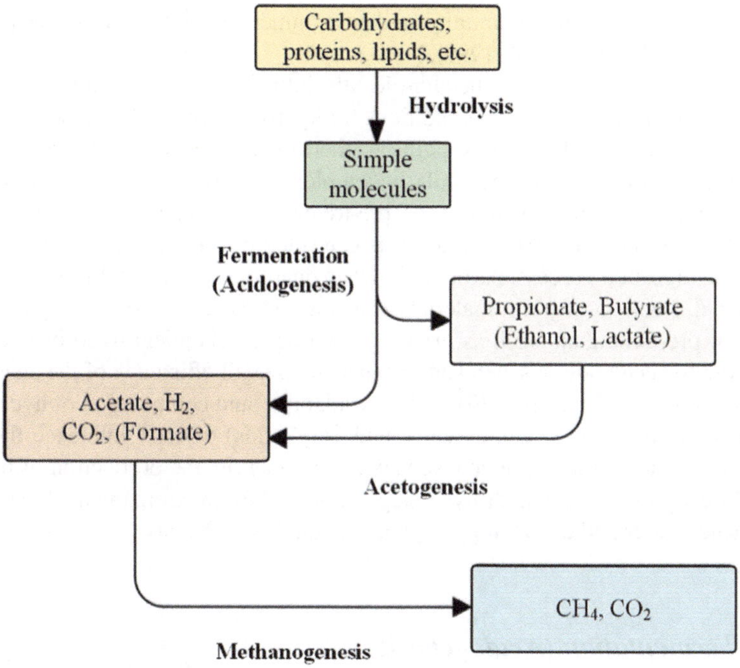

Fig. 2 Hydrolysis, fermentation, acetogenesis, and methanogenesis of organic substrates during the syntrophic CH_4 production

hydrogen (H_2), and CO_2. In the acetogenesis stage, products of VFAs and alcohols are further converted into acetate, H_2, and CO_2. These substances are finally utilized by methanogens to produce CH_4.

2.1 Hydrolysis

Hydrolysis is the first step in the anaerobic digestion process. The primary targets of hydrolysis are complex organic compounds, including carbohydrates (such as cellulose and starch), proteins, and lipids. These macromolecules are typically found in the organic matter being treated, such as agricultural waste, sewage sludge, and food waste (Zhang et al. 2005). Hydrolysis is an enzymatic process (Lu et al. 2006). Each type of macromolecule requires specific types of enzymes for breakdown: carbohydrases for carbohydrates, proteases for proteins, and lipases for lipids. The hydrolytic bacteria secrete extracellular enzymes that catalyze the breakdown of these complex molecules. For instance, cellulose is broken down into simpler sugars like glucose, proteins into amino acids, and lipids into long-chain fatty acids (LCFAs) and glycerol (Lu et al. 2006). The efficiency of this enzymatic breakdown depends on various factors, including the nature and concentration of the substrate, the composition and activity of the microbial community, and environmental conditions such as pH and temperature (Mansfield et al. 1999).

The rate of hydrolysis can be significantly influenced by the conditions within the anaerobic digester. Optimal conditions for hydrolysis typically include a neutral to slightly alkaline pH and mesophilic to thermophilic temperatures (35–55 °C). However, certain substrates, particularly complex carbohydrates like cellulose, may require more stringent conditions or pre-treatment to enhance hydrolysis rates (Castilla-Archilla et al. 2021). Hydrolysis can often be the rate-limiting step in the anaerobic digestion process, particularly when dealing with lignocellulosic biomass or other difficult-to-degrade materials (Sayara and Sánchez 2019). Physical and chemical pre-treatments such as grinding, heating, or chemical hydrolysis can be employed to accelerate this step and enhance the overall efficiency of the anaerobic digestion process (Zhen et al. 2017). The availability and composition of hydrolysis products (simple sugars, amino acids, and fatty acids) directly influence the efficiency of fermentation (acidogenesis) and, subsequently, the production of biogas. Therefore, optimizing the hydrolysis step is essential for maximizing the breakdown of organic matter and enhancing the yield and quality of biogas.

2.2 *Fermentation (Acidogenesis)*

Following hydrolysis, the anaerobic digestion process progresses to fermentation, also known as acidogenesis. This stage is pivotal as it involves the transformation of the monomeric products from hydrolysis into a range of intermediate compounds, primarily VFAs, alcohols, H_2, and CO_2 (Zhang et al. 2005). In the acidogenesis phase, acidogenic bacteria ferment the simple monomers—such as sugars, amino acids, and fatty acids produced during hydrolysis. This step is characterized by the production of short-chain VFAs like acetate, propionate, and butyrate, along with other by-products such as ethanol, lactate, H_2, and CO_2. The specific types and proportions of these fermentation products vary based on the substrate composition and the environmental conditions within the anaerobic digester.

A wide variety of bacteria participate in acidogenesis, each favoring different substrates and producing different end products. Microbial diversity is important for the robustness of the anaerobic digestion process, as it ensures the breakdown of a wide range of organic materials (Joyce et al. 2018). The metabolic activities of these bacteria are also influenced by factors such as temperature, pH, and the presence of inhibitory substances (Li et al. 2020b; Zhang et al. 2005). The optimal conditions for acidogenesis differ slightly from those of hydrolysis, although there is considerable overlap. Generally, acidogenic bacteria thrive in a neutral to slightly acidic pH and at mesophilic to thermophilic temperatures (Calicioglu et al. 2018). However, the process is adaptable and can occur under a wide range of conditions. The dynamics of acidogenesis, including the rates of substrate breakdown and product formation, significantly impact the downstream processes of acetogenesis and methanogenesis.

Balance among the different fermentation products is critical. For instance, the accumulation of certain VFAs or alcohols can inhibit methanogenic archaea if not properly balanced and converted in subsequent stages (Wang et al. 2009). Thus, maintaining an optimal environment for acidogenesis is essential for ensuring that the products of fermentation are suitable substrates for acetogens and methanogens. Fermentation serves as an intermediary step in anaerobic digestion. By converting the products of hydrolysis into compounds that can be utilized by acetogenic and methanogenic microorganisms, acidogenesis ensures the continuity and efficiency of the biogas production process. Disruptions or inefficiencies in this stage can lead to reduced biogas yields and process instability.

2.3 *Acetogenesis*

Acetogenesis is the third critical stage in the anaerobic digestion process. This stage bridges the conversion of organic acids and alcohols produced during acidogenesis into acetic acid, H_2, and CO_2, which are direct precursors for CH_4 production in the final methanogenesis step. In acetogenesis, acetogenic bacteria play a central role, converting VFAs such as propionate and butyrate, along with alcohols and other intermediates produced during fermentation, into acetic acid, H_2, and CO_2 (Wang et al. 2018a). This transformation is vital because methanogens, the archaea responsible for CH_4 production, predominantly utilize acetic acid and H_2.

Acetogenic bacteria are strictly anaerobic and are sensitive to pH, temperature, and toxic substances (Müller 2003). The efficiency of acetogenesis depends on maintaining optimal conditions within anaerobic digester, as disruptions can lead to the buildup of fermentation products, inhibiting the overall process. Acetogenesis often occurs in a syntrophic relationship with H_2-consuming methanogens (Demirel and Scherer 2008). This relationship is crucial for maintaining a low partial pressure of H_2, which is necessary for the energetically favorable conversion of VFAs to acetic acid. If H_2 accumulates, it can halt the acetogenic process; thus, the presence and activity of hydrogenotrophic methanogens are essential for continuous acetogenesis.

Acetogenesis can be a rate-limiting step in anaerobic digestion due to the slower growth rates of acetogens compared to acidogens and the sensitivity of acetogens to environmental conditions (Müller 2003). The efficiency of acetogenesis directly influences the overall effectiveness and yield of CH_4. Successful conversion of VFAs and other intermediates into acetate and H_2 ensures a steady supply of substrates for the methanogenic archaea, thereby maximizing biogas production.

2.4 *Methanogenesis*

Methanogens utilize two main pathways for CH_4 production: acetoclastic methanogenesis and hydrogenotrophic methanogenesis (Zeikus 1977). In acetoclastic methanogenesis, methanogens convert acetic acid directly into CH_4 and CO_2. *Methanosarcina* and *Methanosaeta* are the primary genera responsible for this process (Weimer and Zeikus 1978). Acetoclastic methanogenesis consumes a significant portion of the substrates generated from acidogenesis and acetogenesis. This process typically accounts for about 70% of the CH_4 produced in most anaerobic digestion systems (Li et al. 2020a).

The hydrogenotrophic methanogenesis, on the other hand, involves the reduction of CO_2 with H_2, producing CH_4. This reaction is catalyzed by a different group of methanogens, including *Methanobacterium*, *Methanococcus*, and *Methanospirillum* (Demirel and Scherer 2008). Hydrogenotrophic methanogenesis is essential for the overall balance of anaerobic digestion as it consumes H_2, preventing its accumulation which could otherwise inhibit other biological processes within the anaerobic digester (Ripoll et al. 2020).

Methanogenic archaea are a specialized group of microorganisms distinct from bacteria and eukaryotes. They are obligate anaerobes, meaning that they can only survive and function in environments that absence of oxygen (Garcia et al. 2000). Their unique biological mechanisms enable them to thrive in extreme conditions, such as high temperatures or saline environments (Valentine 2007). Methanogens possess specialized enzymes and coenzymes that enable the production of CH_4, a metabolic capability not found in other microbial groups. Methanogens are sensitive to environmental conditions, such as pH, temperature, and the presence of toxic substances (Yin et al. 2020b). Optimal conditions for methanogenesis generally include a neutral pH and a mesophilic to thermophilic temperature range. Methanogens are also sensitive to high concentrations of sulfide, ammonia, and heavy metals, which can inhibit their activities (Karhadkar et al. 1987; Sprott and Patel 1986; Wang et al. 2018b).

Methanogenesis is tightly linked with the preceding stages of anaerobic digestion, particularly acetogenesis. The syntrophic relationship between acetogenic bacteria and methanogenic archaea is essential for the efficient breakdown of organic material and the production of biogas (Schink 1997). Acetogens convert organic acids and alcohols into acetic acid and H_2, which are then utilized by methanogens. This interdependency ensures the stability and efficiency of the anaerobic digestion process.

3 Substrate Utilization

3.1 Typical Organic Substrates and Their Degradation

Substrates refer to organic matter or waste materials that are broken down by microorganisms to produce biogas. These substrates can come from various sources and are composed of complex organic compounds that need to be decomposed into simpler molecules for microbial consumption. Common substrates include proteins, carbohydrates, lipids, organic acids, and alcohols.

Proteins, prevalent in substrates like manure, food waste, and municipal wastewater, provide both a carbon and nitrogen source for microbial growth in anaerobic digestion systems. The breakdown of proteins begins with their hydrolysis into amino acids, which are subsequently deaminated to produce ammonia, CO_2, and other sulfides during acidogenesis (Ramsay and Pullammanappallil 2001). Although the ammonia generated can act as a nutrient for microorganisms, excessive amounts can lead to ammonia toxicity, inhibiting methanogenic activity (Kayhanian 1999; Koster and Koomen 1988; Sprott and Patel 1986). Therefore, managing the protein content and ensuring balanced nitrogen levels within the anaerobic digester are necessary for maintaining a healthy microbial ecosystem and ensuring efficient methanogenesis.

Carbohydrates in the anaerobic digestion process are typically derived from plant-based materials such as food waste, agricultural residues, and sewage sludge. These substrates mainly consist of polysaccharides like cellulose, hemicellulose, and simple sugars, which are energy-rich and generally broken down quickly by microorganisms (Noike et al. 1985). During hydrolysis, enzymes decompose these complex carbohydrates into simpler sugars, which are then transformed into VFAs, H_2, and CO_2 during the fermentation phase (Yang et al. 2015). Efficient conversion of carbohydrates to VFAs and their subsequent utilization ensures a steady production of CH_4, highlighting the importance of carbohydrates as a significant energy source in the anaerobic digestion process.

Lipids, found in wastes, such as cooking oils, grease, and industrial effluents, are high-energy substrates that contribute significantly to the CH_4 yield due to their high biogas potential. The anaerobic digestion of lipids involves their initial breakdown into glycerol and LCFAs during hydrolysis (Neves et al. 2009). These LCFAs are further degraded through β-oxidation into short-chain VFAs, which are then available for acetogenesis and methanogenesis. While lipids can significantly increase biogas production, they also pose challenges, including the risk of forming scum layers and inhibiting microbial activity if not properly managed (Rasit et al. 2015). Nevertheless, with adequate pre-treatment and process control, lipids can be effectively utilized to enhance CH_4 production in anaerobic digestion systems (Holohan et al. 2022).

Organic acids and alcohols typically originate from industrial waste streams and are characterized by their rapid degradation rates (Kong et al. 2019). Organic acids and alcohols are easily assimilated by microorganisms, leading to a swift production of VFAs, H_2, and CO_2. The ease of conversion of these substrates into VFAs makes them excellent candidates for accelerating the anaerobic digestion process. When managed correctly, organic acids and alcohols can significantly enhance the efficiency and speed of CH_4 production, contributing positively to the overall biogas yield (Christy et al. 2014).

3.2 *Stoichiometry of CH_4 Production*

To estimate the CH_4 production from substrates, many methods have been proposed. Without considering the degradation of organic substrates for microbial synthesis and maintenance, the CH_4 production from organic substrates can be described by Eq. 1 (Buswell and Mueller 1952). However, since organic carbon is comprised of biodegradable and non-biodegradable matter, and only the biodegradable matter can be metabolized into CH_4, the CH_4 yield calculated by this method is overestimated. If nitrogen is included, this can be revised as Eq. 2. In addition, Boyle (1977) proposed the revised equation by including nitrogen and sulfur in the composition of organic matter, which can produce ammonia and hydrogen sulfide (Eq. 3). For solids waste based on volatile solids (VS), theoretical methane production (TMP) could be obtained based on elemental compositions of organic substrates (expressed as TMP_{ele}), such as using the Buswell formula as shown in Eqs. 2 and 4 (Buswell and Mueller 1952; Li et al. 2013). While for liquid organic matter, TMP based on the chemical oxygen demand can be calculated.

$$C_aH_bO_c+\left(a-\frac{b}{4}-\frac{c}{2}\right)H_2O \rightarrow \left(\frac{a}{2}+\frac{b}{8}-\frac{c}{4}\right)CH_4+\left(\frac{a}{2}-\frac{b}{8}+\frac{c}{4}\right)CO_2 \quad (1)$$

$$C_nH_aO_bN_c+\left(n-\frac{a}{4}-\frac{b}{2}+\frac{3c}{4}\right)H_2O \rightarrow \left(\frac{n}{2}+\frac{a}{8}-\frac{b}{4}-\frac{3c}{8}\right)CH_4 + \left(\frac{n}{2}-\frac{a}{8}+\frac{b}{4}+\frac{3c}{8}\right)CO_2+cNH_3 \quad (2)$$

$$C_aH_bO_cN_dS_e+\left(a-\frac{b}{4}-\frac{c}{2}+\frac{3d}{4}+\frac{e}{2}\right)H_2O \rightarrow \left(\frac{a}{2}+\frac{b}{8}-\frac{c}{4}-\frac{3d}{8}-\frac{e}{4}\right)CH_4+\left(\frac{a}{2}-\frac{b}{8}+\frac{c}{4}+\frac{3d}{8}+\frac{e}{4}\right)CO_2 + dNH_3+eH_2S \quad (3)$$

$$TMP_{ele}\left(\frac{mL\ CH_4}{g\ VS}\right)=\frac{22.4\times1000\times\left(\frac{n}{2}+\frac{a}{8}-\frac{b}{4}-\frac{3c}{8}\right)}{12n+a+16b+14c} \quad (4)$$

The theoretical CH_4 production can also be calculated based on the organic composition (expressed as TMP_{org}). For example, when VFA ($C_2H_4O_2$), protein ($C_5H_7NO_2$), lipids ($C_{57}H_{104}O_6$), carbohydrates ($C_6H_{10}O_5$), and lignin ($C_{10}H_{13}O_3$) are expressed as percentages of VS (Angelidaki and Sanders 2004; Kaparaju et al. 2009; Li et al. 2013; Triolo et al. 2011), TMP_{org} can be calculated using Eq. 5.

$$TMP_{org}\left(\frac{mLCH_4}{g\ VS}\right)=(373VFA+496Protein+1014Lipids + 415Carbohydrates+727Lignin)/100 \quad (5)$$

Anaerobic biodegradability (BD) of the substrate could be calculated based on the observed (measured) CH_4 production (OMP) and TMP as follows (Elbeshbishy et al. 2012): BD = OMP/TMP. In addition, comparing the measured CH_4 to the predicted CH_4 helps to estimate the methanogenic activity. The synergistic effect (η) was calculated by Eq. 6 during anaerobic digestion of diverse types of organic matter ($n > 1$) (Awais et al. 2016). A η value greater than 1 indicates a synergistic effect, and a η value less than 1 indicates an antagonistic effect.

$$\eta = \frac{\sum_{i=1}^{n} OMP_i}{\sum_{i=1}^{n} TMP_i} \quad (6)$$

Vrede (1998) reported the representative molecular composition of proteins ($C_{16}H_{30}O_8N_4$, 404 g/mol), carbohydrates ($C_6H_{12}O_6$, 180 g/mol), and lipids ($C_{47}H_{96}O_9$, 804 g/mol). Representative molecular structures for biomass are $C_{75}H_{105}O_{30}N_{15}P$ or $C_5H_9O_3N$ (Mankad and Bungay 1988). Gerber and Span (2008) proposed that biogas production potential was 700 L biogas per kg organics with 71% CH_4 for proteins, 790 L biogas per kg organics with 50% of CH_4 for carbohydrates, and 1250 L biogas per kg organics with 68% CH_4 for lipids.

4 Electron Transfer

Electron transfer is a fundamental aspect of the anaerobic digestion process, underpinning the metabolic reactions that convert organic matter into biogas. This process involves the transfer of electrons from donor molecules (such as organic substrates) to acceptor molecules (such as protons or CO_2) through the metabolic activities of microorganisms. During anaerobic digestion, microorganisms employ various mechanisms to transfer electrons, integral to their energy metabolism. The primary electron transfers include interspecies H_2 transfer (the use of gaseous mediators), interspecies formate transfer (the use of soluble mediators), and direct interspecies electron transfer (DIET; the use of solid mediators). These mechanisms facilitate the reduction of substrates in a series of biochemical reactions that ultimately lead to the production of CH_4.

4.1 Interspecies H_2 and Formate Transfer

Interspecies H_2 transfer was first reported by Wolin (1982) (Fig. 3a). The generation of H_2 by syntrophic bacteria is associated with the valence-state conversion of redox mediators such as nicotinamide adenine dinucleotide couple (NAD^+/NADH), flavin adenine dinucleotide (FAD/$FADH_2$), and ferredoxin couple ($Fd_{(ox)}/Fd_{(red)}$). However, Gibbs free energies of these coupled reactions are positive due to the low redox potential of H^+/H_2 ($E^{0'} = -414$ mV) (Table 1), indicating that these reactions are thermodynamically unfavorable (Thauer et al. 1977). Concentrations of electron carriers H_2 should be maintained at a very low level ($H_2 < 10^{-4}$ atm) for uninhibited hydrogenase activity (McInerney et al. 2008). Methanogens are responsible for the consumption of H_2 so that acetogenic reactions can take place (Pan et al. 2016). For methanogens, the utilization of H_2 is associated with the valence-state conversion of $Fd_{(ox)}/Fd_{(red)}$ and coenzyme F_{420} couple (F_{420}/F_{420}-H_2). From the aspect of a syntrophic methanogenic system, when the H_2 partial pressure is high, the activity of H_2 generation by syntrophic bacteria is inhibited while the utilization of H_2 by methanogens will be activated. On the contrary, when the H_2 partial pressure is low enough, the generation of H_2 will be activated and the utilization by methanogens is inhibited. Therefore, syntrophic bacteria and methanogens should cooperate to obtain energy by adjusting the H_2 partial pressure of the system.

Apart from H_2, formate is another important diffusive electron carrier for syntrophic methanogenesis (Fig. 3a). Thiele and Zeikus (1988) first reported that in the mixed cultures containing *Desulfovibrio vulgaris* and *Methanobacterium formicicum*, more than 90% of the syntrophic CH_4 production from ethanol was mediated via interspecies formate transfer while only less than 10% was accomplished via interspecies H_2 transfer. In this system, H_2 neither inhibited the ethanol oxidation by syntrophic bacteria nor improved the CH_4 production by methanogens. Boone et al. (1989) calculated the diffusion rate of H_2 and formate in the *Syntrophomonas wolfei* and *Methanobacterium formicicum* co-culture, and found that H_2 could not diffuse

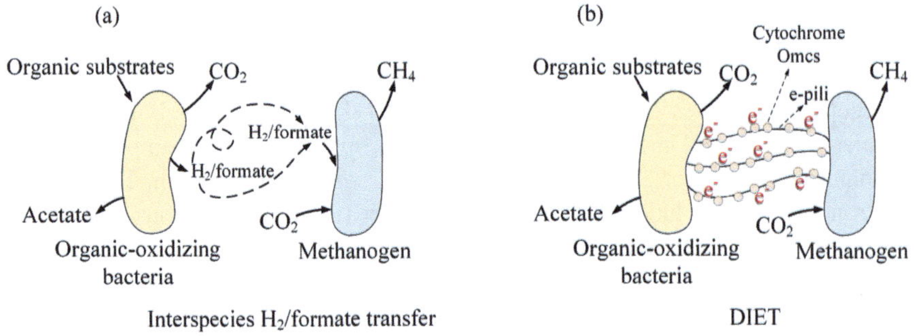

Fig. 3 **a** Interspecies H_2/formate transfer, and **b** DIET between syntrophic bacteria and methanogens

Table 1 Energetics of intracellular redox reactions

Reaction	$\Delta G^{0\prime}$ (kJ/mol)	$\Delta G'$ (kJ/mol) (at 1 Pa H_2)
H_2 formation by anaerobic bacteria		
$2Fd_{(red)} + 2H^+ \rightarrow 2Fd_{(ox)} + H_2$	+3	−26
$Formate^- + H^+ \rightarrow CO_2 + H_2$	−3	−32
$NADH + H^+ \rightarrow NAD^+ + H_2$	+18	−11
$FADH_2 \rightarrow FAD + H_2$	+37	+8
H_2 consumption by methanogens		
$H_2 + F_{420} \rightarrow F_{420}H_2$	−11	+18
$2Fd_{(ox)} + H_2 \rightarrow 2Fd_{(red)}$	−3	+26

$\Delta G^{0\prime}$: represents the standard Gibbs free energy change under standard conditions (pH 7, 25 °C, 1 M concentration for solutes, and 1 atm pressure for gases); $\Delta G'$: represents the Gibbs free energy change under specified non-standard conditions, particularly with respect to H_2 pressure at 1 Pa.

fast enough for CH_4 synthesis while formate could. The existence of formate dehydrogenase in these two microorganisms also proved the presence of interspecies formate transfer (Boone et al. 1989). Furthermore, both propionate and butyrate-oxidizing bacteria were found to possess genes encoding formate dehydrogenase, suggesting the important role of formate transfer in some methanogenic systems (Kosaka et al. 2008; McInerney et al. 2007). Due to the higher solubility of formate than H_2, it was suggested that formate transfer might be the dominant mechanism when the distance between syntrophic partners was long, and H_2 might become more favorable in a short distance (de Bok et al. 2004).

4.2 Direct Interspecies Electron Transfer (DIET)

Recently, DIET has been proposed as an alternative electron exchange strategy for syntrophic CH_4 production (Lovley 2011; Morita et al. 2011; Shrestha and Rotaru 2014). In defined co-cultures grown with ethanol, electrons produced during acetogenesis by *Geobacter* species could be transferred directly to *Methanosarcina barkeri* or *Methanosaeta harundinacea* through electrically conductive pili (e-pili) or outer membrane cytochrome (Rotaru et al. 2014a, b). *Geobacter* cannot efficiently obtain energy from metabolizing ethanol into H_2 or formate, so it could not grow well without the help of methanogens that use H_2 or formate (*Methanospirillum hungatei* or *Methanobacterium formicicum*). However, *G. metallireducens* and *M. harundinacea* successfully grew together by converting ethanol to CH_4 with high expression of genes encoding CO_2 reduction in *M. harundinacea* (Rotaru et al. 2014b). The inability to produce CH_4 via H_2 and the radiotracer experiments suggested that *M. harundinacea* reduced CO_2 by directly accepting electrons from *G. metallireducens*, confirming the existence of DIET (Rotaru et al. 2014b).

Compared with interspecies H_2 or formate transfer, fewer enzymatic steps are involved in DIET because syntrophic bacteria do not need to produce diffusive redox mediators which can save energy and have higher efficiency (Baek et al. 2018; Storck et al. 2016). Besides, e-pili facilitates long-range electron transfer, whose efficiency does not rely on the diffusion of H_2/formate (Fig. 3b), thus making it a faster and potentially more energy-conserving approach for methanogenesis (Cruz Viggi et al. 2014; Lovley 2017a, b). Mei et al. (2018) suggested that 30 kJ/mol more energy was available via DIET compared to complete H_2 production during ethanol oxidation through thermodynamic analysis. Cruz Viggi et al. (2014) calculated the electron transfer rate of DIET and found that it could be 10^6 times higher than that of interspecies H_2 transfer. In addition, DIET can alleviate the negative effect of the H_2 partial pressure (Cruz Viggi et al. 2014), and therefore provide a more stable environment for methanogenesis.

Electrically conductive pili and outer membrane cytochrome are required for extracellular electron transfer (El-Naggar et al. 2010; Shi et al. 2016, 2009). Compared with H_2 or formate transfer, the metal-like conductivity of e-pili allows long-range (centimeter scale) electron transport between microorganisms with a more direct and faster option (Lovley 2017a; Morita et al. 2011). The overlapping π–π orbitals of aromatic amino acids in the pilin monomer might be responsible for the conductivity (Lampa-Pastirk et al. 2016; Tan et al. 2016). Morita et al. (2011) first reported that methanogenic wastewater digester aggregates treating brewery waste were electrically conductive and the conductivity was correlated to the abundance of *Geobacter* species which could generate e-pili for DIET. Shrestha et al. (2014) found that all 14 up-flow anaerobic sludge blanket granules were electrically conductive, and the granule conductivity was moderately correlated with only the abundance of *Geobacter* species. Subsequently, *M. barkeri-G. metallireducens* and *M. harundinacea-G. metallireducens* co-cultures were shown to exchange electrons via DIET (Rotaru et al. 2014a, b). A pilin-deficient *G. metallireducens* failed to form co-cultures with *M. barkeri*, suggesting that e-pili played an important role in DIET for syntrophic methanogenesis (Rotaru et al. 2014a). The transcriptomic and metagenomic analysis further revealed that the functional genes regulating e-pili and accessory proteins were highly assigned and expressed by *Geobacter* species in methanogenic rice paddy soils and anaerobic reactors (Holmes et al. 2017; Mei et al. 2018).

Extracellular cytochrome-to-cytochrome electron transfer can also achieve DIET without e-pili. Multiheme cytochromes on the outer cell surfaces of syntrophic partners were responsible for the electron exchange when they were positioned close enough (McGlynn et al. 2015). Recently, *Geobacter* nanowires were reported to be made up of polymerization of multiheme *c*-type cytochrome (OmcS), which were able to provide continuous electron flow paths for extracellular electron transfer (Wang et al. 2019). However, other studies suggested that outer surface cytochromes were not sufficient to achieve long-range cytochrome-to-cytochrome connections (Malvankar et al. 2012; Ueki et al. 2018). Co-cultures of Aro-5 strains of *G. metallireducens* and *G. sulfurreducens* with poorly conductive pili could not be established via DIET even if the existence of abundant outer surface cytochromes, suggesting the necessity of e-pili for DIET (Ueki et al. 2018). Instead, the OmcS localized on e-pili of *G. sulfurreducens* was proposed to facilitate electron transfer between e-pili and other electron carriers (Lovley 2017b). The deletion of genes encoding OmcS prevented the mutations growing on DIET, suggesting its essential function for direct electron exchange (Summers et al. 2010).

Unlike *G. metallireducens*, methanogens (*M. barkeri* and *M. harundinacea*) that had been confirmed to take part in DIET were found to lack outer surface *c*-type cytochromes and e-pili (Holmes et al. 2018). Therefore, how methanogens accept electrons directly from syntrophic partners and the electron flux for CH_4 production via DIET remain unknown (Cheng and Call 2016). Metagenomic and metatranscriptomic analyses revealed that the CO_2 reduction pathway in *Methanosaeta* was highly expressed during DIET (Holmes et al. 2017; Park et al. 2018), but the difference between methanogenic pathways with interspecies H_2 transfer and DIET cannot be clarified because *Methanosaeta* cannot utilize H_2 for CH_4 production. On the other hand, although the addition of ferroferric oxide (Fe_3O_4) induced genes involved in the CO_2 reduction pathway in *Methanosarcina*, the participation of H_2 could not be completely ruled out because some species of *Methanosarcina* are capable of using

H_2 as the electron donor (Van Steendam et al. 2019). Therefore, it is important to distinguish differences in electron flux between H_2-based and DIET-based pathways.

Thauer et al. (2008) proposed that methanogens with and without cytochromes might reduce CO_2 for CH_4 production with different strategies. During CO_2 reduction with H_2, ferredoxin (Fd), coenzyme F_{420}, coenzyme B, and methanophenazine (MP) were electron carriers for methanogens with cytochromes (Fig. 4). For methanogens with cytochromes (e.g., *Methanosarcina, Methanosaeta* and *Methanolobus* genera), the first (the reduction of CO_2 with H_2 to formyl-methanofuran, endergonic) and last (the production of Coenzyme B-Coenzyme M heterodisulfide (CoB-S-S-CoM), exergonic) steps in methanogenesis via interspecies H_2 transfer are coupled chemiosmotically through methanophenazine-reducing [NiFe] hydrogenase (VhoACG) and heterodisulfide reductase (HdrDE), both containing a subunit of *b*-type cytochrome (VhoC and HdrE) (Beckmann et al. 2016; Deppenmeier 1995). However, for methanogens without cytochromes, VhoACG, HdrDE, and MP are not involved in the reduction of the CoB-S-S-CoM with H_2. Instead, in methanogens without cytochromes, the first and last steps are coupled by flavin-based electron bifurcation through [NiFe] hydrogenase (MvhADG)-heterodisulfide reductase (HdrABC) complex. The HdrABC complex does not contain cytochromes, whereas harbors flavin adenine dinucleotide (FAD) in the HdrA subunit (Thauer et al. 2008). The cytoplasmic HdrABC enzymes (both HdrA1B1C1 and homologous $HdrA_2B_2C_2$) were also found to be specifically involved in the methylotrophic methanogenesis in *Methanosarcina acetivorans* by coupling the oxidation of reduced ferredoxin (−500 mV) with the reduction of both F_{420} (−360 mV) and CoB-S-S-CoM (−140 mV) (Buan and Metcalf 2010; Yan et al. 2017).

Recently, Holmes et al. (2018) have developed a new model for CO_2 reduction in *M. barkeri* during DIET based on the transcriptomic comparison between *M. barkeri-G. metallireducens* co-cultures via DIET and *M. barkeri-Pelobacter carbinolicus* co-cultures via interspecies H_2 transfer. Transcripts for the $F_{420}H_2$ dehydrogenase (Fpo) and HdrABC were abundant during growth with DIET. In this model (Fig. 4), electrons transferred from DIET were first utilized to reduce MP, and then the produced MPH_2 acted as the electron donor for Fpo to reduce F_{420}. Half of the generated $F_{420}H_2$ participates in the flavin-based electron bifurcation with HdrABC to further generate reduced Fd, which is required for the first step in CO_2 reduction. This electron bifurcation simultaneously triggered the reduction of CoM-S-S-CoB necessary for the last step of CH_4 production (Holmes et al. 2018). Besides, this model also balanced the proton flux by assuming that protons produced during the reduction of F_{420} with Fpo are consumed equally by the reduction of MP, and protons released from the oxidation of $F_{420}H_2$ by HdrABC can be utilized in other reactions (Holmes et al. 2018).

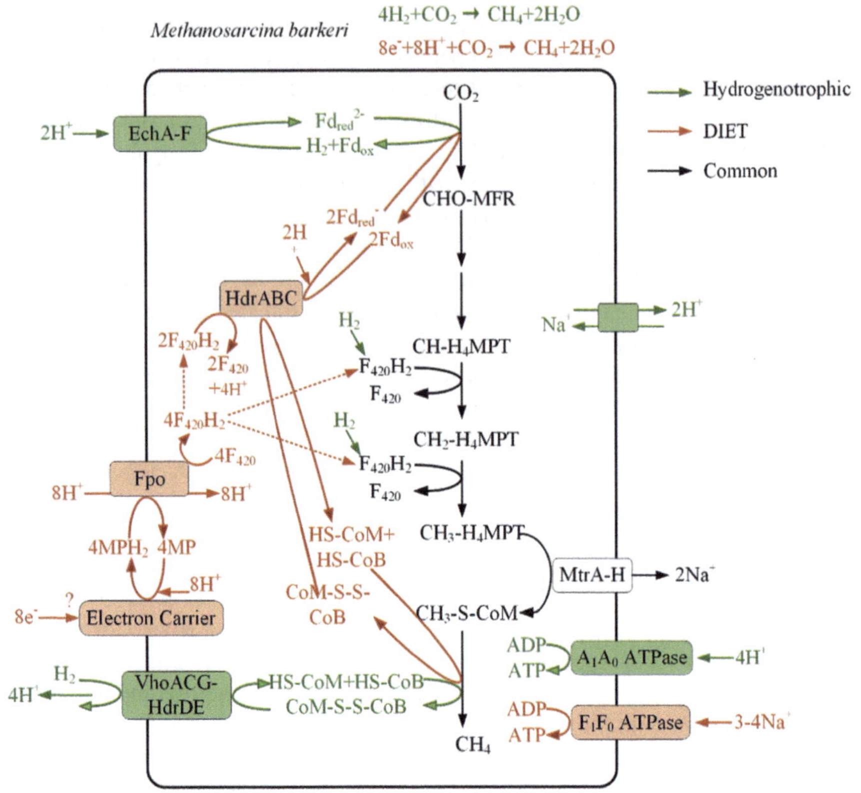

Fig. 4 Model for methanogenic pathways from the CO_2 reduction in *Methanosarcina barkeri* during H_2-based (green) or DIET-based (orange) growth. Note: The H_2-based pathway and DIET-based pathway are cited with adaptation from Thauer et al. (2008) and Holmes et al. (2018), respectively. F_{420}: coenzyme F_{420}; Fd: ferredoxin; MP: methanophenazine; H_4MPT: tetrahydrosarcinapterin; HS-CoB: coenzyme B; HS-CoM: coenzyme M; MFR: methanofuran

5 Energy Production

Energy production during anaerobic digestion is a critical aspect of the metabolic processes undertaken by microorganisms to convert organic materials into biogas. This conversion is not just pivotal for biogas generation but is also essential for sustaining the life and activities of the microbial communities within the anaerobic digester. Microorganisms in anaerobic environments produce energy primarily through the process of fermentation and methanogenesis (Liu et al. 2021; Müller 2003). During fermentation, organic compounds are broken down into simpler molecules such as VFAs, alcohols, and H_2. This breakdown releases energy, which cells capture in the form of adenosine triphosphate (ATP) through substrate-level phosphorylation, a process where a phosphate group is directly transferred to adenosine diphosphate (ADP) to form ATP. As the digestion process progresses toward methanogenesis, methanogenic archaea utilize substrates like H_2, acetate, and CO_2 to produce CH_4. In this stage, energy is produced through different pathways, depending on the type of methanogens and the substrates they use. For acetoclastic methanogens, ATP is generated through the cleavage of acetate (Smith and Ingram-Smith 2007). Hydrogenotrophic methanogens, on the other hand, produce ATP by reducing CO_2 with H_2, a process that involves electron transport chains and chemiosmotic generation of ATP, similar to oxidative phosphorylation in aerobic organisms (Thauer et al. 2008).

The generation of ATP in anaerobic digestion occurs primarily through two mechanisms: substrate-level phosphorylation and chemiosmosis. In substrate-level phosphorylation, ATP is produced directly in the metabolic pathways during the conversion of carbohydrates or other substrates to fermentation end products (Buckel 2021). In contrast, chemiosmosis involves the generation of a proton gradient across a membrane, which drives the synthesis of ATP via ATP synthase (Thauer et al. 2008). This method is particularly relevant in methanogenic pathways where energy conservation is linked to the transfer of electrons and the generation of a proton motive force.

Energy conservation in methanogens is associated with the translocation of protons or sodium ions across the cytoplasmic membrane, which triggers ATP synthase to convert ADP to ATP (Buckel and Thauer 2013). For methanogens with cytochrome, the energy conservation in the CO_2 reduction pathway with H_2 is dependent on methyl-H_4MPT-coenzyme M methyltransferase (Mtr), energy-converting [NiFe] hydrogenase (EchA-F), VhoACG, HdrDE, A_1A_0-ATP, and Na^+/H^+ antiporter (Thauer et al. 2008). In some *Methanosarcina* species, the reduction of MP with H_2 can be alternatively catalyzed by the combination of cytoplasmic F_{420}-reducing hydrogenase (FrhABC) and energy-conserving Fpo, which is also coupled to the translocation of two protons (Bäumer et al. 2000). The energy conservation in methanogens via DIET is still unclear, but the fact that the reaction catalyzed by HdrABC is not coupled to proton or sodium ion translocation across the cytoplasmic membrane, suggesting that the energy conservation via DIET may be different from that via interspecies H_2 transfer in *M. barkeri*. According to the models (Holmes et al. 2018; Thauer et al. 2008), for *M. barkeri*, one main difference was that during H_2-based growth, EchA-F, VhoACG, and HdrDE were responsible for protons translocation, while these membrane-associated protein complexes were replaced by Fpo when growing via DIET. In addition, genes for F_1F_0 ATP synthase which can translocate sodium were highly expressed during the growth of *M. barkeri* via DIET, suggesting the important role of F_1F_0-ATP synthase (Holmes et al. 2018).

6 Information Flow

Information flow within microbial communities, especially in environments like anaerobic digesters, influences the efficiency and effectiveness of the anaerobic digestion process. This flow of information is largely facilitated through mechanisms such as quorum sensing, allowing microorganisms to communicate and coordinate their metabolic activities.

Quorum sensing depends on the excretion of signal substances such as acyl-homoserine lactone (AHL), autoinducer-2 (AI-2), diffusible signal factor (DSF), and 3′–5′ cyclic diguanosine monophosphate (c-di-GMP) (Deng et al. 2011; Hengge 2009; Waters and Bassler 2005). Reaching the threshold concentration of signal substances may result in the regulation of particular gene expression (Platt and Fuqua 2010). This cell–cell communication strategy enables bacteria to cooperatively response to specific environmental niches. To date, many phenotypes have been found to be regulated by quorum sensing such as antibiotic (Brackman et al. 2011), biofilm formation (Jenal et al. 2017), degradation of organic pollutants (Huang et al. 2016), and virulence factor (Ji et al. 1995).

In anaerobic digestion, some specific syntrophic partners were also suggested to be involved in quorum sensing. For instance, the signal molecule N-butyryl-L-homoserine lactone was detected in *Geobacter* co-cultured aggregates, suggesting its role in syntrophic growth and aggregate formation (Wei et al. 2017). A previous study working on metagenomic analysis of 12 anaerobic sludge samples found the existence of diverse quorum sensing genes (Yin et al. 2020a). Genes involved in the regulation of AHLs, AI-2, DSF, and c-di-GMP signals were detected, suggesting the widespread distribution of quorum sensing signals in anaerobic systems. AHLs were able to adjust community structures of bacteria and methanogens (Lv et al. 2018), while AI-2, c-di-GMP, and DSF might regulate extracellular polymeric substances production and biofilm formation (Feng et al. 2014; van Wolferen et al. 2018). Another study collected quorum sensing genes, classified them into seven groups based on their signal molecules, and used them to construct a hierarchical quorum sensing database containing 39,981 quorum sensing genes (Du et al. 2021). This database was further applied to analyze quorum sensing in lab-scale anaerobic reactors and found that AHLs were the most abundant group of quorum sensing genes (Fig. 5).

The ability of microorganisms to communicate and coordinate their activities through quorum sensing is particularly important in anaerobic digestion due to the diversity of microbial species and metabolic pathways involved. Effective communication ensures that different microbial groups act synergistically, optimizing the breakdown of organic matter and enhancing biogas production. However, the complexities of microbial interactions and the vast diversity of species involved in anaerobic digestion present numerous challenges and unknowns. Currently, the specific roles, mechanisms, and impacts of quorum sensing in this context are not fully understood, particularly regarding how these signaling processes influence the overall efficiency and stability of anaerobic digestion systems. Furthermore, the translation of quorum sensing studies from laboratory scales to full-scale anaerobic digestion operations remains a significant hurdle, requiring more in-depth research and practical experiments.

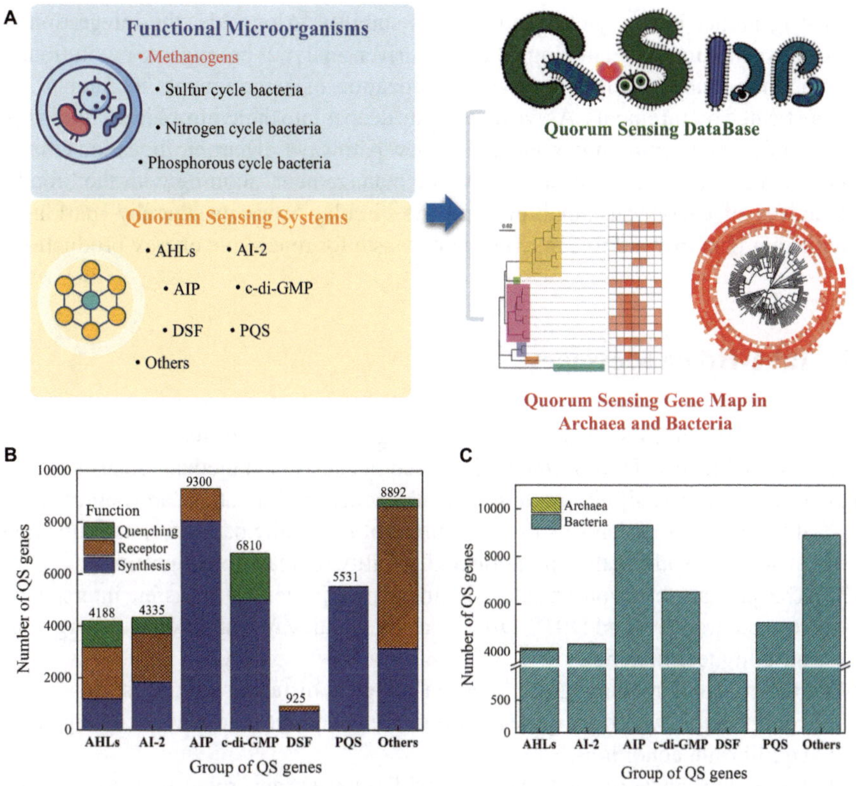

Fig. 5 Construction of the database, and the number of quorum sensing genes in the database. **a** Schematic diagram of quorum sensing gene collection and database construction. Quorum sensing genes were divided into seven groups based on the signal molecules (AHLs, AI-2, AIP, c-di-GMP, DSF, PQS, and others); **b** Number of synthesis, receptor, and quenching genes in each quorum sensing group; **c** Host (bacteria or archaea) of the quorum sensing genes in the database (Du et al. 2021)

7 Future Perspectives

As we continue to explore and understand the intricacies of anaerobic digestion, the future holds promising advancements that could significantly enhance the efficiency and sustainability of this process. Emerging research potentially uncovering new metabolic pathways and expanding our understanding of microbial communities is needed to enhance the efficiency and versatility of the anaerobic digestion process. The discovery of new metabolic pathways and the strategic engineering of microbial consortia are expected to enable the breakdown of a broader array of substrates, including recalcitrant compounds, ushering in more robust and adaptable anaerobic systems. This evolution will be complemented by deeper understanding of microbial functions, aimed at optimizing microbial interactions and metabolic efficiency, thus ensuring higher biogas yields and process stability. Alongside, the integration of innovative monitoring technologies and AI-driven analytics promises to revolutionize operational control, facilitating the customization of environments to cater to the sophisticated requirements. As we investigate deeper into the syntrophic relationships within microbial communities and unlock new pathways, anaerobic digestion is set to become a cornerstone of sustainable waste management, aligning with the broader objectives of energy sustainability and the circular economy, thereby marking a significant leap forward to convert organic waste for renewable energy production.

8 Take-Home Messages

(1) Anaerobic digestion can be broadly categorized into four main stages: hydrolysis, acidogenesis (or fermentation), acetogenesis, and methanogenesis.
(2) Carbohydrates, lipids, protein, organic acids, and alcohols are degraded by different microorganisms through different metabolic pathways, and the types of substrates affect the operation and stability of anaerobic digestion.
(3) The primary electron transfers include interspecies H_2 transfer, interspecies formate transfer, and DIET. DIET is a novel pathway that may save energy and have higher efficiency.
(4) Bacteria and methanogenic archaea adopted different strategies for energy conservation, which are quite important for anaerobic reactions occurring under equilibrium conditions.
(5) Signal substances of quorum sensing widely exist in anaerobic digestion but their critical roles in shaping and maintaining the anaerobic microbial community need further clarification.

References

Adekunle KF, Okolie JA (2015) A review of biochemical process of anaerobic digestion. Adv Biosci Biotechnol 6(3):205–212

Angelidaki I, Sanders W (2004) Assessment of the anaerobic biodegradability of macropollutants. Rev Environ Sci Biotechnol 3:117–129

Awais M, Alvarado-Morales M, Tsapekos P, Gulfraz M, Angelidaki I (2016) Methane production and kinetic modeling for co-digestion of manure with lignocellulosic residues. Energy Fuels 30(12):10516–10523

Baek G, Kim J, Kim J, Lee C (2018) Role and potential of direct interspecies electron transfer in anaerobic digestion. Energies 11(1):107

Bäumer S, Ide T, Jacobi C, Johann A, Gottschalk G, Deppenmeier U (2000) The $F_{420}H_2$ dehydrogenase from *Methanosarcina mazei* is a Redox-driven proton pump closely related to NADH dehydrogenases. J Biol Chem 275(24):17968–17973

Beckmann S, Welte C, Li X, Oo YM, Kroeninger L, Heo Y, Zhang M, Ribeiro D, Lee M, Bhadbhade M, Marjo CE (2016) Novel phenazine crystals enable direct electron transfer to methanogens in anaerobic digestion by redox potential modulation. Energy Environ Sci 9(2):644–655

Boone DR, Johnson RL, Liu Y (1989) Diffusion of the interspecies electron carriers H_2 and formate in methanogenic ecosystems and its implications in the measurement of Km for H_2 or formate uptake. Appl Environ Microbiol 55(7):1735–1741

Boyle WC (1977) Energy recovery from sanitary landfills-a review. In: Schlegel HG, Barnea J (eds) Microbial energy conversion. Seminar Göttingen/FRG, pp 119–138

Brackman G, Cos P, Maes L, Nelis HJ, Coenye T (2011) Quorum sensing inhibitors increase the susceptibility of bacterial biofilms to antibiotics in vitro and in vivo. Antimicrob Agents Chemother 55(6):2655–2661

Buan NR, Metcalf WW (2010) Methanogenesis by *Methanosarcina acetivorans* involves two structurally and functionally distinct classes of heterodisulfide reductase. Mol Microbiol 75(4):843–853

Buckel W (2021) Energy conservation in fermentations of anaerobic bacteria. Front Microbiol 12:703525

Buckel W, Thauer RK (2013) Energy conservation via electron bifurcating ferredoxin reduction and proton/Na^+ translocating ferredoxin oxidation. Biochimica et Biophysica Acta (BBA)-Bioenergetics 1827(2):94–113

Buswell AM, Mueller HF (1952) Mechanism of methane fermentation. Ind Eng Chem 44(3):550–552

Calicioglu O, Shreve MJ, Richard TL, Brennan RA (2018) Effect of pH and temperature on microbial community structure and carboxylic acid yield during the acidogenic digestion of duckweed. Biotechnol Biofuels 11:275

Castilla-Archilla J, Papirio S, Lens PN (2021) Two step process for volatile fatty acid production from brewery spent grain: hydrolysis and direct acidogenic fermentation using anaerobic granular sludge. Process Biochem 100:272–283

Cheng Q, Call DF (2016) Hardwiring microbes via direct interspecies electron transfer: mechanisms and applications. Environ Sci Process Impacts 18(8):968–980

Christy PM, Gopinath LR, Divya D (2014) A review on anaerobic decomposition and enhancement of biogas production through enzymes and microorganisms. Renew Sustain Energy Rev 34:167–173

Cruz Viggi C, Rossetti S, Fazi S, Paiano P, Majone M, Aulenta F (2014) Magnetite particles triggering a faster and more robust syntrophic pathway of methanogenic propionate degradation. Environ Sci Technol 48(13):7536–7543

De Bok FA, Plugge CM, Stams AJ (2004) Interspecies electron transfer in methanogenic propionate degrading consortia. Water Res 38(6):1368–1375

Demirel B, Scherer P (2008) The roles of acetotrophic and hydrogenotrophic methanogens during anaerobic conversion of biomass to methane: a review. Rev Environ Sci Bio/Technol 7:173–190

Deng Y, Wu JE, Tao F, Zhang LH (2011) Listening to a new language: DSF-based quorum sensing in gram-negative bacteria. Chem Rev 111(1):160–173

Deppenmeier U (1995) Different structure and expression of the operons encoding the membrane-bound hydrogenases from *Methanosarcina mazei Gö1*. Arch Microbiol 164:370–376

Du Q, Mu Q, Wu G (2021) Metagenomic and bioanalytical insights into quorum sensing of methanogens in anaerobic digestion systems with or without the addition of conductive filter. Sci Total Environ 763:144509

El-Naggar MY, Wanger G, Leung KM, Yuzvinsky TD, Southam G, Yang J, Lau WM, Nealson KH, Gorby YA (2010) Electrical transport along bacterial nanowires from *Shewanella oneidensis MR-1*. Proc Natl Acad Sci 107(42):18127–18131

Elbeshbishy E, Nakhla G, Hafez H (2012) Biochemical methane potential (BMP) of food waste and primary sludge: influence of inoculum pre-incubation and inoculum source. Biores Technol 110:18–25

Feng H, Ding Y, Wang M, Zhou G, Zheng X, He H, Zhang X, Shen D, Shentu J (2014) Where are signal molecules likely to be located in anaerobic granular sludge? Water Res 50:1–9

Garcia JL, Patel BK, Ollivier B (2000) Taxonomic, phylogenetic, and ecological diversity of methanogenic Archaea. Anaerobe 6(4):205–226

Gerber M, Span R (2008) An analysis of available mathematical models for anaerobic digestion of organic substances for production of biogas. In: International gas union research conference, Paris

Hengge R (2009) Principles of c-di-GMP signalling in bacteria. Nat Rev Microbiol 7(4):263–273

Holmes DE, Shrestha PM, Walker DJ, Dang Y, Nevin KP, Woodard TL, Lovley DR (2017) Metatranscriptomic evidence for direct interspecies electron transfer between *Geobacter* and *Methanothrix* species in methanogenic rice paddy soils. Appl Environ Microbiol 83(9):e00223-e317

Holmes DE, Rotaru AE, Ueki T, Shrestha PM, Ferry JG, Lovley DR (2018) Electron and proton flux for carbon dioxide reduction in *Methanosarcina barkeri* during direct interspecies electron transfer. Front Microbiol 9:416780

Holohan BC, Duarte MS, Szabo-Corbacho MA, Cavaleiro AJ, Salvador AF, Pereira MA, Ziels RM, Frijters CT, Pacheco-Ruiz S, Carballa M, Sousa DZ (2022) Principles, advances, and perspectives of anaerobic digestion of lipids. Environ Sci Technol 56(8):4749–4775

Huang J, Shi Y, Zeng G, Gu Y, Chen G, Shi L, Hu Y, Tang B, Zhou J (2016) Acyl-homoserine lactone-based quorum sensing and quorum quenching hold promise to determine the performance of biological wastewater treatments: an overview. Chemosphere 157:137–151

Jenal U, Reinders A, Lori C (2017) Cyclic di-GMP: second messenger extraordinaire. Nat Rev Microbiol 15(5):271–284

Ji G, Beavis RC, Novick RP (1995) Cell density control of staphylococcal virulence mediated by an octapeptide pheromone. Proc Natl Acad Sci 92(26):12055–12059

Joyce A, Ijaz UZ, Nzeteu C, Shirran SL, O'Flaherty V, Abram F (2018) Linking microbial community structure and function during the acidified anaerobic digestion of grass. Front Microbiol 9:319861

Kaparaju P, Serrano M, Thomsen AB, Kongjan P, Angelidaki I (2009) Bioethanol, biohydrogen and biogas production from wheat straw in a biorefinery concept. Biores Technol 100(9):2562–2568

Karhadkar PP, Audic JM, Faup GM, Khanna P (1987) Sulfide and sulfate inhibition of methanogenesis. Water Res 21(9):1061–1066

Kayhanian M (1999) Ammonia inhibition in high-solids biogasification: an overview and practical solutions. Environ Technol 20(4):355–365

Kong Z, Li L, Xue Y, Yang M, Li YY (2019) Challenges and prospects for the anaerobic treatment of chemical-industrial organic wastewater: a review. J Clean Prod 231:913–927

Kosaka T, Kato S, Shimoyama T, Ishii S, Abe T, Watanabe K (2008) The genome of *Pelotomaculum thermopropionicum* reveals niche-associated evolution in anaerobic microbiota. Genome Res 18(3):442–448

Koster IW, Koomen E (1988) Ammonia inhibition of the maximum growth rate (μ_m) of hydrogenotrophic methanogens at various pH-levels and temperatures. Appl Microbiol Biotechnol 28:500–505

Lampa-Pastirk S, Veazey JP, Walsh KA, Feliciano GT, Steidl RJ, Tessmer SH, Reguera G (2016) Thermally activated charge transport in microbial protein nanowires. Sci Rep 6(1):23517

Lettinga G (1995) Anaerobic digestion and wastewater treatment systems. Antonie Van Leeuwenhoek 67:3–28

Li Y, Zhang R, Liu G, Chen C, He Y, Liu X (2013) Comparison of methane production potential, biodegradability, and kinetics of different organic substrates. Biores Technol 149:565–569

Li Q, Liu Y, Yang X, Zhang J, Lu B, Chen R (2020a) Kinetic and thermodynamic effects of temperature on methanogenic degradation of acetate, propionate, butyrate and valerate. Chem Eng J 396:125366

Li Z, Lou Y, Ding J, Liu BF, Xie GJ, Ren NQ, Xing D (2020b) Metabolic regulation of ethanol-type fermentation of anaerobic acidogenesis at different pH based on transcriptome analysis of *Ethanoligenens harbinense*. Biotechnol Biofuels 13:101

Liu C, Ren L, Yan B, Luo L, Zhang J, Awasthi MK (2021) Electron transfer and mechanism of energy production among syntrophic bacteria during acidogenic fermentation: a review. Biores Technol 323:124637

Lovley DR (2011) Reach out and touch someone: potential impact of DIET (direct interspecies energy transfer) on anaerobic biogeochemistry, bioremediation, and bioenergy. Rev Environ Sci Bio/Technol 10:101–105

Lovley DR (2017a) Electrically conductive pili: biological function and potential applications in electronics. Curr Opin Electrochem 4(1):190–198

Lovley DR (2017b) Syntrophy goes electric: direct interspecies electron transfer. Annu Rev Microbiol 71:643–664

Lu Y, Zhang YH, Lynd LR (2006) Enzyme–microbe synergy during cellulose hydrolysis by *Clostridium thermocellum*. Proc Natl Acad Sci 103(44):16165–16169

Lv L, Li W, Zheng Z, Li D, Zhang N (2018) Exogenous acyl-homoserine lactones adjust community structures of bacteria and methanogens to ameliorate the performance of anaerobic granular sludge. J Hazard Mater 354:72–80

Malvankar NS, Tuominen MT, Lovley DR (2012) Lack of cytochrome involvement in long-range electron transport through conductive biofilms and nanowires of *Geobacter sulfurreducens*. Energy Environ Sci 5(9):8651–8659

Mankad T, Bungay HR (1988) Model for microbial growth with more than one limiting nutrient. J Biotechnol 7(2):161–166

Mansfield SD, Mooney C, Saddler JN (1999) Substrate and enzyme characteristics that limit cellulose hydrolysis. Biotechnol Prog 15(5):804–816

McGlynn SE, Chadwick GL, Kempes CP, Orphan VJ (2015) Single cell activity reveals direct electron transfer in methanotrophic consortia. Nature 526(7574):531–535

McInerney MJ, Rohlin L, Mouttaki H, Kim U, Krupp RS, Rios-Hernandez L, Sieber J, Struchtemeyer CG, Bhattacharyya A, Campbell JW, Gunsalus RP (2007) The genome of *Syntrophus aciditrophicus*: life at the thermodynamic limit of microbial growth. Proc Natl Acad Sci 104(18):7600–7605

McInerney MJ, Struchtemeyer CG, Sieber J, Mouttaki H, Stams AJ, Schink B, Rohlin L, Gunsalus RP (2008) Physiology, ecology, phylogeny, and genomics of microorganisms capable of syntrophic metabolism. Ann N Y Acad Sci 1125(1):58–72

Mei R, Nobu MK, Narihiro T, Yu J, Sathyagal A, Willman E, Liu WT (2018) Novel *Geobacter* species and diverse methanogens contribute to enhanced methane production in media-added methanogenic reactors. Water Res 147:403–412

Morita M, Malvankar NS, Franks AE, Summers ZM, Giloteaux L, Rotaru AE, Rotaru C, Lovley DR (2011) Potential for direct interspecies electron transfer in methanogenic wastewater digester aggregates. MBio 2(4):e00159-e111

Müller V (2003) Energy conservation in acetogenic bacteria. Appl Environ Microbiol 69(11):6345–6353

Neves L, Pereira MA, Mota M, Alves MM (2009) Detection and quantification of long chain fatty acids in liquid and solid samples and its relevance to understand anaerobic digestion of lipids. Biores Technol 100(1):91–96

Noike T, Endo G, Chang JE, Yaguchi JI, Matsumoto JI (1985) Characteristics of carbohydrate degradation and the rate-limiting step in anaerobic digestion. Biotechnol Bioeng 27(10):1482–1489

Pan X, Angelidaki I, Alvarado-Morales M, Liu H, Liu Y, Huang X, Zhu G (2016) Methane production from formate, acetate and H_2/CO_2; focusing on kinetics and microbial characterization. Biores Technol 218:796–806

Park JH, Park JH, Seong HJ, Sul WJ, Jin KH, Park HD (2018) Metagenomic insight into methanogenic reactors promoting direct interspecies electron transfer via granular activated carbon. Biores Technol 259:414–422

Platt TG, Fuqua C (2010) What's in a name? The semantics of quorum sensing. Trends Microbiol 18(9):383–387

Ramsay IR, Pullammanappallil PC (2001) Protein degradation during anaerobic wastewater treatment: derivation of stoichiometry. Biodegradation 12:247–257

Rasit N, Idris A, Harun R, Ghani WA (2015) Effects of lipid inhibition on biogas production of anaerobic digestion from oily effluents and sludges: an overview. Renew Sustain Energy Rev 45:351–358

Ripoll E, López I, Borzacconi L (2020) Hydrogenotrophic activity: a tool to evaluate the kinetics of methanogens. J Environ Manage 270:110937

Rotaru AE, Shrestha PM, Liu F, Markovaite B, Chen S, Nevin KP, Lovley DR (2014a) Direct interspecies electron transfer between *Geobacter metallireducens* and *Methanosarcina barkeri.* Appl Environ Microbiol 80(15):4599–4605

Rotaru AE, Shrestha PM, Liu F, Shrestha M, Shrestha D, Embree M, Zengler K, Wardman C, Nevin KP, Lovley DR (2014b) A new model for electron flow during anaerobic digestion: direct interspecies electron transfer to *Methanosaeta* for the reduction of carbon dioxide to methane. Energy Environ Sci 7(1):408–415

Sayara T, Sánchez A (2019) A review on anaerobic digestion of lignocellulosic wastes: pretreatments and operational conditions. Appl Sci 9(21):4655

Schink B (1997) Energetics of syntrophic cooperation in methanogenic degradation. Microbiol Mol Biol Rev 61(2):262–280

Shi L, Dong H, Reguera G, Beyenal H, Lu A, Liu J, Yu HQ, Fredrickson JK (2016) Extracellular electron transfer mechanisms between microorganisms and minerals. Nat Rev Microbiol 14(10):651–662

Shi L, Richardson DJ, Wang Z, Kerisit SN, Rosso KM, Zachara JM, Fredrickson JK (2009) The roles of outer membrane cytochromes of *Shewanella* and *Geobacter* in extracellular electron transfer. Environ Microbiol Rep 1(4):220–227

Shrestha PM, Malvankar NS, Werner JJ, Franks AE, Elena-Rotaru A, Shrestha M, Liu F, Nevin KP, Angenent LT, Lovley DR (2014) Correlation between microbial community and granule conductivity in anaerobic bioreactors for brewery wastewater treatment. Biores Technol 174:306–310

Shrestha PM, Rotaru AE (2014) Plugging in or going wireless: strategies for interspecies electron transfer. Front Microbiol 5:237

Smith KS, Ingram-Smith C (2007) *Methanosaeta*, the forgotten methanogen? Trends Microbiol 15(4):150–155

Sprott GD, Patel GB (1986) Ammonia toxicity in pure cultures of methanogenic bacteria. Syst Appl Microbiol 7(2–3):358–363

Storck T, Virdis B, Batstone DJ (2016) Modelling extracellular limitations for mediated versus direct interspecies electron transfer. ISME J 10(3):621–631

Summers ZM, Fogarty HE, Leang C, Franks AE, Malvankar NS, Lovley DR (2010) Direct exchange of electrons within aggregates of an evolved syntrophic coculture of anaerobic bacteria. Science 330(6009):1413–1415

Tan Y, Adhikari RY, Malvankar NS, Pi S, Ward JE, Woodard TL, Nevin KP, Xia Q, Tuominen MT, Lovley DR (2016) Synthetic biological protein nanowires with high conductivity. Small 12(33):4481–4485

Thauer RK, Jungermann K, Decker K (1977) Energy conservation in chemotrophic anaerobic bacteria. Bacteriol Rev 41(1):100–180

Thauer RK, Kaster AK, Seedorf H, Buckel W, Hedderich R (2008) Methanogenic archaea: ecologically relevant differences in energy conservation. Nat Rev Microbiol 6(8):579–591

Thiele JH, Zeikus JG (1988) Control of interspecies electron flow during anaerobic digestion: significance of formate transfer versus hydrogen transfer during syntrophic methanogenesis in flocs. Appl Environ Microbiol 54(1):20–29

Triolo JM, Sommer SG, Møller HB, Weisbjerg MR, Jiang XY (2011) A new algorithm to characterize biodegradability of biomass during anaerobic digestion: influence of lignin concentration on methane production potential. Biores Technol 102(20):9395–9402

Ueki T, Nevin KP, Rotaru AE, Wang LY, Ward JE, Woodard TL, Lovley DR (2018) *Geobacter* strains expressing poorly conductive pili reveal constraints on direct interspecies electron transfer mechanisms. MBio 9(4):e01273-e1318

Valentine DL (2007) Adaptations to energy stress dictate the ecology and evolution of the Archaea. Nat Rev Microbiol 5(4):316–323
Van Steendam C, Smets I, Skerlos S, Raskin L (2019) Improving anaerobic digestion via direct interspecies electron transfer requires development of suitable characterization methods. Curr Opin Biotechnol 57:183–190
van Wolferen M, Orell A, Albers SV (2018) Archaeal biofilm formation. Nat Rev Microbiol 16(11):699–713
Vrede T (1998) Elemental composition (C:N:P) and growth rates of bacteria and *Rhodomonas* grazed by *Daphnia*. J Plankton Res 20(3):455–470
Wang Y, Zhang Y, Wang J, Meng L (2009) Effects of volatile fatty acid concentrations on methane yield and methanogenic bacteria. Biomass Bioenerg 33(5):848–853
Wang P, Wang H, Qiu Y, Ren L, Jiang B (2018a) Microbial characteristics in anaerobic digestion process of food waste for methane production—a review. Biores Technol 248:29–36
Wang R, Zhang J, Liu J, Yu D, Zhong H, Wang Y, Chen M, Tong J, Wei Y (2018b) Effects of chlortetracycline, Cu and their combination on the performance and microbial community dynamics in swine manure anaerobic digestion. J Environ Sci 67:206–215
Wang F, Gu Y, O'Brien JP, Sophia MY, Yalcin SE, Srikanth V, Shen C, Vu D, Ing NL, Hochbaum AI, Egelman EH (2019) Structure of microbial nanowires reveals stacked hemes that transport electrons over micrometers. Cell 177(2):361–369
Waters CM, Bassler BL (2005) Quorum sensing: cell-to-cell communication in bacteria. Annu Rev Cell Dev Biol 21:319–346
Wei W, Zhang Y, Komorek R, Plymale A, Yu R, Wang B, Zhu Z, Liu F, Yu XY (2017) Characterization of syntrophic *Geobacter* communities using ToF-SIMS. Biointerphases 12(5):05G601
Weimer PJ, Zeikus JG (1978) One carbon metabolism in methanogenic bacteria: cellular characterization and growth of *Methanosarcina barkeri*. Arch Microbiol 119:49–57
Wolin M (1982) Hydrogen transfer in microbial communities. In: Bull AT, Slater JH (eds) Microbial interactions and communities, vol 1. Academic Press, London, pp 323–356
Yan Z, Wang M, Ferry JG (2017) A ferredoxin-and $F_{420}H_2$-dependent, electron-bifurcating, heterodisulfide reductase with homologs in the domains bacteria and archaea. Mbio 8(1):e02285-e2316
Yang G, Zhang P, Zhang G, Wang Y, Yang A (2015) Degradation properties of protein and carbohydrate during sludge anaerobic digestion. Biores Technol 192:126–130
Yin Q, Gu M, Hermanowicz SW, Hu H, Wu G (2020a) Potential interactions between syntrophic bacteria and methanogens via type IV pili and quorum-sensing systems. Environ Int 138:105650
Yin Q, Gu M, Wu G (2020b) Inhibition mitigation of methanogenesis processes by conductive materials: a critical review. Biores Technol 317:123977
Zeikus JG (1977) The biology of methanogenic bacteria. Bacteriol Rev 41(2):514–541
Zhang B, Zhang LL, Zhang SC, Shi HZ, Cai WM (2005) The influence of pH on hydrolysis and acidogenesis of kitchen wastes in two-phase anaerobic digestion. Environ Technol 26(3):329–340
Zhen G, Lu X, Kato H, Zhao Y, Li YY (2017) Overview of pretreatment strategies for enhancing sewage sludge disintegration and subsequent anaerobic digestion: current advances, full-scale application and future perspectives. Renew Sustain Energy Rev 69:559–577

Cross-feeding in Anaerobic Digestion Ecosystems

Jin Du, Bing Guo, and Guangxue Wu

Abstract Metabolic cross-feeding is an interaction between microbial communities, where a compound produced by the metabolism of one microorganism is utilized by another. There are two types of microbial cross-feeding within anaerobic digestion ecosystems: cascade and reciprocal. Cascade cross-feeding is unidirectional, typically involving catabolic processes and byproducts such as organic acids. In contrast, reciprocal cross-feeding is bidirectional or multidirectional, usually involving anabolic processes and cooperative interactions, exchanging metabolites like amino acids. Cross-feeding interactions based on amino acids and vitamins are common in microbial communities. The adoption of cross-feeding in anaerobic digestion can advance the mechanistic understanding of anaerobic digestion processes, improve better exploration of metabolic pathways, and enhance system efficiency for better-recovering resources.

Keywords Amino acids · Cross-feeding · Microbial interactions · Syntrophy · Vitamins

J. Du
Shenzhen International Graduate School, Tsinghua University, Shenzhen 518055, Guangdong, China

B. Guo
School of Sustainability, Civil and Environmental Engineering, University of Surrey, Guildford GU2 7XH, UK

G. Wu (✉)
Civil Engineering, School of Engineering, College of Science and Engineering, University of Galway, Galway H91 TK33, Ireland
e-mail: guangxue.wu@universityofgalway.ie

G. Wu (ed.), *Anaerobic Digestion*, Green Energy and Technology,
https://doi.org/10.1007/978-3-031-69378-6_3

1 Introduction

Anaerobic digestion is an engineered ecosystem characterized by high substrate loading rates, complex organic compositions, and taxonomically diverse microbial communities. During anaerobic digestion, microorganisms with complementary metabolic functions work together to make biological reactions thermodynamically feasible and consequently carry out efficient transformations from organic wastes to biogas. The feedstock and operational conditions shape microbial communities, and in turn, metabolic activities of these organisms transform the environment that they are living in. This transformation further affects the metabolism of other co-occurring microorganisms. In addition, competition among microorganisms for limiting resources, such as nutrients, minerals, and spaces creates a strong selection pressure for microbial survival.

Both environment-to-organism and organism-to-organism pressures induce a variety of microbial interactions. These complex interactions can improve system performance and enhance the robustness to environmental stressors. Specifically, microbial differentiation promotes division of labor and induces cross-feeding, thereby increasing metabolic efficiency because microorganisms with different physiologies vary in their sensitivities to environmental stresses (Antoniewicz 2020; Evans et al. 2020). Furthermore, metabolic exchange can promote syntrophic interactions, and these cooperative interactions can further strengthen systems against environmental and ecological stresses, facilitating the evolution of more complex traits (Cavaliere et al. 2017). Therefore, understanding anaerobic digestion with the incorporation of cross-feeding among microbial interactions can advance the development of novel technologies and techniques for regulating microbial communities.

2 Definition, Classification, and Outcomes of Cross-feeding

2.1 *Definition of Cross-feeding*

In anaerobic digestion ecosystems, one common consortium interaction motif is cross-feeding, in which essential metabolites are exchanged and shared among community members. Although it is common that microorganisms often excrete various metabolites that may impact the behavior of their neighbors, these metabolites are not always involved in cross-feeding. The terminology to describe metabolic cross-feeding is varied across the literature. Some publications only use cross-feeding to refer to interactions involving two-way exchange of nutrients (Sun et al. 2019; West et al. 2006). However, a broader definition of metabolic cross-feeding is now accepted. According to Smith et al. (2019), metabolic cross-feeding is defined as an interaction between microbial strains where the resulting metabolites from the metabolism of one strain are further utilized by another strain. The following criteria for cross-feeding were proposed by Fritts et al. (2021) and McKinlay (2023): (i) metabolites including molecules, electrons, and protons should be transported from a producer to a recipient; (ii) those transferred metabolites should either be assimilated or participate in the energy conservation in the recipient and/or producer; (iii) the assimilation or energy metabolism related to transferred metabolites should affect the fitness between the producer and/or recipient; and (iv) the producer and recipient should be genotypically or phenotypically distinct populations.

2.2 *Classification of Cross-feeding*

Generally, two types of microbial cross-feedings, i.e., cascade (unidirectional, and usually catabolism and byproducts related, such as organic acids) and reciprocal (bidirectional or multidirectional, and usually anabolism and cooperation related, such as amino acids), exist within anaerobic digestion ecosystems (Fig. 1; Boza et al. 2023; D'Souza et al. 2018).

The biodegradation of organic carbon sequentially to methane can be considered as the cascade cross-feeding involving one-way consumption, where the product of one species is used as the food for the following species. The cascade interaction in anaerobic digestion is also named as syntrophy, referring to all microbial cooperative metabolisms that have a positive impact on the growth and survival of another species. The cascade degradation of substrate requires the cooperation of serial species, thereby promoting biodiversity.

On the other hand, reciprocal cross-feeding means the consumption of metabolic products by all species through cooperative interactions (Escalante et al. 2015). Reciprocal cross-feeding occurs when two or more distinct lineages secrete the same

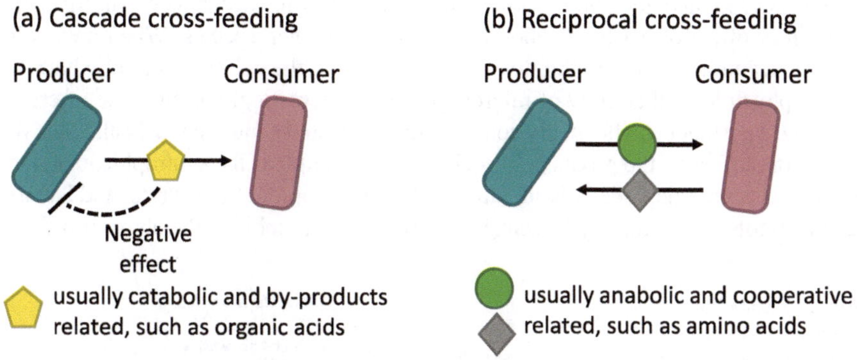

Fig. 1 Two types of microbial cross-feedings including **a** cascade cross-feeding (unidirectional, and also named as syntrophy in anaerobic digestion), and **b** reciprocal cross-feeding (bidirectional or multidirectional)

extracellular metabolite or different metabolites, as a public good, rendering an otherwise inaccessible resource accessible, facilitating all partners' growth (Boza et al. 2023).

2.3 Types of Cross-feeding Metabolites

Metabolites participating in cross-feeding are regarded as biosynthetic intermediates other than communicating signals or toxins (Fig. 2; Fritts et al. 2021). Cross-feeding metabolites can broadly be divided into two categories: the first group covers those that are utilized directly in central metabolism, e.g., organic matters, electron donors, and electron acceptors; and the second group covers essential biosynthetic nutrients, e.g., amino acids, vitamins, and cofactors (Culp and Goodman 2023).

According to whether cross-feeding metabolites are useful to the producer, there are two types of metabolites: metabolic wastes and communally valuable compounds (Fritts et al. 2021). Although metabolic wastes provide no further benefits to the producer, they may serve as electron sources for the recipient microbes. This phenomenon is especially prominent in anaerobic microbial communities because anaerobic fermenting bacteria present a more diverse lifestyle in terms of varied coupling of electron donors and acceptors, thus establishing more cross-feeding interactions (Pacheco et al. 2019). On the other hand, communally valuable compounds are essential metabolites, or public goods, that would benefit both the producer and recipient, including ammonia, amino acids, nucleobases, and vitamins. The evolution of cross-feeding of communally valuable compounds is highly vital for the survival of auxotrophic microbes who cannot produce those costly compounds, like amino acids and vitamins.

Cross-feeding of metabolic wastes can be a detoxifying mechanism to mitigate the inhibitory effect of waste metabolites or reactive oxygen species. When metabolic wastes accumulate to the inhibitory concentration in the cells, it would be harmful to the physiological activity of microorganisms. Interestingly, in the case of cross-feeding, a partner can effectively consume the waste and simultaneously alleviate the inhibitory effect on the producer. A well-known example is the syntrophic methanogenesis: bacteria degrade volatile fatty acids (e.g., propionate, butyrate, and lactate) and alcohols, producing hydrogen (H_2) or formate that inhibits the above processes

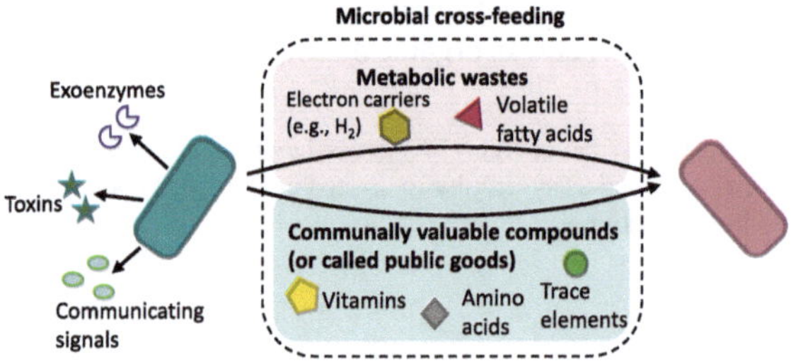

Fig. 2 Metabolites associated with microbial cross-feeding and other secreted metabolites

unless methanogens consume those molecules such as for carbon dioxide (CO_2) reduction to methane. The utilization of product can induce an energy gradient for benefiting the syntrophic activities. Consequently, the unfavorable reaction will be pulled forward.

2.4 *Outcomes of Cross-feeding*

Cross-feeding can induce competition, mutualism, commensalism, or exploitation (Fig. 3), which can be achieved through the production of a metabolite by one species that benefits or harms another species but can also occur when a harmful metabolite to one species is consumed by and benefits another species (Culp and Goodman 2023). Cross-feeding has different ecological outcomes depending on the directionality, where unidirectional cross-feeding is equivalent to commensalism and bidirectional cross-feeding can be considered as mutualism.

The occurrence of cross-feeding may depend on the microbial concentration (Kelly et al. 2021). Usually, a low biomass density induces collaborative interactions while a high biomass density causes competition. Competitive resource allocation can induce overflow metabolism, and the overflow byproducts can stabilize cooperative interactions and lead to cross-feeding (Henson and Heys 2018). Therefore, the benefit of cooperative cross-feeding applies only in the range of intermediate cell densities, because resource concentrations are too low to offset the cost of cooperation at the low density while resources become limiting causing microbial competition at the high density (Bull and Harcombe 2009).

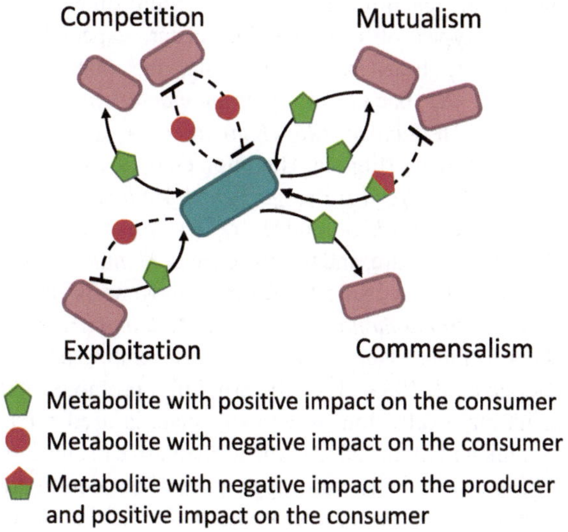

Fig. 3 Ecological outcomes of cross-feeding depending on the types of interactions. Adapted from Culp and Goodman (2023)

Cross-feeding can shape microbial communities and their functions (Gottstein et al. 2016). For example, cross-feeding can alter cellular metabolic fluxes within individual species. In iron-limited *Escherichia coli* cultures, more than 70% of the catabolized glucose carbon was secreted as organic byproducts, primarily lactate; in ammonium-limited *E. coli* chemostat cultures, approximately 40% of the catabolized glucose carbon was secreted as organic byproducts, primarily acetate; in contrast, glucose-limited cultures secreted up to 4% of the glucose as organic byproducts, primarily formate (Henson and Heys 2018). Antoniewicz (2020) showed that although 46 strains alone could not grow with glucose as the sole substrate, 17% of the paired cocultures could, indicating that cross-feeding may benefit microbial growth.

Dynamics of cross-feeding interactions can be affected by (i) the individual enzymes involved in the biosynthesis of the exchanged metabolites, (ii) the cellular allocation of limited resources (e.g., nutrients, expression machinery, and space), (iii) the biotic composition of the microbial community, as well as features of the ecological environment like (i) chemical diffusion, (ii) nutrient availability, or (iii) spatial structure (D'Souza et al. 2018). Some studies found that cross-feeding may be an important driver for maintaining metabolism under nutrient-limited conditions, benefiting microbial survival (Fritts et al. 2021).

3 Cross-feeding for Degrading Organic Matter

Cross-feeding among microorganisms can make unfeasible reactions possible, especially in anaerobic digestion systems where energy production in these biological reactions is close to equilibrium. In anaerobic digestion systems, various types of cross-feeding occur, especially through the interspecies H_2 or formate transfer. For example, the main drivers in microbial interaction between the bacterium *Syntrophobacter fumaroxidans* and the methanogenic archaeon *Methanospirillum hungatei* were attributed to the concurring exchange of H_2 and formate (Fig. 4a; Hamilton et al. 2015). Within the cooperative interactions between *Desulfovibrio vulgar*is and *Methanococcus maripaludis*, the oxidation of lactate to acetate, CO_2, and H_2 by *Desulfovibrio vulgar*is can be inhibited by the high H_2 partial pressure, and the presence of *Methanococcus maripaludis* can alleviate the H_2 inhibition by using H_2 for methanogenesis (Hillesland and Stahl 2010). Moreover, *Syntrophomonas wolfei* and *Methanobacterium formicicum* exchange either H_2 or formate as an electron carrier depending on the spatial distance between cells (Boone et al. 1989). Usually, proximity promotes electron transfer via H_2, because of its rapid diffusion, whereas formate is used for long-distance electron transfer (Schink and Stams 2013). The fermentative bacterium *Pelotomaculum thermopropionicum* was shown to form aggregates when cocultured with the methanogen *Methanothermobacter thermautotrophicus* for facilitating the transfer of H_2 (Ishii et al. 2005).

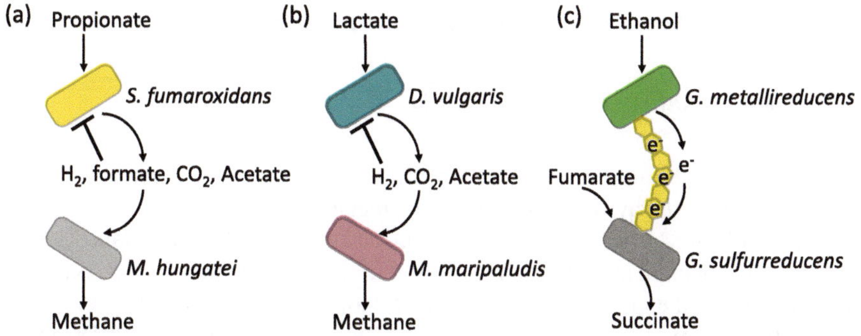

Fig. 4 Syntrophic association for the degradation of organic matters between cocultures of **a** *Syntrophobacter fumaroxidans* and *Methanospirillum hungatei*, **b** *Desulfovibrio vulgar*is and *Methanococcus maripaludis*, and **c** *Geobacter metallireducens* and *Geobacter sulfurreducens*

A cross-feeding example that has been established in laboratories is the synthetic community composed of a sulfate-reducing bacteria *Desulfovibrio vulgaris* and a methanogen *Methanococcus maripaludis* (Fig. 4b; Hillesland et al. 2014). When the environmental condition is deprived of sulfate, the lactate degradation by *D. vulgaris* on its own is thermodynamically infeasible. But the metabolism of *M. maripaludis* could support the growth of both species because *M. maripaludis* could consume the product of lactate degradation, hence pulling forward of the overall reaction. More importantly, evolved *D. vulgaris* was found to lose its ability to utilize sulfate, indicating that it became specialized for obligating syntrophic growth with *M. maripaludis*. This experiment illustrated the evolution of obligate cross-feeding which was initially imposed by environmental conditions and could be ultimately fixed as genes (in other words, the metabolic abilities of microorganisms might be altered) over time (Fritts et al. 2021).

The recently discovered direct interspecies electron transfer (DIET) is also one type of cross-feeding. DIET between two *Geobacter* species could be stimulated by optimizing their growth rates (Nagarajan et al. 2013). For example, a DIET syntrophy between *Geobacter metallireducens* and *Geobacter sulfurreducens* was confirmed, where electrons were transferred from *G. metallireducens* to *G. sulfurreducens*, relieving the thermodynamic barrier on ethanol fermentation and providing electrons to drive fumarate respiration (Fig. 4c; Summers et al. 2010). The DIET syntrophy could be promoted by forming multicellular aggregates, which was dependent on *G. sulfurreducens* nanowire components, i.e., pili and an outer membrane cytochrome (Summers et al. 2010).

4 Cooperative Cross-feeding of Amino Acids and Others

4.1 Cross-feeding of Amino Acids

Within the anaerobic digestion ecosystem, a large proportion of microorganisms may lack metabolic pathways to produce essential metabolites such as amino acids, vitamins, and cofactors. The trait that microorganisms cannot synthesize certain essential metabolites is named as auxotrophy (Fritts et al. 2021). Auxotrophs are organisms fully relying on the environmental supply of certain compounds required for their growth (Cavaliere et al. 2017). For example, Zengler and Zaramela (2018) found that more than 98% of all sequenced microorganisms are auxotrophs, depending on external nutrients and/or byproducts from other microorganisms. Since the synthesis of some microbial compounds costs a lot of energy, auxotrophs could gain great advantages by using external metabolite resources, which can improve their ecological fitness (Fig. 5; Oña and Kost 2022). From the perspective of the whole community, the evolution of cross-feeding can allocate the biosynthetic burden of energetically costly metabolites for benefiting the entire community. Therefore, natural selection may favor auxotrophs by saving the costs of metabolite production (D'Souza et al. 2018).

Amino acids are building blocks of proteins as well as important currency in microbial communities. They can be taken up as energy and nitrogen sources and serve as essential nutrients for amino acid auxotrophs (Culp and Goodman 2023). Amino acid auxotrophs are prevalent, and it is estimated through metabolic pathway analysis that 24–98% of bacteria are unable to synthesize at least one kind of amino acid. Moreover, amino acid auxotrophs are identified in almost all communities by the Earth Microbiome Project, which investigated more than 12,000 free-living and host-associated natural communities (Culp and Goodman 2023).

In microbial ecosystems, microorganisms may evolve to cross-feed biosynthetically costly amino acids, such as methionine, lysine, isoleucine, arginine, and

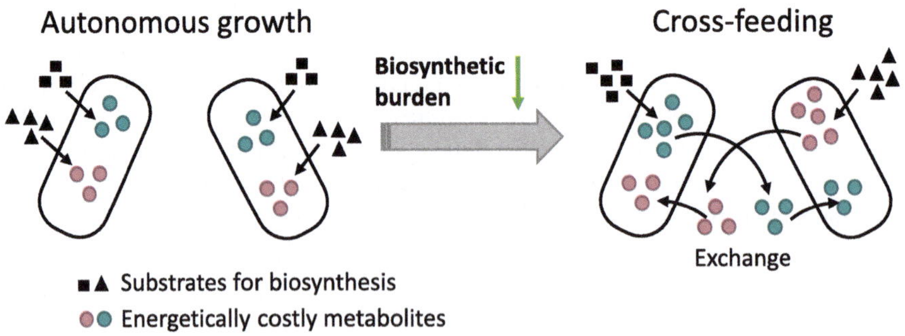

Fig. 5 Schematic illustrations of the benefit of cross-feeding in comparison with autonomous growth from the perspective of energy saving. Adapted from D'Souza et al. (2018)

aromatics. Previous studies have paired amino acid auxotrophs and observed cross-feeding between *E. coli* and *Zymomonas mobilis* by exchanging essential amino acids (Kosina et al. 2016). Furthermore, Kelly et al. (2021) investigated spontaneous obligate cross-feeding between two industrially relevant bacteria, *E. coli* and *Bacillus megaterium* and found that cross-feeding between auxotrophs is a viable method for controlling community composition within synthetic communities. Embree et al. (2015) quantified the simultaneous exchange of several metabolites among the five most abundant members in a complex methanogenic community and revealed the existence of an intertwined network of amino acid auxotrophs. Microfluidic and modeling experiments with closely packed amino acid cross-feeding auxotrophs showed that a single cell might only cross-feed to a relatively small neighborhood of one to four cell lengths (Dal Co et al. 2020). The distance over which cross-feeding can occur is likely further decreased in the presence of competitors.

Recently, the cross-feeding function was assessed by analyzing extracellular polymeric substances-associated amino acids in microbial aggregates collected from 14 lab-scale anaerobic digesters, as well as deciphering their genetically biosynthetic potential by syntrophic bacteria and methanogens (Du et al. 2022). The total concentration of essential amino acids ranged from 1.2 mg/g VSS to 174.0 mg/g VSS. The percentages of glutamic acid (8.5–37.6%), lysine (2.7–22.6%), alanine (5.6–13.2%), and valine (3.0–10.4%) to the total amount of detected amino acids were the highest in most samples. Through metagenomics analysis, several investigated syntrophs (i.e., *Smithella, Syntrophobacter*, *Syntrophomonas*, and *Mesotoga*) and methanogens (i.e., *Methanothrix* and *Methanosarcina*) were auxotrophs, but the genetic ability of syntrophs and methanogens to synthesize some essential amino acids could be complementary, implying potential cross-feeding partnership. For example, as shown in Fig. 6, several syntrophs (i.e., *Smithella* and *Mesotoga*) did not have the genetic capability to produce alanine (Ala) but possessed genes encoding proline (Pro) synthesis. On the contrary, both investigated methanogens *Methanothrix* and *Methanobacterium* had the genetic potential to produce Ala rather than Pro. Besides, there were positive correlations between the syntrophs and the methanogens according to microbial network analysis. Therefore, these microorganisms might benefit each other by complementing the deficiency of Ala and Pro synthesis through potential amino acids cross-feeding.

Cooperative cross-feeding can reduce the biosynthetic burden of amino acids in microbiomes, which may cause the loss of amino acid biosynthetic genes. This metabolic change can be explained by the Black Queen Hypothesis, which proposes that the loss of genes can induce a selective advantage when the essential function or nutrient is provided in the environment, such as through public goods (Heizer et al. 2006).

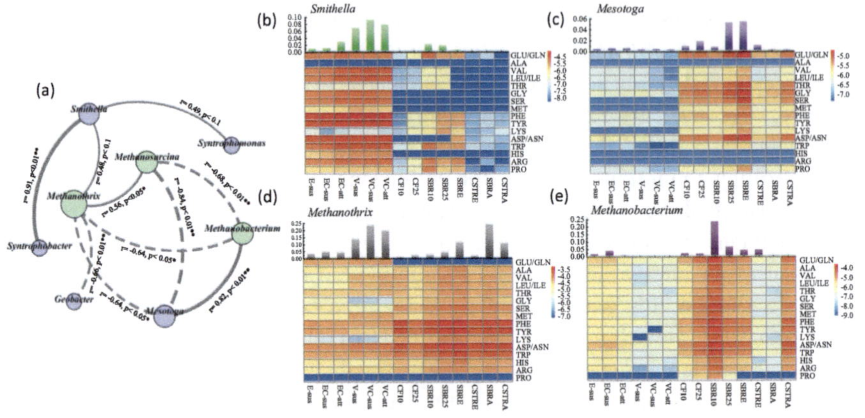

Fig. 6 **a** Potential microbial network among dominant syntrophic bacteria and methanogens; **b–e** genetic ability of some co-occurred microorganisms for amino acids biosynthesis based on metagenomics data from 14 anaerobic digestion systems (the Y coordinate of each bar chart above the heatmap represents the relative abundance of the genus. Adapted from Du et al. (2022)

4.2 *Cross-feeding of Vitamins*

Vitamins are essential compounds for enzymatic function in the cellular metabolism, and vitamins-based cross-feeding interactions also widely exist in microbial communities. In contrast to major organic substrates and amino acids, an extremely low concentration of vitamins at the nanomolar level is required for growth. Meanwhile, vitamins are energetically expensive to synthesize, and auxotrophs lacking the biosynthetic ability of vitamins are common. In this way, vitamins are commonly imported from the environment via inter-microbial cross-feeding. Microbial cross-feeding using B vitamins as substrates (e.g., thiamine [B_1], riboflavin [B_2], niacin [B_3], pantothenate [B_5], pyridoxine [B_6], biotin [B_7], folate [B_9], and corrinoids [B_{12}]) has been well-studied (Putnam and Goodman 2020; Uebanso et al. 2020). Vitamin B_{12} belongs to a group of closely related compounds known as corrinoids. They are ideal cross-fed metabolites because (1) they are essential for microbial growth as about 75% of bacteria encode corrinoid-dependent enzymes; (2) biosynthetic processes of corrinoids are highly expensive, requiring about 30 enzymatic steps, and roughly half of bacteria are auxotrophs; and (3) the import of corrinoids and their biosynthetic intermediates is common, achieved through a dedicated scavenging/transport system (Putnam and Goodman 2020; Uebanso et al. 2020). Furthermore, vitamin B_{12} can also be in exchange with other metabolites. A well-studied example is the cross-feeding between *Eubacterium hallii* (corrinoid producer and 1,2-propanediol consumer) and *Akkermansia muciniphila* (corrinoid consumer and 1,2-propanediol producer) (Culp and Goodman 2023).

5 Possible Practical Applications in Anaerobic Digestion Systems

5.1 The Application of Cross-feeding in Granules

Microbial aggregation, named as granules or biofilm without carriers, is a common phenomenon within anaerobic digesters, which can create environmental gradients and further lead to cross-feeding (McKinlay 2023). Within the spatial structured biofilm, the heterogeneous distribution of chemical gradient causes different accessibility to resources for microorganisms. This can enrich cooperative microbial communities to be segmented for achieving stable system function and inducing the selection of complex functionalities (Cavaliere et al. 2017), depending on the location of microorganisms within biofilm. Microorganisms performing cross-feeding may live closely within the biofilm, and the cooperative interactions among different types of microorganisms can benefit system performance (Fig. 7).

Within the biofilm, spatially structured, aggregative community possesses a high cell density, strengthening competition for resources such as space and nutrients, and finally, colonies need to disperse to populate new substrates (D'Souza et al. 2018). For example, dual-species biofilms possessed higher biomass and cell numbers than mono-species biofilms (Sztajer et al. 2014). Pande et al. (2015) found that cell attachment was contingent on the nutritional status of auxotrophs. Spatially structured environments and limited nutrient availability should facilitate the evolution of metabolic interactions, including cross-feeding, which can help to maintain genotypic diversity within natural microbial populations (Germerodt et al. 2016).

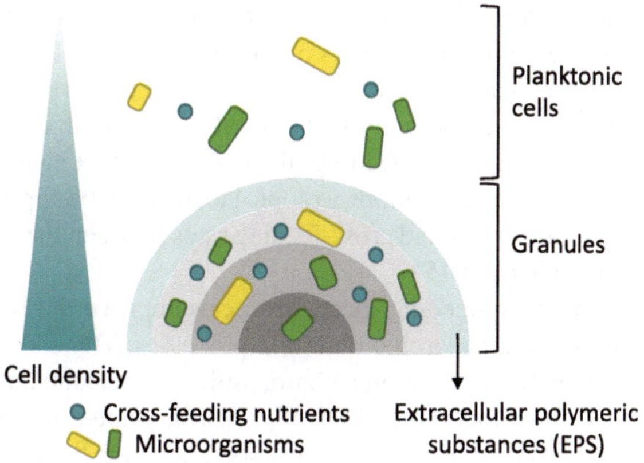

Fig. 7 Microbial cross-feeding in the spatially structured granules and planktonic phase

5.2 The Application of Cross-feeding in Regulating Community Compositions

Some functional microorganisms with desired metabolic capabilities, such as pollutants degradation or bioproduction of value-added products, are difficult to be enriched and kept stabilized in anaerobic digestion systems. This challenge can be overcome by synthetic ecological approaches in laboratory and industrial settings (Kelly et al. 2021). The cross-feeding between auxotrophic pairs can be engineered and used to enable interspecies cooperative growth through a mutualistic ecological interaction.

Moreover, auxotrophy can be used to control the community composition (Ziesack et al. 2019). For example, by selectively supplementing the media for microbial growth, the cell density of an auxotroph can be controlled, allowing for fine-tuning of the community composition (Losoi et al. 2019). These approaches offer engineers huge opportunities to harness the potential of synthetic communities by selectively choosing and combining diverse microorganisms, and further to optimize the community composition for achieving a specific function.

6 Future Perspectives

The adoption of cross-feeding in anaerobic digestion can advance the mechanistic understanding of anaerobic digestion processes. Specifically, understanding microbial ecology of anaerobic digestion through multi-dimensional interactions including cross-feeding can broaden the ecological theory about stability of mutualistic interactions. Microbial interactions can induce synergistic responses that cannot be predicted through activities of individual species (Culp and Goodman 2023). A new mathematical model should be developed to clarify metabolite cross-feeding interactions and community metabolic fluxes (Antoniewicz 2020). For example, when incorporating cross-feeding, the competitive exclusion principle breaks down (Vet et al. 2018). Furthermore, isotopic labeling can be applied to quantify species-specific metabolic fluxes and cross-feeding rates between the populations (Gebreselassie and Antoniewicz 2015).

The cross-feeding mechanism can improve the better exploration of metabolic pathways in anaerobic digestion processes. The diversity of functions and division of labor enable microbial communities to cycle the nutrients and to perform complicated functions more efficiently than individual populations (Che and Men 2019). In anaerobic digestion systems, operation near the equilibrium can provide a greater energetic efficiency by harnessing more of the energy released by central metabolism (Park et al. 2019). A canonical example is overflowing metabolism, where cells incompletely oxidize substrates to metabolites that could, in principle, be further oxidized to generate additional ATP (Dolinšek et al. 2022). However, operating central metabolism near the equilibrium involves reversible reactions, translating to a slow metabolism and a low growth rate (Fritts et al. 2021). At low growth rates, cells shrink to maximize the surface area, which allows for increasing the membrane proteome capacity with respect to the global proteome capacity; while at high growth rate, cells tend to be bigger in volume as a way of lowering the area-to-volume ratio and maximizing the proteome capacity in the cytoplasm, since transport is no longer the growth limiting factor (Regueira et al. 2021).

Cross-feeding can enhance system efficiency for better-recovering resources through anaerobic digestion processes. In addition to the conventional top-down design, nowadays, bottom-up synthetic microbiology has been applied and developed rapidly. While sharing common concepts and methodologies, the creation of synthetic communities differs fundamentally from single-strain engineering by focusing on the coordination of species and the division of labor across members. Specifically, compared with single strains, the community-based organization reduces individual metabolic burden by assigning tasks into multiple strains, thereby helping cells to fulfill designed function with robustness and predictability (Deter and Lu 2022). Although the bottom-up approach facilitated by a variety of synthetic biology tools is a simple and common method to construct synthetic microbial consortia, the top-down strategy offers naturally occurring microbial interdependence that might be missing from an artificial combination of engineered microorganisms using the bottom-up strategy (Che and Men 2019). Therefore, the integration of both the bottom-up approach and the top-down strategy would advance synthetic microbiology in anaerobic digestion ecosystems.

7 Take-Home Messages

(1) Metabolic cross-feeding is defined as an interaction between microbial communities where the resulting molecules from the metabolism of one strain are further utilized by another strain.
(2) Two types of microbial cross-feedings, i.e., cascade (unidirectional, and usually catabolic and byproducts related, such as organic acids) or reciprocal (bidirectional or multidirectional, and usually anabolic and cooperative related, such as amino acids), exist within anaerobic digestion ecosystems.
(3) Amino acids-based and vitamins-based cross-feeding interactions widely exist in microbial communities. The recently discovered DIET is also one type of cross-feeding.
(4) The adoption of cross-feeding in anaerobic digestion can advance the mechanistic understanding of anaerobic digestion processes, improving the better exploration of metabolic pathways and enhancing system efficiency for better-recovering resources.

References

Antoniewicz MR (2020) A guide to deciphering microbial interactions and metabolic fluxes in microbiome communities. Curr Opin Biotechnol 64:230–237

Boone DR, Johnson RL, Liu Y (1989) Diffusion of the interspecies electron carriers H_2 and formate in methanogenic ecosystems and its implications in the measurement of Km for H_2 or formate uptake. Appl Environ Microbiol 55(7):1735–1741

Boza G, Barabás G, Scheuring I, Zachar I (2023) Eco-evolutionary modelling of microbial syntrophy indicates the robustness of cross-feeding over cross-facilitation. Sci Rep 13:907

Bull JJ, Harcombe WR (2009) Population dynamics constrain the cooperative evolution of cross-feeding. PLoS ONE 4(1):e4115

Cavaliere M, Feng S, Soyer OS, Jiménez JI (2017) Cooperation in microbial communities and their biotechnological applications. Environ Microbiol 19(8):2949–2963

Che S, Men Y (2019) Synthetic microbial consortia for biosynthesis and biodegradation: promises and challenges. J Ind Microbiol Biotechnol 46(9–10):1343–1358

Culp EJ, Goodman AL (2023) Cross-feeding in the gut microbiome: ecology and mechanisms. Cell Host Microbe 31(4):485–499

Dal Co A, van Vliet S, Kiviet DJ, Schlegel S, Ackermann M (2020) Short-range interactions govern the dynamics and functions of microbial communities. Nat Ecol Evol 4(3):366–375

D'Souza G, Shitut S, Preussger D, Yousif G, Waschina S, Kost C (2018) Ecology and evolution of metabolic cross-feeding interactions in bacteria. Nat Prod Rep 35(5):455–488

Deter HS, Lu T (2022) Engineering microbial consortia with rationally designed cellular interactions. Curr Opin Biotechnol 76:102730

Dolinšek J, Ramoneda J, Johnson DR (2022) Initial community composition determines the long-term dynamics of a microbial cross-feeding interaction by modulating niche availability. ISME Commun 2(1):77

Du J, Yin Q, Zhou X, Guo Q, Wu G (2022) Distribution of extracellular amino acids and their potential functions in microbial cross-feeding in anaerobic digestion systems. Biores Technol 360:127535

Embree M, Liu JK, Al-Bassam MM, Zengler K (2015) Networks of energetic and metabolic interactions define dynamics in microbial communities. Proc Natl Acad Sci 112(50):15450–15455

Escalante AE, Rebolleda-Gómez M, Benítez M, Travisano M (2015) Ecological perspectives on synthetic biology: insights from microbial population biology. Front Microbiol 6:143

Evans CR, Kempes CP, Price-Whelan A, Dietrich LEP (2020) Metabolic heterogeneity and cross-feeding in bacterial multicellular systems. Trends Microbiol 28(9):732–743

Fritts RK, McCully AL, McKinlay JB (2021) Extracellular metabolism sets the table for microbial cross-feeding. Microbiol Mol Biol Rev 85(1):e00135-e220

Gebreselassie NA, Antoniewicz MR (2015) ^{13}C-metabolic flux analysis of co-cultures: a novel approach. Metab Eng 31:132–139

Germerodt S, Bohl K, Lück A, Pande S, Schröter A, Kaleta C, Schuster S, Kost C (2016) Pervasive selection for cooperative cross-feeding in bacterial communities. PLoS Comput Biol 12(6):e1004986

Gottstein W, Olivier BG, Bruggeman FJ, Teusink B (2016) Constraint-based stoichiometric modelling from single organisms to microbial communities. J R Soc Interface 13(124):20160627

Hamilton JJ, Calixto Contreras M, Reed JL (2015) Thermodynamics and H_2 transfer in a methanogenic, syntrophic community. PLoS Comput Biol 11(7):e1004364

Heizer EM Jr, Raiford DW, Raymer ML, Doom TE, Miller RV, Krane DE (2006) Amino acid cost and codon-usage biases in 6 prokaryotic genomes: a whole-genome analysis. Mol Biol Evol 23(9):1670–1680

Henson MA, Heys JJ (2018) Competitive resource allocation to metabolic pathways contributes to overflow metabolisms and emergent properties in cross-feeding microbial consortia. Biochem Soc Trans 46(2):269–284

Hillesland KL, Stahl DA (2010) Rapid evolution of stability and productivity at the origin of a microbial mutualism. Proc Natl Acad Sci 107(5):2124–2129

Hillesland KL, Lim S, Flowers JJ, Turkarslan S, Pinel N, Zane GM, Elliott N, Qin Y, Wu L, Baliga NS, Zhou J (2014) Erosion of functional independence early in the evolution of a microbial mutualism. Proc Natl Acad Sci 111(41):14822–14827

Ishii S, Kosaka T, Hori K, Hotta Y, Watanabe K (2005) Coaggregation facilitates interspecies hydrogen transfer between *Pelotomaculum thermopropionicum* and *Methanothermobacter thermautotrophicus*. Appl Environ Microbiol 71(12):7838–7845

Kelly EE, Fischer AM, Collins CH (2021) Drawing up a collaborative contract: amino acid cross-feeding between interspecies bacterial pairs. Biotechnol Bioeng 118(8):3138–3149

Kosina SM, Danielewicz MA, Mohammed M, Ray J, Suh Y, Yilmaz S, Singh AK, Arkin AP, Deutschbauer AM, Northen TR (2016) Exometabolomics assisted design and validation of synthetic obligate mutualism. ACS Synth Biol 5(7):569–576

Losoi PS, Santala VP, Santala SM (2019) Enhanced population control in a synthetic bacterial consortium by interconnected carbon cross-feeding. ACS Synth Biol 8(12):2642–2650

McKinlay JB (2023) Are bacteria leaky? Mechanisms of metabolite externalization in bacterial cross-feeding. Annu Rev Microbiol 77:277–297

Nagarajan H, Embree M, Rotaru AE, Shrestha PM, Feist AM, Palsson BØ, Lovley DR, Zengler K (2013) Characterization and modelling of interspecies electron transfer mechanisms and microbial community dynamics of a syntrophic association. Nat Commun 4:2809

Oña L, Kost C (2022) Cooperation increases robustness to ecological disturbance in microbial cross-feeding networks. Ecol Lett 25(6):1410–1420

Pacheco AR, Moel M, Segrè D (2019) Costless metabolic secretions as drivers of interspecies interactions in microbial ecosystems. Nat Commun 10:103

Pande S, Shitut S, Freund L, Westermann M, Bertels F, Colesie C, Bischofs IB, Kost C (2015) Metabolic cross-feeding via intercellular nanotubes among bacteria. Nat Commun 6:6238

Park JO, Tanner LB, Wei MH, Khana DB, Jacobson TB, Zhang Z, Rubin SA, Li SH, Higgins MB, Stevenson DM, Amador-Noguez D, Rabinowitz JD (2019) Near-equilibrium glycolysis supports metabolic homeostasis and energy yield. Nat Chem Biol 15(10):1001–1008

Putnam EE, Goodman AL (2020) B vitamin acquisition by gut commensal bacteria. PLoS Pathog 16(1):e1008208

Regueira A, Lema JM, Mauricio-Iglesias M (2021) Microbial inefficient substrate use through the perspective of resource allocation models. Curr Opin Biotechnol 67:130–140

Schink B, Stams AJM (2013) Syntrophism among prokaryotes. In: Rosenberg E, DeLong EF, Lory S, Stackebrandt E, Thompson E (eds) The prokaryotes: prokaryotic communities and ecophysiology. Springer, Berlin, Heidelberg, pp 471–493

Smith NW, Shorten PR, Altermann E, Roy NC, McNabb WC (2019) The classification and evolution of bacterial cross-feeding. Front Ecol Evol 7:153

Summers ZM, Fogarty HE, Leang C, Franks AE, Malvankar NS, Lovley DR (2010) Direct exchange of electrons within aggregates of an evolved syntrophic coculture of anaerobic bacteria. Science 330(6009):1413–1415

Sun Z, Koffel T, Stump SM, Grimaud GM, Klausmeier CA (2019) Microbial cross-feeding promotes multiple stable states and species coexistence, but also susceptibility to cheaters. J Theor Biol 465:63–77

Sztajer H, Szafranski SP, Tomasch J, Reck M, Nimtz M, Rohde M, Wagner-Döbler I (2014) Cross-feeding and interkingdom communication in dual-species biofilms of *Streptococcus mutans* and *Candida albicans*. ISME J 8(11):2256–2271

Uebanso T, Shimohata T, Mawatari K, Takahashi A (2020) Functional roles of B-vitamins in the gut and gut microbiome. Mol Nutr Food Res 64(18):2000426

Vet S, de Buyl S, Faust K, Danckaert J, Gonze D, Gelens L (2018) Bistability in a system of two species interacting through mutualism as well as competition: Chemostat vs. Lotka-Volterra equations. Plos One 13(6):e0197462

West SA, Griffin AS, Gardner A, Diggle SP (2006) Social evolution theory for microorganisms. Nat Rev Microbiol 4(8):597–607

Zengler K, Zaramela LS (2018) The social network of microorganisms—how auxotrophies shape complex communities. Nat Rev Microbiol 16(6):383–390

Ziesack M, Gibson T, Oliver JKW., Shumaker AM, Hsu BB, Riglar DT, Giessen TW, DiBenedetto NV, Bry L, Way JC, Silver PA, Gerber GK (2019) Engineered interspecies amino acid cross-feeding increases population evenness in a synthetic bacterial consortium. mSystems 4(4):e00352

Microbial Ecology of Anaerobic Digestion

Qin Cao, Simon Mills, and Guangxue Wu

Abstract Anaerobic digestion contains diverse microbial communities with functions complementary to each other at different trophic levels for degrading organic carbon. A comprehensive understanding of microbial communities from taxonomic compositions, microbial diversity, and microbial interactions can benefit biogas production and maintain process stability. Application of ecological theories such as community assembly mechanisms or the r/K selection theory can be utilized to improve system management. Community assembly is mainly dependent on the deterministic processes, implying that community responses to operational adjustments may be predictable. However, stochastic processes can have contributions under some scenarios. In anaerobic digestion systems, the r/K ecological selection theory can be applied to enrich specific types of microorganisms, such as electroactive *Geobacter*, and to remove low-concentration pollutants by enriching K-strategists.

Keywords Community assembly · Microbial community · Microbial ecology · Microbial network · r/k selection theory

---

Q. Cao
CAS Key Laboratory of Environmental and Applied Microbiology, Environmental Microbiology Key Laboratory of Sichuan Province, Chengdu Institute of Biology, Chinese Academy of Sciences, Chengdu 610041, China

S. Mills
Department of Microbiology, School of Biological and Chemical Sciences and Ryan Institute, University of Galway, Galway H91 TK33, Ireland
e-mail: simon.mills@universityofgalway.ie

G. Wu (✉)
Civil Engineering, School of Engineering, College of Science and Engineering, University of Galway, Galway H91 TK33, Ireland
e-mail: guangxue.wu@universityofgalway.ie

G. Wu (ed.), *Anaerobic Digestion*, Green Energy and Technology,
https://doi.org/10.1007/978-3-031-69378-6_4

1 Introduction

During anaerobic digestion, organic carbon is sequentially degraded by syntrophic microorganisms within a complex ecosystem (Fig. 1). This typically occurs in four main steps: (1) hydrolysis of complex organics into proteins, carbohydrates, and long-chain fatty acids; (2) acidification of these products into shorter volatile fatty acids (VFAs) such as propionate and butyrate; (3) acetogenesis resulting in the generation of acetate along with H_2 and CO_2; and (4) methanogenesis of the products of acetogenesis into methane. There are often a multitude of species present which can carry out each step, conferring a high degree of functional redundancy to the community (Louca et al. 2018). Functional redundancy is often linked to system stability (Paulo et al. 2020) and thought to make anaerobic digestion communities more robust. Therefore, a high microbial diversity, which may impart more functional redundancy, is often thought to improve system stability of anaerobic digestion systems (Carballa et al. 2015; McCann 2000).

Understanding anaerobic digestion through the lens of established ecological theories can benefit system design, regulation, and optimization to improve efficiency and stability. Diverse theories can be adopted from non-microbial ecology (e.g., zoology and botany) or even other disciplines to understand the anaerobic digestion ecosystem. For example, Werner et al. (2014) proposed that the adoption of the market theory from economics for characterizing resource investment strategies (a key concept in functional ecology) can be applied to explain microbial cooperative

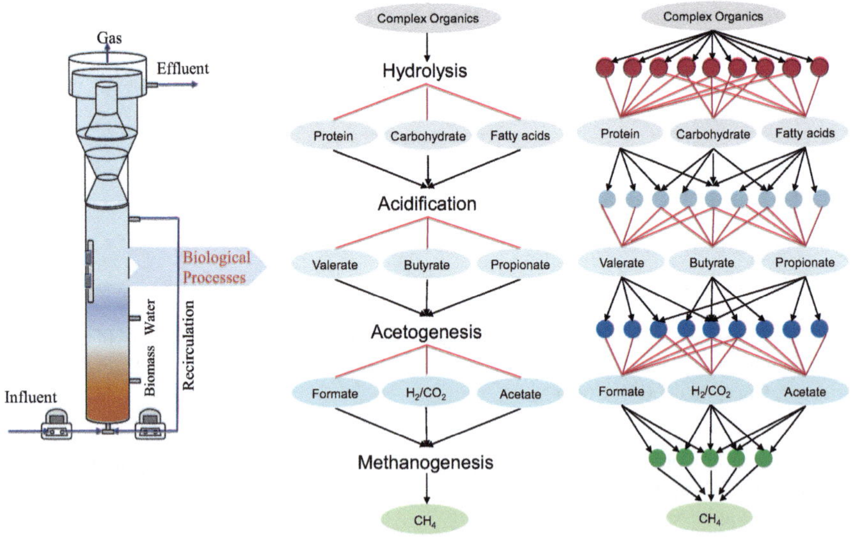

Fig. 1 Microbial degradation of organic carbon in anaerobic digestion ecosystems: four main reaction steps and the diverse responsible microbial communities. Circles with different colors indicate different functional groups of microorganisms

behaviors. The transition to microbial-based management of anaerobic digestion ecosystems cannot be achieved without integration and verification of ecological theory in these engineered microbial systems (Prosser et al. 2007).

2 Microbial Communities in Anaerobic Digestion Ecosystems

The process stability and effectiveness of anaerobic digestion depends on multiple syntrophic interactions among bacteria and archaea (Ziganshina et al. 2017). To enhance biogas production and maintain process stability, it is crucial to comprehensively understand microbial communities in terms of their taxonomic compositions, microbial diversity, interspecies interactions, ecological roles, and the influence of environmental and operational factors (Zhang et al. 2019). Specifically, monitoring syntrophic and methanogenic communities could be an important tool for describing the stability and resilience of anaerobic digestion systems and contribute to better practical applications (De Jonge et al. 2020).

Due to the syntrophic relationship among diverse types of microorganisms in anaerobic digestion microbial communities, conventional culture-based techniques can only isolate a limited number of microorganisms. Nowadays, culture-independent techniques based on high-throughput sequencing and bioinformatic analysis can provide vastly more information on anaerobic digestion microbial communities than culture-based techniques. For example, amplicon-based sequencing of 16S rRNA genes is commonly applied to study the taxonomic and phylogenetic composition of anaerobic digestion microbial communities (Dueholm et al. 2022). This allows ecological aspects of community composition such as diversity, community assembly, microbial networks, and the r/K selection theory to be studied. Additionally, shotgun metagenomics where total community genomic DNA is sequenced rather than short amplicons can be applied to determine potential functions and metabolic pathways in a microbial community (Campanaro et al. 2016). When combined with assembly-based approaches (Campanaro et al. 2020), metagenomics can even be used to recover whole genomes. However, there are some limitations to culture-independent approaches, e.g., polymerase chain reaction biases in amplicon-based studies or inaccuracies in databases (Lupo et al. 2021). Therefore, a combination of culture-dependent approaches (e.g., isolation and sequencing) and culture-independent approaches (e.g., metabarcoding) can be complementary in providing robust and comprehensive characterization of microbial communities within anaerobic digestion ecosystems (Ottoni et al. 2022). However, for studying purely ecological (rather than functional) aspects of community composition, e.g., diversity and community assembly, amplicon-based approaches are the best.

2.1 Community Composition

Bacterial populations in anaerobic digestion systems mainly include hydrolytic/fermentative bacteria, acetogenic bacteria, and homoacetogenic bacteria. Each group includes a complex mix of bacteria, many of which are evolutionarily distantly related but perform the same function. The archaeal populations of anaerobic digesters are primarily made up of methanogens, which have a mutualistic relationship with bacterial populations.

Hydrolytic or fermentative bacteria are mainly obligate and facultative anaerobic heterotrophic bacteria, which have a short doubling time and are greatly affected by substrate, temperature, and other operational parameters. This group is often made up of species from the Bacteroidetes, Firmicutes, and Proteobacteria (Calusinska et al. 2018; Campanaro et al. 2020). The main species are cellulose decomposing bacteria (*Bacteroides succinogenes, Bacteroides cellulosolvens, Butyrivibrio fibrisolvens, Ruminococcus flavefaciens, Clostridium cellobioparus,* and *Clostridium cellulovorans*), hemicellulose decomposing bacteria (*Bacteroides ruminocola, Bacteroides cellulosolvens, Bacteroides xylanisolvens,* and *Butyrivibrio fibrisolvens*), pectin decomposing bacteria (*Bacteroides succinogenes, Clostridium pectinovorum,* and *Bacteroides galacturonicus*), starch decomposing bacteria (*Bacteroides amylophilus*), protein decomposing bacteria (*Clostridium putrificum* and *Clostridium thermoputrificum*), and lipid decomposing bacteria (*Anaerovibrio lipolytica*) (Collins et al. 1994; Hamlin and Hungate 1956; Honda et al. 2013; Koeck et al. 2014; Robert et al. 2007).

Acetogenic bacteria are a class of strictly anaerobic syntrophic bacteria, which generate acetate, H_2 and CO_2 via oxidation of organic acids. This reaction becomes thermodynamically unfavorable at high H_2 partial pressures, and acetogens rely on a syntrophic interaction with hydrogenotrophic methanogens to reduce the H_2 partial pressure. Therefore, acetogens occupy a key niche in anaerobic digestion systems and act as a link between fermentative bacteria and methanogens, allowing electrons to flow from more complex molecules into CH_4. However, they are susceptible to environmental disturbances and their inhibition may lead to the accumulation of VFAs, resulting in the destabilization of the system. The main acetogenic species include syntrophic butyrate-degrading bacteria (*Syntrophomonas, Syntrophococcus, Syntrophus, Syntrophospora, Syntrophothermus,* and *Thermosyntropha*), and syntrophic propionate-degrading bacteria (*Syntrophobacter, Smithella, Pelotomaculum, Desulfotomaculum,* and *Syntrophosphaera*) (de Bok et al. 2001, 2002; Dyksma and Gallert 2019; Ishii et al. 2005; Liu et al. 1999; McInerney et al. 2007; Plugge et al. 2002; Visser et al. 2013; Westerholm et al. 2010; Wu et al. 2006).

Like acetogens, homoacetogens also produce acetate, albeit via reductive acetogenesis, where H_2 is oxidized to reduce CO_2, rather than via fermentative acetogenesis. Therefore, homoacetogens can act as an alternative H_2 sink and keep the hydrogen partial pressure low. The main homoacetogenic bacteria are *Acetobacterium woodii, Acetobacterium wieringae, Clostridium aceticum, Clostridium thermoautotrophicum*, and *Butyribacterium methylotrophicum* (Andreesen et al. 1970; Kerby et al. 1983; Mayer et al. 1977; Poehlein et al. 2016).

The diversity of methanogenic archaea is much lower than the bacteria involved in anaerobic digestion. Methanogens are divided into five major orders, i.e., *Methanobacteriales, Methanococcales, Methanomicrobiales, Methanosarcinales*, and *Methanopyrales*, which can be further divided into 10 families and 31 genera. However, based on utilizable substrates, methanogens can be divided into four major groups, i.e., acetoclastic methanogens (converting acetate), hydrogenotrophic methanogens (converting H_2 and CO_2), methylotrophic methanogens (converting methylated C1 compounds), and alkylotrophic methanogens (converting long-chain alkanes) (Zhou et al. 2022). Among these groups, the first two types are the main pathways responsible for CH_4 production in anaerobic digestion ecosystems. The typical methanogens responsible for acetoclastic methanogenesis are *Methanosaeta* and *Methanosarcina*, while those for hydrogenotrophic methanogenesis include *Methanobacterium, Methanothermobacter, Methanospirillum, Methanobrevibacter*, and *Methanoculleus*.

Environmental and operational factors can shape microbial community structure in anaerobic digestion ecosystems. The core factors include substrates (types and concentrations), organic loading rate (OLR), temperature, ammonia concentration, VFAs, pH, and additives. Substrate can determine the microbial community composition by selecting for organisms which can preferentially degrade it. For example, during anaerobic digestion of straw, the populations of *Ruminococcus, Bacteroides ruminocola*, and *Butyrivibrio fibrisolvens* are often elevated (Chen et al. 2017; Jin et al. 2020; Ma et al. 2019), while *Clostridium, Sedimentibacter, Anaerovibrio*, and *Aminobacterium* are abundant during anaerobic digestion of kitchen waste (De Vrieze et al. 2013a, b; Zheng et al. 2018). Anaerobic co-digestion usually enriches more diverse microbial communities, due to the wider variety of substrates, which can improve system stability. For example, during co-digestion of food waste and waste activated sludge, biodiversity and activity of *Methanosphaera* were promoted (Zhang et al. 2016).

Diverse inhibitors (e.g., ammonia, the accumulation of VFAs, and phenol) can also affect microbial communities. For example, when the ammonia concentration was increased from 4051 to 5734 mg N/L during the anaerobic digestion of thin stillage, the methanogenic population decreased by 56.5% (Moestedt et al. 2016). Among methanogens, *Methanospirillum hungatei* is more sensitive to ammonia, while *Methanosarcina barkei, Methanobacterium thermoautotrophicum*, and *Methanobacteirum formicicum* can tolerate ammonia concentration up to 10000 mg N/L (Jarrell et al. 1987; Koster and Koomen 1988). The accumulation of VFAs can enrich the relative abundances of syntrophic VFA-oxidizing bacteria. For example, mesophilic anaerobic digestion systems with a high butyrate concentration had increased relative abundances of *Syntrophomonas* (Nozhevnikova et al. 2020). *Syntrophobacter, Smithella*, and *Pelotomaculum* were previously enriched in mesophilic anaerobic digestion systems with propionate accumulation, whereas *Desulfotomaculum* and *Syntrophosphaera* were more abundant in thermophilic anaerobic digestion systems with propionate accumulation (Chen et al. 2005; Dyksma and Gallert 2019; Ishii et al. 2005; Liu et al. 1999; Wallrabenstein and Schink 1994). When treating phenol-containing wastewater, Poirier et al. (2016) found that *Methanoculleus* sp. replaced *Methanosarcina sp.* and maintained system stability despite the phenol concentration of 2 g/L. Syntrophic relationships also shifted from *Methanosarcina* coupled with *Syntrophomonadaceae* to *Methanoculleus* coupled with *Synergistaceae*, indicating that the variation of archaeal and bacterial communities might be the early warning microbial indicators associated with phenol inhibition (Poirier et al. 2016). Some more in-depth review articles on inhibition of anaerobic digestion are available such as Chen et al. (2008) or Chen et al. (2014).

2.2 Microbial Diversity in Anaerobic Digestion Systems

In anaerobic digestion ecosystems, operational conditions have a large influence on taxonomic diversity. Diversity can be described in terms of alpha-diversity (α-diversity) and beta-diversity (β-diversity). α-diversity describes the variation of microbial community composition within a sample. The most basic measurement of α-diversity is richness, i.e., the number of taxa at a given taxonomic level that are present in a given sample. Evenness is another commonly applied measure of α-diversity, which describes how the homogeneity of taxa relative abundance in a sample. Several diversity indexes are also available which account for richness and evenness, such as the widely used Shannon index (Shannon 1948) or the Simpson index (Simpson 1949). Other indices also include phylogenetic information such as the Faith's phylogenetic diversity (Faith 1992). β-diversity is a comparison of diversity between samples and is calculated based on a variety of distance metrics, which allows the composition of samples to be compared. These metrics can account for differences in taxa abundances, e.g., the Bray–Curtis distance metric (Bray and Curtis 1957), but can also incorporate phylogenetic distances, e.g., unifrac (Lozupone et al. 2011). Additional metrics which combine both approaches are also available such as the weighted unifrac (Lozupone et al. 2007).

In anaerobic digestion systems, bacterial richness is generally higher than archaea (Nelson et al. 2011). α-diversity can be influenced by different operational parameters, such as substrate, OLR, temperature, VFA concentrations, pH, and ammonia concentrations. Generally, substrates with a high content of easily degradable organic matter increase α-diversity, while more recalcitrant substrates or a high content of toxic substances decrease α-diversity (Sun et al. 2015; Zhang et al. 2021). For example, microbial inhibition was observed with a feedstock consisting of 30%–60% fats, oils, and grease, while food waste addition resulted in increased community diversity (Amha et al. 2017). De Jonge et al. (2020) also found that co-digestion of other wastes such as wastewater and manure with food waste may improve the functional diversity of anaerobic digesters. The community α-diversity may increase when the OLR increases. However, when the OLR is higher than the degradation ability of the system, the substrate may induce microbiome pressure, which could reduce α-diversity of community (Wei et al. 2022; Ziganshina et al. 2015).

Another operational parameter which influences α-diversity is temperature. Generally, α-diversity of mesophilic anaerobic digestion systems is higher than that of thermophilic anaerobic digestion systems (Ao et al. 2021; Liu and Whitman 2008; Wu et al. 2020). Sun et al. (2015) also revealed that high digestion temperature decreased both bacterial and archaeal richness and diversity. Lin et al. (2016) found that α-diversity was lowest while the methane metabolism was highest at 50°C, indicating that functions might be mainly carried out by a limited number of functional microorganisms under thermophilic conditions.

In general, microbial communities with a high α-diversity are thought to be more robust and resistant to shocks or inhibition, with better overall performance (Carballa et al. 2015; McCann 2000), due to more functional redundancy in the microbial community (Paulo et al. 2020). However, higher diversity does not always improve reactor stability (Mills et al. 2023), and therefore, other ecological measures, beyond diversity, should be explored to understand the relationship between community composition and bioreactor performance. These include microbial networks, microbial community assembly, microbial life strategies, etc.

While α-diversity describes the microbial community within a sample, β-diversity is the comparison of microbial communities between samples. Distances (e.g., Bray–curtis or unifrac) between samples are often visualized using different ordination analyses, such as principal component analysis (PCA), principal coordinate analysis (PCoA), and non-metric multidimensional scaling. These allow the composition of microbial communities to be reduced to a single point and to be visualized and grouped as required, to determine what are the major influences on microbial community composition and the degree to which samples differ. For example, β-diversity analysis has been used to demonstrate community composition shifts due to pretreatment (Zhang et al. 2021), substrate (Zhang et al. 2014), temperature (De Vrieze et al. 2015), OLR (Ferguson et al. 2016), presence of toxicants (Chen et al. 2020), and conductive additives (Han et al. 2017). When combined with reactor performance metadata, the causes of the observed clustering patterns can be observed (Calusinska et al. 2018). For example, PCoA showed that samples were separated into different clusters based on substrate type (Wang et al. 2021). Huang et al. (2021) found that the chemical oxygen demand loading rate of vacuum toilet blackwater higher than 3.03 g/(L·d) made a clear microbial community shift based on PCoA. Different additives can also lead to different clusters (Han et al. 2017). PCoA showed that the control, nano-carbon powder, and nano-Al_2O_3 groups clustered together, while Nano-ZnO and nano-CuO groups were clustered within the group and differed significantly from the other groups, indicating that heavy metal shifted microbial community structures (Chen et al. 2020).

Stress factors, such as pH, OLR, hydraulic retention time, and VFAs, can also shift microbial community structure. Li et al. (2023) raised propionic acid loading in an anaerobic digestion system up to 2.7 g/(L·d) and the system tended to be unstable; stop-feeding strategy was used to recover the system; and then the propionic acid

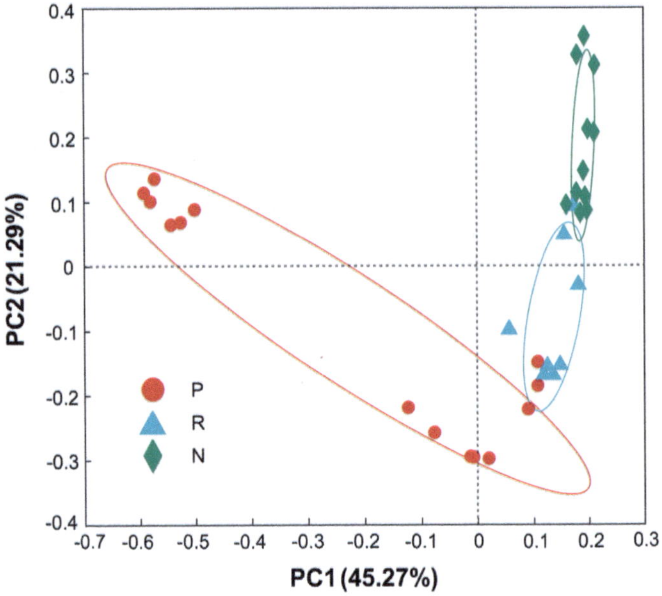

Fig. 2 Principal coordinates analysis (PCoA) demonstrated differences of bacterial communities at various propionic acid loading (PAL) rates. P presents PAL rates up to 2.7 g/(L·d); R demonstrates stop-feeding stage to recover the system; and N includes samples with the PAL rate gradually up to 2.4 g/(L·d) again (adapted from Li et al. 2023)

loading gradually increased to 2.4 g/(L·d). The bacterial communities clearly clustered according to the operational conditions, indicating that propionic acid shaped the bacterial structure by different loading rates (Fig. 2; Li et al. 2023). Poirier et al. (2016) conducted anaerobic digestion with different phenol concentrations up to 5 g/L, and microbial shifts were observed from 0.5 g/L; while PCA showed that the community at 5 g/L was close to other phenol groups after 45 days cultivation, indicating that the community evolved to recover from high phenol concentrations.

2.3 *Microbial Network for Complex Systems*

Due to the diverse microbial interactions in anaerobic digestion ecosystems, the understanding of syntrophic relationships is quite important for system optimization and stability. Microbial network analysis is often used for studying these interactions and determining the statistical relationships between taxa (often known as nodes) and the strength of their connections (often known as edges), which may imply biological relationships. Correlation-based network analysis is most common (e.g., Pearson or Spearman correlations) but networks based on symbiotic distance (such as Jaccard index, Schoener's index, etc.) are also applied (Borcard et al. 2011; Weiss et al. 2016).

Network analysis has been particularly revealing from an ecological point of view in anaerobic digestion. As methanogens and syntrophic bacterial species are difficult to isolate and cultivate, their relationship can be hard to establish. Through network analysis, it can be found that the abundances of some taxa are closely related (Ma et al. 2017). Combined with species annotation, it has been revealed that many of the co-occurring/correlated species were syntrophic bacteria and methanogens (Guo et al. 2021). Even though network analysis does not reflect the real biological interactions, the interaction relationship presented in network analysis can provide some proof of microbial interactions.

Distinct sub-populations within a microbial community can also be identified by finding groups of nodes (often known as modules) with a high degree of interaction. For example, distinct acetolactic and hydrogenotrophic modules within household anaerobic digesters can be identified (Rui et al. 2015). Sub-populations can also change throughout reactor operation, indicating community succession. Wu et al. (2016) found that long-term stability of triplicate anaerobic digesters led to decreased connectivity and increased modularity, suggesting that niche differentiation and community succession occurred over time.

The ecological roles of taxa can also be suggested through network analysis. For example, interspecies interactions such as mutualism, competition, or antagonism can be justified (Braz et al. 2019). Network analysis is useful for identifying keystone species by identifying nodes with significant numbers of positive edges, such as *Methanobacterium, Methanothermus*, and unclassified *Thaumarchaeota* species (Wu et al. 2020a), *Methanobacterium, Methanomicrobiaceae, Alkaliphilus*, and *Petrimonas* species (Thapa et al. 2023), or *Lactococcus* and *Methanosaeta* species (Kim et al. 2015). Despite the advantages of network analysis in deciphering ecological relationships, it is important to note that co-occurrence may be brought about because of preferences for environmental variables rather than interactions.

The degree of "connectedness" within anaerobic digestion microbiome networks has also been demonstrated to be related to performance characteristics of anaerobic digesters, particularly stability. For example, Mercado et al. (2022) found that frequent organic load shocks led to a decrease in microbial interactions and network modularity, which correlated with decreased system performance. Lv et al. (2022) found that temperature shocks also resulted in decreasing modularity but increasing complexity of the networks, which negatively impacted system performance.

Network analysis can also reflect the microbial interactions among different conditions. For example, Zhang et al. (2021) found that the co-digestion of wasted activated sludge (WAS) and thermal pretreatment sludge had more intense network interaction than the mono-digestion of WAS through co-occurrence network analysis. This is perhaps caused by the modification of feedstock parameters, e.g., improved carbon to nitrogen ratios or trace element content (Li et al. 2014; Quiroga et al. 2014; Sawatdeenarunat et al. 2015; Zhang et al. 2013). In general, mesophilic systems have more intricate networks than thermophilic systems (Grimalt-Alemany et al. 2020; Wu et al. 2023; Yu et al. 2023). This is related to the more diverse species in mesophilic systems than in thermophilic systems (Niu et al. 2015; Wu et al. 2020). High OLR can lead to two different situations. While microorganisms must cooperate more intensively to degrade the high concentration of organic matters, the network may be strengthened (Meng et al. 2022) (Fig. 3). Nevertheless, when the OLR is too high to cause the accumulation of VFAs or ammonium, the network intensity may be lowered, as microorganisms are inhibited (Qi et al. 2023). Additives can also increase the

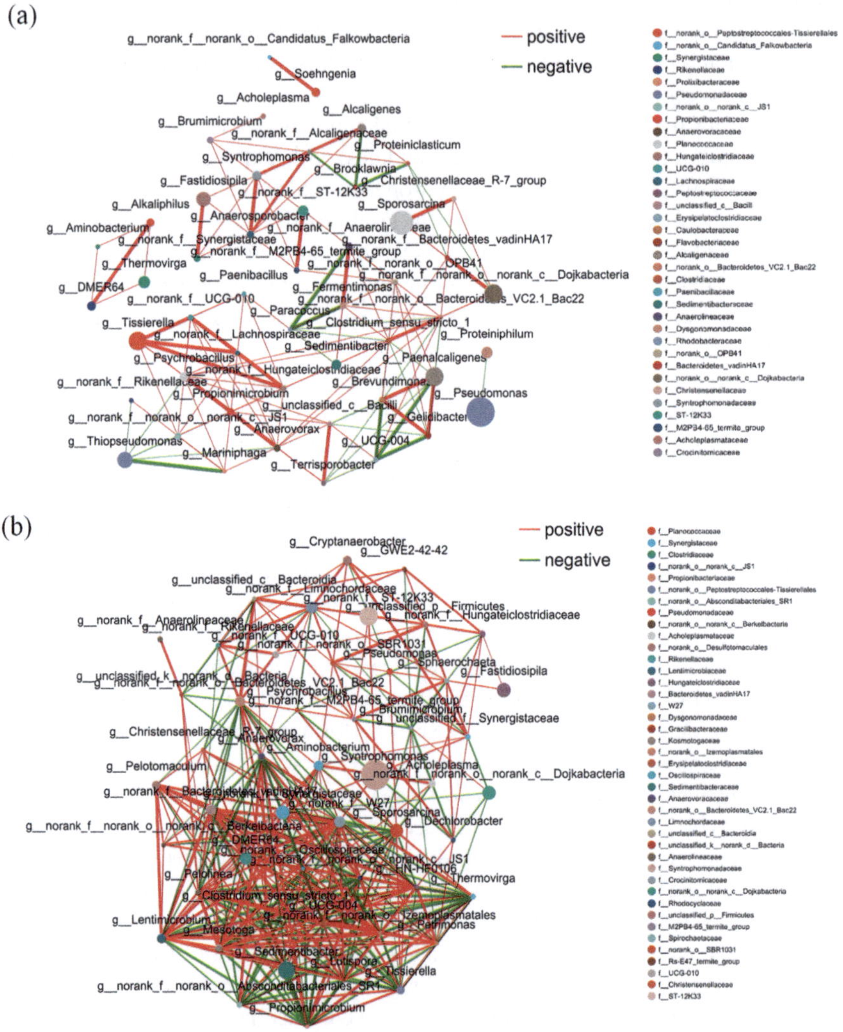

Fig. 3 Correlation networks of different operational taxonomic units (OTUs) in (**a**) cluster I and (**b**) cluster II. OTUs with same color were categorized to the same genera. Red lines indicate positive correlation while green lines indicate negative correlation (Spearman, $r > 0.5$, $p < 0.05$). Cluster I includes samples collected for OLR from 0.2 to 1.6 g/(L·d), and Cluster II contains samples that have OLR ranging from 2.0 to 4.4 g/(L·d) (adapted from Meng et al. 2022)

network intensity. For example, when treating saline wastewater, anaerobic digestion systems with the addition of magnetite demonstrated more complex interactions than the systems without additives, where hydrogenotrophic methanogens showed positive correlation with methane yield (Ma et al. 2021). In addition, the interaction between hydrogenotrophic methanogens and syntrophic bacteria was stronger in the magnetite-supplemented anaerobic digestion system, indicating that magnetite might accelerate interspecies electron transfer (Ma et al. 2021).

3 Mechanisms for Shaping Microbial Communities in Anaerobic Digestion Ecosystems

3.1 Deterministic and Stochastic Processes

Community assembly is the process by which the inoculated species pool is selected through environmental factors and biotic interactions and then evolves into acclimated communities (Fig. 4). Acclimated communities also have feedback effects on the environment and biotic interactions.

Community assembly processes are most commonly described as being either deterministic or stochastic in nature. Deterministic processes include environmental conditions, microbial interactions (i.e., competition, syntrophy, and predation), and microbial traits, while stochastic processes are comprised of mainly random events such as drift, birth, death, immigration, and speciation. These concepts are similar to the concept of niche *vs* neutral theory, where niche processes are deterministic and neutral processes are stochastic. Niche theory predicts that no two species can coexist

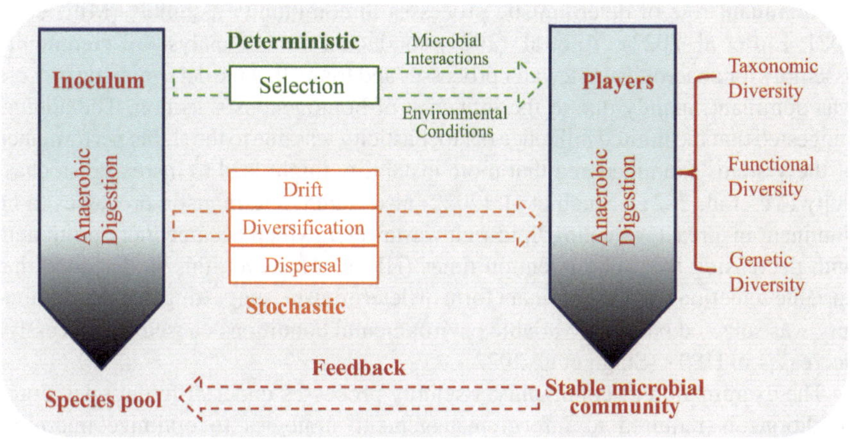

Fig. 4 Process of community assembly within anaerobic digestion ecosystems (adopted from HilleRisLambers et al. 2012)

if they have the same niche and therefore the environment selects the community composition, while the neutral theory describes random drift of microbial communities under identical conditions as all populations are assumed to be ecologically identical (Harpole 2010; Prosser et al. 2007). In reality, microbial communities often experience a mixture of both deterministic and stochastic processes (Liébana et al. 2019; Trego et al. 2021; Zhang et al. 2022b). Therefore, it is important to consider relative change in the contributions of determinism or stochasticity rather than simply defining the community assembly processes as one or the other. Based on the framework described by Vellend (2010), deterministic and stochastic processes can be further categorized into selection, dispersal, diversification, and drift. Selection (deterministic) refers to the modification of the community structure from both biotic and abiotic factors such as the environmental conditions and differences in fitness between individuals and microbial interactions. Selection can be either homogenous (i.e., due to a stable environment) or variable (i.e., due to a variable environment). Dispersal (deterministic and stochastic) refers to the movement and establishment of microorganisms among communities. Diversification (stochastic) refers to the generation of genetic variation. Drift (stochastic) refers to changes resulted from birth, death, reproduction, and colonization. Several methodologies and frameworks have been developed for quantifying the contribution of stochastic or deterministic processes on microbial community assembly (e.g., Stegen et al. 2015). A recent review (Smith et al. 2024) extensively summarizes these in the context of wastewater bioreactors.

Most studies on community assembly in anaerobic digestion to date have identified deterministic processes as being dominant. For example, in a thermal hydrolysis pretreatment coupled anaerobic digestion system, Zhang et al. (2021) found that microbial community assembly was driven by deterministic processes (63.8–81%), such as homogenous selection, but stochasticity increased when SRTs were lengthened (from 19.0% at 10 d to 36.2% at 30 d). Several other studies have also identified the dominant role of deterministic processes in community assembly (Mills et al. 2021; Liu et al. 2023). Yu et al. (2021) conducted a meta-analysis of community assembly for anaerobic wastewater processes and found that the deterministic process was dominant, mainly due to the influence of homogenous selection. The authors suggested that the limited influence of stochasticity was due to the stable performance of the systems examined and that more instability might lead to increased stochasticity (Yu et al. 2021). Singh et al. (2022) also found deterministic processes to be dominant in a reactor treating lipid-rich wastewater, which became more prominent with decreasing hydraulic retention times (HRTs). However, this study found that variable selection was the dominant form of determinism, suggesting that the community was selected based on variable environmental conditions caused by successive decreases in HRTs (Singh et al. 2022).

The examination of community assembly processes under different operational conditions is required to inform management strategies to optimize microbial community development. For example, although they did not use any specific community assembly analysis, Peces et al. (2018) found that different inoculum communities converged to a shared core community in replicated bioreactors and suggested that this was due to deterministic factors. This demonstrates the influence of operational parameters (i.e., SRT and OLR) on microbial communities and indicates that it may be possible to develop a pre-determined community composition by manipulating reactor conditions. Instability caused by shock loads of inhibitory compounds and subsequent recovery often cause a dramatic change in community assembly processes. Cardona et al. (2022) found that deterministic processes increased during an ammonia shock before decreasing again after recovery. The prevalence of deterministic influences during shocks implies that it may be possible to predict community assembly patterns in response to operational conditions.

In contrast however, Zhang et al. (2022a) found that stochasticity increased in unstable systems, indicating that the degree of stochasticity could be used as an indicator for system performance. Increasing contributions of stochasticity have also been shown to correlate with increased diversity (Mills et al. 2021), potentially leading to more functional redundancy, which can improve system stability (Carballa et al. 2015). Therefore, monitoring the balance between stochastic and deterministic processes could be used for managing anaerobic digestion performance. The immigration of influent microorganisms may also affect biodiversity of anaerobic digestion systems, further affecting the system assembly pattern (stronger influence on stochastic processes in dysfunctional systems) (Zhang et al. 2022a, b). The limited number of studies using community assembly analysis and the discrepancies observed between studies demonstrates the need for further work in this area to build up a knowledge base which can inform microbially based management strategies.

3.2 Trait-Based Approach

Trait-based microbial ecology refers to the study of phenotypic characteristics of microbes that influence their fitness in given environmental conditions, regardless of their taxonomy. Application of these approaches can improve the mechanistic understanding of ecological patterns and predict ecological dynamics and function (Lajoie and Kembel 2019). Trait-based approaches can complement taxonomic and phylogenetic approaches, promoting the linkage between microbial diversity and function (Krause et al. 2014).

A trait refers to the phenotype, expressed by an organism that can affect its function or fitness within an ecosystem. Traits are classified in different ways. Simple traits are controlled by single genes, while complex traits are controlled by multiple genes and may also be affected by environmental factors (Falconer and Mackay 1996). Traits can also be classified into continuous traits, such as salt tolerance at different temperatures, and discrete traits, such as the ability of microorganisms to fix nitrogen or carbon dioxide. Additionally, traits can be divided into response traits, which relate to the response of organisms to environmental changes (such as drought tolerance), and effect traits, which refer to the impact of organisms on ecosystem processes (such as photosynthetic capacity) (Webb et al. 2010).

Functional traits are any morphological and physiological characteristics (such as size, reproduction time, or response to a certain antibiotic) that can affect microbial fitness under different environmental conditions, which should be measurable and can be connected to ecological strategies of microbes (Lajoie and Kembel 2019; McGill et al. 2006). Functional traits and the resulting ecological niches determine microbial interactions and ecological function (Krause et al. 2014). Functional traits can be described at species, community, and ecosystem levels and are often conserved despite variations in taxonomic compositions (Louca et al. 2018). The overall ecosystem function may be more affected by the functional trait rather than microbial biodiversity and richness. For example, the relative abundance of methanogens may be quite low in anaerobic digestion systems, but their function is quite important for system stability. Therefore, changes in methanogen populations may not affect total community α-diversity but could greatly impact reactor performance. However, if traits can be conserved during evolution (trait conservatism), phylogenetic diversity may be adopted for assessing trait diversity (Krause et al. 2014). Predicted values for most traits become more accurate over time, indicating that succession is at least partially functionally deterministic (Liang et al. 2022).

Morphology traits can affect microbial communities and their function in anaerobic digestion systems. For example, Vanwonterghem et al. (2014) observed distinct archaeal morphotypes, i.e., cocci (diameter of 1–2 μm), clusters of cocci (diameter 3–7 μm), and rods (3–10 μm); also, rod-shaped cells formed large filaments (15–100 μm) or a chain within a common sheath (up to ~100 μm). Different morphologies can further affect microbial function. The flocculent or filamentous nature of microorganisms may benefit granule formation (Mills et al. 2021).

4 The Application of the r/K Selection Theory in Enriching Functional Microorganisms

The r/K selection theory (i.e., Andrews and Harris 1986) has been used to describe two types of microbial growth strategies in natural and engineered ecosystems and can be linked to oligotrophic or copiotroph life strategies (Ho et al. 2017). r-strategists (copiotrophs) are fast growing opportunists, which utilize resources inefficiently but thrive in a resource-rich environment; while K-strategists (oligotrophs) are slower growing but utilize resources more efficiently and are better adapted to a crowded, resource-limited environment (Erbilgin et al. 2017; Juteau et al. 1999; Rojas-Tirado et al. 2017). Therefore, r-strategists survive in unstable environments with dynamic resource variations (disturbances and/or feast-and-famine intermittent supply), while K-strategists are selected in stable environments with limited resource supply (Cheng et al. 2018). A key genomic difference in these organisms is the number of rRNA genes on their genomes. r-strategists/copiotrophs tend to have more ribosomes, allowing faster growth during an influx of resources, whereas K-strategists/oligotrophs tend to have less rRNA genes which allows for conservation

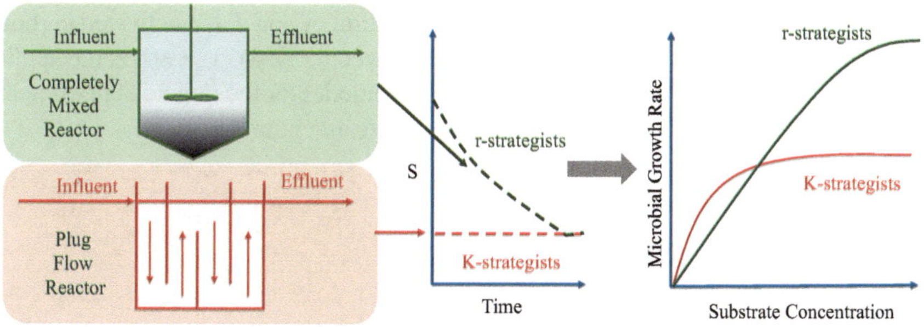

Fig. 5 Reactor configurations have different substrate gradients, with the completely mixed reactor having a low substrate gradient and the plug-flow reactor having a high substrate gradient. The low substrate gradient benefits the enrichment of K-strategists, while the high substrate gradient can enrich both types of microorganisms but usually dominated with r-strategists

of energy but lessens their ability to grow quickly in resource-rich environments (Lankiewicz et al. 2016; Okie et al. 2020).

As shown in Fig. 5, in engineered wastewater treatment systems, there are two typical reactor configurations, i.e., plug-flow reactor (PFR) and completely mixed reactor (CMR). A high substrate gradient exists in PFR, which may enrich mainly r-strategists, while the CMR possesses a relative constant substrate concentration, which can be better for the acclimation of K-strategists.

In anaerobic digestion systems, *Methanosaetaceae* is characterized by high substrate affinity and slow growth rate (K-strategists), while *Methanosarcinaceae* has higher growth rate and low substrate affinity (r-strategists). In anaerobic systems used for biohydrogen production, low levels of initial glucose provided *Lactobacillus* (K-strategist) a competitive advantage over *Clostridium* (r-strategist), and the different substrate utilization was probably determined by the activity rather than the abundance of the two species (Park et al. 2018).

Various types of electrochemically active bacteria such as *Shewanella* and *Geobacter* possess mechanisms of extracellular electron transfer (EET), which can occur in distances from μm to cm and reduce the limitations of cellular size. The EET pathway has been confirmed between *G. metallireducens* and *Methanothrix harundinacea* or *Methanosarcina barkeri* (Rotaru et al. 2014a, b), and the electron/proton flux for reducing CO_2 to CH_4 in *M. barkeri* was further clarified (Holmes et al. 2018). In anaerobic digestion systems, the acclimation of *Geobacter* can enhance CH_4 production by establishing a syntrophic relationship between *Geobacter* and methanogens. Therefore, strategies to enrich and maintain *Geobacter* would benefit anaerobic digestion efficiency. Based on intensive studies, *Geobacter* can be considered as a K-strategist as they have relatively low growth rates and low half-saturation constants when consuming benzoate and acetate (Esteve-Núñez et al. 2005; Marozava et al. 2014). Based on this, Guo et al. (2022) and Du et al. (2024) confirmed that *Geobacter* was notably dominant in constant flow reactors (CFRs) fed with ethanol, where continuously low ethanol concentrations favored the growth of K-strategists (oligotrophs). Remarkably, CFRs with the extended SRT could enrich the relative abundances of *Geobacter* by 71.7% with the addition of powdered activated carbon (PAC) and 70.4% without the addition of PAC (Du et al. 2024). Therefore, the application of the r/K selection theory could be a good model for the targeted enrichment of key populations in environmental wastewater treatment processes.

5 Future Perspectives

The identification of predominant microorganisms is commonly based on the analysis of species abundances, generated using high-throughput sequencing technology. To identify which organisms are present in a given sample and perform various ecological analysis, amplicon-based approaches targeting the 16S rRNA phylogenetic marker gene are typically applied. This requires several steps including DNA extraction, PCR amplification, sequencing library preparation, sequencing, bioinformatic analysis, and classification. All of which can induce certain errors or biases in the final dataset. For example, the specificity and efficiency of different primers vary, and different databases or classification tools may yield different taxonomies. Despite these deficiencies, 16S rRNA-based analysis can provide value, especially when used in combination with physicochemical data and other molecular approaches.

In most amplicon studies, the anaerobic digestion system is operated by regulating operational parameters (environmental factors) and then analyzing the changes in microbial community structure. This can indicate microbial responses to operational parameters (environmental factors). However, the reverse analysis is rarely performed, i.e., using community structure to inform operational parameters. For example, it may be possible to predict system stability by analyzing microbial community structure. One issue with this approach is that amplicon sequencing is not real-time, so the regulation of process parameters based on 16S rRNA gene sequencing has a strong lag time. Combining 16S rRNA gene analysis with real-time monitoring and regulation may be the next issue that researchers are trying to solve. Indeed, the Oxford Nanopore sequencing technology, e.g., MinIon, allows rapid sequencing and analysis (within hours of sampling) (Anderson et al. 2023) and could be a useful tool in process monitoring of anaerobic digestion.

Real-time monitoring of community assembly processes could allow for ecological management of anaerobic digestion. For example, since deterministic processes tend to dominate and are often seen to increase in response to operational changes (Singh et al. 2022), it may be possible to predict microbial community responses and modulate the desired microbial community. However, much more data is needed on different systems responses to defined operational changes to develop this predictive power. The syntrophic nature of anaerobic digestion microbial communities means that monitoring of microbial networks could also be useful for managing processes. For example, the degree of connectedness or modulation could be used as an indicator of process stability and any changes in these characteristics could be used as an early warning of process failure.

It may be also useful to apply established ecological theories to anaerobic digestion ecosystems. However, a fit-for-purpose application strategy is required. For example, under anaerobic conditions, the substrate flux (supply and utilization rates) and substrate concentration are crucial factors for enriching the r-/K-strategists (Yin et al. 2022). Kreft et al. (2020) proposed that anaerobic or anoxic organic matter decomposition is carried out through the cooperation among different types of microbes, which could be due to the high substrate flux. Winkler et al. (2017) found that the selection of r-/K-nitrite oxidizing bacteria not only depended on the substrate concentration, but also the dilution rate, where a low nitrite concentration selected *Nitrobacter* (r-strategist) when the dilution rate was lower than the μ_{max} of *Nitrospira*. Therefore, Yin et al. (2022) proposed the extended r/K selection strategies based on both substrate flux and concentrations. This could be adopted in anaerobic digestion systems for enriching specific types of microorganisms. According to the proposed concept, when the substrate concentration is quite low but with a high substrate supply and utilization rates, r-strategists may still be enriched. Under this condition, the balance between the substrate utilization rate and the supply rate is quite important. Furthermore, the substrate affinity can be substituted by the substrate flux rate under this scenario, and the r/K selection theory would be still working, where r-strategists would have a high substrate utilization rate, while K-strategists have a low substrate utilization rate.

6 Take-Home Messages

(1) Microbial community diversity analysis based on the data from 16S rRNA gene high-throughput sequencing can help understand microbial community structure and interactions in anaerobic digestion systems.

(2) Community structure analysis is an important part of microbial diversity analysis, because it reveals the microbial species that make up the community and the relative proportion of species. Nevertheless, the relative abundance does not represent the absolute number of species in the sample.

(3) Co-occurrence network analysis shows the correlation between samples and species, or the correlation between species. It is often used to study the relationship between microbiomes and samples in complex environments and is strongly related to system stability. Nevertheless, network analysis does not directly reflect the interactions between species.

(4) Deterministic processes are the main process shaping microbial communities in anaerobic digestion ecosystems, while stochastic processes may contribute to the assembly of microbial communities under conditions such as with high influent input. Diverse trait-based properties can be incorporated to understand microbial ecology in anaerobic digestion systems.

(5) The r/K selection can be applied to enrich functional microorganisms in anaerobic digestion systems, which can correspond to the typical reactor configures of plug-flow or completely mixed reactors in engineered wastewater treatment processes.

References

Amha YM, Sinha P, Lagman J, Gregori M, Smith AL (2017) Elucidating microbial community adaptation to anaerobic co-digestion of fats, oils, and grease and food waste. Water Res 123:277–289

Andreesen JR, Gottschalk G, Schlegel HG (1970) *Clostridium formicoaceticum nov.* spec. isolation, description and distinction from *C. aceticum* and *C. thermoaceticum*. Archiv für Mikrobiologie 72:154–174

Andersen MH, Thomsen L, Stokholm-Bjerregaard M, Eriksen S, Hansen SH, Albertsen M, Nielsen PH (2023) Fast DNA-analyses for surveillance of microbial communities in full-scale deammonification tanks: potential for control and troubleshooting. Water Res 236:119919

Andrews JH, Harris RF (1986) R-and K-selection and microbial ecology. In: Marshall KC (ed) Advances in microbial ecology. Springer, New York, NY, USA, pp 99–147

Ao T, Xie Z, Zhou P, Liu X, Wan L, Li D (2021) Comparison of microbial community structures between mesophilic and thermophilic anaerobic digestion of vegetable waste. Bioprocess Biosyst Eng 44:1201–1214

Borcard D, Gillet F, Legendre P (2011) Numerical ecology with R. Springer, New York, NY, USA, pp 59–150

Bray JR, Curtis JT (1957) An ordination of the upland forest communities of southern Wisconsin. Ecol Monogr 27(4):326–349

Braz GHR, Fernandez-Gonzalez N, Lema JM, Carballa M (2019) Organic overloading affects the microbial interactions during anaerobic digestion in sewage sludge reactors. Chemosphere 222:323–332

Calusinska M, Goux X, Fossépré M, Muller EE, Wilmes P, Delfosse P (2018) A year of monitoring 20 mesophilic full-scale bioreactors reveals the existence of stable but different core microbiomes in bio-waste and wastewater anaerobic digestion systems. Biotechnol Biofuels 11:196

Campanaro S, Treu L, Kougias PG, De Francisci D, Valle G, Angelidaki I (2016) Metagenomic analysis and functional characterization of the biogas microbiome using high throughput shotgun sequencing and a novel binning strategy. Biotechnol Biofuels 9:26

Campanaro S, Treu L, Rodriguez-R LM, Kovalovszki A, Ziels RM, Maus I, Zhu X, Kougias PG, Basile A, Luo G, Schlüter A (2020) New insights from the biogas microbiome by comprehensive genome-resolved metagenomics of nearly 1600 species originating from multiple anaerobic digesters. Biotechnol Biofuels 13:25

Carballa M, Regueiro L, Lema JM (2015) Microbial management of anaerobic digestion: exploiting the microbiome-functionality nexus. Curr Opin Biotechnol 33:103–111

Cardona L, Mazéas L, Chapleur O (2022) Deterministic processes drive the microbial assembly during the recovery of an anaerobic digester after a severe ammonia shock. Biores Technol 347:126432

Chen S, Liu X, Dong X (2005) *Syntrophobacter sulfatireducens sp. nov.*, a novel syntrophic, propionate-oxidizing bacterium isolated from UASB reactors. Int J System Evolution Microbiol 55(3):1319–1324

Chen Y, Cheng JJ, Creamer KS (2008) Inhibition of anaerobic digestion process: a review. Biores Technol 99(10):4044–4064

Chen JL, Ortiz R, Steele TW, Stuckey DC (2014) Toxicants inhibiting anaerobic digestion: a review. Biotechnol Adv 32(8):1523–1534

Chen H, Zhang C, Rao Y, Jing Y, Luo G, Zhang S (2017) Methane potentials of wastewater generated from hydrothermal liquefaction of rice straw: focusing on the wastewater characteristics and microbial community compositions. Biotechnol Biofuels 10:140

Chen Y, Yang Z, Zhang Y, Xiang Y, Xu R, Jia M, Cao J, Xiong W (2020) Effects of different conductive nanomaterials on anaerobic digestion process and microbial community of sludge. Biores Technol 304:123016

Cheng Y, Hubbard CG, Zheng L, Arora B, Li L, Karaoz U, Ajo-Franklin J, Bouskill NJ (2018) Next generation modeling of microbial souring–parameterization through genomic information. Int Biodeterior Biodegradation 126:189–203

Collins MD, Lawson PA, Willems A, Cordoba JJ, Fernandez-Garayzabal J, Garcia P, Cai J, Hippe H, Farrow JA (1994) The phylogeny of the genus *Clostridium*: proposal of five new genera and eleven new species combinations. Int J Syst Evol Microbiol 44(4):812–826

de Bok FA, Stams AJ, Dijkema C, Boone DR (2001) Pathway of propionate oxidation by a syntrophic culture of *Smithella propionica* and *Methanospirillum hungatei*. Appl Environ Microbiol 67(4):1800–1804

de Bok FA, Luijten ML, Stams AJ (2002) Biochemical evidence for formate transfer in syntrophic propionate-oxidizing cocultures of *Syntrophobacter fumaroxidans* and *Methanospirillum hungatei*. Appl Environ Microbiol 68(9):4247–4252

De Jonge N, Davidsson Å, la Cour JJ, Nielsen JL (2020) Characterisation of microbial communities for improved management of anaerobic digestion of food waste. Waste Manage 117:124–135

De Vrieze J, De Lathouwer L, Verstraete W, Boon N (2013a) High-rate iron-rich activated sludge as stabilizing agent for the anaerobic digestion of kitchen waste. Water Res 47(11):3732–3741

De Vrieze J, Verstraete W, Boon N (2013b) Repeated pulse feeding induces functional stability in anaerobic digestion. Microb Biotechnol 6(4):414–424

De Vrieze J, Saunders AM, He Y, Fang J, Nielsen PH, Verstraete W, Boon N (2015) Ammonia and temperature determine potential clustering in the anaerobic digestion microbiome. Water Res 75:312–323

Du B, Zhan X, Lens PN, Zhang Y, Wu G (2024) Deciphering anaerobic ethanol metabolic pathways shaped by operational modes. Water Res 249:120896

Dueholm MK, Nierychlo M, Andersen KS, Rudkjøbing V, Knutsson S, Albertsen M, Nielsen PH (2022) MiDAS 4: a global catalogue of full-length 16S rRNA gene sequences and taxonomy for studies of bacterial communities in wastewater treatment plants. Nat Commun 13(1):1908

Dyksma S, Gallert C (2019) *Candidatus Syntrophosphaera thermopropionivorans*: a novel player in syntrophic propionate oxidation during anaerobic digestion. Environ Microbiol Rep 11(4):558–570

Erbilgin O, Bowen BP, Kosina SM, Jenkins S, Lau RK, Northen TR (2017) Dynamic substrate preferences predict metabolic properties of a simple microbial consortium. BMC Bioinf 18:57

Esteve-Núñez A, Rothermich M, Sharma M, Lovley D (2005) Growth of *Geobacter sulfurreducens* under nutrient-limiting conditions in continuous culture. Environ Microbiol 7(5):641–648

Faith DP (1992) Conservation evaluation and phylogenetic diversity. Biol Cons 61(1):1–10

Falconer DS, Mackay TFC (1996) Introduction to quantitative genetics, 4th edn. Longmans Green, Harlow, Essex, UK

Ferguson RM, Coulon F, Villa R (2016) Organic loading rate: a promising microbial management tool in anaerobic digestion. Water Res 100:348–356

Grimalt-Alemany A, Łężyk Mateusz, Kennes-Veiga DM, Skiadas IV, Gavala HN (2020) Enrichment of mesophilic and thermophilic mixed microbial consortia for syngas biomethanation: the role of kinetic and thermodynamic competition. Waste Biomass Valorization 11:465–481

Guo B, Zhang Y, Yu N, Liu Y (2021) Impacts of conductive materials on microbial community during syntrophic propionate oxidization for biomethane recovery. Water Environ Res 93(1):84–93

Guo Q, Yin Q, Du J, Zuo J, Wu G (2022) New insights into the r/K selection theory achieved in methanogenic systems through continuous-flow and sequencing batch operational modes. Sci Total Environ 807:150732

Hamlin LJ, Hungate RE (1956) Culture and physiology of a starch-digesting bacterium (*Bacteroides amylophilus n. sp.*) from the bovine rumen. J Bacteriol 72(4):548–554

Han Z, Chen F, Zhong C, Zhou J, Wu X, Yong X, Zhou H, Jiang M, Jia H, Wei P (2017) Effects of different carriers on biogas production and microbial community structure during anaerobic digestion of cassava ethanol wastewater. Environ Technol 38(18):2253–2262

Harpole W (2010) Neutral theory of species diversity. Nat Educ Knowl 3(10):60

HilleRisLambers J, Adler PB, Harpole WS, Levine JM, Mayfield MM (2012) Rethinking community assembly through the lens of coexistence theory. Annu Rev Ecol Evol Syst 43:227–248

Ho A, Di Lonardo DP, Bodelier PL (2017) Revisiting life strategy concepts in environmental microbial ecology. FEMS Microbiol Ecol 93(3):fix006

Holmes DE, Rotaru AE, Ueki T, Shrestha PM, Ferry JG, Lovley DR (2018) Electron and proton flux for carbon dioxide reduction in *Methanosarcina barkeri* during direct interspecies electron transfer. Front Microbiol 9:3109

Honda T, Fujita T, Tonouchi A (2013) *Aminivibrio pyruvatiphilus gen. nov., sp. nov.*, an anaerobic, amino-acid-degrading bacterium from soil of a Japanese rice field. Int J System Evol Microbiol 63(10):3679–3686

Huang Q, Liu Y, Dhar BR (2021) Pushing the organic loading rate in electrochemically assisted anaerobic digestion of blackwater at ambient temperature: insights into microbial community dynamics. Sci Total Environ 781:146694

Ishii SI, Kosaka T, Hori K, Hotta Y, Watanabe K (2005) Coaggregation facilitates interspecies hydrogen transfer between *Pelotomaculum thermopropionicum* and *Methanothermobacter thermautotrophicus*. Appl Environ Microbiol 71(12):7838–7845

Jarrell KF, Saulnier M, Ley A (1987) Inhibition of methanogenesis in pure cultures by ammonia, fatty acids, and heavy metals, and protection against heavy metal toxicity by sewage sludge. Can J Microbiol 33(6):551–554

Jin S, Jin W, Dong C, Bai Y, Jin D, Hu Z, Huang Y (2020) Effects of rice straw and rice straw ash on rice growth and α-diversity of bacterial community in rare-earth mining soils. Sci Rep 10(1):10331

Juteau P, Rho D, Larocque R, LeDuy A (1999) Analysis of the relative abundance of different types of bacteria capable of toluene degradation in a compost biofilter. Appl Microbiol Biotechnol 52:863–868

Kerby R, Niemczura W, Zeikus JG (1983) Single-carbon catabolism in acetogens: analysis of carbon flow in *Acetobacterium woodii* and *Butyribacterium methylotrophicum* by fermentation and 13C nuclear magnetic resonance measurement. J Bacteriol 155(3):1208–1218

Kim TG, Yun J, Cho KS (2015) The close relation between Lactococcus and Methanosaeta is a keystone for stable methane production from molasses wastewater in a UASB reactor. Appl Microbiol Biotechnol 99:8271–8283

Koeck DE, Pechtl A, Zverlov VV, Schwarz WH (2014) Genomics of cellulolytic bacteria. Curr Opin Biotechnol 29:171–183

Koster IW, Koomen E (1988) Ammonia inhibition of the maximum growth rate (μm) of hydrogenotrophic methanogens at various pH-levels and temperatures. Appl Microbiol Biotechnol 28:500–505

Krause S, Le Roux X, Niklaus PA, Van Bodegom PM, Lennon JT, Bertilsson S, Grossart HP, Philippot L, Bodelier PL (2014) Trait-based approaches for understanding microbial biodiversity and ecosystem functioning. Front Microbiol 5:251

Kreft JU, Griffin BM, González-Cabaleiro R (2020) Evolutionary causes and consequences of metabolic division of labour: why anaerobes do and aerobes don't. Curr Opin Biotechnol 62:80–87

Lajoie G, Kembel SW (2019) Making the most of trait-based approaches for microbial ecology. Trends Microbiol 27(10):814–823

Lankiewicz TS, Cottrell MT, Kirchman DL (2016) Growth rates and rRNA content of four marine bacteria in pure cultures and in the Delaware estuary. ISME J 10(4):823–832

Li Y, Zhang R, He Y, Zhang C, Liu X, Chen C, Liu G (2014) Anaerobic co-digestion of chicken manure and corn stover in batch and continuously stirred tank reactor (CSTR). Biores Technol 156:342–347

Li D, Meng X, Sun Y, Li X, Liu X, Xie Z, Cao Q (2023) The microbial and functional reconstruction of instable syntrophic propionate-oxidizing methanogenesis by system recovering and injection modes changing. Chem Eng J 455:140736

Liang Y, Ma A, Zhuang G (2022) Construction of environmental synthetic microbial consortia: based on engineering and ecological principles. Front Microbiol 13:829717

Liébana R, Modin O, Persson F, Szabó E, Hermansson M, Wilén BM (2019) Combined deterministic and stochastic processes control microbial succession in replicate granular biofilm reactors. Environ Sci Technol 53(9):4912–4921

Liu Y, Whitman WB (2008) Metabolic, phylogenetic, and ecological diversity of the methanogenic archaea. Ann N Y Acad Sci 1125(1):171–189

Lin Q, De Vrieze J, He G, Li X, Li J (2016) Temperature regulates methane production through the function centralization of microbial community in anaerobic digestion. Biores Technol 216:150–158

Liu Y, Balkwill DL, Aldrich HC, Drake GR, Boone DR (1999) Characterization of the anaerobic propionate-degrading syntrophs *Smithella propionica gen. nov., sp. nov.* and *Syntrophobacter wolinii*. Int J System Evol Microbiol 49(2):545–556

Liu J, Xu G, Zhao S, He J (2023) Resilience and functional redundancy of methanogenic digestion microbiome safeguard recovery of methanogenesis activity under the stress induced by microplastics. mLife 2(4):378–388

Louca S, Polz MF, Mazel F, Albright MB, Huber JA, O'Connor MI, Ackermann M, Hahn AS, Srivastava DS, Crowe SA, Doebeli M (2018) Function and functional redundancy in microbial systems. Nat Ecol Evol 2(6):936–943

Lozupone CA, Hamady M, Kelley ST, Knight R (2007) Quantitative and qualitative β diversity measures lead to different insights into factors that structure microbial communities. Appl Environ Microbiol 73(5):1576–1585

Lozupone C, Lladser ME, Knights D, Stombaugh J, Knight R (2011) UniFrac: an effective distance metric for microbial community comparison. ISME J 5(2):169–172

Lupo V, Van Vlierberghe M, Vanderschuren H, Kerff F, Baurain D, Cornet L (2021) Contamination in reference sequence databases: time for divide-and-rule tactics. Front Microbiol 12:755101

Lv Z, Lyu P, Li K, Song F, Zhang Z, Yang Y, Yu H (2022) High temperature shock threatens methane production via disturbing microbial interactions in anaerobic digestion. Sci Total Environ 846:157459

Ma H, Ye L, Hu H, Zhang L, Ding L, Ren H (2017) Determination and variation of core bacterial community in a two-stage full-scale anaerobic reactor treating high-strength pharmaceutical wastewater. J Microbiol Biotechnol 27(10):1808–1819

Ma X, Jiang T, Chang J, Tang Q, Luo T, Cui Z (2019) Effect of substrate to inoculum ratio on biogas production and microbial community during hemi-solid-state batch anaerobic co-digestion of rape straw and dairy manure. Appl Biochem Biotechnol 189:884–902

Ma K, Cao Z, Cui Y, Chen T, Shan S, Shi Y, Wang W, Lv J (2021) Effect of magnetite on anaerobic digestion treating saline wastewater: methane production, biomass aggregation and microbial community dynamics. Biores Technol 341:125783

Marozava S, Röling WF, Seifert J, Küffner R, Von Bergen M, Meckenstock RU (2014) Physiology of *Geobacter metallireducens* under excess and limitation of electron donors. Part I. Batch cultivation with excess of carbon sources. System Appl Microbiol 37(4):277–286

Mayer F, Lurz R, Schoberth S (1977) Electron microscopic investigation of the hydrogen-oxidizing acetate-forming anaerobic bacterium *Acetobacterium woodii*. Arch Microbiol 115:207–213

McCann KS (2000) The diversity-stability debate. Nature 405:228–233

McGill BJ, Enquist BJ, Weiher E, Westoby M (2006) Rebuilding community ecology from functional traits. Trends Ecol Evol 21(4):178–185

McInerney MJ, Rohlin L, Mouttaki H, Kim U, Krupp RS, Rios-Hernandez L, Sieber J, Struchtemeyer CG, Bhattacharyya A, Campbell JW, Gunsalus RP (2007) The genome of *Syntrophus aciditrophicus*: Life at the thermodynamic limit of microbial growth. Proc Natl Acad Sci 104(18):7600–7605

Meng X, Cao Q, Sun Y, Huang S, Liu X, Li D (2022) 16S rRNA genes-and metagenome-based confirmation of syntrophic butyrate-oxidizing methanogenesis enriched in high butyrate loading. Biores Technol 345:126483

Mercado JV, Koyama M, Nakasaki K (2022) Co-occurrence network analysis reveals loss of microbial interactions in anaerobic digester subjected to repeated organic load shocks. Water Res 221:118754

Mills S, Trego AC, Lens PNL, Ijaz UZ, Collins G (2021) A distinct, flocculent, acidogenic microbial community accompanies methanogenic granules in anaerobic digesters. Microbiol Spectrum 9(3):e0078421

Mills S, Nguyen TP, Ijaz UZ, Lens PN (2023) Process stability in expanded granular sludge bed bioreactors enhances resistance to organic load shocks. J Environ Manage 342:118271

Moestedt J, Müller B, Westerholm M, Schnürer A (2016) Ammonia threshold for inhibition of anaerobic digestion of thin stillage and the importance of organic loading rate. Microb Biotechnol 9(2):180–194

Nelson MC, Morrison M, Yu Z (2011) A meta-analysis of the microbial diversity observed in anaerobic digesters. Biores Technol 102(4):3730–3739

Niu Q, Takemura Y, Kubota K, Li YY (2015) Comparing mesophilic and thermophilic anaerobic digestion of chicken manure: microbial community dynamics and process resilience. Waste Manage 43:114–122

Nozhevnikova AN, Russkova YI, Litti YV, Parshina SN, Zhuravleva EA, Nikitina AA (2020) Syntrophy and interspecies electron transfer in methanogenic microbial communities. Microbiology 89:129–147

Okie JG, Poret-Peterson AT, Lee ZM, Richter A, Alcaraz LD, Eguiarte LE, Siefert JL, Souza V, Dupont CL, Elser JJ (2020) Genomic adaptations in information processing underpin trophic strategy in a whole-ecosystem nutrient enrichment experiment. Elife 9:e49816

Ottoni JR, Bernal SP, Marteres TJ, Luiz FN, Dos Santos VP, Mari ÂG, Somer JG, de Oliveira VM, Passarini MR (2022) Cultured and uncultured microbial community associated with biogas production in anaerobic digestion processes. Arch Microbiol 204(6):340

Park JH, Kim DH, Kim SH, Yoon JJ, Park HD (2018) Effect of substrate concentration on the competition between *Clostridium and Lactobacillus* during biohydrogen production. Int J Hydrogen Energy 43(25):11460–11469

Paulo LM, Castilla-Archilla J, Ramiro-Garcia J, Escamez-Picón JA, Hughes D, Mahony T, Murray M, Wilmes P, O'Flaherty V (2020) Microbial community redundancy and resilience underpins high-rate anaerobic treatment of dairy-processing wastewater at ambient temperatures. Front Bioeng Biotechnol 8:192

Peces M, Astals S, Jensen PD, Clarke WP (2018) Deterministic mechanisms define the long-term anaerobic digestion microbiome and its functionality regardless of the initial microbial community. Water Res 141:366–376

Plugge CM, Balk M, Stams AJ (2002) *Desulfotomaculum thermobenzoicum subsp. thermosyntrophicum subsp. nov.*, a thermophilic, syntrophic, propionate-oxidizing, spore-forming bacterium. Int J System Evol Microbiol 52(2):391–399

Poehlein A, Bengelsdorf FR, Schiel-Bengelsdorf B, Daniel R, Dürre P (2016) Genome sequence of the acetogenic bacterium *Acetobacterium wieringae* DSM 1911T. Genome Announc 4(6):e01430-e1516

Poirier S, Bize A, Bureau C, Bouchez T, Chapleur O (2016) Community shifts within anaerobic digestion microbiota facing phenol inhibition: towards early warning microbial indicators? Water Res 100:296–305

Prosser JI, Bohannan BJ, Curtis TP, Ellis RJ, Firestone MK, Freckleton RP, Green JL, Green LE, Killham K, Lennon JJ, Osborn AM, Solan M, van der Gast CJ, Young JP (2007) The role of ecological theory in microbial ecology. Nat Rev Microbiol 5(5):384–392

Qi C, Cao D, Gao X, Jia S, Yin R, Nghiem LD, Li G, Luo W (2023) Optimising organic composition of feedstock to improve microbial dynamics and symbiosis to advance solid-state anaerobic co-digestion of sewage sludge and organic waste. Appl Energy 351:121857

Quiroga G, Castrillón L, Fernández-Nava Y, Marañón E, Negral L, Rodríguez-Iglesias J, Ormaechea P (2014) Effect of ultrasound pre-treatment in the anaerobic co-digestion of cattle manure with food waste and sludge. Biores Technol 154:74–79

Robert C, Chassard C, Lawson PA, Bernalier-Donadille A (2007) *Bacteroides cellulosilyticus sp. nov.*, a cellulolytic bacterium from the human gut microbial community. Int J System Evol Microbiol 57(7):1516–1520

Rojas-Tirado P, Pedersen PB, Pedersen LF (2017) Bacterial activity dynamics in the water phase during start-up of recirculating aquaculture systems. Aquacult Eng 78:24–31

Rotaru AE, Shrestha PM, Liu F, Markovaite B, Chen S, Nevin KP, Lovley DR (2014a) Direct interspecies electron transfer between *Geobacter metallireducens* and *Methanosarcina barkeri*. Appl Environ Microbiol 80(15):4599–4605

Rotaru AE, Shrestha PM, Liu F, Shrestha M, Shrestha D, Embree M, Zengler K, Wardman C, Nevin KP, Lovley DR (2014b) A new model for electron flow during anaerobic digestion: direct interspecies electron transfer to *Methanosaeta* for the reduction of carbon dioxide to methane. Energy Environ Sci 7(1):408–415

Rui J, Li J, Zhang S, Yan X, Wang Y, Li X (2015) The core populations and co-occurrence patterns of prokaryotic communities in household biogas digesters. Biotechnol Biofuels 8:158

Sawatdeenarunat C, Surendra KC, Takara D, Oechsner H, Khanal SK (2015) Anaerobic digestion of lignocellulosic biomass: challenges and opportunities. Biores Technol 178:178–186

Shannon CE (1948) A mathematical theory of communication. Bell Syst Tech J 27(3):379–423

Simpson EH (1949) Measurement of diversity. Nature 163(4148):688

Singh S, Rinta-Kanto JM, Lens PN, Kokko M, Rintala J, O'Flaherty V, Ijaz UZ, Collins G (2022) Microbial community assembly and dynamics in Granular, fixed-biofilm and planktonic microbiomes valorizing long-chain fatty acids at 20 °C. Biores Technol 343:126098

Smith SK, Weaver JE, Ducoste JJ, de Los Reyes III FL (2024) Microbial community assembly in engineered bioreactors. Water Res 255:121495

Stegen JC, Lin X, Fredrickson JK, Konopka AE (2015) Estimating and mapping ecological processes influencing microbial community assembly. Front Microbiol 6:126673

Sun L, Pope PB, Eijsink VG, Schnürer A (2015) Characterization of microbial community structure during continuous anaerobic digestion of straw and cow manure. Microb Biotechnol 8(5):815–827

Thapa A, Park JH, Shin SG, Jo HM, Kim MS, Park Y, Han U, Cho SK (2023) Elucidation of microbial interactions, dynamics, and keystone microbes in high pressure anaerobic digestion. Sci Total Environ 858:159718

Trego AC, McAteer PG, Nzeteu C, Mahony T, Abram F, Ijaz UZ, O'Flaherty V (2021) Combined stochastic and deterministic processes drive community assembly of anaerobic microbiomes during granule flotation. Front Microbiol 12:666584

Vanwonterghem I, Jensen PD, Dennis PG, Hugenholtz P, Rabaey K, Tyson GW (2014) Deterministic processes guide long-term synchronised population dynamics in replicate anaerobic digesters. ISME J 8:2015–2028

Vellend M (2010) Conceptual synthesis in community ecology. Q Rev Biol 85(2):183–206

Visser M, Worm P, Muyzer G, Pereira IA, Schaap PJ, Plugge CM, Kuever J, Parshina SN, Nazina TN, Ivanova AE, Bernier-Latmani R (2013) Genome analysis of *Desulfotomaculum kuznetsovii* strain 17 T reveals a physiological similarity with *Pelotomaculum thermopropionicum* strain SI(T). Stand Genomic Sci 8:69–87

Wallrabenstein C, Schink B (1994) Evidence of reversed electron transport in syntrophic butyrate or benzoate oxidation by *Syntrophomonas wolfei* and *Syntrophus buswellii*. Arch Microbiol 162:136–142

Wang J, Feng K, Lou Y, Lu B, Liu B, Xie G, Ren N, Xing D (2021) The synergistic effect of potassium ferrate and peroxymonosulfate application on biogas production and shaping microbial community during anaerobic co-digestion of a cow manure-cotton straw mixture. Biores Technol 333:125166

Webb CT, Hoeting JA, Ames GM, Pyne MI, LeRoy PN (2010) A structured and dynamic framework to advance traits-based theory and prediction in ecology. Ecol Lett 13(3):267–283

Wei Y, Gao Y, Yuan H, Chang Y, Li X (2022) Effects of organic loading rate and pretreatments on digestion performance of corn stover and chicken manure in completely stirred tank reactor (CSTR). Sci Total Environ 815:152499

Weiss S, Van Treuren W, Lozupone C, Faust K, Friedman J, Deng Y, Xia LC, Xu ZZ, Ursell L, Alm EJ, Birmingham A (2016) Correlation detection strategies in microbial data sets vary widely in sensitivity and precision. ISME J 10(7):1669–1681

Werner GD, Strassmann JE, Ivens AB, Engelmoer DJ, Verbruggen E, Queller DC, Noë R, Johnson NC, Hammerstein P, Kiers ET (2014) Evolution of microbial markets. PNAS 111(4):1237–1244

Westerholm M, Roos S, Schnürer A (2010) *Syntrophaceticus schinkii gen. nov., sp. nov.*, an anaerobic, syntrophic acetate-oxidizing bacterium isolated from a mesophilic anaerobic filter. FEMS Microbiol Lett 309(1):100–104

Winkler MK, Boets P, Hahne B, Goethals P, Volcke EI (2017) Effect of the dilution rate on microbial competition: r-strategist can win over k-strategist at low substrate concentration. PLoS ONE 12(3):e0172785

Wu L, Yang Y, Chen SI, Zhao M, Zhu Z, Yang S, Qu Y, Ma Q, He Z, Zhou J, He Q (2016) Long-term successional dynamics of microbial association networks in anaerobic digestion processes. Water Res 104:1–10

Wu ZL, Lin Z, Sun ZY, Gou M, Xia ZY, Tang YQ (2020) A comparative study of mesophilic and thermophilic anaerobic digestion of municipal sludge with high-solids content: reactor performance and microbial community. Biores Technol 302:122851

Wu ZL, Zhang Q, Xia ZY, Gou M, Sun ZY, Tang YQ (2023) The responses of mesophilic and thermophilic anaerobic digestion of municipal sludge to periodic fluctuation disturbance of organic loading rate. Environ Res 218:114783

Wu C, Liu X, Dong X (2006) *Syntrophomonas erecta subsp. sporosyntropha subsp. nov.*, a spore-forming bacterium that degrades short chain fatty acids in co-culture with methanogens. System Appl Microbiol 29(6):457–462

Wu L, Shan X, Chen S, Zhang Q, Qi Q, Qin Z, Yin H, Zhou J, He Q, Yang Y (2020a) Progressive microbial community networks with incremental organic loading rates underlie higher anaerobic digestion performance. mSystems 5(1):e00357-19

Yin Q, Sun Y, Li B, Feng Z, Wu G (2022) The r/K selection theory and its application in biological wastewater treatment processes. Sci Total Environ 824:153836

Yu J, Tang SN, Lee PK (2021) Microbial communities in full-scale wastewater treatment systems exhibit deterministic assembly processes and functional dependency over time. Environ Sci Technol 55(8):5312–5323

Yu J, Zhao L, Yao Z, Feng J, Yuan X, Wang H, Liang Y, Chen J, Du Y, Shen R (2023) Deeper insights into the synergy of material transformation, microbial network, and energy balance during pilot thermophilic and mesophilic dry anaerobic digestion systems. Sci Total Environ 891:164410

Zhang C, Xiao G, Peng L, Su H, Tan T (2013) The anaerobic co-digestion of food waste and cattle manure. Biores Technol 129:170–176

Zhang W, Werner JJ, Agler MT, Angenent LT (2014) Substrate type drives variation in reactor microbiomes of anaerobic digesters. Biores Technol 151:397–401

Zhang J, Lv C, Tong J, Liu J, Liu J, Yu D, Wang Y, Chen M, Wei Y (2016) Optimization and microbial community analysis of anaerobic co-digestion of food waste and sewage sludge based on microwave pretreatment. Biores Technol 200:253–261

Zhang L, Loh KC, Lim JW, Zhang J (2019) Bioinformatics analysis of metagenomics data of biogas-producing microbial communities in anaerobic digesters: a review. Renew Sustain Energy Rev 100:110–126

Zhang L, Gong X, Wang L, Guo K, Cao S, Zhou Y (2021) Metagenomic insights into the effect of thermal hydrolysis pre-treatment on microbial community of an anaerobic digestion system. Sci Total Environ 791:148096

Zhang L, Gong X, Xu R, Guo K, Wang L, Zhou Y (2022a) Responses of mesophilic anaerobic sludge microbiota to thermophilic conditions: Implications for start-up and operation of thermophilic THP-AD systems. Water Res 216:118332

Zhang L, Guo K, Wang L, Xu R, Lu D, Zhou Y (2022b) Effect of sludge retention time on microbial succession and assembly in thermal hydrolysis pretreated sludge digesters: deterministic versus stochastic processes. Water Res 209:117900

Zheng X, Zhou W, Wan R, Luo J, Su Y, Huang H, Chen Y (2018) Increasing municipal wastewater BNR by using the preferred carbon source derived from kitchen wastewater to enhance phosphorus uptake and short-cut nitrification-denitrification. Chem Eng J 344:556–564

Zhou Z, Zhang CJ, Liu PF, Fu L, Laso-Pérez R, Yang L, Bai LP, Li J, Yang M, Lin JZ, Wang WD (2022) Non-syntrophic methanogenic hydrocarbon degradation by an archaeal species. Nature 601(7892):257–262

Ziganshina EE, Belostotskiy DE, Ilinskaya ON, Boulygina EA, Grigoryeva TV, Ziganshin AM (2015) Effect of the organic loading rate increase and the presence of zeolite on microbial community composition and process stability during anaerobic digestion of chicken wastes. Microb Ecol 70:948–960

Ziganshina EE, Ibragimov EM, Vankov PY, Miluykov VA, Ziganshin AM (2017) Comparison of anaerobic digestion strategies of nitrogen-rich substrates: performance of anaerobic reactors and microbial community diversity. Waste Manage 59:160–171

Anaerobic Digestion Under Environmentally Stressed Conditions

Guangxue Wu, Qidong Yin, and Zhongzhong Wang

Abstract Anaerobic digestion can tackle both challenges of environmental pollution and energy crisis simultaneously and has been commonly applied for renewable energy recovery in the form of biogas from organic waste. However, the performance of anaerobic digestion can be affected by environmental stressors, including both disturbance and inhibition. Typically, three types of inhibition can occur, including the osmotic pressure induced by ions, surface layer formation preventing efficient substrate diffusion, and enzyme inhibition by transferred inhibitors. Microorganisms in anaerobic digestion systems can adopt different strategies responding to stressed conditions, such as microbial morphology adaptation, microbial physiology response, and microbial community acclimatization. The redundancy, resistance, and resilience are proposed as the strategy for improving system stability of anaerobic digestion, including under environmentally stressed conditions.

Keywords Environmental stressor · Inhibition · Microbial morphology · Microbial physiology · Microbial selection · 3R strategy

1 Introduction

Within engineered anaerobic digestion ecosystems, microorganisms may experience diverse environmental stressors. Environmental stressors are referred as those that lead to a sharp reduction in fitness of microbial populations, which includes changing environmental conditions (e.g., disturbance by shock loading) causing a

G. Wu (✉)
Civil Engineering, School of Engineering, College of Science and Engineering, University of Galway, Galway H91 TK33, Ireland
e-mail: guangxue.wu@universityofgalway.ie

Q. Yin
School of Civil Engineering, Sun Yat-Sen University, Guangzhou 51000, Guangdong, China

Z. Wang
Department of Chemical, Biological and Environmental Engineering, Universitat Autònoma de Barcelona Edifici Q, Barcelona, Spain

G. Wu (ed.), *Anaerobic Digestion*, Green Energy and Technology,
https://doi.org/10.1007/978-3-031-69378-6_5

drastic reduction in reproductive output and the persistence of the inhibitors [e.g., free ammonia (NH_3), antibiotics, etc.] leading to marked reductions in the size of populations or repeated cycles of colonization and extinction (Fig. 1; Hoffmann and Hercus 2000). The perturbation can be caused by environmental factors such as salinity and pH, and operational parameters mainly including the shock loading due to the feedstock or the operational change. While the typical inhibitors include the accumulated volatile fatty acids (VFAs), long-chain fatty acids (LCFAs), the produced ammonium and NH_3 from degradation of organic matter such as proteins, urea and nucleic acids, hydrogen sulfide (H_2S) produced from sulfate, the dosed antibiotics during livestock activities, the adsorbed heavy metals, etc.

During anaerobic digestion, diverse syntrophic relationships make the anaerobic reactions feasible, and these microbial interactions are critical to harvest and utilize resources efficiently, resulting in improved system resilience and stability. However, different types of microorganisms (e.g., syntrophic bacteria and methanogenic archaea) in anaerobic digestion systems have different sensitivities to environmental stressors. Therefore, the inhibitory effect of environmental stressors during anaerobic digestion on microorganisms can further break up the syntrophic relationship among microorganisms. Depending on the situation, the inhibition by environmental stressors may induce system failure. While sometimes even under inhibition conditions, anaerobic digestion may still function, which was named as the "inhibited steady state" where the process is running stably but with a low methane yield (Angelidaki and Ahring 1993, 1994).

Generally, the high sensitivity of microbial syntrophy to environmental stressors is the root cause of unstable system performance. Therefore, understanding substrate utilization patterns and community dynamics under stressed conditions can provide valuable information for developing more robust and reliable anaerobic digestion systems (Amha et al. 2018). During system design, it would be better to introduce the concept of redundancy to improve functional diversity, such as the adoption of the nexus concept to establish systems with more diverse microbial communities and metabolic pathways (Wu and Yin 2020). Furthermore, this will improve system resistance under environmentally stressed conditions, including both disturbance and inhibition. Finally, the resilience of anaerobic digestion systems will be achieved through these systematic strategies.

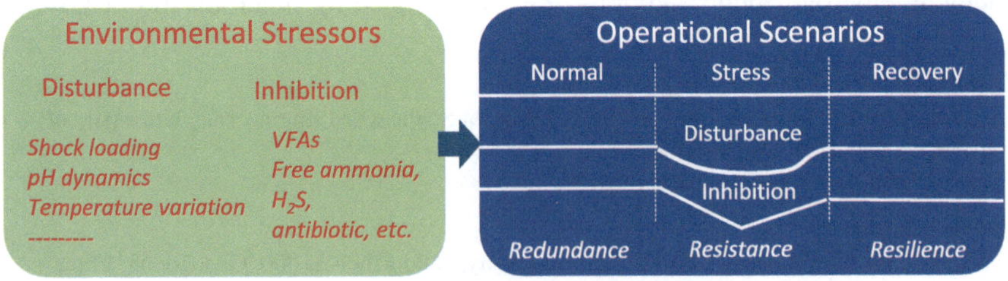

Fig. 1 Stressed phenomenon in anaerobic digestion systems

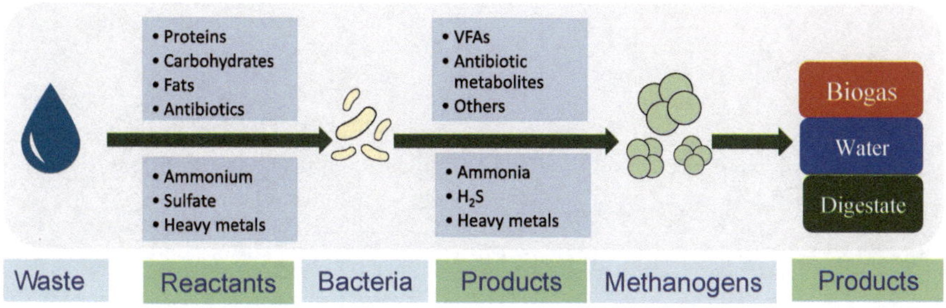

Fig. 2 Inhibition types following degradation pathways during anaerobic digestion

2 Typical Inhibitors and Their Inhibition Mechanisms

2.1 *Types of Inhibitors*

Typical inhibitors in anaerobic digestion systems include those existing in waste sources and those produced during biological reactions (Fig. 2). The former includes antibiotics and heavy metals, and the latter includes intermediates such as the produced VFAs, NH_3, and H_2S. Therefore, the inhibition can be the reactant-based or product-based effect. For the inhibitors, generally, a low concentration can be used as the nutrient (e.g., ammonium and sulfur), benefiting the stable and successful operation of anaerobic digestion systems. Only when their concentrations reach a threshold value, inhibition of microbial activities in anaerobic digestion systems occurs.

2.2 *The Possible Inhibition Mechanisms*

Different inhibition mechanisms can be proposed due to diverse types of interactions such as antagonism, synergism, acclimation, and complexation, all of which could significantly affect the phenomenon of inhibition. The specific inhibition depends on the nature of inhibitors, the acclimation history of biomass, the environmental and operational conditions, and physiology and status of microorganisms. Also, the connections among all inhibitors and environmental conditions should be noticed. For example, the NH_3 concentration is affected by total ammonium nitrogen (TAN) concentration, pH, temperature, and ionic strength (especially for high-loading rate anaerobic digestion systems such as dry anaerobic digestion). The interactions among all inhibitors also affect their inhibitory effects. A typical example is that during anaerobic digestion, the produced high concentration of ammonium can establish a weak buffer system to balance the accumulated high concentration of VFAs, thus mitigating the drastic pH change. This affects the concentration of NH_3 and may enhance system performance under dynamic conditions (or inhibited steady-state conditions).

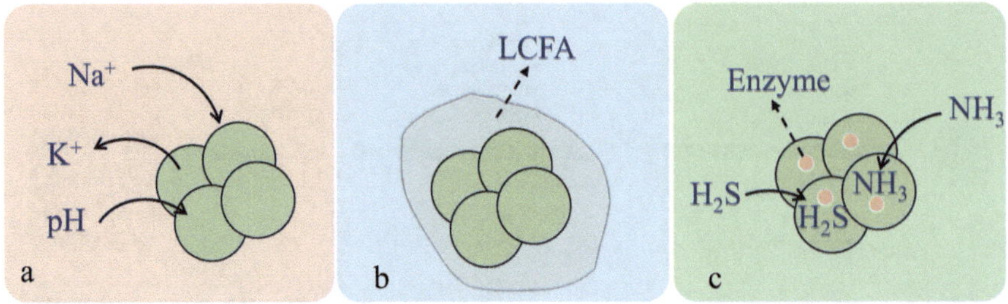

Fig. 3 Typical inhibition mechanisms in anaerobic digestion ecosystems, including the physiological response to pH and salinity (**a**), the formation of inhibitory layer (**b**), and the inhibitor diffusion to the inside of cell causing inhibition on enzymes or metabolic pathways (**c**)

Generally, there are three types of inhibitory mechanisms: microbial physiological response to factors such as ions, the formation of inhibitory layer preventing substrate diffusion and impacting microbial activities, and the penetration of inhibitors to the inside of cell causing inhibition on enzymes or metabolic pathways (Fig. 3).

LCFAs can attach to the surface of microorganisms due to their hydrophobic property, leading to cell membrane damage through solubilizing the lipid bilayer, blocking the accessibility of nutrient due to limited mass transfer, and inhibiting microbial activities due to the failure of homeostasis regulation caused by the formation of acidic environment (Elsamadony et al. 2021; Tian et al. 2018). However, mixed LCFAs may exert different effects on microbial consortia, where acidogenesis could be enhanced due to surfactant effects while acetoclastic methanogenesis was seriously inhibited (Zhu et al. 2019). With similar inhibition effects of LCFAs, the accumulation of surfactants on cell surface can also cause the membrane to swell and leak, disrupting ion gradients and causing cell lysis (Chen et al. 2008).

Previous studies have found that both NH_3 and ammonium are toxic to methanogens, but the inhibitory mechanisms are different (Kayhanian 1999; Rajagopal et al. 2013). NH_3 is usually considered as the active component causing inhibition due to its high permeability to cell membranes and passive diffusion into microbial cells (cytoplasm). The inhibition by the typical NH_3 can be achieved through many pathways, such as a change in intracellular pH caused by converting diffused NH_3 to ammonium through absorbing protons, an increased energy requirement for pumping out potassium (K^+) for pH maintenance (causing potassium deficiency), and the inhibition of activities of specific enzymes. Methanogens need to consume more energy for the potassium antiporter so that intracellular K^+ can be pumped out for protons exchange, maintaining the protons balance and intracellular pH. Ammonium ion, on the other hand, may inhibit specific enzyme reactions which are involved in CH_4 synthesis. A high solution ionic strength caused by the dissolved solutes in raw materials such as livestock manure, as well as adsorption of ammonium on organic matter, can cause reduced chemical activity of both NH_3 and ammonium (Vaddella et al. 2011). Therefore, Ca^{2+} or Mg^{2+} addition could recover ammonia inhibition (Yenigün and Demirel 2013).

As the final product of sulfate reduction, sulfide has toxic effects on methanogens. H_2S can diffuse inside the cell cytoplasm easily and reacts with cellular components. This will further interfere with the assimilation of sulfur and various coenzyme sulfide linkages and denature native proteins by forming sulfide and disulfide cross-links between polypeptide chains (O'Flaherty et al. 1998; Vu et al. 2022). Within anaerobic digesters, the existence of H_2S can be affected by pH, and a high pH can alleviate the inhibition of H_2S on anaerobic digestion. While during anaerobic digestion, oxidation–reduction potential can act as the governing parameter to control the formation of H_2S from sulfate reduction.

Antibiotics are widely used for the treatment and control of human and animal diseases, primarily attributable to their extensive spectrum of antimicrobial activities. However, as the refractory chemical, some antibiotics are discharged into the natural environment due to their limited metabolic transformation or absorption within the physiologies of humans and animals, which can induce lots of environmental risks, such as the propagation of antibiotic resistance genes. For example, as one typical antibiotic, tetracycline can form reversible complexes by binding to 30S subunits of bacterial intracellular ribosomes, thereby inhibiting bacterial protein synthesis (Lins et al. 2015; Lu et al. 2016). Cetecioglu et al. (2013) found that tetracycline exerted a terminal/lethal effect at 8.5 mg/L on the microbial community, which could cause the inhibition of substrate degradation and biogas generation, leading to reactor collapse.

Synergetic effect means that the overall consequence of multiple inhibitors exceeds the sum of their individual effect, and the existence of several inhibitors in an anaerobic digestion system may result in more severe inhibition in methane production due to the potentially formidable joint effects (Yang et al. 2019). For example, LCFAs can have the synergetic co-inhibition combined with ammonia (Tian et al. 2018). The inhibition of methanogens by ammonia can cause the increased accumulation of VFAs and H_2, which renders β-oxidation of LCFA thermodynamically unfavorable; furthermore, the accumulation of LCFA can cause the inhibition of all anaerobic digestion processes (Tian et al. 2018).

3 Microbial Strategies Responding to Inhibition

Under environmentally stressed conditions, microorganisms will develop different responding strategies at molecular, cellular, and community levels. Here, the responding strategies are summarized from aspects of microbial morphology, microbial physiology, and microbial selection (Fig. 4).

3.1 Microbial Morphology

For each type of functional microorganism, different morphologies appear. Usually, filamentous or flocs-forming microorganisms may be acclimated, possessing

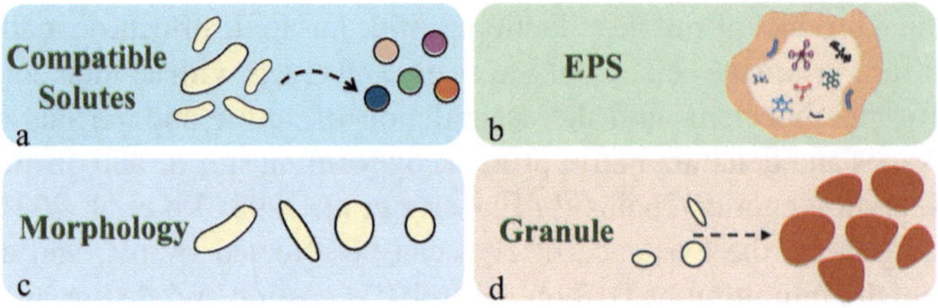

Fig. 4 Microbial responding mechanisms to environmental stressors, such as the formation of metabolites such as secreting organic compatible solutes for protection (**a**), the formation of extracellular polymeric substances (EPS) on the surface (**b**), the adaption of morphology (**c**), and the formation of biofilm or granules (**d**)

different responses to inhibitor diffusion. Regarding acetoclastic methanogens, *Methanosarcina* are large spherical cells possessing a higher volume-to-surface ratio than smaller rod-shaped *Methanosaeta*, which can cause the less diffusion of inhibitors such as NH_3 into the *Methanosarcina* than *Methanosaeta*, resulting in less energy requirement for *Methanosarcina* under inhibited conditions (Wiegant and Zeeman 1986).

Under inhibited conditions, microorganisms may tend to form biofilm-based aggregates, which can alleviate inhibition by limiting substrate diffusion within the aggregates. This is one type of microbial defense mechanisms against unfavorable conditions, and *Methanobacterium thermoformicicum* could grow and form aggregates at the TAN concentration of 9000 mg/L (Hendriksen and Ahring 1991). Synergistic and antagonistic effects may occur during inhibition within microbial aggregates. High biomass concentration or microbial diversity can enhance microbial tolerance to inhibition. For example, microorganisms acclimated at a high sludge retention time (SRT) of 40 days were more tolerant to a TAN concentration of 5000 mg/L in comparison with those acclimated at a low SRT of 25 days (Koster 1986). A high biomass concentration can induce a low inhibitor-to-biomass ratio, which may reduce the inhibition burden for each microorganism.

3.2 Microbial Physiology

The high ammonium concentration may occur simultaneously with the high osmotic pressure. Microorganisms can acclimate to high ammonium concentrations and osmotic pressure by accumulating and secreting low-molecular-weight organic compatible solutes (e.g., betaine) and maintaining proper conformation and activities of enzymes. Yerkes et al. (1997) confirmed that betaine could reduce the toxicity of high concentrations of sodium, resulting in reduced acclimation or lag time and the increased substrate uptake rate and gas production.

Methanosarcina possesses both acetoclastic and hydrogenotrophic methanogenesis and can use diverse substrates (e.g., acetate, hydrogen, and C1 compounds), which may also contribute to their high tolerance to inhibitions such as high temperature and high ammonia stress (Tao et al. 2020). *Methanosarcina* may have a higher acetate uptake rate than *Methanosaeta*, which could be due to that the acetate kinase-phosphoacetyl transferase pathway (*Methanosarcina*) could produce more ATP than the acetyl-CoA synthase pathway (*Methanosaeta*) (Conklin et al. 2006; Rothman et al. 2014). Zhang et al. (2018) found that a high ammonium concentration could inhibit the expressions of methanogenic genes encoding acetyl-CoA decarbonylase and methyl-coenzyme M reductase significantly, and this also indicated the cooperation of the two enzymes in methanogenesis. Microorganisms may revise the metabolic pathways under inhibited conditions. For example, methane production could be achieved through a syntrophic relationship between acetate-oxidizing bacteria and hydrogen-utilizing methanogens under high NH_3 concentrations (Angenent et al. 2002).

Microorganisms can secrete more extracellular polymeric substances (EPS) as a protection mechanism, where EPS can act as the transporting barrier for inhibitors. Because inhibitors must pass through the EPS matrix or react with EPS, microbial activity can be maintained by avoiding the diffusion of inhibitors into cells. Furthermore, polysaccharides were found to play a major role by forming networks to protect microorganisms from inhibition (Yan et al. 2019).

3.3 *Microbial Acclimation*

In anaerobic digestion systems, methanogenesis is more inhibited than other processes of hydrolysis, acidification, and acetogenesis. Generally, for the typical inhibitors of NH_3, LCFAs, H_2S, and antibiotics, acetoclastic methanogens are more sensitive than hydrogenotrophic methanogens, and *Methanosaeta* are more sensitive among the acetoclastics than *Methanosarcina* (Calli et al. 2005; Karakashev et al. 2005).

Due to the diversity of each type of microorganisms, the applied environmental conditions can acclimate specific types of microorganisms. For example, the existence of propionate degraders resistant to high levels of NH_3 nitrogen of 800 mg/L could be acclimated in anaerobic digestion bioreactors (Calli et al. 2005). Also, acclimated microorganisms can have a high tolerance to inhibitor concentrations. For example, Sung and Liu (2003) proposed that acetoclastic methanogens were less sensitive to increased TAN and wider pH ranges after acclimating to high concentrations of TAN. However, Zhu et al. (2021) found that the antibiotic of norfloxacin seriously inhibited the methane production with an 50% inhibitory concentration (IC_{50}) of 0.41 mM, while with the extension of exposure time, the inhibitory effect was strengthened and the IC_{50} values decreased to 0.16 and 0.07 mM in the second and third feeding cycle, respectively.

Most inhibitions during anaerobic digestion are reversible. For example, when exposed to ammonium concentrations of 4000, 8000, and 24,000 mg/L TAN, Parkin et al. (1983) found that the ammonium toxicity was reversible. Pereira et al.

(2005) also confirmed that LCFA inhibition was reversible. Zhu et al. (2021) found that the inhibition of norfloxacin on some bacteria (i.e., *Candidatus Cloacimonas*, *Petrimonas*, *Ercella*, and *Sphaerochaeta*) and hydrogenotrophic methanogens (i.e., *Methanoculleus* and *Methanobacterium*) was reversible, while the inhibition on acetoclastic methanogens of *Methanosaeta* was irreversible.

Some inhibitors can be biodegraded, such as antibiotics. While most of antibiotics can be adsorbed on the surface of biomass or organic solids, their detachment is the rate-limiting step for degradation (Yang et al. 2019). Microplastics in anaerobic digestion systems can also affect the effect of antibiotics. For example, polyamide could enable partial adsorption of ofloxacin, mitigating the inhibitory effect of ofloxacin at low concentrations, but could not effectively alleviate the inhibitory effect of excess ofloxacin on anaerobic digestion (Xiang et al. 2023).

4 Redundancy, Resistance, and Resilience for Anaerobic Digestion Ecosystems

Anaerobic digestion contains diverse types of microbial communities with functional complementary to each other at different trophic levels. The achievement of process stability requires both ecological and engineering principles and technologies. Also, stability can be referred to different aspects, such as ecosystem stability, community stability, and functional stability. The ecosystem stability can be considered from both community and functional stabilities, as not only microbial communities, but also microbe-microbe and microbe-environment interactions are core components of ecosystems. From the engineering aspect, process components such as organic carbon degradation and methane production would be more important, which are mainly kinetic-based processes for the flow of matter and energy.

Redundancy, resistance, and resilience (3R) can be adopted to describe anaerobic digestion performance when facing system disturbance or inhibition. Microbial redundancy is the ability of different types of microorganisms to replace each other's metabolic roles (Free et al. 2018), such as two or more types of microorganisms carrying out the same function. Usually, a high species richness and a high evenness can improve functional redundancy, aiding in recovery of digesters following disturbances (Free et al. 2018). Resistance refers to a community withstand external changes without variations in its original structure and function (Carballa et al. 2015; Liang et al. 2022). Microbial resilience is broadly articulated as a system's ability to recover from external disturbance to its initial or reference state, which can include both ecological (recovery status) and engineering (recovery rate) resilience (Song et al. 2015). As the ability to survive perturbation, robustness can refer to all resilience, resistance, and other complementary properties (Shade et al. 2012), where resilience and resistance are key components of a system's overall robustness (Song et al. 2015).

Within anaerobic digestion ecosystems, hydrolytic-fermentative bacteria are functionally redundant due to their diversity, while syntrophic communities are more resilient with specific metabolic functions (Li et al. 2018). As methanogenesis is more sensitive to stress and less stable, Carballa et al. (2015) proposed *Methanobacteriales* as resistant, *Methanosaeta* as redundant, and *Methanosarcina* and *Methanomicrobiales* as resilient and redundant. Besides, the microbial redundancy of bacteria is generally higher than that of Archaea. For example, Fernandez et al. (1999) found that microbial diversity of bacteria was nearly four times higher than archaea, and despite the stable system function, bacteria possess a higher variation than archaea. Although archaea proportion is relatively low, their function is the core component of anaerobic digestion, and their absence will cause system failure. With applied long SRTs, a high microbial diversity with strong functional redundancy can ensure the system function and stability responding to external disturbances (Zhang et al. 2021).

Although we know that system stability has something to do with redundancy, resilience, and resistance, controlling microbial community dynamics remains challenging. As microbial behavior and dynamic changes are difficult to measure, previous studies have suggested that maintaining stable control conditions can help sustain the abundances of functional microbes, thereby ensuring stable biogas generation (Creamer et al. 2010; Gómez-Quiroga et al. 2022; Jiang et al. 2012). Nevertheless, while microbial ecology of anaerobic digestion system has been widely studied, the microbial community turnover has been observed in the stable performed system (De Vrieze et al. 2013). Therefore, understanding how microbial community turnover and functions respond to certain control conditions is a common challenge in anaerobic digestion research.

The main purpose of ecological engineering is to design anaerobic digestion systems, which can enrich functional microorganisms to remove organic carbon and produce methane efficiently. Generally, ecosystem stability can be ensured by the presence of diverse types of microbial communities for performing the same ecological function (Briones and Raskin 2003). As a dynamic system, different strategies can be applied to enhance system stability. Pulse disturbance can be adopted to enrich microorganisms with high resistance to disturbance. For example, the enrichment of propionate-degrading syntrophic bacteria may be encouraged by using fluctuating propionate levels (Briones and Raskin 2003). In addition, heterogenous systems rather than homogeneous systems would be better applied to enrich both diverse microbial and functional communities, such as that anaerobic baffled reactors with interconnected compartments perform better than completely mixed anaerobic reactors (Bachmann et al. 1985). Another strategy is to increase the organic loading rate (OLR) gradually, so that the acclimation of microbial community may play an important role in maintaining the system stability. Li et al. (2016) reported the functional redundancy of acidogenic bacteria and resilience of *Syntrophomonas* and *Treponema* in a OLR overloading anaerobic digestion system. In an overfed anaerobic digestion system, the resistance of dominant *Bacteroidales* to high VFAs content and low pH could be observed. The severe acidosis made a shift in archaeal populations from acetotrophic to hydrogenotrophic methanogens (Goux et al. 2015). Westerholm et al. (2018) found that the relative abundance of *Synergistia* increased along with the increasing volumetric loading rate in the mesophilic anaerobic digestion system, indicating its redundancy in this system.

5 Future Perspectives

For microbial communities inhibited by diverse inhibitors, predictive understanding of how microbial communities respond to environmental stressors requires mechanistic understanding of relationships and interactions among genes, organisms, environmental chemistry, and physical processes of the anaerobic digestion system (Lui et al. 2021). Two scenarios should be emphasized. One is that for the same inhibitor, diverse forms may exist, such as ammonia-ammonium and sulfide-hydrogen sulfide. The transformation of these forms is affected by factors such as pH and temperature, which makes the understanding of underlying mechanisms difficult. The other one is that different inhibitors may coexist within one ecosystem, and their synergistic or competitive effects should be comprehensively examined. All these studies on different stressed conditions would offer new perspectives on the underlying mechanisms and further guide the development of novel mitigation technologies. Furthermore, the establishment of a framework incorporating all possible mechanisms, scenarios, and mitigation strategies would be quite helpful for practical applications of anaerobic digestion and advancing research in the field.

Recently, many studies have reported the mitigation of methanogenesis inhibition in anaerobic systems amended with conductive materials (Yin et al. 2020). For example, conductive materials can fasten VFAs degradation, avoiding VFA accumulation which may inhibit activities of methanogens (Li et al. 2015). On the other hand, ammonia concentrations are considered as a vital factor that can affect the performance and stability of anaerobic digestion processes (Rajagopal et al. 2013; Yenigün and Demirel 2013), while the presence of magnetite could stimulate syntrophic acetate oxidation in methanogenic systems under high ammonia-stressed conditions (Zhuang et al. 2018). These results highlight the crucial roles of conductive materials on recovering methanogenic activities under stressed conditions, showing a potential extension of their practical applications, which goes beyond the enhancement of direct interspecies electron transfer on CH_4 production. When facing these inhibitory conditions, conductive materials can alleviate the suppression through different mitigation mechanisms, such as avoiding acids accumulation and maintaining intracellular proton and pH balance under ammonia-stressed conditions (Yin et al. 2020).

6 Take-Home Messages

(1) Anaerobic digestion is sensitive to environmental stressors, which can be categorized as perturbation (such as shock loading) and inhibition (inhibitors such as NH_3 and H_2S).
(2) Different inhibitors possess varied inhibition mechanisms (such as being toxic to enzymes, disturbing pH balance, and causing high osmotic pressure), and understanding the mechanistic principles would advance the development of fit-for-purpose alleviation strategies.
(3) Inhibition can not only enrich different microbial communities, but also affect the metabolic pathways of organics degradation.
(4) Acclimation, the formation of aggregates, and enrichment of diverse types of microorganism may enhance system stability responding to environmental stressors.
(5) 3R strategy can be adopted to improve community stability. In general, the higher the diversity of functional microorganisms, the higher the system redundancy, which improves the system's resilience and resistance to withstand perturbation and inhibition.

References

Amha YM, Anwar MZ, Brower A, Jacobsen CS, Stadler LB, Webster TM, Smith AL (2018) Inhibition of anaerobic digestion processes: applications of molecular tools. Bioresour Technol 247:999–1014

Angelidaki I, Ahring BK (1993) Thermophilic anaerobic digestion of livestock waste: the effect of ammonia. Appl Microbiol Biotechnol 38:560–564

Angelidaki I, Ahring BK (1994) Anaerobic thermophilic digestion of manure at different ammonia loads: effect of temperature. Water Res 28(3):727–731

Angenent LT, Sung S, Raskin L (2002) Methanogenic population dynamics during startup of a full-scale anaerobic sequencing batch reactor treating swine waste. Water Res 36(18):4648–4654

Bachmann A, Beard VL, McCarty PL (1985) Performance characteristics of the anaerobic baffled reactor. Water Res 19(1):99–106

Briones A, Raskin L (2003) Diversity and dynamics of microbial communities in engineered environments and their implications for process stability. Curr Opin Biotechnol 14(3):270–276

Calli B, Mertoglu B, Inanc B, Yenigun O (2005) Effects of high free ammonia concentrations on the performances of anaerobic bioreactors. Process Biochem 40(3–4):1285–1292

Carballa M, Regueiro L, Lema JM (2015) Microbial management of anaerobic digestion: exploiting the microbiome-functionality nexus. Curr Opin Biotechnol 33:103–111

Cetecioglu Z, Ince B, Gros M, Rodriguez-Mozaz S, Barceló D, Orhon D, Ince O (2013) Chronic impact of tetracycline on the biodegradation of an organic substrate mixture under anaerobic conditions. Water Res 47(9):2959–2969

Chen Y, Cheng JJ, Creamer KS (2008) Inhibition of anaerobic digestion process: a review. Bioresour Technol 99(10):4044–4064

Conklin A, Stensel HD, Ferguson J (2006) Growth kinetics and competition between *Methanosarcina* and *Methanosaeta* in mesophilic anaerobic digestion. Water Environ Res 78(5):486–496

Creamer KS, Chen Y, Williams CM, Cheng JJ (2010) Stable thermophilic anaerobic digestion of dissolved air flotation (DAF) sludge by co-digestion with swine manure. Bioresour Technol 101(9):3020–3024

De Vrieze J, Verstraete W, Boon N (2013) Repeated pulse feeding induces functional stability in anaerobic digestion. Microb Biotechnol 6(4):414–424

Elsamadony M, Mostafa A, Fujii M, Tawfik A, Pant D (2021) Advances towards understanding long chain fatty acids-induced inhibition and overcoming strategies for efficient anaerobic digestion process. Water Res 190:116732

Fernández A, Huang S, Seston S, Xing J, Hickey R, Criddle C, Tiedje J (1999) How stable is stable? Function versus community composition. Appl Environ Microbiol 65(8):3697–3704

Free A, McDonald MA, Pagaling E (2018) Diversity-function relationships in natural, applied, and engineered microbial ecosystems. Adv Appl Microbiol 105:131–189

Gómez-Quiroga X, Aboudi K, Álvarez-Gallego CJ, Romero-García LI (2022) Successful and stable operation of anaerobic thermophilic co-digestion of sun-dried sugar beet pulp and cow manure under short hydraulic retention time. Chemosphere 293:133484

Goux X, Calusinska M, Lemaigre S, Marynowska M, Klocke M, Udelhoven T, Benizri E, Delfosse P (2015) Microbial community dynamics in replicate anaerobic digesters exposed sequentially to increasing organic loading rate, acidosis, and process recovery. Biotechnol Biofuels 8:122

Hendriksen HV, Ahring BK (1991) Effects of ammonia on growth and morphology of thermophilic hydrogen-oxidizing methanogenic bacteria. FEMS Microbiol Lett 85(3):241–245

Hoffmann AA, Hercus MJ (2000) Environmental stress as an evolutionary force. BioScience 50(3):217–226

Jiang Y, Heaven S, Banks CJ (2012) Strategies for stable anaerobic digestion of vegetable waste. Renew Energy 44:206–214

Karakashev D, Batstone DJ, Angelidaki I (2005) Influence of environmental conditions on methanogenic compositions in anaerobic biogas reactors. Appl Environ Microbiol 71(1):331–338

Kayhanian M (1999) Ammonia inhibition in high-solids biogasification: an overview and practical solutions. Environ Technol 20(4):355–365

Koster IW (1986) Characteristics of the pH-influenced adaptation of methanogenic sludge to ammonium toxicity. J Chem Technol Biotechnol 36(10):445–455

Li H, Chang J, Liu P, Fu L, Ding D, Lu Y (2015) Direct interspecies electron transfer accelerates syntrophic oxidation of butyrate in paddy soil enrichments. Environ Microbiol 17(5):1533–1547

Li L, He Q, Ma Y, Wang X, Peng X (2016) A mesophilic anaerobic digester for treating food waste: process stability and microbial community analysis using pyrosequencing. Microb Cell Fact 15:65

Li L, Peng X, Wang X, Wu D (2018) Anaerobic digestion of food waste: a review focusing on process stability. Bioresour Technol 248:20–28

Liang Y, Ma A, Zhuang G (2022) Construction of environmental synthetic microbial consortia: based on engineering and ecological principles. Front Microbiol 13:829717

Lins P, Reitschuler C, Illmer P (2015) Impact of several antibiotics and 2-bromoethanesulfonate on the volatile fatty acid degradation, methanogenesis and community structure during thermophilic anaerobic digestion. Bioresour Technol 190:148–158

Lu M, Niu X, Liu W, Zhang J, Wang J, Yang J, Wang W, Yang Z (2016) Biogas generation in anaerobic wastewater treatment under tetracycline antibiotic pressure. Sci Rep 6(1):28336

Lui LM, Majumder EL, Smith HJ, Carlson HK, Von Netzer F, Fields MW, Stahl DA, Zhou J, Hazen TC, Baliga NS, Adams PD (2021) Mechanism across scales: a holistic modeling framework integrating laboratory and field studies for microbial ecology. Front Microbiol 12:642422

O'Flaherty V, Mahony T, O'Kennedy R, Colleran E (1998) Effect of pH on growth kinetics and sulphide toxicity thresholds of a range of methanogenic, syntrophic and sulphate-reducing bacteria. Process Biochem 33(5):555–569

Parkin GF, Speece RE, Yang CH, Kocher WM (1983) Response of methane fermentation systems to industrial toxicants. J Water Pollut Control Federat 55(1):44–53

Pereira MA, Pires OC, Mota M, Alves MM (2005) Anaerobic biodegradation of oleic and palmitic acids: evidence of mass transfer limitations caused by long chain fatty acid accumulation onto the anaerobic sludge. Biotechnol Bioeng 92(1):15–23

Rajagopal R, Masse DI, Singh G (2013) A critical review on inhibition of anaerobic digestion process by excess ammonia. Bioresour Technol 143:632–641

Rothman DH, Fournier GP, French KL, Alm EJ, Boyle EA, Cao C, Summons RE (2014) Methanogenic burst in the end-Permian carbon cycle. Proceed Natl Acad Sci 111(15):5462–5467

Shade A, Read JS, Youngblut ND, Fierer N, Knight R, Kratz TK, Lottig NR, Roden EE, Stanley EH, Stombaugh J, Whitaker RJ, Wu CH, McMahon KD (2012) Lake microbial communities are resilient after a whole-ecosystem disturbance. ISME J 6(12):2153–2167

Song HS, Renslow RS, Fredrickson JK, Lindemann SR (2015) Integrating ecological and engineering concepts of resilience in microbial communities. Front Microbiol 6:1298

Sung S, Liu T (2003) Ammonia inhibition on thermophilic anaerobic digestion. Chemosphere 53(1):43–52

Tao Y, Ersahin ME, Ghasimi DS, Ozgun H, Wang H, Zhang X, Guo M, Yang Y, Stuckey DC, van Lier JB (2020) Biogas productivity of anaerobic digestion process is governed by a core bacterial microbiota. Chem Eng J 380:122425

Tian H, Karachalios P, Angelidaki I, Fotidis IA (2018) A proposed mechanism for the ammonia-LCFA synergetic co-inhibition effect on anaerobic digestion process. Chem Eng J 349:574–580

Vaddella VK, Ndegwa PM, Jiang A (2011) An empirical model of ammonium ion dissociation in liquid dairy manure. Trans ASABE 54(3):1119–1126

Vu HP, Nguyen LN, Wang Q, Ngo HH, Liu Q, Zhang X, Nghiem LD (2022) Hydrogen sulphide management in anaerobic digestion: a critical review on input control, process regulation, and post-treatment. Bioresour Technol 346:126634

Westerholm M, Müller B, Singh A, Karlsson Lindsjö O, Schnürer A (2018) Detection of novel syntrophic acetate-oxidizing bacteria from biogas processes by continuous acetate enrichment approaches. Microbial Biotechnol 11(4):680–693

Wiegant WM, Zeeman G (1986) The mechanism of ammonia inhibition in the thermophilic digestion of livestock wastes. Agricult Wastes 16(4):243–253

Wu G, Yin Q (2020) Microbial niche nexus sustaining biological wastewater treatment. NPJ Clean Water 3(1):33

Xiang Y, Xiong W, Yang Z, Xu R, Zhang Y, Wu M, Ye Y, Peng H, Tong J, Wang D (2023) Coexistence of microplastics alters the inhibitory effect of antibiotics on sludge anaerobic digestion. Chem Eng J 455:140754

Yan W, Lu D, Liu J, Zhou Y (2019) The interactive effects of ammonia and carbon nanotube on anaerobic digestion. Chem Eng J 372:332–340

Yang Y, Huang W, Huang W (2019) Antibiotic inhibition on anaerobic digestion of animal manure and controlling strategies: a short review. Clean: Soil, Air, Water 47(1):1700653

Yenigün O, Demirel B (2013) Ammonia inhibition in anaerobic digestion: a review. Process Biochem 48(5–6):901–911

Yerkes DW, Boonyakitsombut S, Speece RE (1997) Antagonism of sodium toxicity by the compatible solute betaine in anaerobic methanogenic systems. Water Sci Technol 36(6–7):15–24

Yin Q, Gu M, Wu G (2020) Inhibition mitigation of methanogenesis processes by conductive materials: a critical review. Bioresour Technol 317:123977

Zhang N, Peng H, Li Y, Yang W, Zou Y, Duan H (2018) Ammonia determines transcriptional profile of microorganisms in anaerobic digestion. Brazil J Microbiol 49:770–776

Zhang L, Guo K, Wang L, Xu R, Lu D, Zhou Y (2021) Effect of sludge retention time on microbial succession and assembly in thermal hydrolysis pretreated sludge digesters: deterministic versus stochastic processes. Water Res 209:117900

Zhu K, Zhang L, Mu L, Ma J, Li C, Li A (2019) Comprehensive investigation of soybean oil-derived LCFAs on anaerobic digestion of organic waste: inhibitory effect and transformation. Biochem Eng J 151:107314

Zhu K, Zhang L, Wang X, Mu L, Li C, Li A (2021) Inhibition of norfloxacin on anaerobic digestion: focusing on the recoverability and shifted microbial communities. Sci Total Environ 752:141733
Zhuang L, Ma J, Yu Z, Wang Y, Tang J (2018) Magnetite accelerates syntrophic acetate oxidation in methanogenic systems with high ammonia concentrations. Microb Biotechnol 11(4):710–720

Ammonia-Induced Inhibition and Its Alleviation Strategies in Anaerobic Digestion Ecosystems

Zhongzhong Wang, Qidong Yin, and Guangxue Wu

Abstract Ammonia, while an essential component in anaerobic digestion, can become toxic at high concentrations, adversely affecting microbial communities and biogas production. Ammonia exists in various forms in anaerobic digestion systems, and different forms of ammonia induce different toxic mechanisms and toxicity levels. Free ammonia is more toxic to anaerobic microorganisms, especially methanogens. The mechanisms of ammonia inhibition on anaerobes encompass both direct chemical interactions between ammonia and microbial cells, and indirect effects of ammonia on the biochemical pathways crucial for energy generation. Microbial communities may employ diverse physiological strategies to tolerate toxic ammonia, such as balancing the intracellular pH, producing osmotic-compounds, and adjusting energy metabolic pathways. Various approaches can be adopted to alleviate ammonia inhibition, such as the control of suitable environmental conditions and the development of advanced biological methods (such as acclimation and bioaugmentation). All these knowledge and approaches can benefit the management of ammonia inhibition in anaerobic digestion and the improvement of system performance.

Keywords Ammonia inhibition · Anaerobic digestion · Methanogens · Microbial community · Mitigation strategies

Z. Wang (✉)
Department of Chemical, Biological and Environmental Engineering, Universitat Autònoma de Barcelona, Barcelona 08193, Spain
e-mail: wang.zhongzhong@uab.cat

Q. Yin
School of Civil Engineering, Sun Yat-Sen University, Guangzhou 51000, Guangdong, China

G. Wu
Civil Engineering, School of Engineering, College of Science and Engineering, University of Galway, Galway H91 TK33, Ireland

G. Wu (ed.), *Anaerobic Digestion*, Green Energy and Technology,
https://doi.org/10.1007/978-3-031-69378-6_6

1 Introduction

The complex biological processes in anaerobic digestion are susceptible to inhibition induced by specific chemicals (e.g., ammonia, hydrogen sulfide, and antibiotics) (Yenigün and Demirel 2013). Ammonia is considered one of the most significant inhibitors (Yenigün and Demirel 2013), which is produced during anaerobic digestion of nitrogenous organic compounds (e.g., proteins, amino acids, urea, and nucleic acids). Particularly, when treating high-nitrogen content organic wastes (e.g., pig slurry), the produced ammonia can significantly impede efficiency and stability of anaerobic digestion systems.

At appropriate concentrations, ammonia serves as a requisite nitrogen source for the synthesis of amino acids, proteins, and nucleic acids, thereby facilitating the growth and metabolism of microorganisms (McCarty 1964). Simultaneously, ammonia contributes to the buffering capacity, assisting in maintaining relatively stable pH by neutralizing the organic acids generated from anaerobic digestion (Jiang et al. 2019; McCarty 1964). However, an undesirably high concentration of ammonia can lead to inhibitory effects, particularly impacting the activity of methanogenic archaea (Capson-Tojo et al. 2020; Wang et al. 2023). This inhibition can result in reduced biogas production, decreased processing efficiency, and even imbalanced system performance (Capson-Tojo et al. 2020; Kalamaras et al. 2020; Liu and Sung 2002).

2 Ammonia Forms in Anaerobic Digestion

The total ammoniacal nitrogen (TAN) in anaerobic digestion systems exists in the form of ammonium (NH_4^+) and free ammonia (FA; NH_3). Both forms can contribute to ammonia-induced inhibition (Yellezuome et al. 2022). However, FA is more toxic to anaerobic microorganisms due to its lipophilic nature, which allows it to diffuse across cellular membrane more easily than NH_4^+. This property makes FA the primary cause of ammonia-induced inhibition in anaerobic digestion ecosystems (Müller et al. 2006; Rajagopal et al. 2013).

2.1 Free Ammonia Calculation

In aquatic environments, the dynamic equilibrium between NH_3 and NH_4^+ can be described by the following reaction: $NH_3(aq) + H_2O(l) \leftrightarrow NH_4^+(aq) + OH^-(aq)$. Free ammonia nitrogen (FAN) refers to the un-ionized form of ammonia nitrogen, while ammonium nitrogen (NH_4^+–N) refers to the ammonium ion nitrogen (Rajagopal et al. 2013). Ideally, the proportion of FAN to TAN is calculated according to Eqs. 1 and 2 (Hansen et al. 1998; Koster 1986):

$$\text{FAN} = \frac{\text{TAN}}{1 + 10^{(\text{pK}_a - \text{pH})}} \tag{1}$$

$$K_a = K_{a\_25} e^{\frac{51965}{R} \times \left(\frac{1}{298.15} - \frac{1}{T}\right)} \tag{2}$$

where R is the ideal gas constant (8.314 J/mol/K), T is the temperature in Kelvin (K), K_a is the acid dissociation constant at temperature T, and K_{a_25} is the acid dissociation constant at 25 °C ($10^{-9.25}$).

When dealing with concentrated feedstock in anaerobic digestion systems, such as animal manure with the existence of high strength ions, the above approach for FAN calculation is inappropriate, as it is mainly based on the ideal solution. Similar to that the concentration of hydrogen ions is not the same as the activity of hydrogen ions (Nielsen et al. 2008), the ionic strength effect on the chemical activity of the ions from concentrated feedstocks should be considered. Generally, the lower dissociation constant of NH_4^+ under these conditions, opposing to the pure water-ammonium solution, is due to the higher ionic strength resulting from dissolved solutes and the adsorption of TAN onto organic matter (Rajagopal et al. 2013), resulting in a reduced chemical activity of both NH_3 and NH_4^+. For these non-ideal solution conditions, the Davies equation (Eq. 3) is typically applied to calculate FAN, incorporating the ionic strength (I) (Eq. 6; Rajagopal et al. 2013). The Davies equation introduces an activity coefficient (f) (Eq. 4) as the correction factor to the ideal equilibrium equation.

$$\text{FAN} = \frac{f \cdot \text{TAN}}{f + 10^{(\text{pK}_a - \text{pH})}} \tag{3}$$

$$f = 10^{\left(-A \cdot z_i^2 \cdot \left(\left(\frac{\sqrt{I}}{1+\sqrt{I}}\right) - \lambda \cdot I\right)\right)} \tag{4}$$

$$A = 1.82 \cdot 10^6 \cdot (\varepsilon \cdot T)^{-\frac{3}{2}} \tag{5}$$

$$I = \frac{1}{2} \sum_{i=1}^{n} C_i \cdot z_i^2 \tag{6}$$

where λ is an empirically determined constant (0.2), ε is the dielectric constant of water at the working temperature (74.828 and 68.345 at 35 and 55 °C, respectively), C_i is the concentration of the species i (M), and z_i is the corresponding charge. Among the different possibilities for calculating f (for instance, Stumm and Morgan 1995), λ of 0.2 was selected for its applicability across a wide range of ionic strengths (up to 0.5 M).

2.2 *Effects of pH and Temperature*

The relative concentrations of NH_3 and NH_4^+ are primarily influenced by pH and temperature (Fig. 1; Koster 1986), with NH_3 prevalent under alkaline conditions. For instance, at 35 °C (equivalent to 308.15 K), the NH_3 proportion increases markedly from 10.1% at pH 8.0 to 52.9% at pH 9.0. This demonstrates the sensitivity of the ammonia–ammonium equilibrium to pH variations and the significant effect of alkalinity on the dominance of FA.

3 Mechanisms of Ammonia Inhibition

The mechanisms of ammonia inhibition on anaerobes encompass both direct chemical interactions between ammonia and microbial cells, and the indirect effects of ammonia on the biochemical pathways crucial for energy generation (Gaspari et al. 2024; Jiang et al. 2019). Generally, methanogens are notably susceptible to ammonia inhibition (Capson-Tojo et al. 2020; Kalamaras et al. 2020).

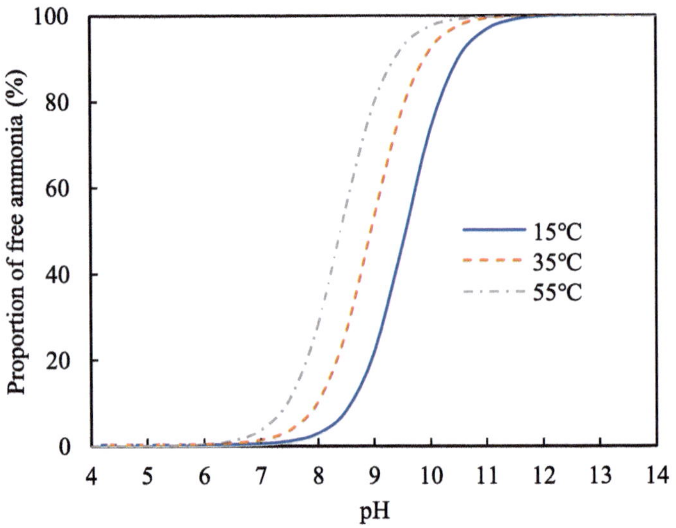

Fig. 1 Variation of the proportion of FA under different pHs and temperatures (temperatures referring to psychrophilic, mesophilic and thermophilic ones usually applied in anaerobic digestion systems)

3.1 Mechanisms of Ammonia Inhibition

The FAN concentration is crucial in ammonia inhibition (Müller et al. 2006; Rajagopal et al. 2013). Following the diffusion of FA into cell, two possible mechanisms of ammonia toxicity have been postulated (Kadam and Boone 1996): direct inhibition of cytosolic enzymes by FA, and intracellular accumulation of NH_4^+ and its consequence inhibitory effects.

Kadam and Boone (1996) illuminated the inhibitory impact of FA on the activity of key ammonia assimilating enzymes: glutamate dehydrogenase, glutamine synthetase, and alanine dehydrogenase. Three methanogen species within the *Methanosarcinaceae* family possessed different enzymatic responses to elevated ammonia levels (Kadam and Boone 1996), indicating that the direct interaction between FA and these enzymes might alter their functionality, thereby affecting the cell's metabolic processes.

The diffusion of FA into the cell and its subsequent conversion to NH_4^+ is another toxicity mechanism. This process captures protons, necessitating increased energy expenditure by the cell to maintain the intracellular pH through pumping potassium (K^+), thereby inhibiting specific enzymatic reactions (Sprott et al. 1984; Wittmann et al. 1995; Yin et al. 2020). When the K^+ pump cannot work efficiently to compensate for the excessive consumption of intracellular protons by FA, an intracellular pH imbalance occurs, leading to cytotoxicity (Sprott and Patel 1986). Sprott et al. (1984) demonstrated the vulnerability of *Methanospirillum hungatei* to ammonia in a K^+-free buffer, observing significant cytoplasmic K^+ loss through an ammonia/K^+ exchange reaction, and highlighting NH_3 as the principal agent of cellular toxicity. Additionally, ammonia affects the cellular retention of other essential cations like Mg^{2+} and Na^+ and disrupts cellular ion homeostasis, potentially leading to deficiencies in micro-nutrients essential for cellular functions (Kadam and Boone 1996).

At elevated levels of TAN, the FAN concentration does not comprehensively account for all inhibition (Akindele and Sartaj 2018; Nakakubo et al. 2008). Lay et al. (1998) demonstrated that the NH_4^+ concentration had a more pronounced inhibitory impact on methanogenesis than FA when the microbial community was adequately acclimated. Furthermore, Sprott and Patel (1986) elucidated that the function of a critical enzyme involved in methane synthesis, which spans the cytoplasmic membrane, depended on its interactions with Ca^{2+} and Mg^{2+}. However, high concentrations of NH_4^+ disrupt these interactions at the Ca^{2+} and Mg^{2+} binding sites, thereby impairing the activity of methane-producing enzymes (Fig. 2).

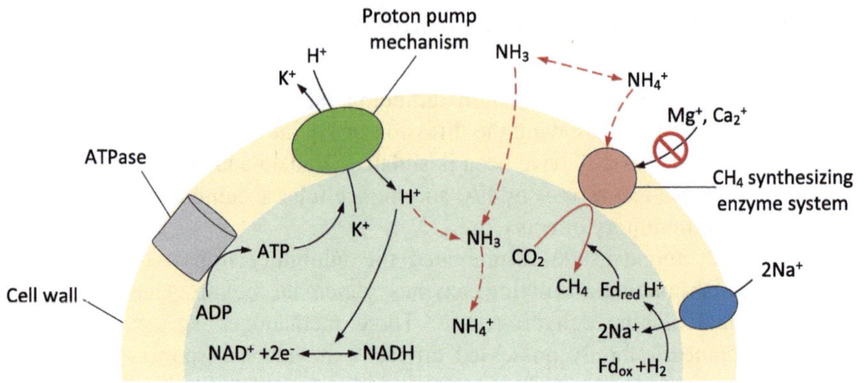

Fig. 2 Proposed mechanism of ammonia inhibition in anaerobic digestion systems (Modified from Yin et al. 2020; Kayhanian 1999)

3.2 *Mechanisms of Ammonia Inhibition Connecting with Energy Maintenance*

For most microorganisms involved in anaerobic digestion, energy generation is facilitated by proton translocating ATPases, relying on a proton motive force (PMF, Δp) across the cell membrane (Jiang et al. 2019; Unden and Kim 2021). PMF (Eq. 7) comprises the combined energies from the transmembrane pH gradient (ΔpH) and the transmembrane electrical gradient ($\Delta\psi$) (Harold 1987; Kadam and Boone 1996).

$$\Delta p = \Delta\psi - 2.303\text{RT}\Delta\text{pH} \quad (\text{kJ/mol}) \tag{7}$$

where Δp is the proton motive force (kJ/mol); ΔpH is the transmembrane pH gradient ($\text{pH}_{\text{in}} - \text{pH}_{\text{out}}$); R is the gas constant (kJ/K·mol); and T is the temperature (K).

The above energy production process is adversely affected by elevated ammonia concentrations outside of the cells (Jiang et al. 2019). Methanogens with a more negative ΔpH compared to their environment can have higher intracellular NH_4^+ concentrations, contributing to a significant portion of the intracellular cations in cells with very negative ΔpH (Jiang et al. 2019; Wolin 1982). Experimental evidence supporting these dynamics is scarce due to the challenges of measuring the intracellular pH and cation concentrations. However, the methanogen strain *Methanolobus taylorii* could increase $\Delta\psi$ by actively extruding potassium, allowing growth in near-neutral cytosol conditions even when the external pH exceeding 7.0 (Jeris and McCarty 1965). This organism's active potassium pump has been proposed as a mechanism to increase $\Delta\psi$ and enable growth under low or negative ΔpH conditions (Koster 1986).

4 Factors Affecting Ammonia Inhibition

Ammonia inhibition in anaerobic digestion significantly impacts the efficiency and stability of biogas production by altering the microbial community structure, gas production rate, and overall process stability. A comprehensive review (Rajagopal et al. 2013) summarized the effects of ammonia levels on anaerobic digestion without sludge adaptation, as shown in Table 1. Capson-Tojo et al. (2020) confirmed that the TAN concentration above 1500 mg/L are considered inhibitory, with the FAN concentration above 50 mg/L posing significant risks to the methanogenic activity. However, after acclimation, TAN concentrations reaching up to 8000–9000 mg/L did not reduce methane production significantly (Miller and Parkin 1983).

Complete ammonia inhibition and "inhibited steady state" exist in anaerobic digestion systems. For complete inhibition, microbial activity is entirely suppressed, resulting in a cessation of metabolic functions. While "inhibited steady state" allows stable methane production despite the presence of inhibitory ammonia. Complete ammonia inhibition typically occurs when TAN concentrations exceed 14 g/L (Cai et al. 2022). However, in most practical anaerobic digestion systems, without an additional nitrogen source, TAN concentrations merely exceed 15 g/L. Hence, "inhibited steady state" is more frequently observed (Zheng et al. 2021).

The ammonia toxicity in anaerobic digestion processes is not only determined by its chemical forms (NH_3 and NH_4^+) but also significantly influenced by environmental conditions such as pH, temperature, the presence of other ions, and substrate composition. These factors can impact the dynamic equilibrium between NH_3 and NH_4^+, subsequently modulating the toxicity levels experienced by methanogens and other microorganisms involved in anaerobic digestion.

4.1 pH

The mechanisms of ammonia inhibition indicate that ammonia toxicity is significantly influenced by the chemical form of ammonia, with FA being more toxic to methanogenic archaea and other microorganisms within anaerobic digestion ecosystems (Müller et al. 2006; Rajagopal et al. 2013). The significant increase in the

Table 1 Effects of ammonia levels on the anaerobic digestion process without any sludge adaption (Rajagopal et al. 2013)

Effect on anaerobic digestion	TAN (mg/L)	Reference
Beneficial	50–200	McCarty (1964)
No antagonistic effect	200–1000	Hobson and Shaw (1976)
Inhibition (especially at high pH)	1500–3000	Angelidaki and Ahring (1993)
Complete inhibition or toxic at any pH	>3000	Procházka et al. (2012), Sung and Liu (2003)

proportion of NH_3 with rising pH highlights the increased risk of ammonia toxicity under alkaline conditions, especially in the presence of high TAN concentrations (Rajagopal et al. 2013).

High pH conditions not only elevate the fraction of toxic NH_3, but also potentially compel methanogens to establish a more negative ΔpH to maintain a near-neutral cytosol pH. This physiological adaptation may inadvertently lead to an accumulation of NH_4^+ inside the cell, contributing to an additional toxicity (Kadam and Boone 1996).

4.2 Temperature

Elevated temperatures induce more toxic FA, thereby potentially increasing its toxicity to microorganisms involved in anaerobic digestion. Moreover, an increase in operational temperature is often associated with enhanced microbial growth and metabolic rates, which, while beneficial for the anaerobic digestion efficiency, can exacerbate the toxicity of ammonia due to the higher susceptibility of more active microbial communities (Nie et al. 2021). For example, thermophilic anaerobic digestion is more prone to inhibition and instability in the presence of high ammonia concentrations compared to mesophilic anaerobic digestion (Rajagopal et al. 2013), highlighting the complex relationship between temperature, microbial activity, and ammonia toxicity. Besides, Massé et al. (2003, 2010) suggested that psychrophilic anaerobic digesters might exhibit greater tolerance to high ammonia concentrations compared to mesophilic or thermophilic anaerobic digesters. This indicates that psychrophilic anaerobic digestion could offer advantages in handling nitrogen-rich organic wastes by mitigating the inhibitory effect of ammonia.

4.3 Salinity and Presence of Other Ions

Salinity levels and the specific ionic composition of the anaerobic digestion medium are critical factors in determining the extent of ammonia toxicity, through their impacts on osmotic balance and the rates of NH_3 diffusion. High concentration salts exacerbate ammonia's toxic effects by altering the osmotic equilibrium, thereby affecting the diffusion of NH_3 into microbial cells (Hafner and Bisogni 2009).

The specific ionic composition of the digestion medium plays a significant role in modulating ammonia toxicity. Cations such as K^+ and Na^+ are particularly influential, as they directly affect the osmotic balance and ion exchange within methanogenic cells. These ions can either exacerbate or mitigate ammonia's toxic effects, depending on their concentrations and the overall ionic balance within the medium. The interaction between these cations and ammonia is a delicate one, with the potential to significantly influence the stability and efficiency of the anaerobic digestion process (Chen et al. 2008).

Moreover, the presence of divalent cations like Mg^{2+} and Ca^{2+} in the digestion medium can have profound effects on the microbial community, particularly in terms of the aggregation and stability of microbial flocs. These cations contribute to the structural integrity of microbial aggregates, which can enhance the resilience of the microbial community to ammonia toxicity by affecting the physical and chemical environment of the digestion medium (Capson-Tojo et al. 2020).

4.4 Substrate Composition

The substrate composition also determines ammonia toxicity in anaerobic digestion systems. The type and composition of the feedstock directly influences the ammonia concentration generated during anaerobic digestion, affecting the system's susceptibility to ammonia inhibition. Organic wastes rich in proteins and urea produce a relatively high ammonia concentration upon decomposition, posing significant toxicity risks to microorganisms (Capson-Tojo et al. 2020; Wang et al. 2023).

The carbon-to-nitrogen (C/N) ratio of the substrate is a critical parameter in managing ammonia levels within the anaerobic digester. A balanced C/N ratio ensures optimal microbial growth and metabolic activity by providing adequate carbon for energy production and nitrogen for protein synthesis without excessive ammonia production. Typically, substrates with a C/N ratio below 15 pose a risk of ammonia inhibition (Cai et al. 2021). Adjusting the substrate composition, such as through co-digestion with carbon-rich materials, can mitigate ammonia toxicity by diluting the nitrogen content and adjusting the C/N ratio to a more favorable one. For example, Xie et al. (2011) demonstrated that co-digestion of pig manure and food waste improved the methanogenic process by providing a more balanced C/N ratio, preventing the accumulation of volatile fatty acids and maintaining an optimal pH for methanogens.

5 Strategies for Alleviation of Ammonia Inhibition

Current strategies for reducing ammonia inhibition in anaerobic digestion systems include: (1) ammonia concentration reduction, (2) microbial activity enhancement, (3) augmentation of microorganisms resisting to ammonia, and (4) the addition of conductive materials (Fig. 3).

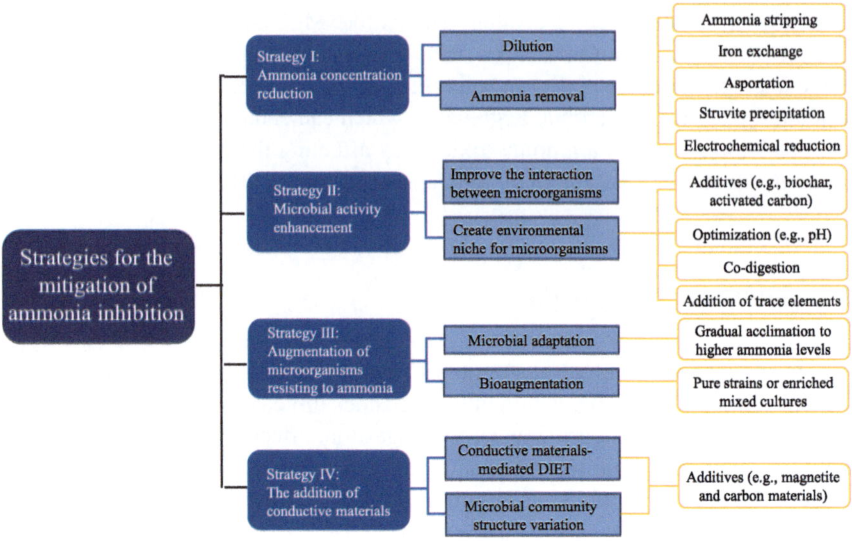

Fig. 3 Strategies for alleviating ammonia inhibition in anaerobic digestion ecosystems (DIET: direct interspecies electron transfer)

5.1 Optimization of Environmental Conditions

Several strategies, including substrate dilution, pH and temperature adjustment, and balancing the C/N ratio, are often used in optimizing the microbial environment to ensure robust and sustainable biogas production.

Diluting the substrate is a common method to control ammonia concentrations (Yenigün and Demirel 2013). This is achieved by either adding water or incorporating substrates with a low nitrogen content, thereby reducing the overall ammonia concentration. For instance, chicken manure is typically diluted to maintain a total solids (TS) content of about 10%, resulting in a TAN concentration ranging from 5.0 to 7.0 g/L (Fuchs et al. 2018). However, dilution lowers the concentration of organic matter, compromising digestion efficiency and necessitating larger digester volumes to achieve the required hydraulic retention time. Additionally, it increases the volume of digestate, complicating post-processing and increasing transportation costs (Fuchs et al. 2018). Therefore, maintaining an optimal level of TS is crucial, especially when processing nitrogen-rich substrates.

Proper management of pH is crucial in anaerobic digestion to mitigate the toxicity of ammonia on microorganisms. Maintaining a controlled pH range of 6.5–8.2 is imperative (Wang et al. 2023). However, the control of a stable pH is challenging, and the use of chemicals significantly increases the operational cost.

Temperature also significantly affects ammonia inhibition and microbial digestion kinetics. Anaerobic digestion at high TAN concentrations is more susceptible to inhibition and less stable under thermophilic conditions compared to mesophilic conditions (Gallert et al. 1998). Operating reactors at psychrophilic temperatures (below 20 °C) can counteract ammonia inhibition, achieving stable performance even at TAN levels up to 8.2 g/L (Massé et al. 2014). However, lower temperatures slow down the digestion process, necessitating a longer retention time and a larger reactor volume (Fuchs et al. 2018).

Balancing the C/N ratio through co-digestion helps mitigate ammonia inhibition, enhancing the stability and efficiency of anaerobic digestion systems. An optimal C/N ratio (typically ranging from 20 to 30) ensures that enough carbon is available to balance the nitrogen in the feedstock, preventing excessive ammonia generation and high pH conditions (Wang et al. 2023). Blending waste streams for co-digestion is one of the most cost-effective and simplest approaches for reducing ammonia toxicity (Rajagopal et al. 2013). Co-digestion strategies involve combining nitrogen-rich materials with carbon-rich organic wastes such as yard trimmings, food waste, or fats, oils, and greases, which can not only dilute the nitrogen content but also introduce additional microbial populations that enhance the breakdown of complex substrates, improving the overall stability and biogas output.

5.2 *Microbial Adaptation and Tolerance*

Microbial consortia in anaerobic digestion systems show substantial adaptability and resilience to ammonia toxicity, which is crucial for maintaining stability and efficiency when processing nitrogen-rich substrates like food waste and animal manure. Methanogenic archaea, a key group within these communities, exhibit significant evolutionary adaptations that enable them to function optimally under elevated ammonia conditions. This capacity highlights the robust nature of these microorganisms and provides valuable insights for enhancing the operational strategies of anaerobic digestion facilities.

5.2.1 Adaptation Mechanisms

Anaerobic microorganisms exhibit a range of adaptation mechanisms to cope with ammonia inhibition, including modulation of the intracellular pH, synthesis or transport of osmo-protectants, and alterations in the membrane permeability (Bucci et al. 2023; Sprott and Patel 1986; Yan et al. 2020b). These adaptive responses enable methanogenic archaea and other microorganisms to maintain functionality even under conditions of elevated ammonia concentrations.

Anaerobic microorganisms can adjust their intracellular pH to prevent the accumulation of toxic FA, by producing extracellular buffering compounds or by transporting protons across the cell membrane (Sprott et al. 1984; Sprott and Patel 1986). Notably, Sprott and Patel (1986) observed that elevated ammonia levels led to a significant efflux of cytoplasmic K^+ in *Methanospirillum hungatei*, *Methanosarcina barkeri*, and *Methanothrix concilii*.

Moreover, anaerobic microorganisms develop osmotic balance by synthesizing or transporting osmo-protectants to cope with elevated ammonia levels. The production or uptake of osmo-protectants such as glutamate, glutamine, phosphate, Nε-acetyl-L-lysine, and glycine betaine could achieve osmotic balance and mitigate ammonia inhibition (Grammann et al. 2002; Vyrides et al. 2010; Yan et al. 2020b).

Certain anaerobic bacteria and archaea can modify the lipid composition of their cell membrane to reduce the permeability of FA, thereby limiting its entry into the cell. These changes often involve adjusting the ratio of saturated to unsaturated fatty acids in the membrane lipids, affecting both the fluidity and permeability of the membrane (Liu et al. 2023).

5.2.2 Tolerance Variability

Different microbial groups exhibit varied degree of tolerance to ammonia, influenced by a range of physiological and genetic factors (Liu et al. 2023). This variability impacts the overall efficiency of biogas production. Methanogenic archaea are generally more susceptible to ammonia inhibition compared to bacterial species involved in the hydrolysis and acidogenesis phases (Capson-Tojo et al. 2020). The high sensitivity of methanogens can be traced back to their unique cellular structures and metabolic requirements. The lack of peptidoglycan in the cell wall of methanogens diminishes their ability against the disruptive effects of high ammonia concentrations (Yan et al. 2020a). Additionally, these archaea rely on specific cofactors that are susceptible to oxidation in the presence of ammonia, further impairing their functionality and survival.

Bacterial populations involved in hydrolysis and acidogenesis exhibit higher resilience to ammonia (Capson-Tojo et al. 2020; Kalamaras et al. 2020). For instance, some syntrophic bacteria, such as *Syntrophaceticus schinkii*, *Tepidanaerobacter acetatoxydans*, and *Clostridium ultunense*, which work closely with methanogens by converting the products of acidogenesis into substrates for methane production, show better tolerance to ammonia (Westerholm et al. 2011, 2016).

5.2.3 Development of Tolerant Strains

The development of tolerant strains to counteract ammonia toxicity in anaerobic digestion represents a forward-thinking approach to enhance the resilience and efficiency of biogas production systems. For this, recent advancements have focused on the selective adaptation and genetic engineering of microbial strains that exhibit enhanced tolerance to high ammonia concentrations.

Selective adaptation involves gradually acclimatizing microbial communities to increasing ammonia levels, enriching the population with naturally tolerant strains to resist higher levels of ammonia without succumbing to toxicity (Westerholm et al.

2015; Yan et al. 2019a, b). This gradual adaptation can be achieved through stepwise increments in the ammonia concentration of the feedstock or by incrementally introducing ammonia directly into the digester (Rajagopal et al. 2013). Over time, this exposure selects ammonia-tolerant microbial strains, which proliferate and become dominant within the microbial community (Westerholm et al. 2012). These adapted strains possess enhanced mechanisms for coping with ammonia stress, such as improved ion exchange capabilities, enhanced ammonia assimilation pathways, and alterations in cell membrane composition to reduce ammonia permeability (Xiao et al. 2019).

The effectiveness of the acclimatization process is influenced by several factors, including the increasing rate of ammonia concentration, the initial microbial community composition, and the operational conditions of the anaerobic digester, such as temperature and pH (Jiang et al. 2019). Optimal acclimatization protocols balance the need for rapid adaptation with the risk of overwhelming the microbial community, thereby maximizing the resilience of the system to ammonia toxicity while maintaining efficient biogas production. By fostering a microbial community adept at handling elevated ammonia levels, the acclimatization process enhances the stability and performance of anaerobic digestion systems, supporting sustainable biogas production (Rajagopal et al. 2013).

Genetic engineering, on the other hand, offers a more targeted approach by modifying specific genes or metabolic pathways in microorganisms to enhance their ammonia tolerance. Techniques such as CRISPR-Cas9 and other gene editing tools have been applied to alter the genetic makeup of methanogenic archaea, enabling them to withstand higher ammonia concentrations without compromising their metabolic efficiency. For instance, the CRISPR-Cas9 system has been utilized in *Methanococcus maripaludis*, a methanogenic archaeon, to perform precise genome edits through gene knockouts, gene tagging, and single-nucleotide mutagenesis (Li et al. 2022). This gene editing approach focuses on altering genes involved in ammonia assimilation, osmoregulation, and stress response, thereby improving the overall resilience of methanogenic archaea to ammonia inhibition (Li et al. 2022).

5.3 Bioaugmentation

Bioaugmentation, which involves introducing specific strains or consortia of microorganisms into anaerobic digestion systems, has emerged as a promising strategy to mitigate the inhibitory effects of ammonia on microbial communities (Li et al. 2023). Both pure strains and enriched mixed cultures used in bioaugmentation have been proven effective in mitigating ammonia inhibition in the anaerobic digestion system, despite the diverse underlying mechanisms involved (Fotidis et al. 2014, 2017).

The introduction of syntrophic acetate-oxidizing bacteria (SAOB) such as *Syntrophaceticus schinkii*, *Tepidanaerobacter acetatoxydans*, and *Clostridium ultunense* has shown promising in alleviating ammonia inhibition (Westerholm et al. 2016). These bacteria convert acetate into hydrogen and carbon dioxide, which are then utilized by hydrogenotrophic methanogens to produce methane, thereby maintaining the flow of intermediates essential for methanogenesis under ammonia stressed conditions (Schnürer et al. 1996; Westerholm et al. 2010). Moreover, the strategic inoculation of hydrogenotrophic methanogens, which exhibit higher tolerance to ammonia compared to their acetoclastic counterparts, has been beneficial. These methanogens, including strains such as *Methanoculleus* and *Methanobacterium*, can utilize the hydrogen produced by SAOB, effectively bypassing the direct inhibitory effects of ammonia on acetoclastic methanogenesis (Fotidis et al. 2014; Koster and Lettinga 1984; Li et al. 2023).

Bioaugmentation can be coupled with other process optimization strategies such as pH control, co-digestion with carbon-rich substrates, and the use of support media to enhance microbial retention and activity (Carabeo-Pérez et al. 2024; Li et al. 2023). These combined approaches have been shown to stabilize anaerobic digestion, improve methane yields, and reduce the risk of reactor failure due to ammonia toxicity (Fotidis et al. 2014).

5.4 Addition of Conductive Materials

In recent years, the incorporation of conductive materials (e.g., magnetite and carbon materials) into anaerobic digestion systems has emerged as an innovative strategy to mitigate ammonia inhibition and enhance overall process stability (Lü et al. 2016; Yan et al. 2019a, b; Zhuang et al. 2018). The improvement of the methane production rate induced by the presence of conductive materials under ammonia-stressed conditions approximately ranges from 10 to 58%. For example, Zhuang et al. (2018) observed a 36–58% increase in the CH_4 production rate with the supplementation of magnetite nanoparticles in batch digesters containing 5 g/L of ammonium (about 20 mg/L of FAN). Poirier et al. (2016) also found that activated carbon reduced the lag phase of CH_4 production by 25% in the presence of 19 g/L of TAN (about 426–511 mg/L of FAN) in batch digesters.

Unlike the dilution of reactor contents which mitigates the ammonia inhibition by reducing the TAN concentration, many conductive materials (e.g., activated carbon and magnetite) do not have physicochemical properties to absorb ammonia. In these cases, the fluctuations of TAN levels were not obvious throughout methanogenic processes, suggesting that conductive materials did not affect the concentration of ammonia but reduced its inhibition (Florentino et al. 2019; Poirier et al. 2016). One exception is that biochar which exhibited large specific surfaces was reported to reduce TAN and FAN concentrations, therefore benefiting the CH_4 production (Pan et al. 2019).

Nevertheless, the mitigating effects of conductive materials are mainly attributed to three aspects. First, conductive materials-mediated direct interspecies electron transfer (DIET) was suggested to increase the efficiency of electron transfer between syntrophic bacteria and methanogens, accelerating biogas production (Poirier et al. 2016; Zhuang et al. 2018). Second, conductive materials could affect microbial community structure. The existence of conductive materials provides a better ecological niche for microorganisms that can utilize conductive materials to transfer electrons. Thus, the relative abundance of syntrophic bacteria and methanogens with the DIET ability was increased by the supplementation of conductive materials (Florentino et al. 2019; Zhuang et al. 2018). For instance, *Geobacteraceae* and *Methanosarcinaceae* were enriched by the supplementation of magnetite in the anaerobic reactor with 5.0 g/L of NH_4^+–N (Zhuang et al. 2018). Florentino et al. (2019) also observed that the relative abundance of *Geobacteraceae* family in the GAC-addition group was enhanced under high ammonia concentrations. Finally, the positive effect may be due to the energy conservation advantage of DIET which although is still largely inconclusive. Conductive materials-mediated DIET is proposed to be more energy conserving and therefore may alleviate the energy-shortage problem induced by ammonia inhibition. More energy obtained might also be the reason for the dominance of some DIET-capable microbes in ammonia-stressed conditions.

The concentration of ammonia is crucial in determining the stimulating effects of conductive materials. Improvements are minor under low ammonia concentrations but significant under high ammonia conditions. Zhuang et al. (2018) found that the addition of magnetite did not substantially affect methanogenesis in both the ammonia-free and low ammonia (0.5 g/L of NH_4^+–N) bioreactors. One possible explanation of the controversial results is that microorganisms may change their metabolic pathways in response to environmental changes. It was mentioned that electroactive microbes can express different electron transfer pathways responding to major changes in environmental variables (Ishii et al. 2018). Also, electroactive microbes may lose their extracellular electron transfer capabilities when growing under non-specific conditions such as the absence of conductive solid surface (Yee et al. 2020).

High ammonia concentrations can shift CH_4 formation from acetoclastic to syntrophic pathways, as acetoclastic methanogens (especially *Methanothrix*) are sensitive to high TAN concentrations, whereas SAOB are more tolerant (Schnürer and Nordberg 2008). This is because acetoclastic methanogens (especially *Methanothrix*) are sensitive to high TAN concentration (Calli et al. 2005), while SAOB are tolerant under ammonia-stressed conditions (Karakashev et al. 2006). Therefore, *Methanothrix* species and acetoclastic metabolism are often found to be dominant with low ammonia concentrations (Zhuang et al. 2018). In this case, although *Methanothrix* species are capable of DIET, the lack of syntrophic partners may limit the possibility of benefiting from the addition of conductive materials (Yin and Wu 2019). On the contrary, SAOB is frequently detected as the dominant pathway in anaerobic digester treating wastes containing high concentrations of ammonia (Sun et al. 2014). Also, *Methanosarcina* species which can utilize versatile substrates for CH_4 production including DIET can be enriched under ammonia-stressed conditions (Zhuang et al. 2018), suggesting that conductive materials supplementation might promote DIET between SAOB and methanogens, accelerating CH_4 production efficiency (Zhuang et al. 2018). In addition, other environmental variables such as the type of carbon substrates and conductive materials could also lead to different stimulating results under ammonia-stressed conditions (Pan et al. 2019; Poirier et al. 2016).

6 Future Perspectives

Ammonia-induced inhibition is a critical challenge in the field of anaerobic digestion, directly impacting the efficiency of biogas production. The complexity of microbial interactions within anaerobic digestion systems presents both challenges and opportunities for innovation. Future research efforts are expected to focus intensively on the microbial ecology of anaerobic digestion processes, employing high-throughput sequencing and metagenomic techniques.

Operational optimization represents another pivotal strategy for mitigating ammonia inhibition in full-scale applications. By finely tuning parameters such as temperature, pH, and the organic loading rate, it is possible to foster a more ammonia-tolerant microbial consortium. The advent of advanced process control technologies, leveraging real-time data analysis and machine-learning algorithms, is poised to significantly improve the adaptability and performance of anaerobic digestion systems in the face of ammonia challenges.

Pretreatments to reduce ammonia concentration in the feedstocks include scalable and cost-effective methods such as enzymatic hydrolysis and selective membrane technologies. These techniques promise to lower the ammonia content in high-nitrogen feedstocks, thereby alleviating inhibitory effects on the microbial digestion process and enhancing overall system efficiency.

Integrating anaerobic digestion with nitrogen recovery technologies offers a comprehensive approach to manage ammonia levels while promoting nutrient recycling. Techniques such as ammonia stripping and ion exchange are under exploration for their potential to recover nitrogenous compounds from digestate, turning a waste management challenge into an opportunity for creating value-added products.

7 Take-Home Messages

(1) While ammonia is essential for anaerobic digestion, maintaining its concentration within a specific threshold is crucial to prevent toxicity and ensure optimal microbial activity and biogas production.
(2) Ammonia toxicity primarily affects methanogenic archaea by disrupting intracellular processes and membrane functions, leading to decreased methane output and process inefficiencies.
(3) Environmental factors such as pH, temperature, and the organic loading rate significantly influence ammonia's form and its toxicity in anaerobic digestion systems.
(4) Microbial communities in anaerobic digesters can adapt to elevated levels of ammonia through various mechanisms, including changes in community composition and metabolic adaptations. Leveraging these natural adaptive processes is key for enhancing tolerance.

(5) Employing mitigation strategies can effectively reduce ammonia concentrations and mitigate its inhibitory effects, thus improving the overall efficiency and stability of anaerobic digestion processes.

References

Akindele AA, Sartaj M (2018) The toxicity effects of ammonia on anaerobic digestion of organic fraction of municipal solid waste. Waste Manag 71:757–766

Angelidaki I, Ahring BK (1993) Thermophilic anaerobic digestion of livestock waste: the effect of ammonia. Appl Microbiol Biotechnol 38:560–564

Bucci L, Ghiotto G, Zampieri G, Raga R, Favaro L, Treu L, Campanaro S (2023) Adaptation of anaerobic digestion microbial communities to high ammonium levels: insights from strain-resolved metagenomics. Environ Sci Technol 58(1):580–590

Cai Y, Zheng Z, Wang X (2021) Obstacles faced by methanogenic archaea originating from substrate-driven toxicants in anaerobic digestion. J Hazard Mater 403:123938

Cai Y, Zhu M, Meng X, Zhou JL, Zhang H, Shen X (2022) The role of biochar on alleviating ammonia toxicity in anaerobic digestion of nitrogen-rich wastes: a review. Bioresour Technol 351:126924

Calli B, Mertoglu B, Inanc B, Yenigun O (2005) Effects of high free ammonia concentrations on the performances of anaerobic bioreactors. Process Biochem 40(3–4):1285–1292

Capson-Tojo G, Moscoviz R, Astals S, Robles Á, Steyer JP (2020) Unraveling the literature chaos around free ammonia inhibition in anaerobic digestion. Renew Sustain Energy Rev 117:109487

Carabeo-Pérez A, López MI, Rivera GG, Henderson D, Jiménez J (2024) Rice straw and swine manure anaerobic co-digestion enhancement through bioaugmentation: effect on the microbial community. BioEnergy Res 17(1):756–767

Chen Y, Cheng JJ, Creamer KS (2008) Inhibition of anaerobic digestion process: a review. Bioresour Technol 99(10):4044–4064

Florentino AP, Sharaf A, Zhang L, Liu Y (2019) Overcoming ammonia inhibition in anaerobic blackwater treatment with granular activated carbon: the role of electroactive microorganisms. Environ Sci Water Res Technol 5(2):383–396

Fotidis IA, Wang H, Fiedel NR, Luo G, Karakashev DB, Angelidaki I (2014) Bioaugmentation as a solution to increase methane production from an ammonia-rich substrate. Environ Sci Technol 48(13):7669–7676

Fotidis IA, Treu L, Angelidaki I (2017) Enriched ammonia-tolerant methanogenic cultures as *Bioaugmentation inocula* in continuous biomethanation processes. J Clean Product 166:1305–1313

Fuchs W, Wang X, Gabauer W, Ortner M, Li Z (2018) Tackling ammonia inhibition for efficient biogas production from chicken manure: status and technical trends in Europe and China. Renew Sustain Energy Rev 97:186–199

Gallert C, Bauer S, Winter JJ (1998) Effect of ammonia on the anaerobic degradation of protein by a mesophilic and thermophilic biowaste population. Appl Microbiol Biotechnol 50:495–501

Gaspari M, Ghiotto G, Centurion VB, Kotsopoulos T, Santinello D, Campanaro S, Treu L, Kougias PG (2024) Decoding microbial responses to ammonia shock loads in biogas reactors through metagenomics and metatranscriptomics. Environ Sci Technol 58(1):591–602

Grammann K, Volke A, Kunte HJ (2002) New type of osmoregulated solute transporter identified in halophilic members of the bacteria domain: TRAP transporter TeaABC mediates uptake of ectoine and hydroxyectoine in *Halomonas elongata* DSM 2581T. J Bacteriol 184(11):3078–3085

Hafner SD, Bisogni JJ (2009) Modeling of ammonia speciation in anaerobic digesters. Water Res 43(17):4105–4114

Hansen KH, Angelidaki I, Ahring BK (1998) Anaerobic digestion of swine manure: inhibition by ammonia. Water Res 32(1):5–12

Harold FM (1987) The vital force: a study of bioenergetics. WH Freeman, New York

Hobson PN, Shaw BG (1976) Inhibition of methane production by *Methanobacterium formicicum*. Water Res 10(10):849–852

Ishii SI, Suzuki S, Tenney A, Nealson KH, Bretschger O (2018) Comparative metatranscriptomics reveals extracellular electron transfer pathways conferring microbial adaptivity to surface redox potential changes. ISME J 12(12):2844–2863

Jeris JS, McCarty PL (1965) The biochemistry of methane fermentation using C^{14} tracers. Journal 37(2):178–192

Jiang Y, McAdam E, Zhang Y, Heaven S, Banks C, Longhurst P (2019) Ammonia inhibition and toxicity in anaerobic digestion: a critical review. J Water Process Eng 32:100899

Kadam PC, Boone DR (1996) Influence of pH on ammonia accumulation and toxicity in halophilic, methylotrophic methanogens. Appl Environ Microbiol 62(12):4486–4492

Kalamaras SD, Vasileiadis S, Karas P, Angelidaki I, Kotsopoulos TA (2020) Microbial adaptation to high ammonia concentrations during anaerobic digestion of manure-based feedstock: biomethanation and 16S rRNA gene sequencing. J Chem Technol Biotechnol 95(7):1970–1979

Karakashev D, Batstone DJ, Trably E, Angelidaki I (2006) Acetate oxidation is the dominant methanogenic pathway from acetate in the absence of *Methanosaetaceae*. Appl Environ Microbiol 72(7):5138–5141

Kayhanian M (1999) Ammonia inhibition in high-solids biogasification: an overview and practical solutions. Environ Technol 20(4):355–365

Koster IW (1986) Characteristics of the pH-influenced adaptation of methanogenic sludge to ammonium toxicity. J Chem Technol Biotechnol 36(10):445–455

Koster I, Lettinga G (1984) The influence of ammonium-nitrogen on the specific activity of pelletized methanogenic sludge. Agricult Wastes 9(3):205–216

Lay JJ, Li YY, Noike T (1998) The influence of pH and ammonia concentration on the methane production in high-solids digestion processes. Water Environ Res 70(5):1075–1082

Li J, Zhang L, Xu Q, Zhang W, Li Z, Chen L, Dong X (2022) CRISPR-Cas9 toolkit for genome editing in an autotrophic CO_2-fixing methanogenic archaeon. Microbiol Spect 10(4):e01165-e1222

Li ZY, Inoue D, Ike M (2023) Mitigating ammonia-inhibition in anaerobic digestion by bioaugmentation: a review. J Water Process Eng 52:103506

Liu T, Sung S (2002) Ammonia inhibition on thermophilic aceticlastic methanogens. Water Sci Technol 45(10):113–120

Liu C, Zhang X, Chen C, Yin Y, Zhao G, Chen Y (2023) Physiological responses of *Methanosarcina barkeri* under ammonia stress at the molecular level: the unignorable lipid reprogramming. Environ Sci Technol 57(9):3917–3929

Lü F, Luo C, Shao L, He P (2016) Biochar alleviates combined stress of ammonium and acids by firstly enriching *Methanosaeta* and then *Methanosarcina*. Water Res 90:34–43

Masse DI, Masse L, Croteau F (2003) The effect of temperature fluctuations on psychrophilic anaerobic sequencing batch reactors treating swine manure. Bioresour Technol 89(1):57–62

Massé DI, Masse L, Xia Y, Gilbert Y (2010) Potential of low-temperature anaerobic digestion to address current environmental concerns on swine production. J Animal Sci 88(13):e112–e120

Massé DI, Rajagopal R, Singh G (2014) Technical and operational feasibility of psychrophilic anaerobic digestion biotechnology for processing ammonia-rich waste. Appl Energy 120:49–55

McCarty PL (1964) Anaerobic waste treatment fundamentals, Part III, toxic materials and their control. Public Works 95:91–94

Miller SW, Parkin GF (1983) Response of methane fermentation to continuous addition of selected industrial toxicants. In: Proceedings of the 37th purdue industrial waste conference on Ann Arbor science publishers, Ann Arbor, pp 729–743

Müller T, Walter B, Wirtz A, Burkovski A (2006) Ammonium toxicity in bacteria. Curr Microbiol 52:400–406

Nakakubo R, Møller HB, Nielsen AM, Matsuda J (2008) Ammonia inhibition of methanogenesis and identification of process indicators during anaerobic digestion. Environ Eng Sci 25(10):1487–1496

Nie E, He P, Zhang H, Hao L, Shao L, Lü F (2021) How does temperature regulate anaerobic digestion? Renew Sustain Energy Rev 150:111453

Nielsen AM, Spanjers H, Volcke EI (2008) Calculating pH in pig manure taking into account ionic strength. Water Sci Technol 57(11):1785–1790

Pan J, Ma J, Liu X, Zhai L, Ouyang X, Liu H (2019) Effects of different types of biochar on the anaerobic digestion of chicken manure. Bioresour Technol 275:258–265

Poirier S, Desmond-Le Quéméner E, Madigou C, Bouchez T, Chapleur O (2016) Anaerobic digestion of biowaste under extreme ammonia concentration: identification of key microbial phylotypes. Bioresour Technol 207:92–101

Procházka J, Dolejš P, Máca J, Dohányos M (2012) Stability and inhibition of anaerobic processes caused by insufficiency or excess of ammonia nitrogen. Appl Microbiol Biotechnol 93:439–447

Rajagopal R, Massé DI, Singh G (2013) A critical review on inhibition of anaerobic digestion process by excess ammonia. Bioresour Technol 143:632–641

Schnürer A, Nordberg Å (2008) Ammonia, a selective agent for methane production by syntrophic acetate oxidation at mesophilic temperature. Water Sci Technol 57(5):735–740

Schnürer A, Schink B, Svensson BH (1996) *Clostridium ultunense* sp. nov., a mesophilic bacterium oxidizing acetate in syntrophic association with a hydrogenotrophic methanogenic bacterium. Int J Syst Evolut Microbiol 46(4):1145–1152

Sprott GD, Patel GB (1986) Ammonia toxicity in pure cultures of methanogenic bacteria. Syst Appl Microbiol 7(2–3):358–363

Sprott GD, Shaw KM, Jarrell KF (1984) Ammonia/potassium exchange in methanogenic bacteria. J Biol Chem 259(20):12602–12608

Stumm W, Morgan JJ (1995) Aquatic chemistry: chemical equilibria and rates in natural waters, 3rd edn. Wiley, New York

Sun L, Müller B, Westerholm M, Schnürer A (2014) Syntrophic acetate oxidation in industrial CSTR biogas digesters. J Biotechnol 171:39–44

Sung S, Liu T (2003) Ammonia inhibition on thermophilic anaerobic digestion. Chemosphere 53(1):43–52

Unden G, Kim OB (2021) Bioenergetics theory and components: energy conservation and growth by anaerobic bacteria. In: Jez J (ed) Encyclopedia of biological chemistry III. Elsevier, Amsterdam, pp 15–22

Vyrides I, Santos H, Mingote A, Ray MJ, Stuckey DC (2010) Are compatible solutes compatible with biological treatment of saline wastewater? Batch and continuous studies using submerged anaerobic membrane bioreactors (SAMBRs). Environ Sci Technol 44(19):7437–7442

Wang Z, Hu Y, Wang S, Wu G, Zhan X (2023) A critical review on dry anaerobic digestion of organic waste: characteristics, operational conditions, and improvement strategies. Renew Sustain Energy Rev 176:113208

Westerholm M, Roos S, Schnürer A (2010) *Syntrophaceticus schinkii* gen. nov., sp. nov., an anaerobic, syntrophic acetate-oxidizing bacterium isolated from a mesophilic anaerobic filter. FEMS Microbiol Lett 309(1):100–104

Westerholm M, Roos S, Schnürer A (2011) *Tepidanaerobacter acetatoxydans* sp. nov., an anaerobic, syntrophic acetate-oxidizing bacterium isolated from two ammonium-enriched mesophilic methanogenic processes. Syst Appl Microbiol 34(4):260–266

Westerholm M, Levén L, Schnürer A (2012) Bioaugmentation of syntrophic acetate-oxidizing culture in biogas reactors exposed to increasing levels of ammonia. Appl Environ Microbiol 78(21):7619–7625

Westerholm M, Müller B, Isaksson S, Schnürer A (2015) Trace element and temperature effects on microbial communities and links to biogas digester performance at high ammonia levels. Biotechnol Biofuels 8:154

Westerholm M, Moestedt J, Schnürer A (2016) Biogas production through syntrophic acetate oxidation and deliberate operating strategies for improved digester performance. Appl Energy 179:124–135

Wittmann C, Zeng AP, Deckwer WD (1995) Growth inhibition by ammonia and use of a pH-controlled feeding strategy for the effective cultivation of *Mycobacterium chlorophenolicum*. Appl Microbiol Biotechnol 44:519–525

Wolin MJ (1982) Interspecies hydrogen transfer: 15 years later. ASM News 48:561–565

Xiao J, Li QY, Tu JP, Chen XL, Chen XH, Liu QY, Liu H, Zhou XY, Zhao YZ, Wang HL (2019) Stress response and tolerance mechanisms of ammonia exposure based on transcriptomics and metabolomics in *Litopenaeus vannamei*. Ecotoxicol Environ Saf 180:491–500

Xie S, Lawlor PG, Frost JP, Hu Z, Zhan X (2011) Effect of pig manure to grass silage ratio on methane production in batch anaerobic co-digestion of concentrated pig manure and grass silage. Bioresour Technol 102(10):5728–5733

Yan M, Fotidis IA, Tian H, Khoshnevisan B, Treu L, Tsapekos P, Angelidaki I (2019a) Acclimatization contributes to stable anaerobic digestion of organic fraction of municipal solid waste under extreme ammonia levels: focusing on microbial community dynamics. Bioresour Technol 286:121376

Yan W, Lu D, Liu J, Zhou Y (2019b) The interactive effects of ammonia and carbon nanotube on anaerobic digestion. Chem Eng J 372:332–340

Yan M, Treu L, Campanaro S, Tian H, Zhu X, Khoshnevisan B, Tsapekos P, Angelidaki I, Fotidis IA (2020a) Effect of ammonia on anaerobic digestion of municipal solid waste: Inhibitory performance, bioaugmentation and microbiome functional reconstruction. Chem Eng J 401:126159

Yan M, Treu L, Zhu X, Tian H, Basile A, Fotidis IA, Campanaro S, Angelidaki I (2020b) Insights into ammonia adaptation and methanogenic precursor oxidation by genome-centric analysis. Environ Sci Technol 54(19):12568–12582

Yee MO, Deutzmann J, Spormann A, Rotaru AE (2020) Cultivating electroactive microbes—from field to bench. Nanotechnology 31(17):174003

Yellezuome D, Zhu X, Wang Z, Liu R (2022) Mitigation of ammonia inhibition in anaerobic digestion of nitrogen-rich substrates for biogas production by ammonia stripping: a review. Renew Sustain Energy Rev 157:112043

Yenigün O, Demirel B (2013) Ammonia inhibition in anaerobic digestion: a review. Process Biochem 48(5–6):901–911

Yin Q, Wu G (2019) Advances in direct interspecies electron transfer and conductive materials: electron flux, organic degradation and microbial interaction. Biotechnol Adv 37(8):107443

Yin Q, Gu M, Wu G (2020) Inhibition mitigation of methanogenesis processes by conductive materials: a critical review. Bioresour Technol 317:123977

Zheng Z, Cai Y, Zhang Y, Zhao Y, Gao Y, Cui Z, Hu Y, Wang X (2021) The effects of C/N (10–25) on the relationship of substrates, metabolites, and microorganisms in "inhibited steady-state" of anaerobic digestion. Water Res 188:116466

Zhuang L, Ma J, Yu Z, Wang Y, Tang J (2018) Magnetite accelerates syntrophic acetate oxidation in methanogenic systems with high ammonia concentrations. Microb Biotechnol 11(4):710–720

Sulfur Metabolism and Its Inhibition Alleviation in Anaerobic Digestion Ecosystems

Qigui Niu and Guangxue Wu

Abstract Sulfate reduction mediated by sulfate-reducing bacteria (SRB) significantly impacts the efficiency of anaerobic digestion, particularly due to the inhibitory toxicity of its metabolite, hydrogen sulfide, on the microbial community of anaerobic digestion. Sulfate reduction occurs through the assimilatory sulfate reduction and/or the dissimilatory sulfate reduction, and the produced hydrogen sulfide can inhibit methanogenesis. Generally, two types of SRB exist in anaerobic digestion systems, with incomplete SRB converting organic carbon to acetate and complete SRB converting organic carbon to carbon dioxide. The competition between SRB and methanogens is affected by diverse factors, especially the carbon to sulfate ratio. Furthermore, alleviation strategies for the inhibition of hydrogen sulfide were proposed. Specifically, the potential of optimizing the substrate ratio and the addition of adsorbent conductive materials (such as biochar and zero-valent iron) to regulate the competition and collaboration between SRB and methanogens were detailed.

Keywords Anaerobic digestion · Methanogens · Microbial interactions · Sulfate-reducing bacteria · Sulfate reduction

1 Introduction

Sulfur possesses a wide range of valences with values from −2 (the complete reduction state) to +6 (the complete oxidation state) and can be mutually transformed through both chemical and biological processes. Sulfur is one of the necessary elements for living organisms, such as for the synthesis of cysteine and methionine (Block and Henry 1961). Sulfur also plays a key role in biogeochemical cycles,

Q. Niu (✉)
School of Environmental Science and Engineering, Shandong University, Qingdao 266237, Shandong, China
e-mail: niuqg@sdu.edu.cn

G. Wu
Civil Engineering, School of Engineering, College of Science and Engineering, University of Galway, Galway H91 TK33, Ireland

G. Wu (ed.), *Anaerobic Digestion*, Green Energy and Technology,
https://doi.org/10.1007/978-3-031-69378-6_7

with pyrite (FeS_2) and gypsum ($CaSO_4$) frequently detected in rocks and sediments (Grein et al. 2013). Furthermore, sulfur-containing chemical products, such as sulfur soap and methyl mercaptan, are important for our daily life.

Through anthropogenic activities, sulfate-rich wastewater is generated from diverse industries, such as distilleries, pharmaceuticals, food processing, and pulp and paper, with sulfur concentrations of 100–15,000 mg/L (Li et al. 2015; Oliveira et al. 2020; Sun et al. 2018; Yan et al. 2018). Anaerobic digestion has already been widely applied for the treatment of sulfate-rich wastewater, and hydrogen sulfide (H_2S) can be produced through dissimilatory sulfate reduction during anaerobic treatment. A high H_2S concentration can inhibit microbial activities, especially of methanogens, causing instability or even failure of anaerobic digestion systems. Therefore, understanding the potential biological transformation mechanisms and then mitigation of sulfide inhibition of methanogenesis are major challenges for anaerobic digestion.

2 Metabolic Pathways of Sulfate-Reducing Bacteria

2.1 Sulfate Reduction

Sulfate reduction is a biochemical process in which sulfate is used as the terminal electron acceptor for the degradation of organic compounds (Zhang et al. 2022). The pathways of sulfate reduction include assimilatory sulfate reduction (ASR) and dissimilatory sulfate reduction (DSR). The most significant difference between ASR and DSR is that in ASR, sulfate is used for the synthesis of sulfur-containing amino acids, whereas in DSR, sulfate is used to generate energy and act as an electron acceptor to produce H_2S, without further synthesizing sulfur-containing amino acids (Kushkevych et al. 2020).

As depicted in Fig. 1, the canonical microbial reduction of sulfate includes the following three core steps (Bell et al. 2020): (i) the reduction of sulfate to produce adenosine 5′-phosphosulfate (APS) by sulfate adenylyltransferase (*sat*), (ii) the reduction of APS to sulfite by adenylylsulfate reductase alpha and beta subunits (*aprAB*), and (iii) sulfite reduction to sulfide mediated by dissimilatory sulfite reductase alpha and beta subunits (*dsrAB*) and D subunit (*dsrD*). The reduction of 1 mol sulfate consumes 8 mol of electrons and 8.5 mol of protons (Zhang et al. 2022).

2.2 Sulfate-Reducing Bacteria

SRB occupy a diverse range of ecological niches in nature, inhabiting various environments such as seawater, river and lake sediments, urban sewers, and others. The majority of SRB are primarily distributed across various genera,

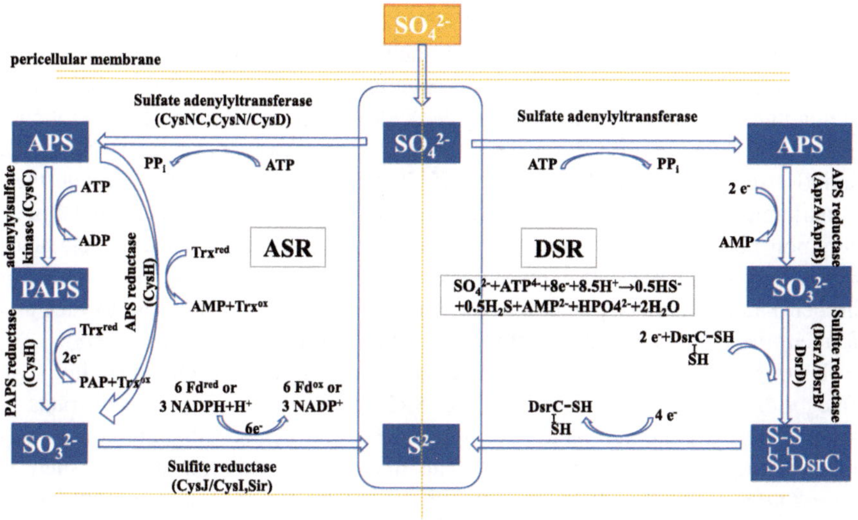

Fig. 1 Assimilatory and dissimilatory sulfate reduction pathways [Left: assimilatory sulfate reduction (ASR); Right: dissimilatory sulfate reduction (DSR); Modified from Zhang et al. 2022]

such as *Archaeoglobus*, *Desulfovibrio*, *Desulfobacterium*, *Desulfotomaculum*, *Desulfobacca*, and *Desulfobacter* (Novair et al. 2024).

Different SRB possess distinct physiologies. *Desulfocapsa sulfexigens* possesses all the genes for dissimilatory sulfate reduction, but it cannot respire sulfate (Finster et al. 2013). *Desulfitobacterium dehalogenans* was not seen as SRB because it could not reduce sulfate but sulfite, thiosulfate, and sulfur into sulfide (Utkin et al. 1994). Generally, *dsrD* can be used as the biomarker to classify SRB, which is the rate-limiting steps in the global sulfur cycle (Anantharaman et al. 2018). *dsrD* can be conducted by four types of processes, i.e., siroheme-dependent dissimilatory sulfite reductase (dsr), siroheme-dependent anaerobic sulfite reductase (asr) genes, octaheme cytochrome *c* sulfite reductase (*mccA*), and reverse dissimilatory sulfite reductase genes (rdsr), while rdsr can also participate in sulfur oxidation (Anantharaman et al. 2018). Some SRB, such as *Desulfurivibrio alkaliphilus*, have been confirmed to be able to oxidize sulfide (Thorup et al. 2017). SRB from *Deltaproteobacteria* can perform thiosulfate disproportionation through thiosulfate reductase (*phsA*) (Jørgensen and Bak 1991; Krämer and Cypionka 1989). Fang et al. (2020) proposed that SRB may use the more thermodynamically favorable sulfate-to-ZVS pathway rather than the sulfate-to-sulfide pathway to alleviate the inhibition of sulfide.

For the growth of SRB, the presence of an adequate amount of sulfate and carbon source is crucial for providing SRB with an optimal anaerobic environment and enabling them to vigorously engage in sulfate reduction activities (Baumgartner et al. 2006). SRB can utilize hydrogen and a diverse range of organic substrates, such as short-chain volatile fatty acids (VFAs; such as acetate, propionate, etc.), long-chain fatty acids (LCFAs), aromatic compounds (e.g., benzoate and phenol), amino acids, one-carbon compounds (such as methanol, carbon monoxide, and methyl mercaptan), and alkanes (Aeckersberg et al. 1998; Bak and Pfennig 1987; Böttcher et al. 2005; Davidova et al. 2006; Grossi et al. 2007; Kniemeyer et al. 2007; Nanninga and Gottschal 1987; Parshina et al. 2005; Tanimoto and Bak 1994). Despite their remarkable ability to utilize organic carbon sources, SRB cannot directly metabolize common polymeric macromolecules such as starch, fibrin, proteins, nucleic acids (DNA and RNA), and fats. This suggests that SRB rely on other microorganisms in their natural habitats, potentially forming symbiotic relationships to degrade and ferment these macromolecules into SRB-usable substrates (Zhao et al. 2008).

Distinguished by their metabolic pathways, SRB can be classified into two categories: (i) the incomplete SRB (IO-SRB) that degrade organic substrates to acetate; and (ii) the complete SRB (CO-SRB) that degrade organic compounds to CO_2 (Muyzer and Stams 2008). The first group, including species such as *Desulfovibrio* and *Desulfotomaculum*, incompletely degrades organic compounds to usually acetate. In contrast, the second group, represented by *Desulfobacter postgatei* and *Desulfobacca acetoxidans*, is capable of completely degrading organic compounds to CO_2. When degrading acetate, CO-SRB may use either the citric acid cycle (CAC) (such as in *Desulfobacter*) or the oxidative carbon monoxide dehydrogenase (OCMD) pathway (such as *Desulfotomaculum* and *Desulfobacterium*) (Goevert and Conrad 2008; Qian et al. 2019; Thauer 1988). The OCMD pathway is more advantage as all oxidation steps proceed as exergonic reactions, while the oxidation of succinate to fumarate in the CAC pathway is an endergonic reaction (+33 mV) (Qian et al. 2019; Thauer et al. 2009).

SRB growing on H_2 as the energy source include several incomplete (e.g., *Desulfovibrio* sp.) and complete (e.g., *Desulfosarcina variabilis*, *Desulfonema limicola*, and *Desulfobacterium autotrophicum*) SRB (Colleran et al. 1995). The CO-SRB of H_2 utilizing species are capable of true autotrophic growth on H_2 and CO_2. The growth of the IO-SRB *Desulfovibrio* sp. requires, in addition to CO_2, at least a 2-carbon compound, such as acetate as the sole carbon source, for assimilation and cell carbon synthesis (Colleran et al. 1995; Widdel 1988). Badziong et al. (1979) demonstrated that approximately two-thirds of the cell material was derived from acetate with the remaining third coming from CO_2. Thus, besides competing H_2 with *Methanobacterium* sp., *Desulfovibrio* sp. also somewhat suppressed *Methanosaeta* sp. through consuming acetate.

SRB possess remarkable environmental adaptability, enabling them to autonomously regulate their metabolic activity in response to changing environmental conditions. Marietou et al. (2022) found that *Desulfobacterium autotrophicum* had a key adaptive mechanism for low energy in marine sediments, which can switch from the complete to incomplete lactate oxidation mode based on the abundance of sulfate substrates. Hausmann et al. (2016) found that the rare *Desulfosporosinus* species in the peatland maintained a low-abundance population but could increase their ribosome content and metabolic activity responding to favorable environmental conditions. It was confirmed that *Desulfosporosinus* species contain a high number of ribosomal (*rrn*) operons of 8–10 copies in their genomes, which can enable them to react quickly by increasing ribosomes (Hausmann et al. 2016; Pester et al. 2012).

2.3 *Effects of Metabolites Produced by SRB*

H_2S is notoriously recognized for its rotten egg odor, asphyxiating properties, and high toxicity. Being a metabolite of SRB, H_2S may pose health risks to workers operating in certain environments, resulting in acute central neurotoxicity, pulmonary edema, conjunctivitis, and other adverse health effects (Guidotti 2010; Kilburn 1993). H_2S also exerts a profound influence on the metabolic activity of anaerobic microorganisms. Free-state H_2S can diffuse into cell membranes, resulting in protein denaturation (Speece 1983). Furthermore, H_2S reacts with metal ions to form insoluble metal sulfides, leading to the precipitation of trace metal components of cofactors involved in the methanogenesis process. This precipitation indirectly hinders the activity of methanogens.

The thresholds for the inhibition of anaerobic digestion are in wide ranges of 150–1100 mg/L for dissolved sulfide and 50–250 mg/L for free sulfide (Omil et al. 1995). Paula and Foresti (2009) found that 100–500 mg/L dissolved sulfide showed no inhibition in anaerobic digestion reactors. Stucki et al. (1993) reported that anaerobic digestion was sensitive to undissociated H_2S above 50 mg/L when pure CO-SRB was used in sulfate reduction. In the study of Kaksonen et al. (2003) on sulfide toxicity with batch kinetic experiments, they found that the inhibition concentrations of dissolved sulfide for ethanol and acetate oxidation were 248 and 356 mg/L, while the corresponding values of free sulfide were 84 and 124 mg/L, respectively. In contrast, microorganisms exhibit greater tolerance toward S^{2-} and HS^-, which exist in an ionic state and are more difficult to penetrate the intracellular methanogenic compartment.

As a highly corrosive gas, H_2S can cause technical problems, such as the corrosion of pipes and other facilities (e.g., mixers and pumps). The existing format of sulfides depends on pH. When pH was about 8, it is mainly in the ionized HS^- form, while most are in the H_2S form at pH 6 (Fig. 2). When dissolved in water, it creates a highly acidic environment, thereby shortening the lifespan of equipment. The dissolution and ionization of H_2S generates H^+, which undergoes cathodic depolarization. This process leads to the continuous removal and utilization of hydrogen atoms on the metal surface, generating a steady stream of sulfur ions and ferrous sulfide precipitates, thus perpetuating the corrosion process (von Wolzogen Kühr and van der Vlugt 1934). Alternatively, H_2S causes corrosion of reactors through the chemical and microbially influenced corrosion mechanisms.

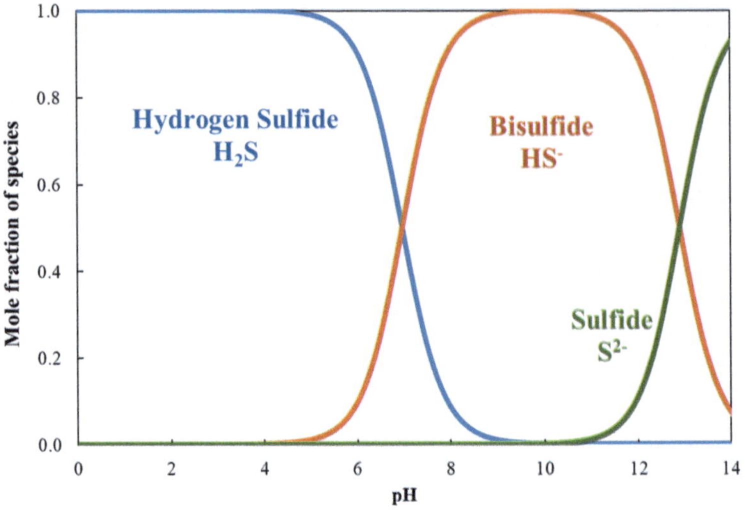

Fig. 2 Equilibrium diagram of the distribution of sulfide species under different pHs at 298 K with pK1 of 6.97 and pK2 of 12.90 (Dean 1999)

3 Microbial Interactions Between SRB and Methanogens

3.1 Competition and Collaboration Between SRB and Methanogens

SRB can use diverse types of substrates, such as H_2, formate, VFAs, LCFAs, alcohols (ethanol, methanol, etc.), and some aromatic compounds, while methanogens can only use limited types of substrates, such as H_2/CO_2, formate, methanol, and acetate. Generally, when anaerobic digestion is applied to treat sulfate-containing wastewater, SRB out-competed methanogens (Dar et al. 2008). Kinetically, SRB holds a competitive edge over methanogens due to its higher affinity for substrates like H_2 (Colleran et al. 1995; Cord-Ruwisch et al. 1988). Thermodynamically, SRB possess a competitive advantage due to its lower Gibbs free energy. The utilization of hydrogen and acetic acid by SRB yields an energy that is 4.08 and 16.6 kJ lower per mol, respectively, when compared to methanogens (Paulo et al. 2015). Physiologically, the produced sulfide from sulfate reduction has a higher toxic effect on methanogens than on SRB (Jung et al. 2022). These factors give SRB a leading position in competition with methanogens for easily degradable carbon sources. However, considering CO- and IO-SRB, IO-SRB may cooperate with methanogens, while CO-SRB may compete with methanogens for H_2 and acetate (Fig. 3).

The interaction between SRB and methanogens is substantially affected by the types of SRB and environmental variables such as sulfate concentrations, carbon substrates, and chemical oxygen demand (COD)/sulfate ratios (Chen et al. 2019;

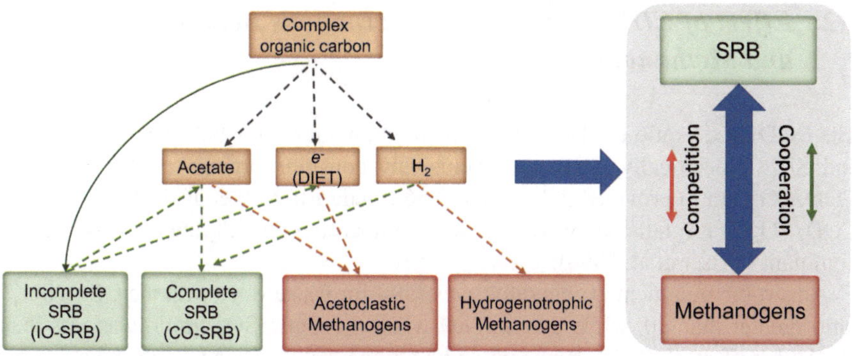

Fig. 3 Schematic diagram of the relationships between SRB and methanogens incorporating both IO- and CO-SRB

Muyzer and Stams 2008; Wu et al. 2018). For example, studies have found the coexistence of acetoclastic/hydrogenotrophic methanogens and SRB in the presence of sulfate (Dar et al. 2008; Kuivila et al. 1990), and this pattern of coexistence can persist or adapt to sulfate perturbations and environmental changes (Ma et al. 2017; Omil et al. 1998). When there is no sulfate or low concentration of sulfate, SRB may act as the syntrophs such as the function of acetogens, and their syntrophic cooperation with methanogens (both hydrogenotrophic and acetotrophic methanogens, depending on the end-products from activities of SRB) would benefit their activities. For example, under conditions without sulfate, SRB still can be acclimated, especially for *Desulfovibrio* species, indicating that SRB can act as acetogens (Raskin et al. 1994, 1995). Therefore, the diverse metabolic flexibility can enhance their survival under different dynamic conditions. Even with a high sulfate concentration, IO-SRB may also have syntrophic relationship with methanogens. For example, *Desulfarculus baarsii* can convert acetate to CO_2, which can be further used to produce methane by hydrogenotrophic methanogens (Cetecioglu et al. 2019).

Under carbon-limited conditions, the relationship between SRB and methanogens gradually evolves toward a cooperative symbiosis. SRB can decompose complex carbon sources into small molecule substances for methanogens to utilize. For example, Shi et al. (2020) found that favorable SRB conditions facilitated the creation of a crucial medium for methane metabolism (methyl coenzyme M), leading to a gradual increase in the diversity of methanogen in urban sewers. Conversely, methanogens enhance sulfide consumption through the metabolism of L-cysteine and methionine, resulting in a positive sulfide metabolism reaction and facilitating sulfide production (Shi et al. 2020). Zan et al. (2021) demonstrate that the addition of difficultly degradable carbon sources like food waste enhances the diversity of fermentative bacteria and modifies the methanogen types in biofilms, leading to an elevation in the expression of key enzymes involved in sulfate reduction and methanogenesis.

3.2 *Effect of COD/SO_4^{2-} on the Interactions Between SRB and Methanogens*

The COD/SO_4^{2-} ratio is a key factor affecting the interaction between methanogens and SRB. Specifically, SRB can exclusively consume organic matter when the COD/SO_4^{2-} ratio provided is less than 0.66 (Oude Elferink et al. 1998). When the COD/SO_4^{2-} ratio falls below 1.0, the cell number and growth rate of SRB become dominant (Visser et al. 1993).

Anaerobic treatment is generally successful for wastewater with the COD/SO_4^{2-} ratio greater than 10, while deteriorated when the COD/SO_4^{2-} ratio was less than 10 (Hulshoff et al. 2001; Rinzema and Lettinga 1988). Using a lab-scale upflow anaerobic sludge blanket (UASB) reactor for treating starch wastewater, when the COD/SO_4^{2-} ratio was above 2, stable biogas production was obtained with efficient simultaneous removal of total COD (73.5–80.3%) and sulfate (82.6 ± 6.4%), while when the COD/SO_4^{2-} ratio was below 2, methanogenesis was suppressed through electrons competition and sulfide inhibition, deteriorating methane conversion (Lu et al. 2016). During anaerobic digestion of a very high strength and sulfate rich vinasse, the COD/SO_4^{2-} ratios below 0.10 enhanced the propionic acid degradation (up to 90%) by propionate degrading SRB; a COD/SO_4^{2-} ratio of 0.10 caused inhibition by H_2S_{aq}, $[H_2S]_{free}$ and propionic acid to SRB, methanogens or both; while a COD/SO_4^{2-} ratio of 0.20 severely inhibited methanogens and SRB, leading to reactor failure (Barrera et al. 2014). During thermophilic anaerobic digestion of vinasse, CH_4 production was 35% lower at the COD/SO_4^{2-} ratio of 7.5 compared with the ratio of 12.0, with a diversion of approximately 13.6% of the electron flow to sulfidogenesis, while the diversion of electrons to sulfidogenesis was negligible at COD/SO_4^{2-} ratios higher than 25 (Kiyuna et al. 2017).

When using specific types of organic carbon, the interaction is also determined by the COD/SO_4^{2-} ratio. When using the UASB reactors treating a VFA mixture (with an acetate:propionate:butyrate ratio of 5:3:2 on the COD basis) or acetate as the sole substrate at different COD/SO_4^{2-} ratios, the COD/SO_4^{2-} ratio was a key factor in the partitioning of acetate utilization between methanogens and SRB; with excess of sulfate (COD/SO_4^{2-} ratio lower than 0.67), SRB became predominant over methanogens after prolonged reactor operation: 250 and 400 days were required to increase the amount of acetate used by SRB from 50 to 90% in the reactor treating the VFA mixture or acetate, respectively (Omil et al. 1998). During anaerobic treatment of sulfate-rich methanol wastewater in a lab-scale UASB reactor at the COD/SO_4^{2-} ratio of 0.5, more than 85% of influent COD was confirmed to be utilized for methanogenesis, revealing that methanogens prevailed over sulfidogens and dominated significantly in the COD utilization (Wu et al. 2019).

The examination of relationships between SRB and methanogens can be achieved by analyzing the electron flow. With the influent COD/SO_4^{2-} ratios ranging from 0.5 to 20 (Fig. 4), Hu et al. (2015) revealed that COD removal decreased from 87.8 to 79.2% as the COD/SO_4^{2-} ratio decreased, and microbial electron flow accounting indicated that methanogens utilized over 79.4% of the total microbial electron flow

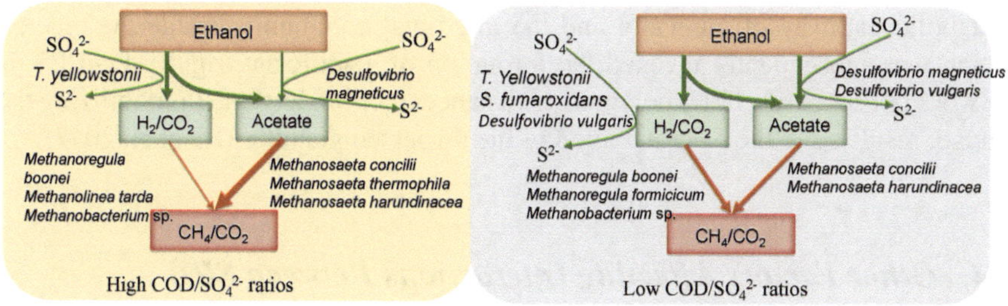

Fig. 4 Degradation pathways of ethanol, acetate, and sulfate with corresponding functional microorganisms at varying COD/SO_4^{2-} ratios (Hu et al. 2015)

across all ratios. Although no significant sulfide inhibition was observed throughout the experimental cycle, SRB species dominated ethanol degradation (Hu et al. 2015). By treating simulated wastewater constituents from a fiber manufacturing plant at a COD/SO_4^{2-} ratio of 1.0 with acetate and ethanol as carbon sources, Wu et al. (2018) obtained that the stabilized microbial electron flow distribution was 30% for methanogens and 70% for SRB after 330 days' operation, differing from the initial distribution of 60 and 40%, respectively.

3.3 Effect of Growth Modes on the Interactions Between SRB and Methanogens

In anaerobic reactors with better biomass retention, the competition between SRB and methanogens is determined not only by growth kinetics, but also by immobilization properties, substrate diffusion limitations inside biofilms, and environmental conditions (Colleran et al. 1995; Oude Elferink et al. 1998). The microbial community's ability to form biofilms or granules, as well as the actual water flow velocity in the reaction zone, are non-negligible factors. SRB were observed to have a lower attachment capacity compared to methanogens in fixed-film reactors (Isa et al. 1986). Additionally, the water flow velocity in the reactor determines the frequency of microbial regeneration, with higher upflow velocities (4–6 m/h) in a UASB reactor leading to increased methanogenic activity (Omil et al. 1996).

With acetate as the main substrate in a laboratory scale anaerobic fluidized bed, SRB out-competed methanogens in the biofilm at lower acetate concentrations, while methanogens were able to form a biofilm faster than SRB at higher acetate concentrations presumably due to methanogens' higher ability to adhere carrier surfaces compared with SRB (Yoda et al. 1987). When using methanol as the substrate and at a low COD/SO_4^{2-} ratio, Lu et al. (2017) found that the conversion of methanol to methane depended on both low levels of acetoclastic and hydrogenotrophic methanogens and IO-SRB species (e.g., *Desulfovibrio* sp.) that utilize H_2–CO_2 with acetate to mineralize the methanol. The sulfidogenesis served to diversify the metabolic pathway of methanol and the microbial community inside the reactor, which was subsequently invoked the formation of a sufficient, rigid [–Fe–EPS–]$_n$ network (EPS: extracellular polymeric substances) for binding and immobilizing the sludge, resulting in the re-granulation of the dispersed granules (Lu et al. 2017).

3.4 *Other Factors Affecting Interactions Between SRB and Methanogens*

Both SRB and methanogens have specific optimal temperatures for growth, and changes in ambient temperature can shift the dominant microbial communities and their competitive interactions. Madden et al. (2014) found that at lower temperatures of 15 °C, methanogenesis was only affected at COD/SO_4^{2-} ratios of 1:2 or less. Gas production decreased rapidly when the temperature dropped to 25 °C, while increasing the reactor temperature from 25 to 35 °C led to increased sulfate reduction rates but decreased methane generation rates, indicating that SRB are less sensitive to temperature variations than methanogens (Shin et al. 1996).

The change of pH has an important effect on the functional microbial community, and SRB and methanogens also have different responses to the change of pH. Under alkaline conditions, activities of SRB are reduced due to the alkalinization of the cytoplasm by inhibiting specific enzymes and destroying cell homeostasis, while under acidic conditions, microbial growth is limited due to that more energy will be used to pump out protons from the membrane for maintaining intracellular pH. Thus, the optimal pH for SRB is 6–7.5, and microbial activity will be decreased at pH 8 (Hu et al. 2020).

In addition, different biological processes require different redox potentials (ORP). For example, ORPs of −100 to −200 mV can enhance VFAs production, ORPs of −275 to −265 mV can increase sulfide removal, and ORPs of −50 to −300 mV are suitable for SRB growth (Huan et al. 2013; Yin et al. 2016). Methanogens are strictly anaerobic organisms that can only grow and metabolize under conditions where the ORP is below −320 mV. A reduction in ORP can stimulate the growth of syntrophic bacteria and enhance the methanogenesis process, leading to improved activities of these microorganisms (Yao et al. 2023).

4 Alleviation Strategies for the Inhibition of H_2S

Another concern during sulfate reduction is that the produced H_2S can cause inhibition of methanogens. The strategies for the alleviation of H_2S inhibition are mainly achieved through the prevention of H_2S formation or the efficient removal of the produced H_2S.

By increasing pH from 6.5 to 8.0, the methane production increased by 64.1%, while the H_2S content in biogas decreased by 44.7%; because more sulfate and organic sulfur were transferred to solid and liquid rather than H_2S formation at a high initial pH (Yan et al. 2018). O'Flaherty et al. (1998) found that sulfide inhibition for all the groups of microorganisms was related to the unionized sulfide concentrations in the pH range of 6.8–7.2 and to the total sulfide concentrations at pH above 7.2; and a high pH benefitted the growth of methanogens and a low pH benefitted SRB growth. Hilton and Oleszkiewicz (1988) showed that the inhibition of SRB and methanogens was correlated with the total sulfide and unionized sulfide concentration, respectively. Therefore, increasing pH may be an effective approach to alleviate the sulfur inhibition on the growth of methanogens (Yang et al. 2016).

Lupitskyy et al. (2018) found that an inorganic reactive adsorbent (zinc oxide nanowires, dosage 1 g/L) could effectively remove soluble sulfide (HS^-) and prevent H_2S formation, with the removal capability of 625 mg S^{2-}/L per gram of ZnO. Oliveira et al. (2020) found that biochar could remove H_2S from biogas (>98%), alleviate sulfide toxicity to microorganisms by removing 89% of unionized sulfide (H_2S_{aq}), and improve system stability, where the relative abundance of methanogens (*Methanobacterium* and *Methanosaeta*) increased from 0.7 to 3.7% and that of SRB (*Desulfovibrio*) decreased from 9.3 to 0.5%. Surface modification can improve the adsorption capacity of biochar, and the iron modified biochar increases the H_2S adsorption capacity by 2.5–3.9 times, trapping sulfur within the pores of the biochar (Choudhury and Lansing 2021).

Recently, the presence of conductive materials shows positive effects on mitigating sulfate effects on methanogenic systems. The mitigating effect of conductive materials under sulfate-stressed conditions is primarily attributed to their capability to enhance the electron competitive ability of methanogens (Jin et al. 2019; Li et al. 2017; Liu et al. 2019). With the dosage of conductive stainless steel into anaerobic digesters for treating sulfate-containing wastewater with the COD/SO_4^{2-} ratio of 1.23, the production of methane was increased by 7.5–24.6%, where the dissimilatory Fe (III) reducers such as *Geobacter* species were enriched on the surface of the stainless steel, possible establishing the direct interspecies electron transfer (DIET) with methanogenesis for outcompeting the sulfate reduction (Li et al. 2017). While with the dosage of magnetite, the methane production can be increased by about 3–10 folds at each COD/SO_4^{2-} ratio, and *Veillonella* species, the Fe(III)-reducing genus were found to reduce sulfate to H_2S (Jin et al. 2019). The addition of conductive materials (Fe_3O_4) promoted methanogenesis by re-enriching syntrophic partners dominant in the sulfate-free system and recovering their syntrophic interactions via DIET (Liu et al. 2019). Furthermore, for conductive iron oxides such as Fe_3O_4, a dissimilatory iron reduction may occur, which could convert sulfate to sulfur by incorporating a possible DIET between the sulfur-oxidizing bacteria and methanogens; thereby mitigating methanogenesis inhibition due to sulfate reduction (Jung et al. 2020). Specifically, when magnetite was added, H_2S production decreased remarkably, with the H_2S content in biogas below 100 ppm at magnetite doses of 8 mM Fe or higher, and the extracellular accumulation of S^0 occurred, where the reduced H_2S production appears to be due to the anaerobic oxidation of sulfide to S^0 (Jung

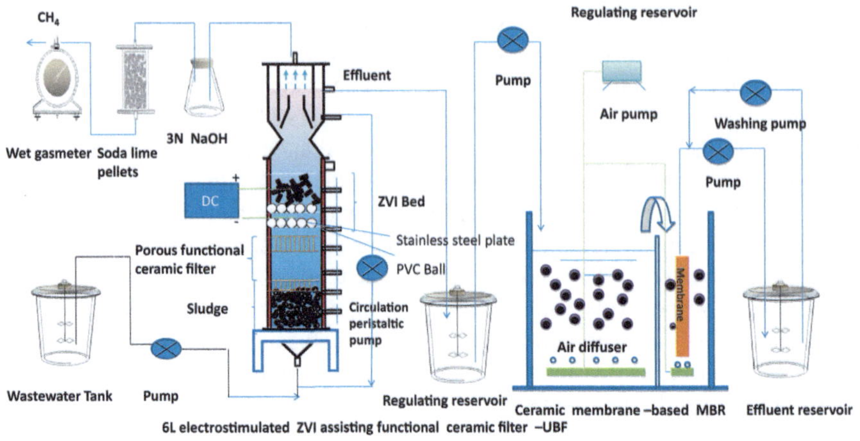

Fig. 5 Schematic diagram of electrically stimulated ZVI-assisted functional ceramic filter-UBF/ ceramic membrane MBR setup (Wang et al. 2020)

et al. 2020). The addition of biochar enhances DIET between potential electroactive microorganisms (*Syntrophomonas*) and methanogens, with possible involvement of *Desulfovibrio* in the syntrophic metabolism of electron transfer (Chen et al. 2023).

Zero-valent iron (ZVI) can accelerate granular sludge formation, mitigate sulfide inhibition, and enhance methane production. Li et al. (2019) achieved efficient methane yields at COD/SO_4^{2-} ratios greater than 0.9 by establishing a microelectrode-field-ZVI UASB system. ZVI functioned as an electron donor for methanogenesis and sulfate reduction, mitigating the inhibitory effects of SRB and its H_2S production on the anaerobic digestion process. Wang et al. (2020) explored the anaerobic synergistic degradation of para-ester and sulfate by dosing ZVI, with the introduction of a novel upflow biofilter (UBF) ceramic membrane within a membrane reactor (MBR) (Fig. 5). Notably, the UBF demonstrated satisfactory performance even when the COD/SO_4^{2-} ratio was as low as 1.57 and OLR was high at 6.8 g COD/ L/d, while at the ratio as low as 0.88, UBF operation was hindered by SRB inhibition, which could be restored by elevating the COD/SO_4^{2-} ratio (Wang et al. 2020).

5 Future Perspectives

5.1 *Mechanism Understanding*

In anaerobic digestion ecosystems, SRB and methanogens maintain a complex relationship within the symbiotic ecological niche. Improper regulation of this relationship can lead to consequences such as reduced methane production, metal corrosion, and inadequate wastewater treatment and disposal. However, further research is needed to elucidate the scientific and rigorous regulation process underlying this mechanism, particularly in the removal of high COD and high-strength sulfate from wastewater. Additionally, it is crucial to consider factors such as temperature, pH, COD/SO_4^{2-} ratio, and the practical application of coexisting microbial communities. Depending on the availability of easily and difficultly degradable carbon sources, appropriate strategies must be implemented to ensure the robustness of the microbial community's metabolic network while achieving production goals.

In microbial ecology systems, each biological reaction includes substrate degradation, electron transfer, and energy conversion. If only one aspect is considered, the underlying mechanism cannot be comprehensively explored. For example, Bell et al. (2020) found that by applying geochemical monitoring alone, limited microbial sulfate-reducing activity was obtained, while the examination of the microbial community at genome and proteome levels revealed active sulfur cycles of sulfate reduction and possibly sulfur disproportionation.

5.2 *The Extension of Function of SRB and Methanogens*

SRB and methanogens are not only exclusively found in the anaerobic treatment of industrial wastewater but also reside in domestic wastewater, urban sewers, and soil contaminated with heavy metals. Under various environmental conditions, the types of SRB and methanogens, and their metabolic pathways, would be different. Li et al. (2023) conducted a comprehensive investigation of methane production, microbial communities, and functional genes in sewers exhibiting different COD/SO_4^{2-} ratios at 33 sampling sites. Functional gene prediction revealed that the primary methane production pathways in sewers with COD/SO_4^{2-} ratios of 5.8 and 2.5 were acetate- and hydrogen-trophic, respectively. Conversely, in sewers with a COD/SO_4^{2-} ratio of 0.3, methyl-trophic methane production dominated. The genes expressed by methylotrophic methanogens in sewers with low COD/SO_4^{2-} ratios and high sulfate concentrations avoided competitive inhibition by SRB, thus facilitating methane production in these environments (Li et al. 2023).

During anaerobic biological treatment of wastewater containing heavy metals, these metals tend to bind with microorganisms, leading to the rupture of cell membranes and organelles. This binding process also results in the denaturation and inactivation of enzymes. The ability of SRB to transform sulfate into sulfur ions for the immobilization of heavy metals is widely favored. Sulfide reacts with various heavy metals to form insoluble metal sulfides (MeS), which precipitate rapidly. Mori et al. (2000) reported that sulfide production protects methanogenesis from the toxic effects of high concentrations of Cd or Cu (2 mM), and the presence of sulfide even induced the recovery of methanogenesis in cultures exposed to varying concentrations of heavy metals. For Zn and Ni, sulfide amendment could completely alleviate their toxic effects, while for Cu, sulfide amendment alleviated its toxicity only when the Cu^{2+} concentration was below 0.47 mM but failed to do so when the Cu^{2+} concentration rose to 0.8 mM (Zayed and Winter 2000).

5.3 Strategies to Establish Interspecies Mutualistic Relationships

The core of microbial interspecies symbiotic relationships lies in the transfer of electrons from one microorganism to another. The newly discovered DIET allows for the coupling of metabolic pathways between two distinct microorganisms, fostering interspecies symbiosis (Nozhevnikova et al. 2020). The bacterial-archaeal interspecies symbiosis based on DIET plays a pivotal role in microbial-driven geochemical transformations and the cycling of carbon and sulfur. Magnetite effectively promotes DIET, serving as an electronic conduit for electron transfer and enabling methanogens to compensate for external surface c-type cytochromes. Jin et al. (2019) significantly enhanced the community's resistance to the toxicity of H_2S by incorporating magnetite, while simultaneously maintaining the stability of methanogenic metabolism. By establishing interspecies mutualistic relationships among species, microbial communities can demonstrate enhanced resilience in responding to diverse and intricate environmental pressures.

6 Take-Home Messages

(1) When anaerobic digestion of sulfate-rich wastewater, the better competition of SRB for organic carbon accompanied with the inhibition by the produced H_2S can reduce the methanogenesis efficiency and even cause system failure.
(2) With diverse metabolic pathways of both sulfate (the electron acceptor) and organic carbon or H_2 (the electron donor), various types of SRB can be enriched under different environmental conditions with varied metabolic physiology.
(3) The competition between SRB and methanogens depends mainly on the COD/SO_4^{2-} ratio, and the formation of aggregate can affect the substrate utilization and further the competition or cooperation between SRB and methanogens.
(4) The alleviation of H_2S inhibition can be achieved through the removal of H_2S by adjusting pH or adding adsorbents (such as zinc oxide nanowires and biochar) or the enhancement of methanogenesis by dosing conductive materials such as biochar and ZVI.

References

Aeckersberg F, Rainey FA, Widdel F (1998) Growth, natural relationships, cellular fatty acids and metabolic adaptation of sulfate-reducing bacteria that utilize long-chain alkanes under anoxic conditions. Arch Microbiol 170:361–369

Anantharaman K, Hausmann B, Jungbluth SP, Kantor RS, Lavy A, Warren LA, Rappé MS, Pester M, Loy A, Thomas BC, Banfield JF (2018) Expanded diversity of microbial groups that shape the dissimilatory sulfur cycle. ISME J 12(7):1715–1728

Badziong W, Ditter B, Thauer RK (1979) Acetate and carbon dioxide assimilation by *Desulfovibrio vulgaris* (Marburg), growing on hydrogen and sulfate as sole energy source. Arch Microbiol 123:301–305

Bak F, Pfennig N (1987) Chemolithotrophic growth of *Desulfovibrio sulfodismutans* sp. nov. by disproportionation of inorganic sulfur compounds. Arch Microbiol 147(2):184–189

Barrera EL, Spanjers H, Romero O, Rosa E, Dewulf J (2014) Characterization of the sulfate reduction process in the anaerobic digestion of a very high strength and sulfate rich vinasse. Chem Eng J 248:383–393

Baumgartner LK, Reid RP, Dupraz C, Decho AW, Buckley DH, Spear JR, Przekop KM, Visscher PT (2006) Sulfate reducing bacteria in microbial mats: changing paradigms, new discoveries. Sedim Geol 185(3–4):131–145

Bell E, Lamminmäki T, Alneberg J, Andersson AF, Qian C, Xiong W, Hettich RL, Frutschi M, Bernier-Latmani R (2020) Active sulfur cycling in the terrestrial deep subsurface. ISME J 14(5):1260–1272

Block RJ, Henry SM (1961) Metabolism of the sulphur amino-acids and of sulphate in *Blattella germanica*. Nature 191(4786):392–393

Böttcher ME, Thamdrup B, Gehre M, Theune A (2005) $^{34}S/^{32}S$ and $^{18}O/^{16}O$ fractionation during sulfur disproportionation by *Desulfobulbus propionicus*. Geomicrobiol J 22(5):219–226

Cetecioglu Z, Dolfing J, Taylor J, Purdy KJ, Eyice Ö (2019) COD/sulfate ratio does not affect the methane yield and microbial diversity in anaerobic digesters. Water Res 155:444–454

Chen L, Zhang R, Li Y, Zhang Y, Fang W, Zhang P, Zhang G (2023) Internal driving mechanisms of microbial community and metabolism for granular activated carbon enhancing anaerobic digestion of high-strength sulfate organic wastewater. Chem Eng J 476:146648

Chen J, Wade MJ, Dolfing J, Soyer OS (2019) Increasing sulfate levels show a differential imethanogensct on synthetic communities comprising different methanogens and a sulfate reducer. 16:20190129

Choudhury A, Lansing S (2021) Adsorption of hydrogen sulfide in biogas using a novel iron-impregnated biochar scrubbing system. J Environ Chem Eng 9(1):104837

Colleran E, Finnegan S, Lens P (1995) Anaerobic treatment of sulphate-containing waste streams. Antonie Van Leeuwenhoek 67:29–46

Cord-Ruwisch R, Seitz HJ, Conrad R (1988) The capacity of hydrogenotrophic anaerobic bacteria to compete for traces of hydrogen depends on the redox potential of the terminal electron acceptor. Arch Microbiol 149:350–357

Dar SA, Kleerebezem R, Stams AJ, Kuenen JG, Muyzer G (2008) Competition and coexistence of sulfate-reducing bacteria, acetogens and methanogens in a lab-scale anaerobic bioreactor as affected by changing substrate to sulfate ratio. Appl Microbiol Biotechnol 78(6):1045–1055

Davidova IA, Duncan KE, Choi OK, Suflita JM (2006) *Desulfoglaeba alkanexedens* gen. nov., sp. nov., an n-alkane-degrading, sulfate-reducing bacterium. Int J Syst Evolut Microbiol 56(12):2737–2742

Dean JA (1999) Lange's handbook of chemistry. McGraw-Hill, Inc.

Fang W, Gu M, Liang D, Chen GH, Wang S (2020) Generation of zero valent sulfur from dissimilatory sulfate reduction under methanogenic conditions. J Hazard Mater 383:121197

Finster KW, Kjeldsen KU, Kube M, Reinhardt R, Mussmann M, Amann R, Schreiber L (2013) Complete genome sequence of *Desulfocapsa sulfexigens*, a marine deltaproteobacterium specialized in disproportionating inorganic sulfur compounds. Stand Genom Sci 8(1):58–68

Goevert D, Conrad R (2008) Carbon isotope fractionation by sulfate-reducing bacteria using different pathways for the oxidation of acetate. Environ Sci Technol 42(21):7813–7817

Grein F, Ramos AR, Venceslau SS, Pereira IA (2013) Unifying concepts in anaerobic respiration: insights from dissimilatory sulfur metabolism. Biochim Biophys Acta Bioenerget 1827(2):145–160

Grossi V, Cravo-Laureau C, Méou A, Raphel D, Garzino F, Hirschler-Réa A (2007) Anaerobic 1-alkene metabolism by the alkane-and alkene-degrading sulfate reducer *Desulfatibacillum aliphaticivorans* strain CV2803T. Appl Environ Microbiol 73(24):7882–7890

Guidotti TL (2010) Hydrogen sulfide: advances in understanding human toxicity. Int J Toxicol 29(6):569–581

Hausmann B, Knorr KH, Schreck K, Tringe SG, Glavina Del Rio T, Loy A, Pester M (2016) Consortia of low-abundance bacteria drive sulfate reduction-dependent degradation of fermentation products in peat soil microcosms. ISME J 10(10):2365–2375

Hilton BL, Oleszkiewicz JA (1988) Sulfide-induced inhibition of anaerobic digestion. J Environ Eng 114(6):1377–1391

Hu Y, Jing Z, Sudo Y, Niu Q, Du J, Wu J, Li YY (2015) Effect of influent ratios on UASB treatment of a synthetic sulfate-containing wastewater. Chemosphere 130:24–33

Hu K, Xu D, Chen Y (2020) An assessment of sulfate reducing bacteria on treating sulfate-rich metal-laden wastewater from electroplating plant. J Hazard Mater 393:122376

Huan NH, Hai NX, Yem T, Tuan NN (2013) Factors effect to the sulfide generation rate in the To Lich River, Vietnam. ARPN J Eng Appl Sci 8(3):190–199

Hulshoff LW, Lens PN, Weijma J, Stams AJ (2001) New developments in reactor and process technology for sulfate reduction. Water Sci Technol 44(8):67–76

Isa Z, Grusenmeyer S, Verstraete W (1986) Sulfate reduction relative to methane production in high-rate anaerobic digestion: microbiological aspects. Appl Environ Microbiol 51(3):580–587

Jin Z, Zhao Z, Zhang Y (2019) Potential of direct interspecies electron transfer in synergetic enhancement of methanogenesis and sulfate removal in an up-flow anaerobic sludge blanket reactor with magnetite. Sci Total Environ 677:299–306

Jørgensen BB, Bak F (1991) Pathways and microbiology of thiosulfate transformations and sulfate reduction in a marine sediment (Kattegat, Denmark). Appl Environ Microbiol 57(3):847–856

Jung H, Kim D, Choi H, Lee C (2022) A review of technologies for in-situ sulfide control in anaerobic digestion. Renew Sustain Energy Rev 157:112068

Jung H, Baek G, Lee C (2020) Magnetite-assisted in situ microbial oxidation of H_2S to S^0 during anaerobic digestion: a new potential for sulfide control. 397:124982

Kaksonen AH, Franzmann PD, Puhakka JA (2003) Performance and ethanol oxidation kinetics of a sulfate-reducing fluidized-bed reactor treating acidic metal-containing wastewater. Biodegradation 14:207–217

Kilburn KH (1993) Case report: profound neurobehavioral deficits in an oil field worker overcome by hydrogen sulfide. Am J Med Sci 306(5):301–305

Kiyuna LSM, Fuess LT, Zaiat M (2017) Unraveling the influence of the COD/sulfate ratio on organic matter removal and methane production from the biodigestion of sugarcane vinasse. Bioresour Technol 232:103–112

Kniemeyer O, Musat F, Sievert SM, Knittel K, Wilkes H, Blumenberg M, Michaelis W, Classen A, Bolm C, Joye SB, Widdel F (2007) Anaerobic oxidation of short-chain hydrocarbons by marine sulphate-reducing bacteria. Nature 449(7164):898–901

Krämer M, Cypionka H (1989) Sulfate formation via ATP sulfurylase in thiosulfate-and sulfite-disproportionating bacteria. Arch Microbiol 151:232–237

Kuivila KM, Murray JW, Devol AH (1990) Methane production in the sulfate-depleted sediments of two marine basins. Geochim Cosmochim Acta 54(2):403–411

Kushkevych I, Cejnar J, Treml J, Dordević D, Kollar P, Vítězová M (2020) Recent advances in metabolic pathways of sulfate reduction in intestinal bacteria. Cells 9(3):698

Li W, Niu Q, Zhang H, Tian Z, Zhang Y, Gao Y, Li YY, Nishimura O, Yang M (2015) UASB treatment of chemical synthesis-based pharmaceutical wastewater containing rich organic sulfur compounds and sulfate and associated microbial characteristics. Chem Eng J 260:55–63

Li Y, Zhang Y, Yang Y, Quan X, Zhao Z (2017) Potentially direct interspecies electron transfer of methanogenesis for syntrophic metabolism under sulfate reducing conditions with stainless steel. Bioresour Technol 234:303–309

Li W, Niu Q, Wu J, Luan X, Qi W, Zhang Y, Li YY, Gao Y, Yang M (2019) Enhanced anaerobic performance and SMD process in treatment of sulfate and organic S-rich TMBA manufacturing wastewater by micro-electric field—zero valent iron-UASB. J Hazard Mater 379:120695

Li Y, Jiang J, Zhang W, Yang G (2023) Changes in stoichiometric ratio of carbon and sulfate affect methanogenesis pathways in sulfate-rich sewers. J Clean Product 426:139112

Liu Y, Gu M, Yin Q, Wu G (2019) Inhibition mitigation and ecological mechanism of mesophilic methanogenesis triggered by supplement of ferroferric oxide in sulfate-containing systems. Bioresour Technol 288:121546

Lu X, Zhen G, Ni J, Hojo T, Kubota K, Li YY (2016) Effect of influent ratios on biodegradation behaviors of starch wastewater in an upflow anaerobic sludge blanket (UASB) reactor. Bioresour Technol 214:175–183

Lu X, Zhen G, Ni J, Kubota K, Li YY (2017) Sulfidogenesis process to strengthen re-granulation for biodegradation of methanolic wastewater and microorganisms evolution in an UASB reactor. Water Res 108:137–150

Lupitskyy R, Alvarez-Fonseca D, Herde ZD, Satyavolu J (2018) In-situ prevention of hydrogen sulfide formation during anaerobic digestion using zinc oxide nanowires. J Environ Chem Eng 6(1):110–118

Ma TT, Liu LY, Rui JP, Yuan Q, Feng DS, Zhou Z, Dai LR, Zeng WQ, Zhang H, Cheng L (2017) Coexistence and competition of sulfate-reducing and methanogenic populations in an anaerobic hexadecane-degrading culture. Biotechnol Biofuels 10:207

Madden P, Al-Raei AM, Enright AM, Chinalia FA, de Beer D, O'Flaherty V, Collins G (2014) Effect of sulfate on low-temperature anaerobic digestion. Front Microbiol 5:376

Marietou A, Kjeldsen KU, Glombitza C, Jørgensen BB (2022) Response to substrate limitation by a marine sulfate-reducing bacterium. ISME J 16(1):200–210

Mori K, Hatsu M, Kimura R, Takamizawa K (2000) Effect of heavy metals on the growth of a methanogen in pure culture and coculture with a sulfate-reducing bacterium. J Biosci Bioeng 90(3):260–265

Muyzer G, Stams AJ (2008) The ecology and biotechnology of sulphate-reducing bacteria. Nat Rev Microbiol 6(6):441–454

Nanninga HJ, Gottschal JC (1987) Properties of *Desulfovibrio carbinolicus* sp. nov. and other sulfate-reducing bacteria isolated from an anaerobic-purification plant. Appl Environ Microbiol 53(4):802–809

Novair SB, Atigh ZB, Lajayer BA, Shu W, Price GW (2024) The role of sulphate-reducing bacteria (SRB) in bioremediation of sulphate-rich wastewater: focus on the source of electron donors. Process Saf Environ Protect 184:190–207

Nozhevnikova AN, Russkova YI, Litti YV, Parshina SN, Zhuravleva EA, Nikitina AA (2020) Syntrophy and interspecies electron transfer in methanogenic microbial communities. Microbiology 89:129–147

O'Flaherty V, Mahony T, O'Kennedy R, Colleran E (1998) Effect of pH on growth kinetics and sulphide toxicity thresholds of a range of methanogenic, syntrophic and sulphate-reducing bacteria. Process Biochem 33(5):555–569

Oliveira FR, Surendra KC, Jaisi DP, Lu H, Unal-Tosun G, Sung S, Khanal SK (2020) Alleviating sulfide toxicity using biochar during anaerobic treatment of sulfate-laden wastewater. Bioresour Technol 301:122711

Omil F, Méndez R, Lema JM (1995) Anaerobic treatment of saline wastewaters under high sulphide and ammonia content. Bioresour Technol 54(3):269–278

Omil F, Lens P, Pol LH, Lettinga G (1996) Effect of upward velocity and sulphide concentration on volatile fatty acid degradation in a sulphidogenic granular sludge reactor. Process Biochem 31(7):699–710

Omil F, Lens P, Visser A, Hulshoff Pol LW, Lettinga G (1998) Long-term competition between sulfate reducing and methanogenic bacteria in UASB reactors treating volatile fatty acids. Biotechnol Bioeng 57:676–685

Oude Elferink SJ, Boschker HT, Stams AJ (1998) Identification of sulfate reducers and *Syntrophobacter* sp. in anaerobic granular sludge by fatty-acid biomarkers and 16S rRNA probing. Geomicrobiol J 15(1):3–17

Parshina SN, Sipma J, Nakashimada Y, Henstra AM, Smidt H, Lysenko AM, Lens PN, Lettinga G, Stams AJ (2005) *Desulfotomaculum carboxydivorans* sp. nov., a novel sulfate-reducing bacterium capable of growth at 100% CO. Int J Syst Evolut Microbiol 55(5):2159–2165

Paula DR, Foresti E (2009) Sulfide toxicity kinetics of a UASB reactor. Brazil J Chem Eng 26:669–675

Paulo LM, Stams AJ, Sousa DZ (2015) Methanogens, sulphate and heavy metals: a complex system. Rev Environ Sci Bio/technol 14:537–553

Pester M, Brambilla E, Alazard D, Rattei T, Weinmaier T, Han J, Lucas S, Lapidus A, Cheng JF, Goodwin L, Pitluck S, Peters L, Ovchinnikova G, Teshima H, Detter JC, Han CS, Tapia R, Land ML, Hauser L, Kyrpides NC, Ivanova NN, Pagani I, Huntmann M, Wei CL, Davenport KW, Daligault H, Chain PS, Chen A, Mavromatis K, Markowitz V, Szeto E, Mikhailova N, Pati A, Wagner M, Woyke T, Ollivier B, Klenk HP, Spring S, Loy A (2012) Complete genome sequences of *Desulfosporosinus orientis* DSM765T, *Desulfosporosinus youngiae* DSM17734T, *Desulfosporosinus meridiei* DSM13257T, and *Desulfosporosinus acidiphilus* DSM22704T. J Bacteriol 194(22):6300–6301

Qian Z, Tianwei H, Mackey HR, van Loosdrecht MCM, Guanghao C (2019) Recent advances in dissimilatory sulfate reduction: From metabolic study to application. Water Res 150:162–181

Raskin L, Stromley JM, Rittmann BE, Stahl DA (1994) Group-specific 16S rRNA hybridization probes to describe natural communities of methanogens. Appl Environ Microbiol 60(4):1232–1240

Raskin L, Zheng D, Griffin ME, Stroot PG, Misra P (1995) Characterization of microbial communities in anaerobic bioreactors using molecular probes. Antonie Van Leeuwenhoek 68(4):297–308

Rinzema A, Lettinga G (1988) Anaerobic treatment of sulfate-containing waste water. In: Wise DL (ed) Biotreatment systems (Vol III). CRC Press Inc., Boca Raton, pp 65–109

Shi X, Gao G, Tian J, Wang XC, Jin X, Jin P (2020) Symbiosis of sulfate-reducing bacteria and methanogenic archaea in sewer systems. Environ Int 143:105923

Shin HS, Oh SE, Bae BU (1996) Competition between SRB and MPB according to temperature change in the anaerobic treatment of tannery wastes containing high sulfate. Environ Technol 17(4):361–370

Speece RE (1983) Anaerobic biotechnology for industrial wastewater treatment. Environ Sci Technol 17(9):416A-427A

Stucki G, Hanselmann KW, Hürzeler RA (1993) Biological sulfuric acid transformation: reactor design and process optimization. Biotechnol Bioeng 41(3):303–315

Sun R, Zhang L, Zhang Z, Chen GH, Jiang F (2018) Realizing high-rate sulfur reduction under sulfate-rich conditions in a biological sulfide production system to treat metalladen wastewater deficient in organic matter. Water Res 131:239–245

Tanimoto Y, Bak F (1994) Anaerobic degradation of methylmercaptan and dimethyl sulfide by newly isolated thermophilic sulfate-reducing bacteria. Appl Environ Microbiol 60(7):2450–2455

Thauer RK (1988) Citric-acid cycle, 50 years on: modifications and an alternative pathway in anaerobic bacteria. Eur J Biochem 176(3):497–508

Thauer RK, Stackebrandt E, Hamilton WA (2009) Energy metabolism and phylogenetic diversity of sulphate-reducing bacteria. In: Barton LL, Hamilton WA (eds) Sulphate-reducing bacteria environmental and engineered systems. Cambridge University Press, Cambridge, pp 1–38

Thorup C, Schramm A, Findlay AJ, Finster KW, Schreiber L (2017) Disguised as a Sulfate Reducer: Growth of the deltaproteobacterium *Desulfurivibrio alkaliphilus* by sulfide oxidation with nitrate. Mbio 8(4):e00671–e00677

Utkin I, Woese C, Wiegel J (1994) Isolation and characterization of *Desulfitobacterium dehalogenans* gen. nov., sp. nov., an anaerobic bacterium which reductively dechlorinates chlorophenolic compounds. Int J Syst Evolut Microbiol 44(4):612–619

Visser A, Nozhevnikova AN, Lettinga G (1993) Sulphide inhibition of methanogenic activity at various pH levels at 55°C. J Chem Technol Biotechnol 57(1):9–13

Von Wolzogen Kühr CAH, van der Vlugt IS (1934) The graphitization of cast iron as an electrobiochemical process in anaerobic soils. Water 18(16):147–165

Wang C, Wu H, Wu J, Zhu B, Fan W, Lu T, Gao Y, Qi W, Niu Q (2020) Promoted biodegradation of para-ester wastewater by electrostimulated ZVI assisting novel UBF/ceramic membrane MBR and microbial community. J Taiwan Instit Chem Eng 113:285–292

Widdel F (1988) Microbiology and ecology of sulfate- and sulfur-reducing bacteria. In: Zehnder AJB (ed) Biology of anaerobic microorganisms. Wiley, New York, pp 469–585

Wu J, Niu Q, Li L, Hu Y, Mribet C, Hojo T, Li YY (2018) A gradual change between methanogenesis and sulphidogenesis during a long-term UASB treatment of sulfate-rich chemical wastewater. Sci Total Environ 636:168–176

Wu J, Liu Q, Feng B, Kong Z, Jiang B, Li YY (2019) Temperature effects on the methanogenesis enhancement and sulfidogenesis suppression in the UASB treatment of sulfate-rich methanol wastewater. Int Biodeter Biodegrad 142:182–190

Yan L, Ye J, Zhang P, Xu D, Wu Y, Liu J, Zhang H, Fang W, Wang B, Zeng G (2018) Hydrogen sulfide formation control and microbial competition in batch anaerobic digestion of slaughterhouse wastewater sludge: effect of initial sludge pH. Bioresour Technol 259:67–74

Yang G, Zhang G, Zhuan R, Yang A, Wang Y (2016) Transformations, inhibition and inhibition control methods of sulfur in sludge anaerobic digestion: a review. Curr Organ Chem 20(26):2780–2789

Yao Y, Wei Y, Li J, Han R, Jing C, Liu R, Niu Q (2023) Microbial electron flow promotes naphthalene degradation in anaerobic digestion in the presence of nitrate electron acceptor: focus on electron flow regulation and microbial interaction succession. J Hazard Mater 459:132293

Yin J, Yu X, Zhang Y, Shen D, Wang M, Long Y, Chen T (2016) Enhancement of acidogenic fermentation for volatile fatty acid production from food waste: effect of redox potential and inoculum. Bioresour Technol 216:996–1003

Yoda M, Kitagawa M, Miyaji Y (1987) Long term competition between sulfate-reducing and methane-producing bacteria for acetate in anaerobic biofilm. Water Res 21(12):1547–1556

Zan F, Tang W, Jiang F, Chen G (2021) Diversion of food waste into the sulfate-laden sewer: interaction and electron flow of sulfidogenesis and methanogenesis. Water Res 202:117437

Zayed G, Winter J (2000) Inhibition of methane production from whey by heavy metals–protective effect of sulfide. Appl Microbiol Biotechnol 53:726–731

Zhang Z, Zhang C, Yang Y, Zhang Z, Tang Y, Su P, Lin Z (2022) A review of sulfate-reducing bacteria: metabolism, influencing factors and application in wastewater treatment. J Clean Product 376:134109

Zhao Y, Ren N, Wang A (2008) Contributions of fermentative acidogenic bacteria and sulfate-reducing bacteria to lactate degradation and sulfate reduction. Chemosphere 72(2):233–242

Anaerobic Biotransformation of Organoarsenicals and Its Inhibition on Anaerobic Digestion

Rui Tang and Zhenhu Hu

Abstract Extensive use of organoarsenic feed additives has caused organoarsenicals to be detected in animal manure and wastewater, and further within anaerobic digestion systems. Anaerobic organoarsenical biodegradation is initiated with the reduction of electron-withdrawing groups, followed by the cleavage of C–As bonds or the reduction of arsenate group. Electrochemical stimulation can accelerate the degradation of the organoarsenicals with electron-donating groups. The biotransformation of organoarsenicals can be significantly affected by microbial compositions and property, and environmental factors such as co-existing organic matter. Due to the equipment of arsenate or nitro groups in organoarsenicals and their anaerobic degradation products, methanogenic activity, especially the acetoclastic methanogenic activity can be inhibited, and anaerobic sludge characteristics including physiochemical properties and microbial structure are remodeled. In response to the stress of organoarsenicals at environment-relevant levels, anaerobic granular sludge can protect itself through accelerating the excretion of extracellular polymeric substances. Two inhibition alleviation strategies that rely on sulfate reduction and zero-valent iron mediation, respectively, can enhance organoarsenical degradation and immobilize the generated inorganic arsenic simultaneously. The understanding of organoarsenical-induced inhibition on anaerobic digestion and the development of the corresponding mitigation techniques can achieve highly efficient arsenic removal and biogas production during anaerobic digestion of animal manure and wastewater.

Keywords Electrochemical stimulation · Organoarsenical · Sulfate reduction · Transformation pathway · Zero-valent iron

R. Tang
Shanghai Key Laboratory of Materials Protection and Advanced Materials in Electric Power, College of Environmental and Chemical Engineering, Shanghai University of Electric Power, Shanghai 200090, China

Z. Hu (✉)
School of Civil Engineering, Hefei University of Technology, Hefei 230009, Anhui, China
e-mail: zhhu@hfut.edu.cn

G. Wu (ed.), *Anaerobic Digestion*, Green Energy and Technology,
https://doi.org/10.1007/978-3-031-69378-6_8

1 Introduction

Aromatic organoarsenicals have been extensively used as feed additives in animal feeding operations for several decades, aiming to prevent parasitic infection, stimulate animal growth, and improve meat pigmentation (Hu et al. 2019). Commonly used organoarsenicals, i.e., phenylarsonic acid (PAA), roxarsone (ROX, 4-hydroxy-3-nitrophenylarsonic acid), arsanilic acid (ASA, 4-aminophenylarsonic acid), nitrasone (NIT), and 4-hydroxy-phenylarsonic acid (4-HPAA), are the derivatives of phenylarsonic acid with diverse functional groups (i.e., hydroxy-, amino-, and nitro-) at *para* and *meta* positions (Table 1). Since ROX was approved as the animal feed additive by the US Food and Drug Administration in 1944 (Fisher et al. 2015), other aforementioned organoarsenic feed additives subsequently have emerged in animal feeding operations globally (Mafla et al. 2015; Silbergeld and Nachman 2008). Although the use of these organoarsenic feed additives has then been forbidden in European countries, USA, and China due to concerns about arsenic exposure, they are still authorized in many countries, e.g., Argentina, Mexico, Vietnam, and Brazil (Fisher et al. 2015; Hu et al. 2019). Most of these countries are the major producers and exporters of meat all over the world.

The dosage of aromatic organoarsenicals in animal feed typically ranges from 20 to 200 mg/kg (Mangalgiri et al. 2015; Silbergeld and Nachman 2008). For blackhead disease prevention, the dosage of Histostat-50, a commercially prepared NIT, can even reach 375 mg/kg (Mangalgiri et al. 2015). Of the dietary organoarsenic additives, more than 90% is excreted by animals without any chemical changes (Gupta et al. 2018). For example, the arsenic concentration in animal manure ranges from the

Table 1 Physicochemical property and dosage of organoarsenic feed additives

Additives	Molecular formula	Chemical structure	pKa	Dosage (mg/kg)	References
PAA	$C_6H_7AsO_3$	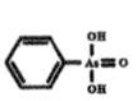	$pK_{a1} = 3.47$ $pK_{a2} = 8.48$	–	Pressman and Brown (1943)
4-HPAA	$C_6H_7AsO_4$	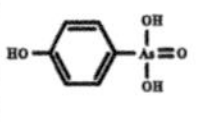	$pK_{a1} = 3.89$ $pK_{a2} = 8.37$ $pK_{a3} = 10.05$	–	Pressman and Brown (1943)
ASA	$C_6H_8AsNO_3$	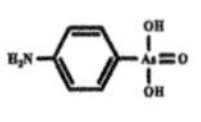	$pK_{a1} = 1.90$ $pK_{a2} = 4.10$ $pK_{a3} = 9.20$	100	Chen and Huang (2012)
NIT	$C_6H_6AsNO_5$	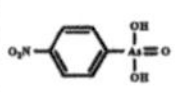	$pK_{a1} = 2.20$ $pK_{a2} = 7.78$	375	Adak et al. (2015)
ROX	$C_6H_6AsNO_6$	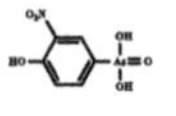	$pK_{a1} = 3.49$ $pK_{a2} = 5.74$ $pK_{a3} = 9.13$	20–50	Qiang and Adams (2004)

Note NIT, 4-nitrophenylarsonic acid; PAA, phenylarsonic acid; 4-HPAA, 4-hydroxy-phenylarsonic acid; ASA, 3-aminophenylarsonic acid; ROX, 4-hydroxy-3-nitrophenylarsonic acid undetectable level to 315.1 mg/kg (Fisher et al. 2015; Huang et al. 2014), and 70–75% of the total arsenic is soluble (Garbarino et al. 2003), which can be easily dissolved in wastewater (Fei et al. 2018). When used for land application, animal manure or wastewater can release arsenic into soil. Consequently, these arsenics are likely to get access to human food chain, e.g., vegetables (Huang et al. 2014), rice (Geng et al. 2017), and aquatic products, exhibiting a serious threat to human health (Zhu et al. 2014) (Fig. 1). In the natural environment, microorganisms (Fu et al. 2017; Stolz et al. 2007) and sunlight irradiation (Meng et al. 2019) stimulate the transformation of organoarsenicals into more toxic forms, As(V) and As(III), thereby increasing the threat of arsenic to ecological security and human health. Regarding the health risk caused by arsenic, World Health Organization recommended that the allowable arsenic level in drinking water should not exceed 10 μg/L (WHO 2017), which has been adopted in many countries, e.g., USA and China. With increasing concerns, it is necessary to remove organic/inorganic arsenic from wastewater before its discharging or application.

Owing to bioenergy recovery, high organic loading, and low operating cost, anaerobic digestion has been applied widely for the treatment of animal manure and wastewater (Cheng et al. 2020). During anaerobic digestion, removal efficiency of conventional pollutants, e.g., chemical oxygen demand (COD), ammonium nitrogen (NH_4^+–N), and phosphorus (P), have received major attention. While recently, the impacts of organoarsenicals on anaerobic digestion systems and its regulations have been investigated for their risk control (Fei et al. 2018; Guzmán-Fierro et al. 2015; Mafla et al. 2015; Shui et al. 2016; Stolz et al. 2007).

Fig. 1 Fate of organoarsenicals supplemented in animal feed

2 Anaerobic Biotransformation of Organoarsenicals

Anaerobic organoarsenical biodegradation commonly involves the reduction of arsenate group and the cleavage of C–As bonds (Tang et al. 2020a). The presence of electron-withdrawing groups such as nitro group can accelerate the organoarsenical transformation, whereas electron-donating groups such as amino group will cause a slow degradation (Cortinas et al. 2006). Therefore, ROX and ASA, two commonly used organoarsenicals equipped with electron-withdrawing and electron-donating groups, respectively, are selected to show the anaerobic organoarsenical biotransformation behavior.

2.1 Biotransformation of ROX

The biotransformation of ROX initiates with the reduction of its nitro group into amino group, thus generating 4-hydroxy-3-aminophenylarsonic acid (HAPA; Cortinas et al. 2006; Shi et al. 2014). This reaction is typically catalyzed by nonspecific nitroreductase through transferring six electrons to nitro group with H_2, exogenous, or endogenous substances as the electron donors (Cortinas et al. 2006). Anaerobic microorganisms participating in this process include *Clostridium* species (e.g., *Clostridium* sp. strain OhILAs, *Clostridium celerecrescens,* and *Clostridium irregulare*), methanogens, and sulfate-reducing bacteria (Cortinas et al. 2006; Stolz et al. 2007; Thomas et al. 2014). Moreover, electroactive bacteria (EEB) such as *Aeromonas hydrophila*, *Geobacter sulfurreducens*, and *Shewanella putrefaciens* CN32, can reduce ROX into HAPA intracellularly and extracellularly (Han et al. 2017). With the contribution of both intracellular and extracellular actions, EEB exhibit a high capability for ROX nitro group reduction.

The further degradation of HAPA can be achieved via two pathways through diverse anaerobic reactions (Fig. 2), such as methanogenesis (Stolz et al. 2007; Tang et al. 2019), sulfate reduction (Cortinas et al. 2006), and denitrification (Liu et al. 2014). In pathway I, HAPA undergoes a microbially catalyzed cleavage of the C–As bond, thereby producing As(V) and *o*-aminophenol (Cortinas et al. 2006; Shui et al. 2016). In pathway II, the intracellular reduction of HAPA to 4-hydroxy-3-aminophenylarsenite [HAPA(III)] via reducing the arsenate group by a molybdopterin oxidoreductase, ArrA, encoded by the *arrA* gene (Thomas et al. 2014), and then the cleavage of the C–As bond, due to a C–As lyase expressed by *arsI* gene carriers, e.g., *Bacillus* sp. MD1 and *S. putrefaciens* 200 (Pawitwar et al. 2017; Yoshinaga and Rosen 2014), generates As(III) and o-aminophenol (Han et al. 2017; Tang et al. 2020a). The difference between pathways I and II is whether arsenate group is reduced or not before the C–As bond cleavage. Compared to microbial reduction of nitro group, C–As bond cleavage is much slower (Stolz et al. 2007; Zhang et al. 2014), thereby becoming the rate-limiting step for biodegradation of ROX into inorganic arsenic. The end products from HAPA reduction, As(III) and As(V),

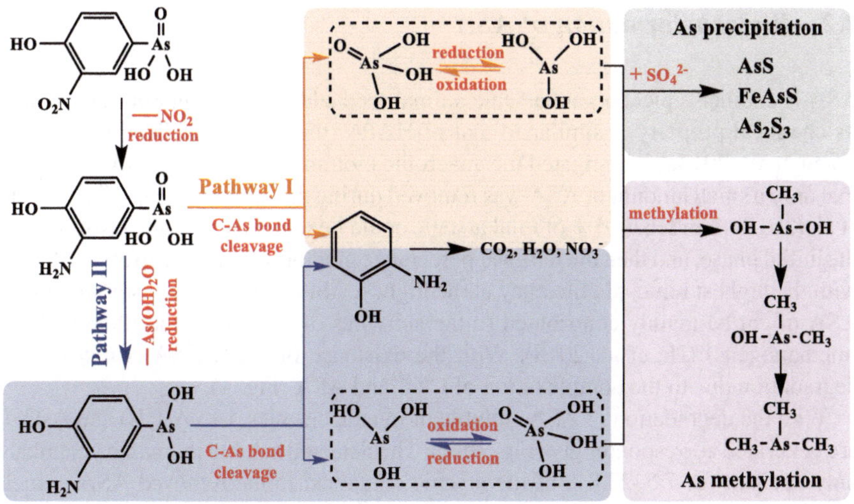

Fig. 2 Anaerobic biotransformation pathways of ROX (Tang et al. 2021c)

can be mutually transformed to each other in anaerobic bioreactors, and thus exist simultaneously. Specifically, As(III) is generated via microbial reduction of As(V) under anaerobic conditions (Cortinas et al. 2006), while As(V) is generated through anaerobic respiration of As(III) for microbial growth (Stolz et al. 2006).

The produced As(III) and As(V) can be further transformed through processes such as methylation and precipitation (Fig. 2). Arsenic methylation and volatilization were initially thought to follow the Challenge pathway (Mestrot et al. 2013a), in which As(III) undergoes an alternation of oxidative methylation and microbial reduction to release volatile arsenic. While Hayakawa et al. (2005) demonstrated a direct non-oxidative arsenic methylation pathway, where As(III) is consecutively methylated to monomethylarsenite [MMA(III)], dimethylarsenite [DMA(III)], and finally trimethylarsenite [TMA(III)]. These methylarsenites can be converted to the corresponding arsines (i.e., AsH_3, monomethylarsine and dimethylarsine) and oxidized to methylarsenates (i.e., monomethylarsenate [MMA], dimethylarsenate [DMA], and trimethylarsenate [TMA]). The arsenic methylation and volatilization are primarily manipulated by microbial *arsM* gene (Qin et al. 2006), which encodes As(III) S-adenosylmethionine methyltransferase. Intensive and continuous release of volatile arsenic from anaerobic digestion threatens human health. For example, the proportion of the volatilized arsenic to the influent ROX can reach to 0.01–0.26‰ in batch digesters (Mestrot et al. 2013b; Tang et al. 2020b) and 0.02‰ in an up-flow anaerobic sludge blanket (UASB) reactor (Tang et al. 2017). Furthermore, inorganic arsenic can precipitate in forms of realgar (As_4S_4), orpiment (As_2S_3), and arsenopyrite (FeAsS) during bio-reduction of As(V) and sulfate (Battaglia-Brunet et al. 2012; Ji et al. 2017; Rodriguez-Freire et al. 2014), which enables effective mitigation of the mobility and bioaccessibility of arsenic.

2.2 Biotransformation of ASA

ASA is another typical organoarsenic animal feed additive without nitro group, and its chemical property is similar to that of HAPA (the product of ROX reduction). Wang et al. (2014a) investigated the anaerobic biotransformation of ASA and found that only a small amount of ASA was removed during the initial 18 days (Fig. 3a). He et al. (2018) found that ~5% of total arsenic could be removed through adsorption in the initial phase, and then the removal percentage gradually increased to 36.9–58.7%, with the highest removal efficiency at the highest ASA addition. The degradation of ASA might be mainly contributed to the activities of bacteria, rather than archaea (methanogens) (He et al. 2018). With the existence of Cu^{2+}, ASA is resistant to degradation due to the complexation of Cu^{2+} and ASA (Fig. 4).

With the degradation of ASA, soluble inorganic arsenics, i.e., As(III) and As(V), are generated correspondingly (Fig. 3b, c). The detected soluble inorganic arsenicals only account for 7.5–23.8% of the arsenic contained in the removed ASA, which could be due to that most of the produced inorganic arsenic is adsorbed and precipitated in the sludge. The inorganic arsenic species seems to be dependent on the initial ASA concentration. At the ASA concentration above 0.46 mM, As(III) becomes the primary species. Furthermore, in addition to As(III) and As(V), methylated arsenics including MMA and DMA are also detected, indicating the methylation of arsenic exists in the anaerobic digestion system.

2.3 Electrochemically Stimulated Biotransformation of Organoarsenicals

Microbial electrochemical systems, i.e., microbial electrolysis cell (MEC) and microbial fuel cell (MFC), can transfer electrons, and have been successfully applied for enhancing the degradation of organoarsenicals (Chen et al. 2018; Shi et al. 2014). In the MEC applied with a voltage of 0.5 V, Shi et al. (2014) found that 76.0 μM ROX was rapidly transformed to HAPA and completely degraded into As(V) within 7 days (Fig. 5a). Without voltage, HAPA is resistant to degradation and persists for even more than 200 days in digesters (Cortinas et al. 2006; Zhang et al. 2014). Therefore, the voltage application can be better applied for the cleavage of the C–As bond.

With MEC, the anode functions as a stable electron acceptor to facilitate microbial degradation of organoarsenicals such as HAPA, with the end products of mainly As(V) rather than As(III) (Fig. 5b) (Shi et al. 2014). In the MFC inoculated with *Shewanella oneidensis* MR-1, ROX is degraded into inorganic arsenic in the anode chamber following the pathway II as described in Fig. 2, and the closed-circuit voltage can reach ~800 mV (Chen et al. 2018). In such a system, direct contact between electrode and microbes largely determines the kinetics of extracellular electron transfer (Torres et al. 2010). Accordingly, microbes attached on electrodes are more efficient to metabolize ROX and generate power than planktonic microbes (Chen et al. 2018).

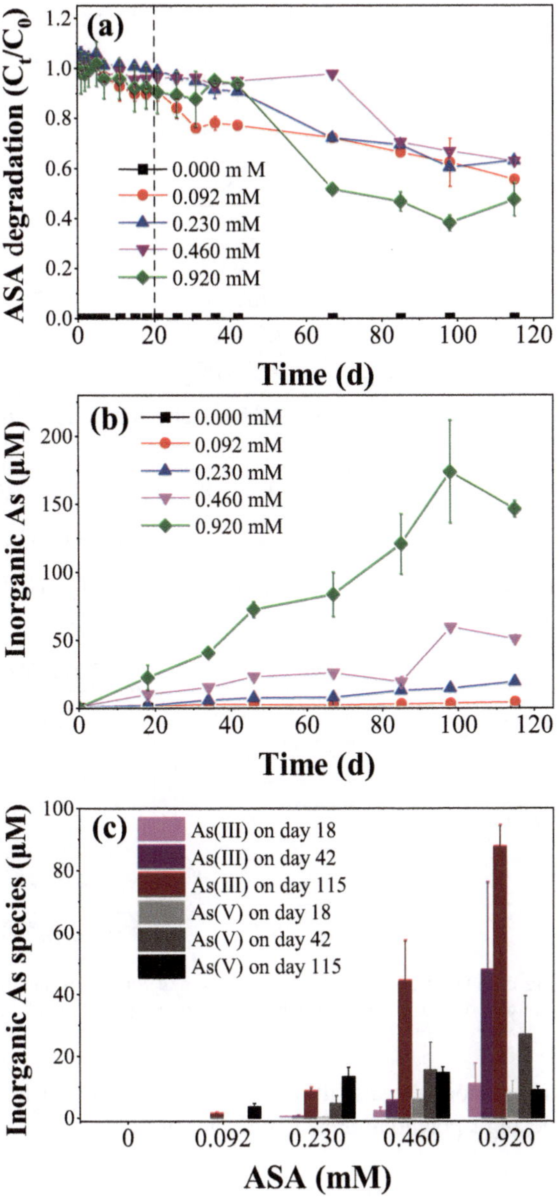

Fig. 3 ASA transformation during anaerobic methanogenesis. **a** ASA degradation, **b** inorganic arsenic generation, **c** inorganic arsenic speciation (Wang et al. 2014a)

The larger-scale operations and complex compositions of real wastewater may cause perturbations on their performance. For example, labile organic carbons such as acetate are the inhibitory factor for HAPA degradation in MEC (Shi et al. 2014). One more concern is that inorganic arsenic released via bioelectrochemical degradation

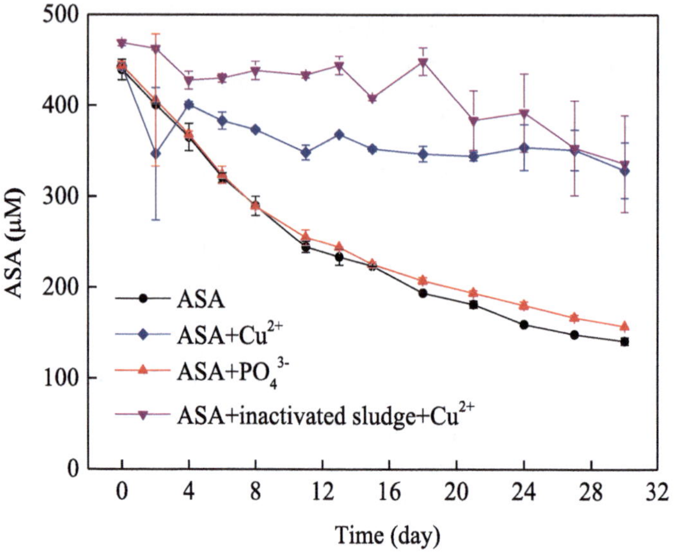

Fig. 4 Biotransformation of ASA during anaerobic digestion of pig manure (He et al. 2018)

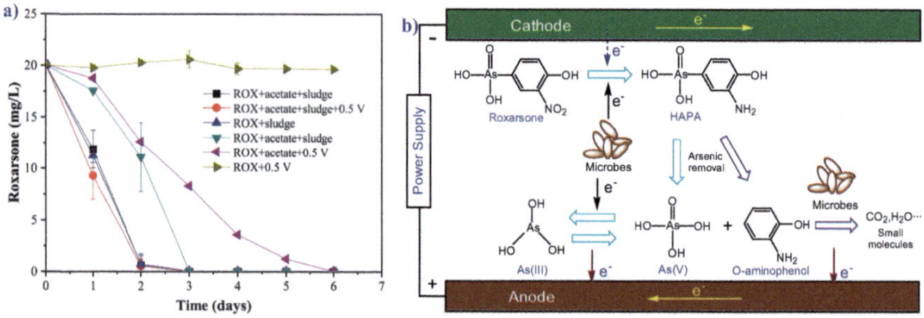

Fig. 5 **a** Degradation of ROX under different conditions, and **b** degradation and biotransformation pathway of ROX with electrochemical stimulation (Shi et al. 2014)

of organoarsenicals still persists in the solution (Chen et al. 2018; Shi et al. 2014), posing a potential threat to effluent safety.

2.4 *Microbial Functions Responsible for Organoarsenic Biotransformation*

Organoarsenic biotransformation is a complex biological process. The functional genes can be categorized into two groups, i.e., those control the uptake of arsenic and the detoxification of arsenic. The potential functional microorganisms are summarized in Table 2.

Three uptake systems that are manipulated by *glpF*, *pit*, and *pst* genes to uptake the specific arsenic species, respectively, have been identified (Fig. 6). Microbial *glpF* gene encodes aquaglyceroporin (GlpF) for the uptake of As(III) (Meng et al. 2004). Due to the similar structure to phosphate, As(V) enters into cells through phosphate transporters, Pit or Pst, encoded by *pit* and *pst* genes, respectively (Kruger et al. 2013; Li et al. 2013). As Pst is much more specific for phosphate transport, it is less efficient for As(V) uptake. With the collaborative actions of these uptake systems, microbes can transform organoarsenicals intracellularly and extracellularly simultaneously. However, transporters for organoarsenicals uptake remain unkown.

The detoxification of adverse arsenic species can be achieved through transforming arsenic into less toxic forms and microbially catalyzed arsenic extrusion. Generally, trivalent arsenicals are commonly more toxic than pentavalent forms (Qin et al. 2006). Organoarsenic oxidase encoded by *arsH* gene stimulates the oxidation of trivalent organoarsenicals including MMA(III), ROX(III), and phenylarsenite (Chen et al. 2015a), while the oxidation of inorganic As(III) to As(V) is catalyzed by Aio/Arx (Warelow et al. 2017). Furthermore, arsenic extrusion can effectively mitigate cellular arsenic accumulation, which is driven by arsenic-resistant genes encoding efflux permeases. Currently, Acr3, ArsB, ArsG, ArsJ, ArsK, and ArsP are the six known arsenic efflux permeases, and each confers resistance to different arsenic species (Chen et al. 2015b, 2019; Shi et al. 2018). Among them, Acr3 and ArsB are widely distributed in arsenic-resistant microorganisms for extruding As(III) exclusively (Shi et al. 2018; Tisa and Rosen 1990). When Acr3 or ArsB is coupled with ArsA, a membrane-bound anion-translocating ATPase, As(III) extrusion becomes more efficient (Tisa and Rosen 1990). Another widespread efflux transporter, ArsK, can also extrude As(III), as well as several trivalent organoarsenicals, i.e., MMA(III) and ROX(III) (Shi et al. 2018). ArsK-catalyzed extrusion of As(III) is more efficient than that of MMA(III) and ROX(III).

2.5 Factors Affecting Organoarsenic Biotransformation

Microbial compositions and the property of sludge exert significant influence on organoarsenic degradation. Methanogenic and sulfate-reducing sludge exhibits a higher capability for reducing nitro group of ROX into amino group and cleaving C–As bond compared to denitrifying sludge (Cortinas et al. 2006). Wherein, the reduction rate of nitro group by methanogenic and sulfate-reducing sludge is up to ~50 times of that by denitrifying sludge. Furthermore, Han et al. (2017) reported that several pure EEB are more efficient to degrade ROX than mixed cultures, e.g., anaerobic and aerobic sludge. Among the EEB, *Shewanella oneidensis* MR-1 is the most efficient, followed by *S. putrefaciens* CN32, *Shewanella decolorationis* S12, *A. hydrophila* ATCC7966, and *G. sulfurreducens* PCA. The growth of *S. oneidensis* MR-1 can even be stimulated by the addition of ROX (Wang et al. 2020). Compared to

Table 2 Summary of microorganisms involved in organoarsenic biotransformation and the corresponding functional genes

Functions	Genes	Functional proteins	Associated microorganisms	References
-NO_2 reduction	*arsE/ arsF*	ArsE/ArsF: nitroreductase	*S. putrefaciens* 200	Chen et al. (2019)
Acetyltion of HAPA	*nhoA1/ nhoA2*	NhoA1/NhoA2: *n*-hydroxyarylamine *o*-acetyltransferases	*Enterobacter* sp. CZ-1	Huang et al. (2020)
C–As cleavage	*arsI*	ArsI: C–As bond lyase	*Bacillus* sp. MD1, *S. putrefaciens* 200	Pawitwar et al. (2017)
As reduction	*arsC*	ArsC: As(V) reductase	*Aeropyrum pernix*, *Agrobacterium tumefaciens*, *Alicaligenes faecalis*, *Cenibacterium arsenoxidans*, *Nitrobacter hamburgensis*, *Rhodoferax ferrireducens*, *Sulfolobus tokodaii*	Stolz et al. (2006)
	arr	Arr: dissimilatory As(V) reductase	*Alkaliphilus metalliredigenes*, *Bacillus areniciselenatis*, *Bacillus selenitireducens*, *Chrysiogenes arsenatis*, *Clostridium* sp. OhILAs, *Desulfosporosinus* sp. Y5, *Sulfurospirillum barnesii*, *Shewanella* sp. *strain* ANA-3, *Wolinella succinogenes*	Stolz et al. (2006)
As oxidation	*aio/arx*	Aio/Arx: As(III) oxidase	*Rhizobium* sp. str. NT-26, *Alkalilimnicola ehrlichii strain* MLHE-1, *Ectothiorhodospira strain* PHS-1	Warelow et al. (2017)
	arsH	ArsH: organoarsenical oxidase	*Pseudomonas putida* KT2440, *Sinorhizobium meliloti* Rm1021	Chen et al. (2015a)
As uptake	*glpF*	GlpF: aquaglyceroporin for As(III) uptake	*Escherichia coli*	Meng et al. (2004)
	Pit/pst	Pit/Pst: unspecific and specific phosphate transporter for As(V) uptake, respectively	*Agrobacterium tumefaciens* 5A, *Agrobacterium tumefaciens* GW4, *E. coli*, *Sinorhizobium* sp. M14	Kruger et al. (2013), Li et al. (2013)
As extrusion	*acr3*	ArsB or Acr3: As(III) efflux permease	*A. tumefaciens* GW4, *A. tumefaciens* 5A, *Halomonas* sp. HAL1, *Rhizobium* sp. str. NT-26	Shi et al. (2018)

(continued)

Table 2 (continued)

Functions	Genes	Functional proteins	Associated microorganisms	References
	arsB		*Bacillus* sp. UWC, *E. coli*, *Ochrobactrum tritic*, *Shewanella* sp. ANA-3, *Staphylococcus aureus*	
	arsJ	ArsJ: As(V) efflux transporter	*Aeromonas media* WS, *Bermanella marisrubri*, *Halomonas* sp. GFAJ-1, *Marinobacter daepoensis*, *Marinimicrobium agarilyticum*, *Oceanimonas* sp. GK1, *Pesudochrobatrum aeruginosa* DK2, *S. putrefaciens* 200	Shi et al. (2018)
	arsG	ArsG, efflux protein for ASA(III)/HAPA(III)	*S. putrefaciens* 200	Chen et al. (2019)
	arsK	ArsK: efflux protein for As(III), MMA(III) and ROX(III)	*A. tumefaciens* GW4, *A. tumefaciens* 5A, *Ensifer adhaerens*, *Mesorhizobium* sp. YR577, *Pesudochrobatrum* sp. B5, *Rhizobium nepotum*, *Rhizobium* sp. str. NT-26, *Sinorhizobium* sp. NFACC03	Shi et al. (2018)
	arsP	ArsP: methylarsenite efflux permease	*Cupriavidus metallidurans* CH34, *Campylobacter sputorum*, *Campylobacter gracilis*, *Campylobacter jejuni*, *Campylobacter lari NCTC*, *S. oneidensis* MR-1	Chen et al. (2015b), Shi et al. (2018)
As methylation	*arsM*	ArsM: As(III) S-adenosylmethionine methyltransferas	Various fungi, bacteria, methanoarchaea, and eukaryotic microorganisms	Wang et al. (2014b)

Note ROX(III), 4-hydroxy-3-nitrophenylarsenite; HAPA(III), 4-hydroxy-3-aminophenylarsenite; ASA(III), 4-aminophenylarsenite; MMA(III), monomethylarsenite

sludge flocs, anaerobic granular sludge (AGS) has obvious advantages in converting ROX into HAPA due to its good cohesion and high microbial activity (Zhang et al. 2014). Furthermore, AGS is normally coated by a layer of extracellular polymeric substances (EPS) composed of proteins, polysaccharides, etc. The EPS can protect the microbes in the AGS from harsh conditions, potentially endowing AGS with high tolerance to arsenic. On the other hand, with long-term exposure to organic/inorganic arsenic, arsenic resistance of sludge can be enhanced, thereby facilitating organoarsenic biotransformation. For instance, AGS with 270-day exposure to ROX enables arsenic volatilization at 9.06 ng As/g VSS/d, ~8 times of arsenic volatilization rate by AGS without the exposure to any arsenic (Tang et al. 2020b).

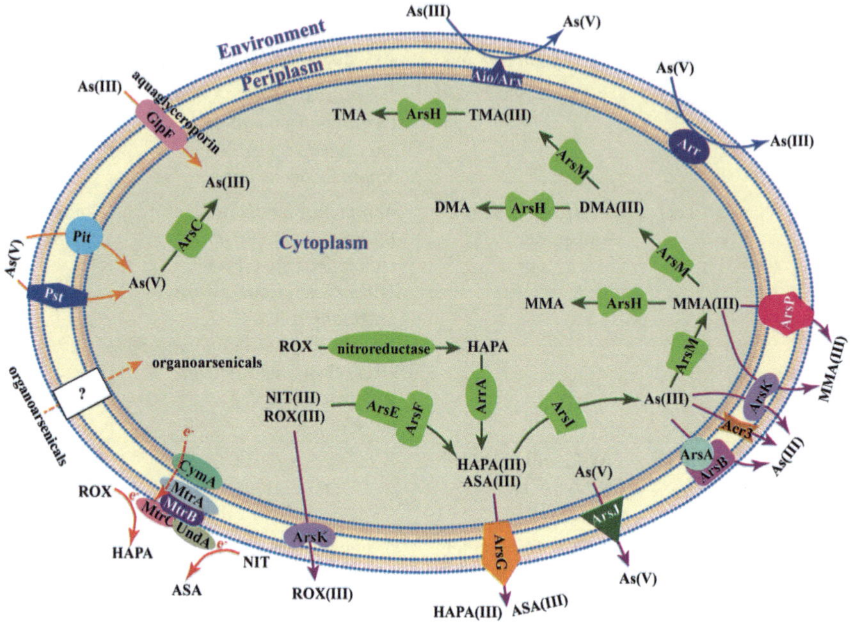

Fig. 6 Mechanism for microbial transformation of organoarsenicals (Tang et al. 2021c). ROX(III), 4-hydroxy-3-nitrophenylarsenite; HAPA(III), 4-hydroxy-3-aminophenylarsenite; ASA(III), 4-aminophenylarsenite; NIT(III), 4-nitrophenylarsenite; MMA, monomethylarsenate; MMA(III), monomethylarsenite; DMA, dimethylarsenate; DMA(III), dimethylarsenite; TMA, trimethylarsenate; TMA(III), trimethylarsenite

Organic substances serve as electron donors or electron shuttles for organoarsenic biotransformation. The microbial reduction of ROX into HAPA can be enhanced by organic substances, and glucose is the most efficient electron donor for ROX reduction, followed by lactate and endogenous substances of inocula (Cortinas et al. 2006). Arsenic volatilization rate is also associated with the type of organic substances, which affects the acclimated microbial communities. For example, amendment of rice straw triggers arsenic volatilization rates at 35–1350 μg/kg/year, i.e., 2–8 times higher than that in the cow dung amended anaerobic system (Huang et al. 2012). However, the presence of organic substances is an obstacle to HAPA biodegradation (Shi et al. 2014; Zhang et al. 2014), and only when organic substances are depleted, HAPA can be utilized as carbon source by anaerobic microbes (Shi et al. 2014). Natural organic matter, e.g., humic substances, and microbially secreted materials, e.g., riboflavin, can work as electron shuttles to accelerate electron transfer from microbe cells to electron acceptors, therefore favoring extracellular reductive reactions (Marsili et al. 2008). The addition of 5 μM anthraquinone-2,6-disulfonate, a humic acid analog, causes the increase of ROX biodegradation rate by ~120% (Han et al. 2017).

Arsenic concentrations and species, temperature, Fe^{3+}, and salt level also affect organoarsenic biotransformation (Chen et al. 2016; Garbarino et al. 2003; Mestrot et al. 2013b). Arsenic concentrations and species play crucial roles in arsenic volatilization. Note that organoarsenicals cannot be directly volatilized before degrading into inorganic arsenic, so arsenic volatilization is less efficient in the organoarsenic-loaded system compared to inorganic or methylated arsenic loaded systems. Mestrot et al. (2013b) reported that the digester spiked with MMA produced the most volatile arsenic (413 ng), which was 2.3 and 8.1 times of that in As(V)-spiked and ROX-spiked digesters, respectively. This indicates that arsenic volatilization is closely dependent on the speciation of arsenic source. Accordingly, sulfate-reducing process that can reduce methylated or inorganic arsenic has demonstrated to be an effective approach to mitigate arsenic volatilization (Tang et al. 2020a). During the compost of poultry litter, ROX degradation is accelerated at 40 °C, but retarded at 15 °C (Garbarino et al. 2003), which could be due to that organoarsenic degradation is biologically catalyzed (Cortinas et al. 2006; Shi et al. 2014) and the microbial activity is highly dependent on temperature (Walker et al. 2018). The existence of Fe^{3+} has been reported to increase biotransformation rate of ROX to HAPA by ~3 times, since *S. oneidensis* MR-1 induces Fe^{3+}/Fe^{2+} redox cycling for accelerating electron transfer (Chen et al. 2016). Salt level has been identified as another crucial factor affecting organoarsenic biotransformation, which can even alter the metabolic pathways of organoarsenicals by *S. putrefaciens* (Chen and Rosen 2016). The demethylation of MMA and the thiolation of MMA and DMA occur only under a high salt condition.

3 Inhibition of Organoarsenicals on Anaerobic Digestion

3.1 Inhibition on Methanogenesis

Organoarsenicals exert adverse influence on methane production. Firstly, the toxicity of organoarsenicals to methanogenic activity is species dependent. Compared to acetotrophic methanogenesis, hydrogenotrophic methanogenesis is less susceptible to organoarsenicals (Sierra-Alvarez et al. 2010). The addition of 1 mM organoarsenicals causes 84% (ROX), 50% (HAPA), 40% (ASA), and 55% (PAA) inhibition on acetotrophic methanogenesis (Sierra-Alvarez et al. 2010), which is mainly induced by arsenate and nitro groups of organoarsenicals rather than phenyl (Donlon et al. 1995; Sierra-Alvarez et al. 2010). Secondly, the inhibitory effect of aromatic organoarsenicals on methanogenesis is concentration dependent. The ROX concentration at 0.25, 0.43, and 0.78 mM caused 20, 50, and 80% inhibition (IC20, IC50, and IC80) on methanogenesis, respectively (Sierra-Alvarez et al. 2004). Yao et al. (2022) found that 0.19 mM ROX caused a 65.7% decline in the cumulative methane production compared to the control without ROX addition (Fig. 7). Similarly, <0.46 mM ASA has no influence on methane production, whereas 0.92 mM ASA completely suppresses

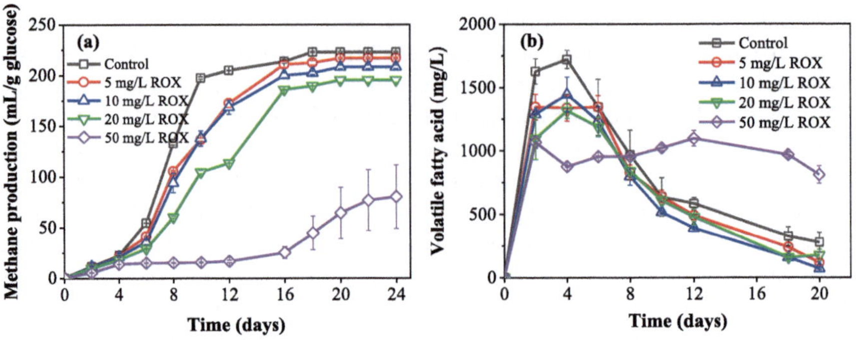

Fig. 7 Effect of ROX on **a** methane production and **b** volatile fatty acids (VFAs) generation (Yao et al. 2022)

methane generation (Wang et al. 2014a). The varying inhibition could probably be associated with the sludge type and the generated inorganic arsenic concentrations.

The inhibition of organoarsenicals toward methanogenesis becomes progressively aggravated with a longer-time exposure of sludge to organoarsenicals (Shui et al. 2016; Sierra-Alvarez et al. 2010). This is probably caused by microbial degradation of organoarsenicals into more toxic arsenic forms, e.g., As(III). Additionally, co-existing contaminants, i.e., Cu^{2+} and sulfadiazine, can dramatically alter effects of organoarsenicals on methanogenic activity. The inhibition of 0.46 mM ASA on methane production decreases from 100% to <40% with the addition of 2 mM Cu^{2+} (He et al. 2018). These detoxification mechanisms might be ascribed to the complexation between Cu^{2+} and ASA, or the obstruction on biotransformation of ASA to As(III) by Cu^{2+}. While the coexistence of 0.23 mM ROX and 0.4 mM sulfadiazine exerts a synergetic effect on methanogenesis, completely inhibiting methane production (Tang et al. 2019).

3.2 Impact on Characteristics of Anaerobic Sludge

Arsenic accumulation in the sludge and further structural damage of AGS have been extensively observed (Ji et al. 2017; Shui et al. 2016; Tang et al. 2017; Zhang et al. 2014). For instance, after 240-day batch anaerobic digestion, arsenic concentration in the AGS reaches >700 mg/kg when 0.08 mM ROX is fed, and 2600 mg/kg when 1.33 mM ROX is fed (Zhang et al. 2014). While in a continuously operated anaerobic bioreactor, arsenic is progressively accumulated in the sludge and the arsenic concentration can even reach >10,000 mg/kg within 260 days (Tang et al. 2017). These accumulated arsenics often exist in different forms including adsorbed, cellular bonded, and precipitated arsenic (Ji et al. 2017). Among them, adsorbed arsenic is the most exchangeable, followed by cellular bonded arsenic and precipitated arsenic

(He et al. 2016). The speciation of arsenic determines the toxicity and bioavailability of arsenic.

Granular size has shown to be a crucial physical parameter that impacts the performance of anaerobic reactors (Wu et al. 2016). A 23-day exposure of AGS to 19 μM ROX has no obvious impact on AGS size (Fig. 8) (Tang et al. 2021a). However, further increasing the exposure time causes a slow decrease in the AGS size from >1.8 mm to 1.26 mm, equivalent to 64.0% of the size found in the reactor without ROX addition (R0) at the corresponding time. After day 352, the granular sizes in the reactors without ROX addition (R0) and with 19 μM ROX addition (R1) become once more stable and remain at a size of ~2.0 and ~1.2 mm, respectively. The 34.1% reduction in granular size in R1 that occurs over the duration of the ROX amendment is very likely connected with the production of ROX metabolites. Small particle sizes enable the AGS to reduce the mass transfer efficiency, thereby reducing the toxic loading imposed by ROX and its metabolites on the AGS. Also, Sierra-Alvarez et al. (2004) reported that fine anaerobic sludge exhibited a higher capability to sequestrate As(III) due to a higher surface area compared to larger-sized AGS (Sierra-Alvarez et al. 2004).

Long-term exposure to organoarsenic-contaminated wastewater confers an obvious damage on the external and internal structure of the AGS. Compared to the AGS without ROX, the AGS with ROX secretes more EPS. For example, AGS is coated with a yellow-green film after 260-day exposure to ROX (Fig. 9) (Shui et al. 2016). Tang et al. (2021a) observed similar phenomena that a gray and rough EPS layer is formed on the surface of AGS after 420-day exposure to ROX.

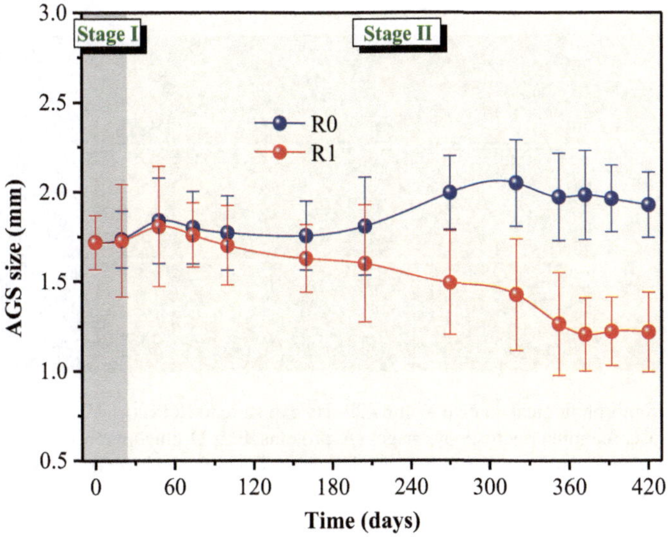

Fig. 8 Evolution of AGS diameter in both reactors (R0: without ROX; R1: with 19 μM ROX) (Tang et al. 2021a)

Fig. 9 Images of AGS surface after exposure to ROX (Shui et al. 2016)

Confocal laser scanning microscopy further confirms that both protein and carbohydrate excretion is enhanced with an increased protein to carbohydrate ratio (Fig. 10). Furthermore, surface spalling of the AGS is observed due to the death of surface microbes caused by ROX-associated toxicity. This behavior is also a self-protective pathway since arsenic might be sequestered by surface EPS. This is consistent with the hypothesis that the stress of toxic arsenical compounds stimulates EPS secretion to protect the microbes themselves (Sheng et al. 2010). Even with the production of dense EPS, ROX and its arsenical metabolites can still get access to the interior of AGS, and exert serious inhibition on the growth of methanogens which inhabit in the center of the AGS (Zhang et al. 2014). This therefore causes the occurrence of cavities in the interior. In addition, ROX loading significantly alters the morphotypes of microbes in the AGS. The thin filaments and rods are dominant on the surface of AGS without ROX, while the cocci and short rods are dominating on that with ROX (Shui et al. 2016; Zhang et al. 2014).

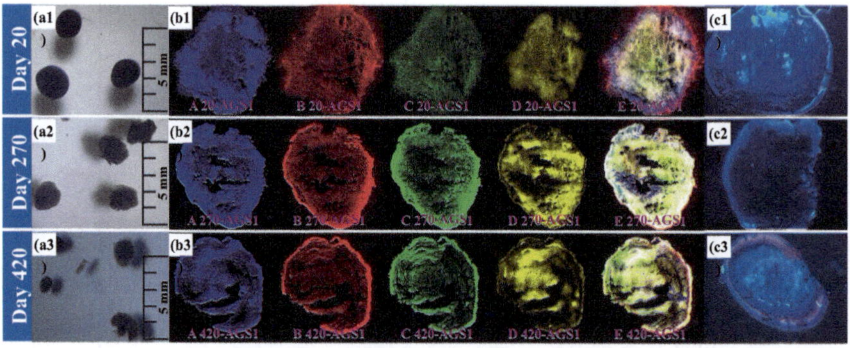

Fig. 10 AGS morphological change within 420-day exposure to ROX: (a1–3) digital photos, (b1–3) confocal laser scanning microscopy images (A: proteins; B: α-D-glucopyranose polysaccharides; C: β-D-glucopyranose polysaccharides; D: nucleic acids; E: combined layers), and (c1–3) live/dead staining images

3.3 *Impact on Microbial Community Structure*

The tolerance to ROX and its metabolites differs between various microbes within anaerobic sludge (Sierra-Alvarez et al. 2004). The 420-day exposure to 19 μM ROX-amended conditions reshapes the microbial community structure (Tang et al. 2021a), where the archaeal operational taxonomic units (OTUs) decrease from 1260 to 906 with ROX addition while remaining at a stable level of ~1200 without ROX addition (Table 3). Furthermore, the decrease in the Shannon Index and the increase in the Simpson Index for archaea under ROX-added conditions are observed, suggesting a reduction in archaea diversity. All these confirm the susceptibility of archaea to arsenical compounds.

As for bacteria (Fig. 11), *Levilinea* (21.7%) and *Sulfurovum* (21.5%), both typical acetogenic bacteria (Amin et al. 2021), are the dominant bacteria in the raw AGS. However, after 270-day exposure to ROX, the bacterial community composition is greatly altered. *Olsenella* (11.0%), *Megasphaera* (8.0%), *Clostridium* sensu stricto (8.0%), *Brooklawnia* (6.0%), and *Alloprevotella* (5.6%) become dominant on day 270. *Clostridium* species are previously demonstrated to be the main contributors for degrading ROX (Stolz et al. 2007) and active in fermentation (Jiang et al. 2011). This might reflect the increase in ROX-degrading activity.

Table 3 Microbial operational taxonomic units (OTUs)

AGS	OTUs		Shannon index		Simpson index	
	Bacteria	Archaea	Bacteria	Archaea	Bacteria	Archaea
Raw-AGS	2821	1260	4.31	1.96	0.08	0.30
270-AGS0	2779	1479	4.77	2.37	0.03	0.21
270-AGS1	2472	1146	4.22	2.33	0.05	0.17
420-AGS0	2829	1203	5.15	2.17	0.01	0.23
420-AGS1	2830	906	4.96	1.30	0.02	0.50

Note 270-AGS0, 270-AGS1, 420-AGS0 and 420-AGS1 stand for the AGS collected on day 270 and 420 in the reactors with ROX and without ROX, respectively

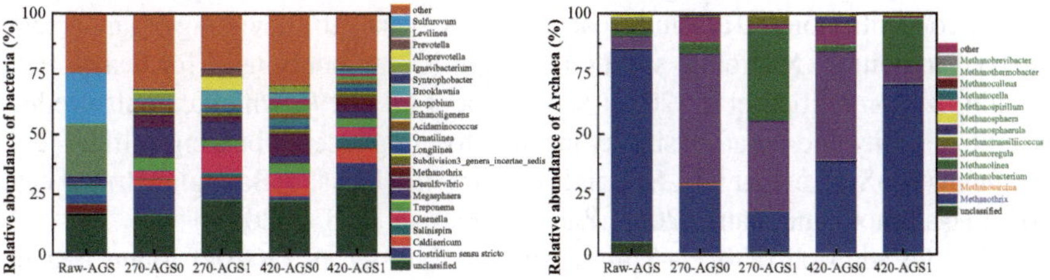

Fig. 11 Microbial community on the genus level (Tang et al. 2021a)

As for archaea, *Methanothrix* is found to be the primary methanogen in the raw AGS, i.e., a typical acetoclastic methanogen. After the 270-day operation, *Methanothrix* abundance decreases from 80.3% to 16.9% in the ROX-added reactor, much lower than that (27.9%) in the ROX-free reactor; hydrogenotrophic methanogens, i.e., *Methanobacterium* (37.5%), *Methanolinea* (37.8%), *Methanomassiliicoccus* (4.0%), and *Methanoregula* (2.0%), dominate in the ROX-added reactor, with a total relative abundance of ~80%. However, after the 420-day operation, the relative abundance of acetoclastic methanogen, *Methanothrix*, has ultimately recovered to 70.0%, whereas hydrogenotrophic methanogen abundance decreases to 28.6%. Generally, acetoclastic methanogens are more susceptible to ROX and its metabolites than hydrogenotrophic methanogens (Sierra-Alvarez et al. 2010). Yao et al. (2022) also reported that ROX exerted obvious inhibition on both acetoclastic and hydrogenotrophic methanogens, and in the 19 μM ROX-loaded environment, the hydrogenotrophic methanogen *Methanobacterium* became dominant with a relative abundance of up to 83%. The results obtained within 270 days and on day 420 show an inconsistent trend, and the one on day 420 is contradictory with the previous reports (Sierra-Alvarez et al. 2010). The different results indicate that the inhibition of ROX and its metabolites on *Methanothrix* is restorable, even under ROX-loaded conditions. One possible reason may be that the progressive restoration of VFA-producing bacteria, e.g., *Syntrophobacter*, allows a competitive advantage of *Methanothrix* over hydrogenotrophic methanogens due to abundant substrate supply. The mechanisms underlying this deserve further investigation.

4 Strategy to Alleviate the Inhibition Associated with Organoarsenicals

4.1 Alleviation Driven by Sulfate Reduction

The degradation products of organoarsenicals, especially As(III) and As(V), are more toxic than organoarsenicals themselves, and the inhibition on anaerobic digestion is mainly induced by inorganic arsenics. Therefore, reducing aqueous inorganic arsenic concentration and arsenic bioaccessibility are key to alleviating organoarsenic induced inhibition. Microbial sulfate reduction is commonly used for heavy metal immobilization (Cibati et al. 2013; Sahinkaya et al. 2019), which can also reduce arsenic mobility and bioaccessibility through forming arsenic-bearing sulfides, e.g., orpiment (As_2S_3), realgar (As_4S_4), and arsenopyrite (FeAsS) (Battaglia-Brunet et al. 2012; Hashimoto and Kanke 2018; Zacarias-Estrada et al. 2020).

With the addition of 0.72–5 mg/L sulfate, Tang et al. (2020a) found that the nearly complete inhibition of acetoclastic methanogenic activity by ROX was slowly restored to the initial level that had not been affected by ROX, i.e., 210.4 mL CH_4/g VSS/d (Fig. 12). This is mainly because the sulfate addition causes a 66.7% decrease in soluble inorganic arsenic and significant arsenic precipitation in the sludge. Based

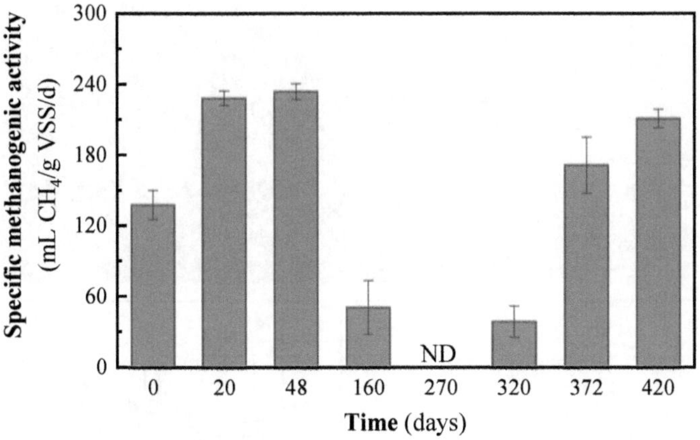

Fig. 12 Acetotrophic methanogenic activity changes (Tang et al. 2020a)

on the results of field-emission scanning electron microscope-energy-dispersive X-ray spectroscopy (FESEM-EDS) and X-ray photoelectron spectroscopy (XPS analysis, sulfate addition induces arsenic precipitation in the AGS through the formation of orpiment (Fig. 13). This demonstrates that microbial sulfate reduction plays an important role in arsenic immobilization.

Furthermore, Tang et al. (2021b) proposed a sulfate-mediated bioelectrochemical system where ROX and its derived organoarsenicals were completely degraded within 13 days under the combination of 0.5 V voltage and 350 μM sulfate. The reason is that organoarsenicals are used as the electron donor for sulfate reduction, and electrochemical stimulation accelerates this process. In addition, up to 83% of total arsenic is immobilized in the sludge due to the formation of arsenic-bearing sulfides, i.e., crystalline alacranite, in the cathode chamber (Fig. 14).

4.2 Alleviation Mediated by Zero-Valence Iron

Zero-valence iron (ZVI) is reported to favor the enrichment of hydrogenotrophic methanogens (Romero-Gueiza et al. 2016) that have higher tolerance to ROX than acetoclastic methanogens (Sierra-Alvarez et al. 2010; Tang et al. 2021a). This provides an opportunity for ZVI to alleviate the methanogenesis inhibition by ROX through enhancing hydrogenotrophic methanogenesis. More importantly, in the micro-ZVI-enhanced anaerobic digesters, Fe^{2+} is often released due to microbial corrosion (Feng et al. 2014). Fe^{2+} together with its oxidation product, Fe^{3+}, exhibits a priority and high potential to co-precipitate with inorganic arsenic in the forms of scorodite ($FeAsO_4$) (Li et al. 2021) and arsenopyrite (FeAsS) (Zhao et al. 2021). Fe-based materials including nano-ZVI and other Fe-modified materials also have an excellent ability to adsorb inorganic arsenic (Pan et al. 2021; Wang et al. 2017).

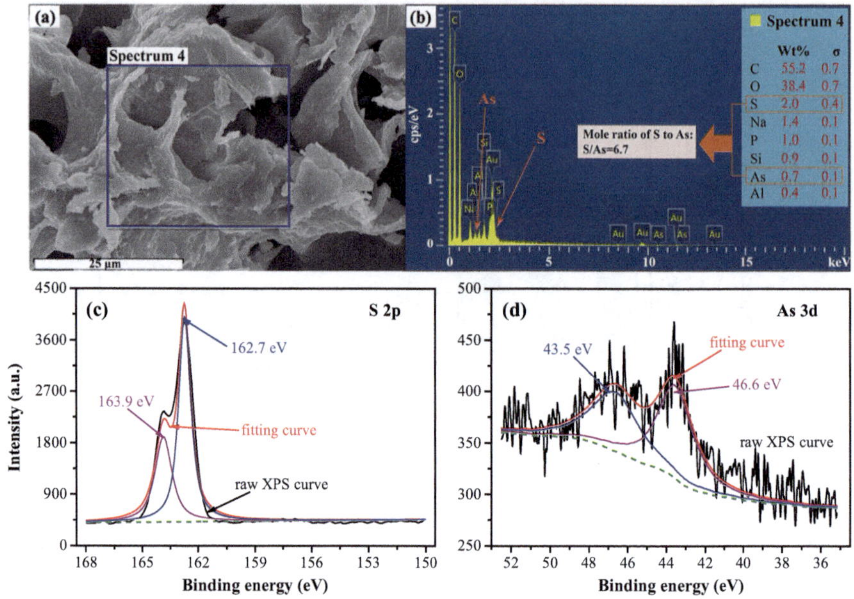

Fig. 13 Characterization of precipitated arsenic species in anaerobic granular sludge: **a**, **b** FESEM-EDS analysis, and the XPS spectra of **c** S 2p and **d** As 3d of the AGS residue after sequential extraction (Tang et al. 2020a)

The ZVI-mediated arsenic immobilization via the coprecipitation or adsorption is a possible way to reduce arsenic bioaccessibility and mobility, thereby alleviating the inhibition on microbial activity.

As reported by Yao et al. (2022), when the ZVI concentration increased from 1.0 to 10.0 g/L, the inhibition of ROX on methanogenesis was decreased from 65.7% to 12.7% and 28.1%, respectively (Fig. 15). The ZVI-mediated alleviation mechanisms are as follows: (i) additional CH_4 production through microbial reduction of CO_2 with Fe as direct or indirect elector donors (Eqs. 1 and 2) (Wei et al. 2018), (ii) 4.3–8.0 times enhancement on hydrogenotrophic methanogenic activity, and (iii) immobilization of up to > 95% arsenic in the form of $FeAsO_4$. However, the regulation of the ZVI-mediated alleviation technique, its techno-economic evaluation and feasibility for practical application require further investigation.

$$4Fe + 8H^+ \rightarrow 4Fe^{2+} + 4H_2 \tag{1}$$

$$4H_2 + CO_2 \rightarrow CH_4 + 2H_2O \tag{2}$$

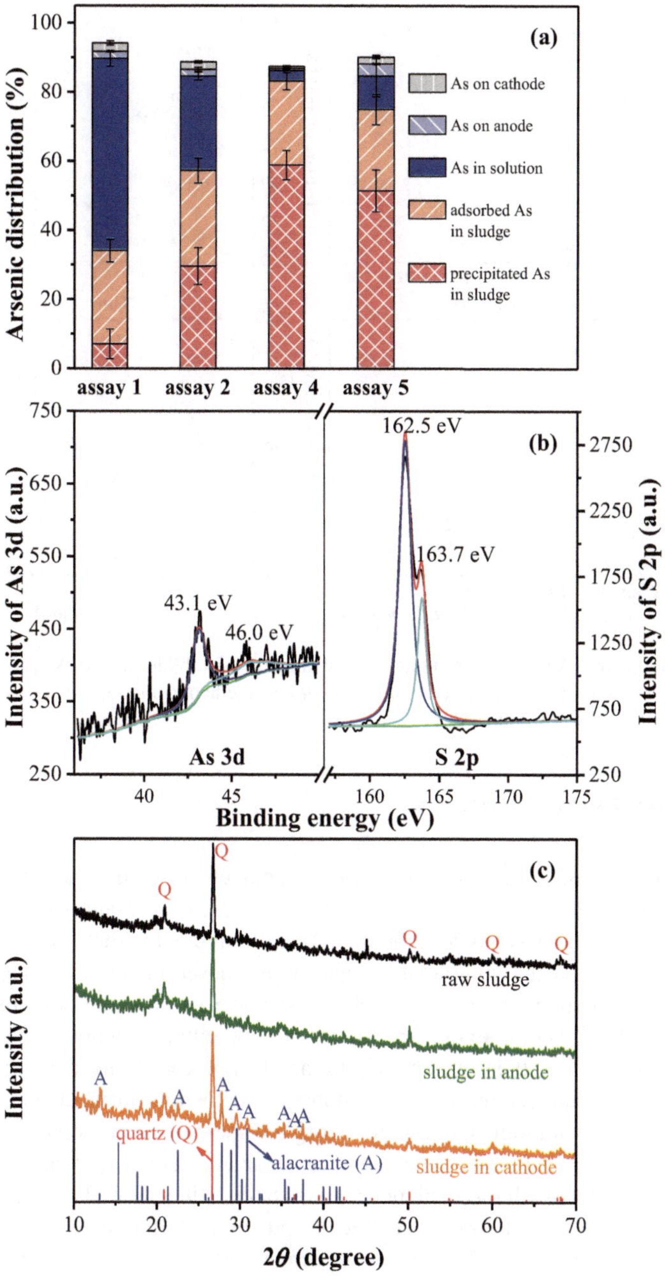

Fig. 14 **a** Arsenic distribution in assay 1 (without sulfate and voltage application), assay 2 (350 μM sulfate, without voltage application), assay 4 (0.5 V voltage, without glucose) and assay 5 (500 mg/L glucose + 0.5 V voltage) after 32 days, **b** XPS spectrum, and **c** X-ray diffraction pattern for precipitated arsenic in the sludge (Tang et al. 2021b)

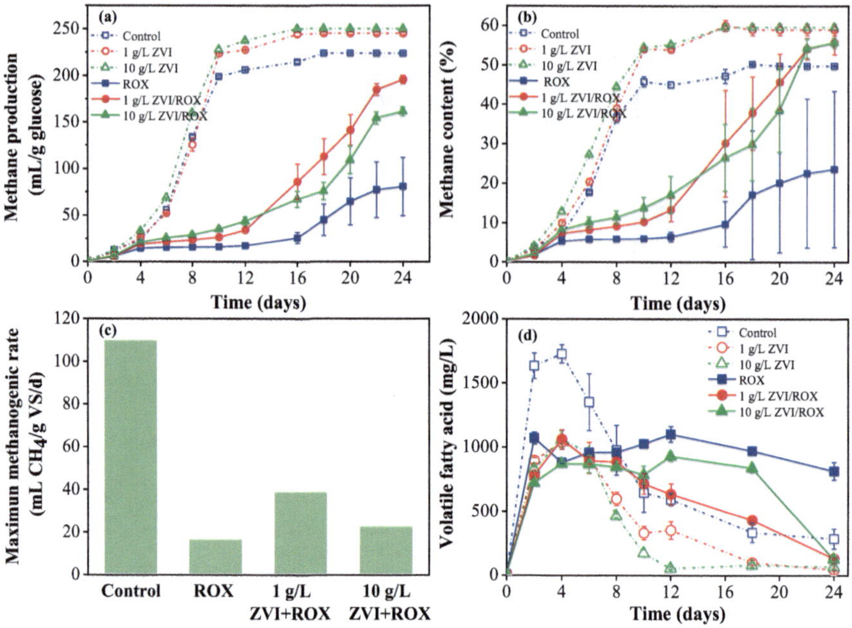

Fig. 15 Effects of ZVI on ROX-induced methanogenesis inhibition and VFAs generation: **a** methane production; **b** methane content; **c** maximum methanogenic rate; **d** VFAs generation ([ROX] = 50 mg/L) (Yao et al. 2022)

5 Future Perspectives

To eliminate/minimize the interference of organoarsenicals with anaerobic digestion and environmental risk associated with arsenic, there exist many issues that need to be figured out. As(III) has been demonstrated to interact with organic substances such as proteins via the complexation and hydrophobic interactions with –OH, –NH, and –C=O (Ma et al. 2021). Animal wastewater and manure contain complex matrices potentially complexing or co-precipitating with organoarsenicals and their arsenical metabolites, thereby detoxifying arsenicals. Clarifying their interactions can provide better understanding on organoarsenic behaviors during anaerobic digestion and even other biological treatment of organoarsenic-bearing waste (water). The anaerobic degradation pathways have been extensively studied; nevertheless, little is known about molecular biological mechanisms behind their degradation, particularly the microbial resistance to organoarsenicals and some of their arsenical derivatives. Clarifying this issue will help us better understand the mechanisms that govern the conversion of pentavalent aromatic arsenicals to trivalent forms, as well as the uptake and extrusion of pentavalent aromatic organoarsenicals and the majority of methylated arsenicals. The elimination or immobilization of organoarsenicals is critical to reduce their potential influence on anaerobic digesters and the natural environment.

Sulfur- or Fe-based approaches have been developed and demonstrated high efficiency at a laboratory scale (Tang et al. 2021b; Yao et al. 2022). The larger-scale application must be tested with real wastewater to ensure practicality. In addition, their techno-economic evaluation is also required. This is the key to fixing the problem encountered in the real application.

6 Take-Home Messages

(1) Anaerobic biotransformation of organoarsenicals is initiated with the reduction of electron-withdrawing groups excluding arsenate group, followed by two different pathways. One is the direct cleavage of C–As bond, and the other is the reduction of arsenate groups before C–As bond cleavage.

(2) Degradation of organoarsenical with electron-donating groups is the rate-limiting step, and electrochemical stimulation is an effective alternative to facilitate the degradation. Microbial compositions and property, co-existing organic matter, and other environmental factors can significantly influence the organoarsenical transformation.

(3) Organoarsenicals exert serious inhibition on methanogenic activity, especially on acetoclastic methanogenic activity because of arsenate and nitro groups. The inhibition induced by environment-relevant organoarsenical levels is restorable. In response to organoarsenical stress, accelerating the excretion of EPS is commonly observed as a protective mechanism for anaerobic sludge.

(4) ZVI mediation and sulfate reduction are the two efficient ways to alleviate the negative effect of organoarsenicals on anaerobic digestion due to the immobilization of arsenic, and the change of metabolic pathways.

References

Adak A, Mangalgiri KP, Lee J, Blaney L (2015) UV irradiation and UV-H_2O_2 advanced oxidation of the roxarsone and nitarsone organoarsenicals. Water Res 70:74–85

Amin FR, Khalid H, El-Mashad HM, Chen C, Liu G, Zhang R (2021) Functions of bacteria and archaea participating in the bioconversion of organic waste for methane production. Sci Total Environ 763:143007

Battaglia-Brunet F, Crouzet C, Burnol A, Coulon S, Morin D, Joulian C (2012) Precipitation of arsenic sulphide from acidic water in a fixed-film bioreactor. Water Res 46(12):3923–3933

Chen WR, Huang CH (2012) Surface adsorption of organoarsenic roxarsone and arsanilic acid on iron and aluminum oxides. J Hazard Mater 227:378–385

Chen J, Rosen BP (2016) Organoarsenical biotransformations by *S. putrefaciens*. Environ Sci Technol 50(15):7956–7963

Chen J, Bhattacharjee H, Rosen BP (2015a) ArsH is an organoarsenical oxidase that confers resistance to trivalent forms of the herbicide monosodium methylarsenate and the poultry growth promoter roxarsone. Mol Microbiol 96(5):1042–1052

Chen J, Madegowda M, Bhattacharjee H, Rosen BP (2015b) ArsP: a methylarsenite efflux permease. Mol Microbiol 98(4):625–635

Chen G, Ke Z, Liang T, Liu L, Wang G (2016) *Shewanella oneidensis* MR-1-induced Fe (III) reduction facilitates roxarsone transformation. PLoS ONE 11(4):e0154017

Chen G, Xu R, Liu L, Shi H, Wang G, Wang G (2018) Limited carbon source retards inorganic arsenic release during roxarsone degradation in *Shewanella oneidensis* microbial fuel cells. Appl Microbiol Biotechnol 102:8093–8106

Chen J, Zhang J, Rosen BP (2019) Role of ArsEFG in roxarsone and nitarsone detoxification and resistance. Environ Sci Technol 53(11):6182–6191

Cheng HH, Narindri B, Chu H, Whang LM (2020) Recent advancement on biological technologies and strategies for resource recovery from swine wastewater. Bioresour Technol 303:122861

Cibati A, Cheng KY, Morris C, Ginige MP, Sahinkaya E, Pagnanelli F, Kaksonen AH (2013) Selective precipitation of metals from synthetic spent refinery catalyst leach liquor with biogenic H_2S produced in a lactate-fed anaerobic baffled reactor. Hydrometallurgy 139:154–161

Cortinas I, Field JA, Kopplin M, Garbarino JR, Gandolfi AJ, Sierra-Alvarez R (2006) Anaerobic biotransformation of roxarsone and related N-substituted phenylarsonic acids. Environ Sci Technol 40(9):2951–2957

Donlon BA, Razo-Flores E, Field JA, Lettinga G (1995) Toxicity of N-substituted aromatics to acetoclastic methanogenic activity in granular sludge. Appl Environ Microbiol 61(11):3889–3893

Fei J, Wang T, Zhou Y, Wang Z, Min X, Ke Y, Hu W, Chai L (2018) Aromatic organoarsenic compounds (AOCs) occurrence and remediation methods. Chemosphere 207:665–675

Feng Y, Zhang Y, Quan X, Chen S (2014) Enhanced anaerobic digestion of waste activated sludge digestion by the addition of zero valent iron. Water Res 52:242–250

Fisher DJ, Yonkos LT, Staver KW (2015) Environmental concerns of roxarsone in broiler poultry feed and litter in Maryland, USA. Environ Sci Technol 49(4):1999–2012

Fu QL, Blaney L, Zhou DM (2017) Natural degradation of roxarsone in contrasting soils: degradation kinetics and transformation products. Sci Total Environ 607:132–140

Garbarino JR, Bednar AJ, Rutherford DW, Beyer RS, Wershaw RL (2003) Environmental fate of roxarsone in poultry litter. I. Degradation of roxarsone during composting. Environ Sci Technol 37(8):1509–1514

Geng A, Wang X, Wu L, Wang F, Chen Y, Yang H, Zhang Z, Zhao X (2017) Arsenic accumulation and speciation in rice grown in arsanilic acid-elevated paddy soil. Ecotoxicol Environ Saf 137:172–178

Gupta SK, Le XC, Kachanosky G, Zuidhof MJ, Siddique T (2018) Transfer of arsenic from poultry feed to poultry litter: a mass balance study. Sci Total Environ 630:302–307

Guzmán-Fierro VG, Moraga R, León CG, Campos VL, Smith C, Mondaca MA (2015) Isolation and characterization of an aerobic bacterial consortium able to degrade roxarsone. Int J Environ Sci Technol 12:1353–1362

Han JC, Zhang F, Cheng L, Mu Y, Liu DF, Li WW, Yu HQ (2017) Rapid release of arsenite from roxarsone bioreduction by exoelectrogenic bacteria. Environ Sci Technol Lett 4(8):350–355

Hashimoto Y, Kanke Y (2018) Redox changes in speciation and solubility of arsenic in paddy soils as affected by sulfur concentrations. Environ Pollut 238:617–623

Hayakawa T, Kobayashi Y, Cui X, Hirano S (2005) A new metabolic pathway of arsenite: arsenic–glutathione complexes are substrates for human arsenic methyltransferase Cyt19. Arch Toxicol 79:183–191

He X, Zhang Y, Shen M, Zeng G, Zhou M, Li M (2016) Effect of vermicomposting on concentration and speciation of heavy metals in sewage sludge with additive materials. Bioresour Technol 218:867–873

He P, Wu G, Tang R, Ji P, Yuan S, Wang W, Hu Z (2018) Influence of arsanilic acid, Cu^{2+}, PO_4^{3-} and their interaction on anaerobic digestion of pig manure. Front Environ Sci Eng 12:9

Hu Y, Cheng H, Tao S, Schnoor JL (2019) China's ban on phenylarsonic feed additives, a major step toward reducing the human and ecosystem health risk from arsenic. Environ Sci Technol 53(21):12177–12187

Huang H, Jia Y, Sun GX, Zhu YG (2012) Arsenic speciation and volatilization from flooded paddy soils amended with different organic matters. Environ Sci Technol 46(4):2163–2168

Huang L, Yao L, He Z, Zhou C, Li G, Yang B, Deng X (2014) Roxarsone and its metabolites in chicken manure significantly enhance the uptake of As species by vegetables. Chemosphere 100:57–62

Huang K, Gao F, Le XC, Zhao FJ (2020) N-hydroxyarylamine o-acetyltransferases catalyze acetylation of 3-amino-4-hydroxyphenylarsonic acid in the 4-hydroxy-3-nitrobenzenearsonic acid transformation pathway of *Enterobacter* sp. strain CZ-1. Appl Environ Microbiol 86(2):e02050-19

Ji P, Tang R, He P, Yuan S, Wang W, Hu ZH (2017) Characterization of arsenic species in the anaerobic granular sludge treating roxarsone-contaminated wastewater. Chem Eng J 327:162–168

Jiang L, Wang J, Liang S, Cai J, Xu Z, Cen P, Yang S, Li S (2011) Enhanced butyric acid tolerance and bioproduction by *Clostridium tyrobutyricum* immobilized in a fibrous bed bioreactor. Biotechnol Bioeng 108(1):31–40

Kruger MC, Bertin PN, Heipieper HJ, Arsène-Ploetze F (2013) Bacterial metabolism of environmental arsenic—mechanisms and biotechnological applications. Appl Microbiol Biotechnol 97:3827–3841

Li H, Li M, Huang Y, Rensing C, Wang G (2013) In silico analysis of bacterial arsenic islands reveals remarkable synteny and functional relatedness between arsenate and phosphate. Front Microbiol 4:347

Li Y, Qi X, Li G, Wang H (2021) Double-pathway arsenic removal and immobilization from high arsenic-bearing wastewater by using nature pyrite as in situ Fe and S donator. Chem Eng J 410:128303

Liu H, Wang G, Ge J, Liu L, Chen G (2014) Fate of roxarsone during biological nitrogen removal process in wastewater treatment systems. Chem Eng J 255:500–505

Ma WJ, Zhang JT, Wang Y, Li GF, Wu XX, Yao YX, Cheng YF, Huang BC, Jin RC (2021) Extracellular polymeric substances excreted by anammox sludge act as a barrier for As(III) invasion: binding property and interaction mechanism. Chemosphere 278:130414

Mafla S, Moraga R, Leon CG, Guzman-Fierro VG, Yanez J, Smith CT, Mondaca MA, Campos VL (2015) Biodegradation of roxarsone by a bacterial community of underground water and its toxic impact. World J Microbiol Biotechnol 31:1267–1277

Mangalgiri KP, Adak A, Blaney L (2015) Organoarsenicals in poultry litter: detection, fate, and toxicity. Environ Int 75:68–80

Marsili E, Baron DB, Shikhare ID, Coursolle D, Gralnick JA, Bond DR (2008) *Shewanella* secretes flavins that mediate extracellular electron transfer. Proceed Natl Acad Sci 105(10):3968–3973

Meng YL, Liu Z, Rosen BP (2004) As (III) and Sb (III) uptake by GlpF and efflux by ArsB in *E. coli*. J Biol Chem 279(18):18334–18341

Meng J, Xu F, Yuan S, Mu Y, Wang W, Hu ZH (2019) Photocatalytic oxidation of roxarsone using riboflavin-derivative as a photosensitizer. Chem Eng J 355:130–136

Mestrot A, Planer-Friedrich B, Feldmann J (2013a) Biovolatilisation: a poorly studied pathway of the arsenic biogeochemical cycle. Environ Sci Processes Impacts 15(9):1639–1651

Mestrot A, Xie WY, Xue X, Zhu YG (2013b) Arsenic volatilization in model anaerobic biogas digesters. Appl Geochem 33:294–297

Pan H, Yang X, Chen H, Sarkar B, Bolan N, Shaheen SM, Wu F, Che L, Ma Y, Rinklebe J, Wang H (2021) Pristine and iron-engineered animal-and plant-derived biochars enhanced bacterial abundance and immobilized arsenic and lead in a contaminated soil. Sci Total Environ 763:144218

Pawitwar SS, Nadar VS, Kandegedara A, Stemmler TL, Rosen BP, Yoshinaga M (2017) Biochemical characterization of ArsI: a novel C–As Lyase for degradation of environmental organoarsenicals. Environ Sci Technol 51(19):11115–11125

Pressman D, Brown DH (1943) Ionization constants of several substituted phenylarsonic acids. J Am Chem Soc 65(4):540–543

Qiang Z, Adams C (2004) Potentiometric determination of acid dissociation constants (pKa) for human and veterinary antibiotics. Water Res 38(12):2874–2890

Qin J, Rosen BP, Zhang Y, Wang G, Franke S, Rensing C (2006) Arsenic detoxification and evolution of trimethylarsine gas by a microbial arsenite S-adenosylmethionine methyltransferase. Proceed Natl Acad Sci 103(7):2075–2080

Rodriguez-Freire L, Sierra-Alvarez R, Root R, Chorover J, Field JA (2014) Biomineralization of arsenate to arsenic sulfides is greatly enhanced at mildly acidic conditions. Water Res 66:242–253

Romero-Güiza MS, Vila J, Mata-Alvarez J, Chimenos JM, Astals S (2016) The role of additives on anaerobic digestion: a review. Renew Sustain Energy Rev 58:1486–1499

Sahinkaya E, Isler E, Yurtsever A, Coban I (2019) Sulfidogenic treatment of acid mine drainage using anaerobic membrane bioreactor. J Water Process Eng 31:100816

Sheng GP, Yu HQ, Li XY (2010) Extracellular polymeric substances (EPS) of microbial aggregates in biological wastewater treatment systems: a review. Biotechnol Adv 28(6):882–894

Shi L, Wang W, Yuan SJ, Hu ZH (2014) Electrochemical stimulation of microbial roxarsone degradation under anaerobic conditions. Environ Sci Technol 48(14):7951–7958

Shi K, Li C, Rensing C, Dai X, Fan X, Wang G (2018) Efflux transporter ArsK is responsible for bacterial resistance to arsenite, antimonite, trivalent roxarsone, and methylarsenite. Appl Environ Microbiol 84(24):e01842-e1918

Shui M, Ji F, Tang R, Yuan S, Zhan X, Wang W, Hu Z (2016) Impact of roxarsone on the UASB reactor performance and its degradation. Front Environ Sci Eng 10(6):149–157

Sierra-Alvarez R, Cortinas I, Yenal U, Field JA (2004) Methanogenic inhibition by arsenic compounds. Appl Environ Microbiol 70(9):5688–5691

Sierra-Alvarez R, Cortinas I, Field JA (2010) Methanogenic inhibition by roxarsone (4-hydroxy-3-nitrophenylarsonic acid) and related aromatic arsenic compounds. J Hazard Mater 175(1–3):352–358

Silbergeld EK, Nachman K (2008) The environmental and public health risks associated with arsenical use in animal feeds. Ann N Y Acad Sci 1140(1):346–357

Stolz JF, Basu P, Santini JM, Oremland RS (2006) Arsenic and selenium in microbial metabolism. Ann Rev Microbiol 60:107–130

Stolz JF, Perera E, Kilonzo B, Kail B, Crable B, Fisher E, Ranganathan M, Wormer L, Basu P (2007) Biotransformation of 3-nitro-4-hydroxybenzene arsonic acid (roxarsone) and release of inorganic arsenic by *Clostridium species*. Environ Sci Technol 41(3):818–823

Tang R, Chen H, Yuan S, Zhan X, Wang W, Hu ZH (2017) Arsenic accumulation and volatilization in a 260-day cultured upflow anaerobic sludge blanket (UASB) reactor. Chem Eng J 311:277–283

Tang R, Yuan S, Chen F, Zhan X, Wang W, Hu Z (2019) Effects of roxarsone and sulfadiazine on biogas production and their degradation during anaerobic digestion. Int Biodeter Biodegrad 140:113–118

Tang R, Wu G, Yue Z, Wang W, Zhan X, Hu ZH (2020a) Anaerobic biotransformation of roxarsone regulated by sulfate: degradation, arsenic accumulation and volatilization. Environ Pollut 267:115602

Tang R, Yuan S, Wang Y, Wang W, Wu G, Zhan X, Hu Z (2020b) Arsenic volatilization in roxarsone-loaded digester: Insight into the main factors and arsM genes. Sci Total Environ 711:135123

Tang R, Luo H, Prommer H, Yue Z, Wang W, Su K, Hu ZH (2021a) Response of anaerobic granular sludge to long-term loading of roxarsone: from macro-to micro-scale perspective. Water Res 204:117599

Tang R, Prommer H, Yuan S, Wang W, Sun J, Jamieson J, Hu ZH (2021b) Enhancing roxarsone degradation and in situ arsenic immobilization using a sulfate-mediated bioelectrochemical system. Environ Sci Technol 55(1):393–401

Tang R, Wang Y, Yuan S, Wang W, Yue Z, Zhan X, Hu ZH (2021c) Organoarsenic feed additives in biological wastewater treatment processes: removal, biotransformation, and associated impacts. J Hazard Mater 406:124789

Thomas JA, Chovanec P, Stolz JF, Basu P (2014) Mapping the protein profile involved in the biotransformation of organoarsenicals using an arsenic metabolizing bacterium. Metallomics 6(10):1958–1969

Tisa LS, Rosen BP (1990) Molecular characterization of an anion pump: the ArsB protein is the membrane anchor for the ArsA protein. J Biol Chem 265(1):190–194

Torres CI, Marcus AK, Lee HS, Parameswaran P, Krajmalnik-Brown R, Rittmann BE (2010) A kinetic perspective on extracellular electron transfer by anode-respiring bacteria. FEMS Microbiol Rev 34(1):3–17

Walker TW, Kaiser C, Strasser F, Herbold CW, Leblans NI, Woebken D, Janssens IA, Sigurdsson BD, Richter A (2018) Microbial temperature sensitivity and biomass change explain soil carbon loss with warming. Nat Clim Change 8(10):885–889

Wang HL, Hu ZH, Tong ZL, Xu Q, Wang W, Yuan S (2014a) Effect of arsanilic acid on anaerobic methanogenic process: kinetics, inhibition and biotransformation analysis. Biochem Eng J 91:179–185

Wang P, Sun G, Jia Y, Meharg AA, Zhu Y (2014b) A review on completing arsenic biogeochemical cycle: microbial volatilization of arsines in environment. J Environ Sci 26(2):371–381

Wang S, Gao B, Li Y, Creamer AE, He F (2017) Adsorptive removal of arsenate from aqueous solutions by biochar supported zero-valent iron nanocomposite: batch and continuous flow tests. J Hazard Mater 322:172–181

Wang G, Han N, Liu L, Ke Z, Li B, Chen G (2020) Molecular density regulating electron transfer efficiency of *S. oneidensis* MR-1 mediated roxarsone biotransformation. Environ Pollut 262:114370

Warelow TP, Pushie MJ, Cotelesage JJ, Santini JM, George GN (2017) The active site structure and catalytic mechanism of arsenite oxidase. Sci Rep 7(1):1757

Wei J, Hao X, van Loosdrecht MC, Li J (2018) Feasibility analysis of anaerobic digestion of excess sludge enhanced by iron: a review. Renew Sustain Energy Rev 89:16–26

WHO (2017) Guidelines for drinking-water quality: fourth edition incorporating the first addendum. World Health Organization, Geneva

Wu J, Afridi ZU, Cao ZP, Zhang ZL, Poncin S, Li HZ, Zuo JE, Wang KJ (2016) Size effect of anaerobic granular sludge on biogas production: a micro scale study. Bioresour Technol 202:165–171

Yao G, Tang R, Luo H, Yuan S, Wang W, Xiao L, Chu X, Hu ZH (2022) Zero-valent iron mediated alleviation of methanogenesis inhibition induced by organoarsenic roxarsone. Sci Total Environ 819:152080

Yoshinaga M, Rosen BP (2014) AC· As lyase for degradation of environmental organoarsenical herbicides and animal husbandry growth promoters. Proceed Natl Acad Sci 111(21):7701–7706

Zacarías-Estrada OL, Ballinas-Casarrubias L, Montero-Cabrera ME, Loredo-Portales R, Orrantia-Borunda E, Luna-Velasco A (2020) Arsenic removal and activity of a sulfate reducing bacteria-enriched anaerobic sludge using zero valent iron as electron donor. J Hazard Mater 384:121392

Zhang FF, Wang W, Yuan SJ, Hu ZH (2014) Biodegradation and speciation of roxarsone in an anaerobic granular sludge system and its impacts. J Hazard Mater 279:562–568

Zhao Z, Wang J, Li Y, Zhu T, Yu Q, Wang T, Liang S, Zhang Y (2020) Why do DIETers like drinking: metagenomic analysis for methane and energy metabolism during anaerobic digestion with ethanol. Water Res 171:115425

Zhao J, Su A, Tian P, Tang X, Collins RN, He F (2021) Arsenic(III) removal by mechanochemically sulfidated microscale zero valent iron under anoxic and oxic conditions. Water Res 198:117132
Zhu YG, Yoshinaga M, Zhao FJ, Rosen BP (2014) Earth abides arsenic biotransformations. Ann Rev Earth Planet Sci 42:443–467

Nanoparticle-Induced Inhibition of Anaerobic Digestion and Its Alleviation

Haining Huang and Yinguang Chen

Abstract The increasing global application of nanoparticles (NPs) inevitably results in the release and accumulation of NPs in environmental media such as wastewater and waste sludge. To date, diverse NPs including zero valent metallic NPs, metal oxide NPs, carbonaceous nanomaterials, and multi-compound NPs have exhibited inhibitory effects on anaerobic digestion, which mainly depend on their elementary composition, size, concentration, and surface properties. Once introduced into anaerobic digesters, NPs can be absorbed onto sludge and change sludge morphology. Subsequently, NPs can inhibit the activities of key enzymes involved in hydrolysis, acidogenesis, and methanogenesis, thereby affecting the efficiencies of these critical steps of anaerobic digestion. In addition, NPs exposure can shift the microbial communities, reducing the abundances of key methanogens. In general, the toxicity of NPs to anaerobic microorganisms is attributed to the production of reactive oxygen species and the release of toxic substances. Hence, strategies targeting these two aspects such as the addition of other NPs, biochar, cationic polyacrylamide, and sulfide can efficiently recover methane yields of anaerobic digesters exposed to NPs.

Keywords Alleviation strategies · Anaerobic digestion · Inhibitory effects · Nanoparticle

1 Introduction

Nanoparticles (NPs) are nanomaterials with a diameter ranging from about 1 to 100 nm. They are widely and increasingly produced and employed in many emerging technologies and consumer industries, such as biomedicine, electronic equipment, and additional commodities for personal care, agriculture, water/soil treatment, and renewable energy, due to their distinct chemical and physical properties (Wang et al. 2024; Wang and Chen 2016). Accordingly, the increasing global application and

H. Huang · Y. Chen (✉)
State Key Laboratory of Pollution Control and Resource Reuse, School of Environmental Science and Engineering, Tongji University, Shanghai 200092, China
e-mail: yinguangchen@tongji.edu.cn; yinchentj@163.com

G. Wu (ed.), *Anaerobic Digestion*, Green Energy and Technology,
https://doi.org/10.1007/978-3-031-69378-6_9

inappropriate disposal of NPs will inevitably result in the release of NPs into the surrounding environment, threatening the ecological security.

Wastewater treatment plants (WWTPs) are the well-known hotspots that collect diverse types of contaminants, including the released NPs. Cervantes-Avilés and Keller (2021) detected 13 metal-based NPs in the influent of a municipal WWTP in Southern California at concentrations ranging from 2.6 ng/L for Cd-based NPs to 10,700 ng/L for Ti-based NPs. However, a significant proportion of NPs is difficult to biodegrade effectively by conventional treatment technologies in WWTPs. Hence, the fate of most NPs will be enriched in activated sludge with high levels by adsorption and sedimentation. Tou et al. (2017) have identified various NPs including Ti-, Fe-, Zn-, Sn-, and Pb-containing NPs in sewage sludge and found that the concentration of all NPs could be up to $10^7 - 10^{11}$ particles/g within the bioavailable fraction of metals. Similarly, Gottschalk et al. (2009) estimated that the concentrations of ZnO NPs and TiO_2 NPs in sludge were about 17.1–23.2 mg/g and 136–211 mg/g, respectively. Anerobic digestion is one of the most widely applied processes for wastewater and sludge treatment. During anaerobic digestion, the contact between accumulated NPs and key microorganisms is inevitable. Therefore, understanding the potential effects (particularly the inhibitory effects) of NPs on anaerobic digestion and then developing mitigation strategies are critical challenges for operating anaerobic digestion systems efficiently.

2 Typical Nanoparticles Inhibiting the Function of Anaerobic Digestion

In general, NPs can be roughly divided into four categories according to their element compositions: zero valent metallic NPs, metal oxide NPs, carbonaceous nanomaterials, and multi-compound NPs. Although the biological effects of NPs might depend on diverse conditions such as the size and dosages of applied NPs as well as the characteristics of substrates for anaerobic digestion, the effects of NPs on anaerobic digestion can generally be divided into three categories: no effect, promoting effect, and inhibiting effect (Baniamerian et al. 2019). The non-influential NPs mainly include NPs with relatively stable character, such as TiO_2 NPs and SiO_2 NPs (Gonzalez-Estrella et al. 2013; Zheng et al. 2015). NPs exhibiting promotion-effects are mainly iron-based conductive NPs or trace element NPs. Following, NPs inhibiting anaerobic digestion are focused.

2.1 *Zero Valent Metallic Nanoparticles*

Zero valent metallic nanoparticles refer to zero valent metals (such as Fe, Ni, Cu, Co, Ag, Au) at nanoscale. They are widely applied in environmental processes due to their significant strong chemical reducibility, high efficiency, and large specific surface (Li et al. 2016a, b). Table 1 summarizes zero valent metallic nanoparticles showing considerable inhibitory effects on the function of anaerobic digestion. García et al. (2012) found that Ag NPs (30 nm) had a toxicity effect and inhibited the biogas production by 33–50%, whereas Au NPs (20 nm) showed no or slight inhibition. Gitipour et al. (2016) suggested that the negative effects of Ag NPs on anaerobic digestion depend on their surface charge. Both citrate-coated Ag NPs (negative charge) and polyvinylpyrrolidone-coated Ag NPs (minimal charge) showed no toxicity and inhibitory effects on biogas production at any concentrations, which might be due to aggregation and subsequent settling or lower attraction between the NPs and the negatively charged bacterial membrane; whereas branched polyethyleneimine-coated Ag NPs (positive charge) at 100 mg/L demonstrated severe toxicity (Gitipour et al. 2016).

Gonzalez-Estrella et al. (2013) tested 11 different types of inorganic nanoparticles and found that Cu NPs had the most toxic effect on methanogenesis, with the inhibited activity of acetoclastic methanogens rather than H_2-utilizing methanogens. Targeting model acetogen *Moorella thermoacetica*, Wu et al. (2021) confirmed that Cu NPs at 1 mg/L could significantly reduce acetic acid production to 40.7% of the control.

nZVI shows promotional effects on anaerobic digestion. Abdelsalam et al. (2017a, b) found that the addition of 20 mg/L of nZVI NPs to bio-digester under mesophilic temperature stimulated the activities of methanogenic archaea, resulting in an increased biogas and methane production up to 45 and 59%, respectively. However, Yang et al. (2013) showed that nZVI NPs (55 nm) with a concentration of 1 mM (56 mg/L) inhibited methanogenesis more than 20% during anaerobic digestion of sludge, due to the disruption of cell integrity in exposure to nZVI NPs.

2.2 *Metal Oxide Nanoparticles*

Metal oxide NPs are widely applied in chemistry, physics, and material science due to their small sizes and high density of corner or edge surface sites (Baniamerian et al. 2019). The irons released from metal oxide NPs are essential for enzymes and cofactors for anaerobic digestion reactions (Jadhav et al. 2021). However, typical metal oxide NPs such as MgO NPs, CuO NPs, ZnO NPs, and CeO_2 NPs are toxic to anaerobic microorganisms in digesters (Table 2).

Wang et al. (2016) found that MgO NPs exposure at dosages above > 100 mg/g TSS (total suspended solids) resulted in significant inhibition on methane generation in anaerobic granular sludge (AGS) reactors, and the yield of methane was dramatically reduced to only 1.1% of the control at the concentration of 500 mg/g TSS. Similarly, Qi et al. (2021) showed that ZnO NPs (5–100 mg/g TSS) promoted the accumulation of soluble chemical oxygen demand (COD), soluble protein and polysaccharide, but inhibited their following fermentation and methane production. This inhibition became stronger as the ZnO NPs concentration increased, leading to the methane production declined by 84.55% (5 mg/ g TSS), 92.39% (30 mg/g

Table 1 Reported zero valent metallic nanoparticles showing inhibiting effect on anaerobic digestion

NPs	Size (nm)	Concentration	Substrate	Temperature	Effects	References
Cu	40–60	10–1500 mg/L	Granular sludge	30 °C, batch	Highly inhibitory to acetoclastic and hydrogenotrophic methanogens with IC50 values of 62 and 68 mg/L	Gonzalez-Estrella et al. (2013)
	20–40	0–1 mg/L	*Moorella thermoacetica*	55 °C, batch	Acetic acid production was inhibited to 40.7% of the control at 1 mg/L of Cu NPs	Wu et al. (2021)
Ag	170	500 mg/g TSS	Waste activated sludge	30 °C, batch	Decrease 27% of methane production	Casals et al. (2014)
	30	130 mg/L	Cellulose	37/55 °C, batch	Toxicity effect, decrease 33–50% biogas production	García et al. (2012)
	10–15	0.5–100 mg/L cationic silver NPs	Biosolid from wastewater treatment plant	37 °C, batch	Up to complete inhibition of the anaerobic biogas production	Gitipour et al. (2016)
	20–40	5–1000 mg/g TS	municipal waste activated sludge	37 °C, batch	12.1% decrease in the methane production at the highest dosage	Ünsar et al. (2016)
Au	20	75 mg/L	Cellulose	37/55 °C, batch	Zero or slight toxicity effect on ordinary heterotrophic organisms, ammonia-oxidizing bacteria, and anaerobic bacteria	García et al. (2012)
nZVI	55	>1 mM	Waste activated sludge	37 °C, batch	Inhibited methanogenesis by more than 20%	Yang et al. (2013)

TS total solids; *TSS* total suspended solids

TSS), and 93.72% (100 mg/g TSS), respectively (Qi et al. 2021). Luna-delRisco et al. (2011) compared the bioeffects of both CuO and ZnO with different sizes and found that: (1) nanoparticles (i.e., CuO NPs of 30 nm and ZnO NPs of 50–70 nm) were more toxic than the bulk ones (i.e., CuO of 5 μm and ZnO of 1 μm); and (2) the half-maximal effective concentrations for methane inhibition were estimated to be 10.7 mg Cu/L for CuO NPs and 57.4 mg Zn/L for ZnO NPs.

Table 2 Reported metal oxide NPs showing inhibiting effect on anaerobic digestion

NPs	Size (nm)	Concentration	Substrate	Temperature	Effects	References
MgO	154	500 mg/g TSS	Anaerobic granular sludge	35 °C, batch	Decrease 99% of methane production	Wang et al. (2016)
CuO	30	7.5–480 mg/L	Cattle manure	36 °C, batch	Inhibition of biogas production up to 96%	Luna-delRisco et al. (2011)
	30–50	11–1100 mg/L	Waste activated sludge	37 °C, batch	Inhibition of biogas production up to 84%	Ünsar et al. (2016)
CeO_2	12	640 mg/L	Cellulose treatment plant	37/55 °C, batch	Toxicity effect, decrease nearly 100% biogas production	García et al. (2012)
	15–30	11–1100 mg/L	Waste activated sludge	37 °C, batch	Caused 9.2% increase in the methane production at the highest dosage	Ünsar et al. (2016)
	-	100–1000 mg/L	Sludge	30 °C, batch	Up to 35% decrease in biogas production	Nguyen et al. (2015)
ZnO	140	10–1500 mg/L	Waste activated sludge	35 °C, batch	Inhibition of methane production up to 75.1%	Mu and Chen (2011)

(continued)

Table 2 (continued)

NPs	Size (nm)	Concentration	Substrate	Temperature	Effects	References
	10–30	10–1500 mg/L	Granular sludge	30 °C, batch	Highly inhibitory to acetoclastic and hydrogenotrophic methanogens with the half-maximal inhibitory concentrations of 87 and 250 mg/L	Gonzalez-Estrella et al. (2013)
	145	5–500 mg/L	Waste activated sludge	37 °C, batch	VFA accumulation in the anaerobic digestion process, 25% reduction on biogas and 50% reduction on methane production	Zhang et al. (2017)

2.3 *Carbon-Based Nanomaterials*

The carbonaceous nanoparticles include graphene, fullerene (C_{60}), single and multi-walled carbon nanotubes (SWNT and MWNT), and nanofibers. Among them, C_{60}, SWNT and MWNT show no negative effects on anaerobic digestion (Ambuchi et al. 2017; Gil et al. 2019; Yang et al. 2017). While the graphene effect on anaerobic digestion was controversial. Lin et al. (2017) observed that the addition of 1.0 g/L graphene resulted in an enhancement of 25% in methane yield and 19.5% in biogas production rate. However, for nano-graphene oxide (NGO), Dong et al. (2019) demonstrated that the methane production dropped with increasing NGO additions, with the cumulative methane production decreased by 7.6% and 12.6% at the NGO dosing rates of 0.054 mg/mg volatile solids (VS) and 0.108 mg/mg VS, respectively.

Table 3 summarized the reported effects of nanoplastic on anaerobic digestion. Generally, nano-polystyrene possesses a common negative impact on methanogenesis, indicated by the decreased methane production, prolonged lag phases, and higher volatile fatty acids accumulation (Azizi et al. 2021). Moreover, particle size and surface charge of polystyrene are important factors. For instance, nano-sized polystyrene particles were more inhibitory to digester microbial communities than micro-sized particles (Zhang et al. 2020). Moreover, compared to anionic polystyrene nanoparticles, a relatively lower concentration (100 vs. 20 μg/mL) of cationic polystyrene nanoparticles could lead to more decline in methane production (17.47 vs. 22.98%) (Feng et al. 2018).

Table 3 Reported nanoplastic showing inhibiting effect on anaerobic digestion

NPs	Size (nm)	Concentration	Substrate	Temperature	Effects	References
Polystyrene	50 ± 6	0–50 μg/L	Synthetic wastewater (glucose)	35 °C, Continuous	20 and 50 μg/L decreased methane production (by 19.0–28.6%) and COD removal (19.3–30.0%)	Wei et al. (2020)
Polystyrene	54.8	0–0.2 g/L	Sewage sludge	37 °C, batch	Methane yields linearly decreased with increasing polystyrene concentrations	Fu et al. (2018)
Polystyrene	60–80	Anionic polystyrene (PS-SO_3H: 0–100 μg/mL), cationic polystyrene (PS-NH_2: 0–20 μg/mL)	Synthetic wastewater (glucose)	37 °C, batch	20 μg/mL of cationic polystyrene could lead to higher inhibition of methane production than 100 μg/mL of anionic polystyrene	Feng et al. (2018)
Polystyrene	80	0–0.25 g/L	Synthetic wastewater	37 °C, batch	80 nm polystyrene reduced CH_4-production by 19.3%, while 5 μm polystyrene decreased by 17.9% during anaerobic digestion	Zhang et al. (2020)
Polystyrene	50	50 mg/g TSS	Waste sludge	35 °C, batch	50 nm NPs resulted in a 15.5% decrease in methane yield	Wang et al. (2022b)

2.4 Multi-compound Nanoparticles

To overcome the shortcomings such as non-uniform dispersion, agglomeration and low recovery of NPs, supporting/coating the NPs onto carrier additives are widely applied (Baniamerian et al. 2019). Magnetic nanoparticles (MNPs) are a major class of nanoscale materials with superparamagnetic behavior, which is easy to separate using an external magnetic force (Sun et al. 2008). Ni ferrite nanoparticles (Fe_2NiO_4) and Ni Zn ferrite nanoparticles (Fe_4NiO_4Zn) are two types of MNPs containing the trace metals Ni and Fe. Chen et al. (2018) compared their effects on biogas (methane) production and found that anaerobic digestion was stimulated by the addition of 100 mg Ni/L in Fe_2NiO_4 NPs, while it was inhibited by the addition of 1–100 mg Ni/L in Fe_4NiO_4Zn NPs. The structure of Fe_4NiO_4Zn NPs has a Zn^{2+} dopant which can be released into solution to result in toxicity. However, Fe_2NiO_4 NPs adopts an inverse spinel structure with Ni^{2+} in the octahedral sites, and Fe^{3+} equally distributed between the octahedral and tetrahedral sites of the O^{2-} face-centered cubic cell, resulting in no toxicity to anaerobic microorganisms.

3 Nanoparticle-Induced Diverse Impacts on Anaerobic Digestion

3.1 Changing Sludge Morphology

Functional microorganisms in anaerobic digestion systems exist in the aggregated forms of sludge flocs. Once added into the anaerobic digester, NPs can adhere to the extracellular polymeric substances (EPS) on the microbial cell envelope or the excreted biomolecules (lipopolysaccharides and lipoteichoic acid) (Wang and Chen 2016). This was confirmed by scanning electron micrographs images (Fig. 1) (Mu and Chen 2011; Wang et al. 2016). Generally, EPS secreted by microorganisms can protect them from toxic substances. For instance, Cervantes-Avilés et al. (2018) found that TiO_2 NPs resulted in the remodel of sludge EPS by increasing the concentration of protein and polysaccharide in the loosely bound EPS to protect anaerobic microorganisms from nanoparticle exposure.

Khan et al. (2011) suggested that EPS can also exacerbate the toxicity of NPs to environmentally bacterial species and aquatic organisms, as the propensity of Ag-NPs to aggregate was reduced due to the increase of negative surface charge in the presence of EPS (10–250 mg/L). Additionally, when the toxicity of NPs exceeds the buffer capacity of EPS, NPs exposure can damage sludge structures. Wang et al. (2016) assessed the impacts of diverse NPs (Ag NPs, MgO NPs, nZVI, Fe_2O_3 NPs) on the surface of AGS, and found that Ag NPs and MgO NPs at high dosages (500 mg/g TSS) broken the sludge surface, which was further confirmed by Lactate dehydrogenase assays. This might be one of the key reasons for the inhibited function of anaerobic digestion under the exposure of Ag NPs and MgO NPs. Wei et al. (2020) found that

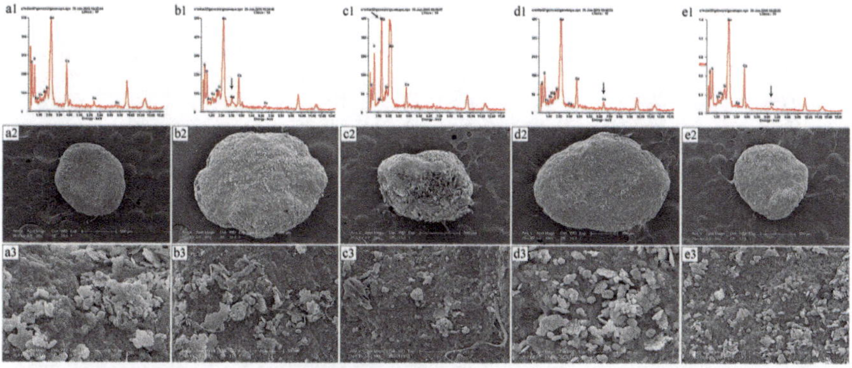

Fig. 1 Scanning electron micrograph (SEM) images (a2–e2, a3–e3) and energy-dispersive X-ray (EDX) analysis (a1–e1) of AGS in the reactors exposed to Ag NPs, MgO NPs, nZVI, and Fe_2O_3 NPs at the indicated concentrations: Control (a1–a3), Ag NPs-500 (b1–b3), MgO NPs-500 (c1–c3), nZVI-10 (d1–d3), Fe_2O_3 NPs-100 (e1–e3) (Wang et al. 2016)

nano-polystyrene at 10 μg/L stimulated the production of proteins, polysaccharides and lipids contents (but not DNA) of EPS to protect the inner microbial cells of AGS from nano-polystyrene exposure. Whereas higher nano-polystyrene exposure levels of 20 and 50 μg/L reduced the corresponding concentrations of EPS key components, which were further confirmed by fluorescent staining experiments and Fourier transform infrared (FTIR) spectroscopy analysis (Fig. 2).

3.2 *Affecting Steps and Key Enzymes of Anaerobic Digestion*

Anaerobic digestion process usually includes solubilization, hydrolysis, acidogenesis, and methanogenesis, converting organic substrates into methane and carbon dioxide (Fig. 3). These steps involve numerous functional microorganisms and key enzymes such as protease, acetate kinase, phosphoric acid transferase, succinyl-CoA transferase, coenzyme F_{420}, and so on (Wang et al. 2024). Under NPs exposure conditions, Mu and Chen (2011) reported that ZnO NPs at higher dosages (30 and 150 mg/g TSS) inhibited hydrolysis (hydrolysis of protein) and methanation steps, but had no significant effects on solubilization and acidification processes. These differences might be attributed to the fact that ZnO NPs inhibited the activties of protease and coenzyme F_{420}, but not acetate kinase. Similarly, Wang et al. (2016) reported that Ag NPs and MgO NPs at 500 mg/g TSS significantly decreased all activities of protease, acetate kinase and coenzyme F_{420}, contributing to the disfunction of anaerobic digestion. Wu et al. (2021) targeted model acetogen *Moorella thermoacetica* and found that Cu NPs at 1 mg/L inhibited the gene expressions and catalytic activities of the key enzymes related to acetic acid production, such as phosphofructokinase, pyruvate kinase, phosphotransacetylase, and acetate kinase.

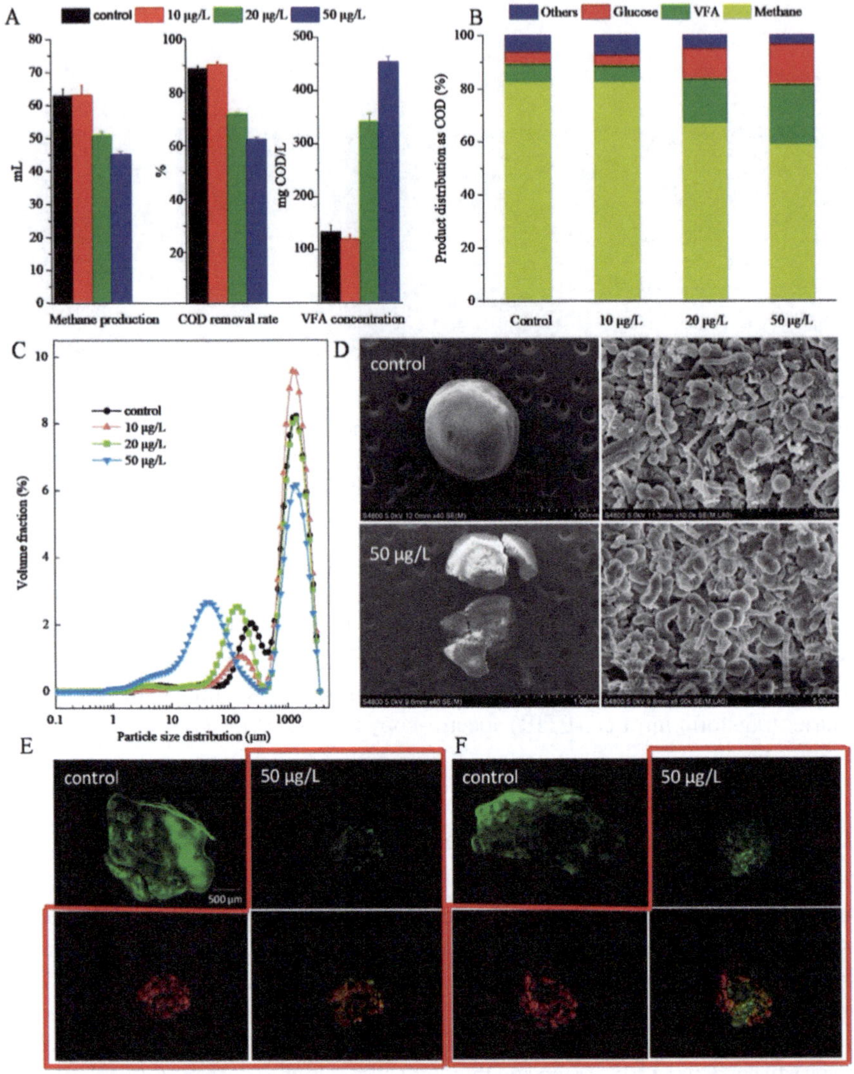

Fig. 2 Fluorescence photos of live cells (green), dead cells (red) and total cells (overlay) in AGS from control (A1–A3) and experimental (A4–A6) reactors (A), compositions of EPS in AGS from control and experimental reactors (B), distribution profiles of EPS in AGS from control (C1–C6) and experimental (C7-C12) reactors with proteins (green, C2, C8), α-d-glucopyranose polysaccharides (red, C3, C9), β-d-glucopyranose polysaccharides (blue, C4, C10), lipids (pink, C5, C11), and overlay images (C6, C12) over the whole Exposure phase (C) and FTIR spectra of EPS from control and experimental AGS samples (D) (Wei et al. 2020)

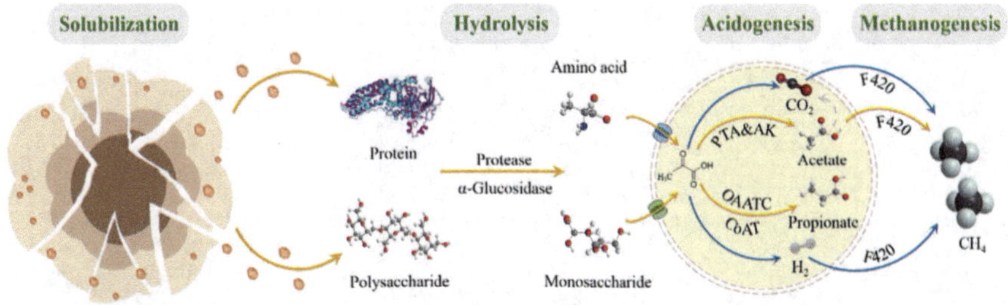

Fig. 3 Metabolic pathways (i.e., solubilization, hydrolysis, acidogenesis, and methanogenesis) of anaerobic digestion. Typically, the metabolites of one phase are the substrates for the next phase. Acetate kinase (AK) and phosphoric acid transferase (PTA) are mainly responsible for acetate formation, while succinyl-CoA transferase (CoAT) and oxaloacetate transferase (OAATC) are mainly involved in propionate synthesis (Wang et al. 2024)

3.3 Shifting Microbial Community Structures

Anaerobic digestion process is carried out by many microorganisms consisting of both bacteria and archaea. Wang et al. (2016) applied fluorescence in situ hybridization and confocal laser-scanning microscopy to assess the shift in the microbial community of AGS to NPs exposure, and they suggested that: (1) Bacteria domain (red, a1-e1) and Archaea domain (green, a2–e2) were the primary AGS components and consisted of the phyla α-Proteobacteria (red, a3–e3), β-Proteobacteria (rose, a4–e4), and Bacteroidetes (blue, a5–e5) as well as the genus *Methanosaeta* (green, a6–e6); (2) either 500 mg/g TSS Ag NPs (b1-b6) or 500 mg/g TSS MgO NPs (c1-c6) caused a significant decrease in the center of AGS, whereas 10 mg/g TSS nZVI (d1–d6) and 100 mg/g TSS Fe_2O_3 NPs (e1–e6) resulted in more microorganisms in AGS (Fig. 4), which were consistent with corresponding methane productions. Similar results were also reported by a previous study, in which the abundance of Archaea was dramatically reduced from 39.5% in the control to 27.1% and 3.5% after the long-term exposure to 30 mg/g TSS and 150 mg/g TSS of ZnO NPs, respectively (Mu and Chen 2011). *Methanospirillum*, a hydrogenotrophic methanogen, is the most sensitive methanogen to ZnO NPs; whereas *Methanosaeta*, a strict acetic acid type methanogen dominant in anaerobic digestion, can responded to external ZnO NPs stimuli (Wang et al. 2021). In addition, ZnO NPs could inhibit the activity of *Clostridium* XI during anaerobic digestion, which is a typical anaerobe that plays an important role in the production of acetic acid and butyrate during the degradation of carbohydrates (Wang et al. 2021).

Wei et al. (2020) confirmed that the higher exposure levels of nano-polystyrene at 20 and 50 μg/L changed microbial communities along the direction against hydrolysis-acidification and methanation. Specifically, nano-polystyrene exposure not only significantly reduced the abundances of key bacteria such as *Longilinea* sp. and *Paludibacter* sp. which have ability of degrading organic substances with volatile fatty acids and hydrogen as their major products, but also dominant methanogens including acetoclastic *Methanosaeta* sp. and hydrogenotrophic *Methanobacterium*

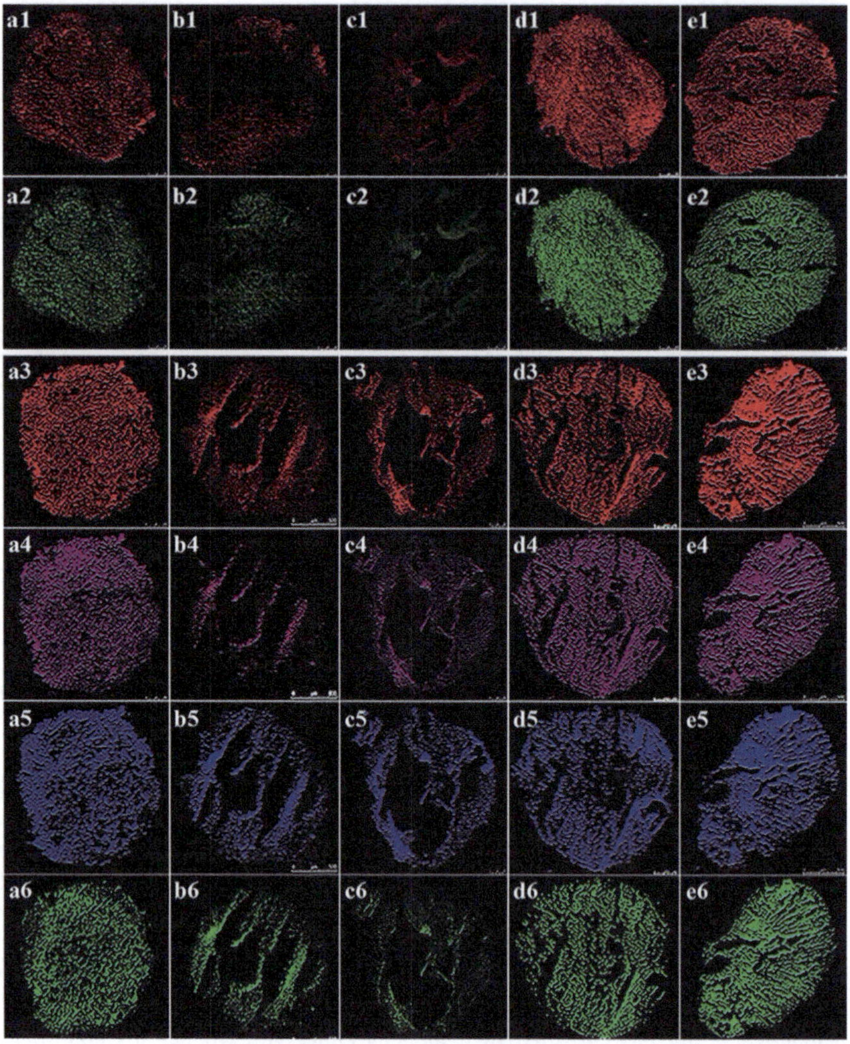

Fig. 4 Fluorescence in situ hybridization of AGS sections of biomass cultured over the long term in the absence of NPs (a1–a6) and with 500 mg/g TSS Ag NPs (b1–b6), 500 mg/g TSS MgO NPs (c1–c6), 10 mg/g TSS nZVI (d1–d6) or 100 mg/g TSS Fe_2O_3 NPs (e1–e6) viewed via CLSM and photographed at higher (i.e., 62×) magnification. The Cy-3-labeled Bacteria domain probe (EUB338) (red, a1–e1) and the FAM-labeled Archaea-domain probe (ARC915) (green, a2–e2) were simultaneously hybridized with the sections. A Cy-3-labeled α-Proteobacteria probe (ALF968) (red, a3–e3), a Cy-5-labeled β-Proteobacteria probe (BET42a) (rose, a4–e4), a HEX-labeled Bacteroidetes probe (CFB719) (blue, a5–e5) and a FAM-labeled *Methanosaeta* probe (MX825) (green, a6-e6) were simultaneously hybridized with the sections (Wang et al. 2016)

sp.. All these variations were consistent with the deteriorating performances of AGS observed in the long-term exposure operation.

4 Underlying Mechanisms Responsible for Nanoparticle-Induced Inhibition of Anaerobic Digestion

4.1 Reactive Oxygen Species

Reactive oxygen species (ROS) generation is the most widely accepted mechanism for nanoparticle-induced bioeffects. Figure 5 summarizes different ROS generation pathways on particle surface such as electrochemical corrosion reactions, Fenton and Haber–Weiss Reactions, and surface catalytic reactions. In general, the ROS formation processes are affected by the characteristics of metallic NPs including size, shape, oxide composition, bulk composition, and degree of crystallinity. For example, the ROS yields of NPs are much higher than micron-sized particles or massive sheets, because NPs have more fraction of surface atoms, which will speed up the rates of corrosion reactions (Kessler et al. 2022).

When micro organisms encounter NPs, the NPs can attach on the surface of cells, penetrate cell membrane, enter inside cells, and produce ROS. Most cells have the ability of neutralizing a certain amount of ROS via enzymatic reactions, but this capability is limited owing to the restricted number of the microbial enzyme molecules. When the ROS production exceeds the buffer capacity of microbes, NPs can cause oxidative stress on the membrane, peroxidation of various organelle constituents and homeostatic imbalance, resulting in cellular metabolic disturbance or even microbial death.

Usually, ROS, including superoxide (O_2^-), hydrogen peroxide (H_2O_2), and the hydroxyl radical (OH·), are produced in the presence of oxygen. However, during anaerobic digestion, ROS was still observed and was associated with the deteriorated anaerobic digestion performance. For instance, Mu and Chen (2011) found that ZnO NPs exposure could inhibit biomass viability and the inhibitory effects were consistent with ROS production. Wang et al. (2022b) suggested that the production of intracellular ROS induced by polystyrene showed a significant size-related increase, which was responsible for the corresponding reduction in cell viability. The exposure of (nano)microplastics (e.g., polyethylene, polystyrene) could considerably increase the intracellular ROS levels, which also aligned with the viability of microbial biomass (Wei et al. 2019a, 2020).

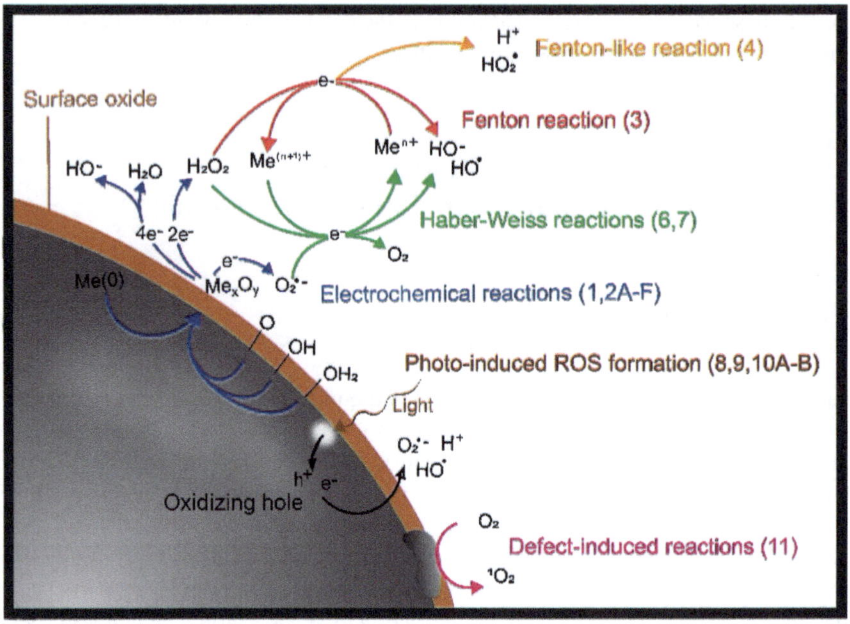

Fig. 5 Schematic illustration of the different routes of ROS generation taking place on a surface of an oxidized metal particle or massive surface and in solution. Note that all reactions take place on or in close proximity to the particle surface with adsorbed, chemisorbed, and desorbed species as intermediate states (Kessler et al. 2022)

4.2 The Release of Toxic Substances

The released toxic substance was another key mechanism responsible for the bioeffects of NPs. For metallic NPs such as ZnO NPs and CuO NPs, the toxic material refers to the released metal ions. Generally, the presence of heavy metal ions (i.e., Cu, Zn, Fe, Ni, Co, Mo) during anaerobic digestion of organic matter is known to be fundamental for numerous enzyme-catalyzed reactions. However, high concentrations of these elements can inhibit the biological degradation in anaerobic reactors. Luna-delRisco et al. (2011) demonstrated that CuO NPs shows a higher solubility compared to bulk CuO, and the concentration of released copper increases with the dosages of CuO NPs, which is the key reason for the nanoparticle toxicity to anaerobic microorganisms. Mu and Chen (2011) also suggested that the toxic effect of ZnO NPs on methane production was mainly due to the release of Zn^{2+} from ZnO NPs, which may cause the inhibitory effects on the hydrolysis and methanation steps of anaerobic digestion.

For nanoplastics, the released toxic substances mainly refer to the leaching of toxic chemicals or additives, which can affect microbial activities in digester by directly damaging microbial cells and by inhibiting activities of key enzymes associated with biochemical reactions and defense against oxidative stress (Azizi et al. 2021).

For instance, Wei et al. (2019b) identified the leaching of bisphenol-A (BPA) from polyvinyl chloride as the key inhibition mechanism that facilitated the rupture of microbial cell walls. The leaching of BPA (0.5–3.6 μg/L) increased with increasing the number of particles from 10 to 60 particles/g TS, and the BPA concentration above 1.8 μg/L hindered methane production. Wei et al. (2020) also found that sodium dodecyl sulfate leached from polystyrene nanoparticles could contribute to methanogenesis inhibition. Some leached compounds from nanoplastics may not inhibit microbial communities. For instance, Wei et al. (2019a) investigated the leaching of potential additives from polyethylene, such as acetyl tri-n-butyl citrate (ATBC), diethylhexyl phthalate, and Irgafos 168 phosphate (plastic antioxidant). Only the leaching of ATBC was observed, while the leaching of other additives was insignificant. Further tests with direct spiking of ATBC in digesters revealed no adverse impact of ATBC on methanogenic communities. Moreover, Chen et al. (2021) found that caprolactam released from polyamide microparticles could increase the activities of key enzymes involved in anaerobic digestion.

5 Strategies for Alleviation of Nanoparticle Inhibition

During anaerobic digestion, NPs will inevitably encounter diverse conditions or substances, which may change their characteristics and in turn affect their bioeffects. Based on this fact, some strategies have been applied to alleviate the inhibitory impacts induced by NPs on anaerobic digestion.

5.1 The Addition of Other NPs

Generally, diverse NPs such as ZnO NPs, TiO_2 NPs, and Fe_2O_3 NPs showed different effects on the performance of anaerobic digestion. However, their co-occurrence can result in intriguingly combined effects. For example, the presence of TiO_2 NPs can reduce the toxicity of ZnO NPs to anaerobic fermentation or anaerobic digestion (Zhang et al. 2019; Zheng et al. 2019). The introduction of TiO_2 NPs in the multi-NPs systems promoted the recovery of enzyme activity, cell viability, microbial abundance, and microbial community structures, which was mainly attributed to the decrease in Zn^{2+} (Zhang et al. 2019). Similarly, Mu et al. (2022) reported that the presence of 50 mg/L Fe_3O_4 NPs entirely mitigated the inhibition caused by 100 mg/L ZnO NPs, which had been decreased by 39.0% compared with the control. Fe_3O_4 NPs could promote the agglomeration of NPs, strengthen EPS resistance, recover enzyme activities (acetate kinase and coenzyme F_{420}), maintain NADH/NAD^+ as well as cell viability, and stimulate direct interspecies electron transfer (DIET), and all these were responsible for the reduced toxicity of ZnO NPs (Mu et al. 2022).

5.2 The Addition of Carbon-Based Materials

Carbon-based materials, including granular activated carbon, biochar and hydrochar, were considered as conductive materials and were applied in anaerobic digestion processes to improve the yield of methane (Lu et al. 2020). Additionally, these carbon-based materials can be applied as a remediation method to recover the function of anaerobic digestion deteriorated by NPs. For instance, Mu et al. (2023) found that the adsorption of granular activated carbon on sludge can weaken the effective contact between ZnO NPs and microbes, and also adsorb some released Zn^{2+} that has contributed primarily to ZnO NPs toxicology. In addition, the presence of granular activated carbon resulted in the enrichment of the tolerant hydrogenotrophic methanogens and DIET-linking partners of *Methanosarcina* with *Geobacter/Syntrophomonas*. These two key mechanisms contribute to the recovery of methane yield to 97% of the control under the exposure of ZnO NPs. Similar alleviation effects were also observed by Azizi et al. (2023), in which granular activated carbon could counteract the adverse effect induced by nanoplastics via adsorbing nanoplastics and enriching DIET-active microbes. With respect to biochar and hydrochar, Wang et al. (2022a, 2023) found that they could competitively adsorb nanoplastics from AGS (Fig. 6), enhance AGS integrity and cell viability by stimulating EPS secretion, and maintain key acidogens as well as methanogens, contributing to the mitigated inhibition of polyethylene-NPs on methane production.

5.3 The Addition of Coagulant

Cationic polyacrylamide (cPAM) is frequently used for wastewater pretreatment and sludge dewatering. Generally, it has been reported that cPAM can inhibit anaerobic digestion (Wang et al. 2018). However, when co-existing with ZnO NPs, cPAM showed an unexpected alleviation effect on anaerobic digestion of waste sludge. Zhang et al. (2021) indicated that the presence of 4.0 mg/g TSS cPAM significantly reduced the inhibition rate of methane yield by 19.3% compared to the individual ZnO NPs during anaerobic digestion, and this behavior was attributed to following aspects: (1) cPAM promoted the solubilization and acidification stages by weakening the contact between ZnO NPs and anaerobes in anaerobic digestion systems, thus providing abundant substance for sequent bio-utilization; (2) the introduction of cPAM into the ZnO NPs-conditioned anaerobic digestion system restored the relative abundance of the *Euryarchaeota* (a well-known methanogen) from 15.3% to 23.3%; and (3) cPAM reduced the production of ROS induced by ZnO NPs.

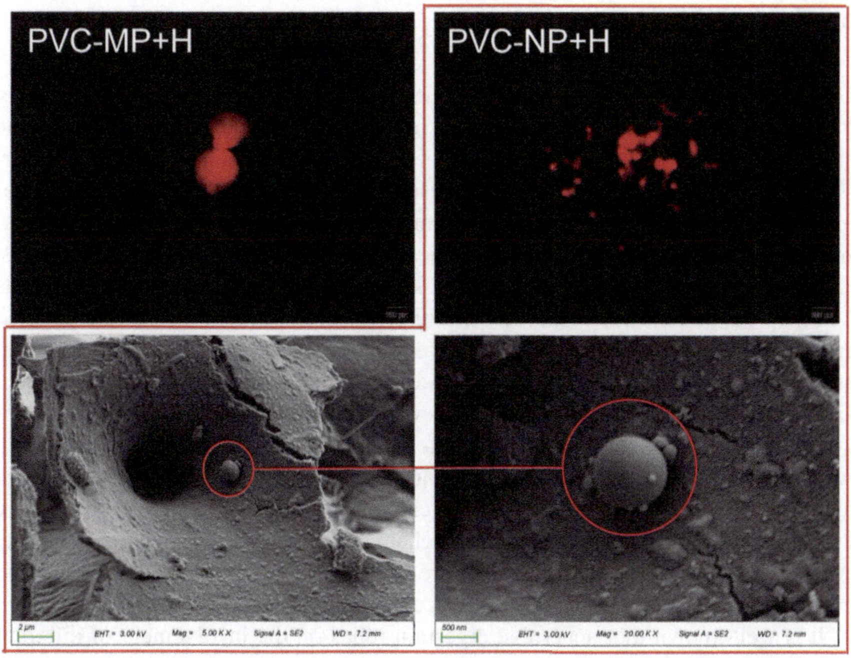

Fig. 6 The fluorescence images of PVC-MPs/PVC-NPs adsorbed on the surface of hydrochar and SEM images of hydrochar surface from "PVC-NP + H" reactors (Wang et al. 2023)

5.4 The Addition of Sulfur-Containing Species

Sulfur-containing species such as hydrogen sulfide (H_2S) is undesirable during anaerobic digestion, because it can result in the corrosion of digesters. However, recently, sulfur-containing species have been proven to play significant roles in regulating the toxicity of NPs. Lombi et al. (2012) investigated the fate of ZnO NPs during anaerobic digestion, and found it can be transformed into the species such as ZnS or Zn-cysteine, which further limited the bioavailability of ZnO NPs. Gonzalez-Estrella et al. (2015) demonstrated that the toxicity of Cu NPs and ZnO NPs to methanogens was respectively sevenfold and 14-fold lower in sulfate-containing than in sulfate-free tests according to the inhibitory constants. The addition of FeS could serve as an antidote to methanogenic toxicity induced by Cu NPs and Zn NPs (Gonzalez-Estrella et al. 2016). The attenuation behavior of H_2S is mainly attributed to two mechanisms: (1) sulfidation, which can lead to the reduced solubility and bioavailability of toxic soluble ions; and (2) passivation, through which the metal-sulfide layer formed on the surface of NPs can prevent the release of toxic ions (Wang et al. 2024). In addition, Huang et al. (2019) confirmed that H_2S can be involved in the biosynthesis of cysteine, which is critical for the buffer ability to oxidative stress.

6 Future Perspectives

Generally, current research related to NPs toxicity to anaerobic digestion mainly focuses on the phenotypic outcomes, such as the yield of methane production, the activity of key enzymes, and the shift of microbial community. To better understand the effect of NPs on microorganisms in anaerobic digesters, more deeper investigations should be carried out from the following aspects: (1) comprehensive evaluation of the responses of typical acidogens, acetoclastic methanogens, and hydrogenotrophic methanogen to NPs exposure at the gene and protein levels will help the better understanding the inhibition mechanisms (such as the targeted enzyme) of NPs and further guide the development of toxicity alleviation strategies; and (2) more insights into mixed cultures relating to metagenome, metatranscriptome, and metaproteomics will provide detailed evidence for understanding microbial adaptation, cooperation, and resistance to the toxicity of NPs during anaerobic digestion. Apart from NPs, several other contaminants such as antibiotics, polycyclic aromatic hydrocarbons, and perfluoroalkyl substances, also accumulate in sludge, co-existing with NPs. They may interact with NPs, changing the characteristics of each other (Wang et al. 2024). In future studies, more attention should be paid to the combined effects of NPs and other contaminants on anaerobic digestion.

It is well-known that anaerobic digestion relies on the involvement of different microorganisms. To effectively alleviate NPs toxicity, future research may need to shift the focus from the NPs themselves to the enhancement of key microbial functions. Based on the known mechanisms, the possible alleviation strategies may be implemented through following aspects: (1) the introduction of microorganisms producing antioxidants or EPS, which can intensify the buffering ability of the entire microbial community to NPs exposure; and (2) the addition of quorum sensing signals, which can promote the communication between microorganisms to survive from external stresses (Lee et al. 2013).

7 Take-Home Messages

(1) The inhibitory effects of NPs on anaerobic digestion mainly depend on their elementary composition, size, concentration, and surface properties.
(2) The hydrolysis and methanogenesis processes are more sensitive to NPs. NPs exposure could significantly reduce the abundance of key methanogens, such as acetoclastic methanogens and hydrogenotrophic methanogens.
(3) The production of ROS and the release of toxic substances are the two main mechanisms responsible for NPs toxicity to anaerobic digestion.
(4) Strategies such as the addition of TiO_2 NPs, biochar, cPAM, and sulfide can efficiently recover methane yields of anaerobic digesters exposed to NPs.

References

Abdelsalam E, Samer M, Attia YA, Abdel-Hadi MA, Hassan HE, Badr Y (2017a) Effects of Co and Ni nanoparticles on biogas and methane production from anaerobic digestion of slurry. Energy Convers Manage 141:108–119

Abdelsalam E, Samer M, Attia YA, Abdel-Hadi MA, Hassan HE, Badr Y (2017b) Influence of zero valent iron nanoparticles and magnetic iron oxide nanoparticles on biogas and methane production from anaerobic digestion of manure. Energy 120:842–853

Ambuchi JJ, Zhang Z, Shan L, Liang D, Zhang P, Feng Y (2017) Response of anaerobic granular sludge to iron oxide nanoparticles and multi-wall carbon nanotubes during beet sugar industrial wastewater treatment. Water Res 117:87–94

Azizi SM, Hai FI, Lu W, Al-Mamun A, Dhar BR (2021) A review of mechanisms underlying the impacts of (nano)microplastics on anaerobic digestion. Biores Technol 329:124894

Azizi SM, Zakaria BS, Haffiez N, Dhar BR (2023) Granular activated carbon remediates antibiotic resistance propagation and methanogenic inhibition induced by polystyrene nanoplastics in sludge anaerobic digestion. Biores Technol 377:128938

Baniamerian H, Isfahani PG, Tsapekos P, Alvarado-Morales M, Shahrokhi M, Vossoughi M, Angelidaki I (2019) Application of nano-structured materials in anaerobic digestion: current status and perspectives. Chemosphere 229:188–199

Casals E, Barrena R, García A, González E, Delgado L, Busquets-Fité M, Font X, Arbiol J, Glatzel P, Kvashnina K, Sánchez A (2014) Programmed iron oxide nanoparticles disintegration in anaerobic digesters boosts biogas production. Small 10(14):2801–2808

Cervantes-Avilés P, Keller AA (2021) Incidence of metal-based nanoparticles in the conventional wastewater treatment process. Water Res 189:116603

Cervantes-Avilés P, Ida J, Toda T, Cuevas-Rodríguez G (2018) Effects and fate of TiO_2 nanoparticles in the anaerobic treatment of wastewater and waste sludge. J Environ Manage 222:227–233

Chen JL, Steele TW, Stuckey DC (2018) The effect of Fe_2NiO_4 and Fe_4NiO_4Zn magnetic nanoparticles on anaerobic digestion activity. Sci Total Environ 642:276–284

Chen H, Tang M, Yang X, Tsang YF, Wu Y, Wang D, Zhou Y (2021) Polyamide 6 microplastics facilitate methane production during anaerobic digestion of waste activated sludge. Chem Eng J 408:127251

Dong B, Xia Z, Sun J, Dai X, Chen X, Ni BJ (2019) The inhibitory impacts of nano-graphene oxide on methane production from waste activated sludge in anaerobic digestion. Sci Total Environ 646:1376–1384

Feng Y, Feng LJ, Liu SC, Duan JL, Zhang YB, Li SC, Sun XD, Wang SG, Yuan XZ (2018) Emerging investigator series: inhibition and recovery of anaerobic granular sludge performance in response to short-term polystyrene nanoparticle exposure. Environ Sci: Water Res Technol 4(12):1902–1911

Fu SF, Ding JN, Zhang Y, Li YF, Zhu R, Yuan XZ, Zou H (2018) Exposure to polystyrene nanoplastic leads to inhibition of anaerobic digestion system. Sci Total Environ 625:64–70

García A, Delgado L, Torà JA, Casals E, González E, Puntes V, Font X, Carrera J, Sánchez A (2012) Effect of cerium dioxide, titanium dioxide, silver, and gold nanoparticles on the activity of microbial communities intended in wastewater treatment. J Hazard Mater 199:64–72

Gil A, Siles JA, Serrano A, Chica AF, Martín MA (2019) Effect of variation in the C/[N+P] ratio on anaerobic digestion. Environ Prog Sustain Energy 38(1):228–236

Gitipour A, Thiel SW, Scheckel KG, Tolaymat T (2016) Anaerobic toxicity of cationic silver nanoparticles. Sci Total Environ 557:363–368

Gonzalez-Estrella J, Sierra-Alvarez R, Field JA (2013) Toxicity assessment of inorganic nanoparticles to acetoclastic and hydrogenotrophic methanogenic activity in anaerobic granular sludge. J Hazard Mater 260:278–285

Gonzalez-Estrella J, Puyol D, Sierra-Alvarez R, Field JA (2015) Role of biogenic sulfide in attenuating zinc oxide and copper nanoparticle toxicity to acetoclastic methanogenesis. J Hazard Mater 283:755–763

Gonzalez-Estrella J, Gallagher S, Sierra-Alvarez R, Field JA (2016) Iron sulfide attenuates the methanogenic toxicity of elemental copper and zinc oxide nanoparticles and their soluble metal ion analogs. Sci Total Environ 548:380–389

Gottschalk F, Sonderer T, Scholz RW, Nowack B (2009) Modeled environmental concentrations of engineered nanomaterials (TiO_2, ZnO, Ag, CNT, fullerenes) for different regions. Environ Sci Technol 43(24):9216–9222

Huang H, Zheng X, Yang S, Chen Y (2019) More than sulfidation: Roles of biogenic sulfide in attenuating the impacts of CuO nanoparticle on antibiotic resistance genes during sludge anaerobic digestion. Water Res 158:1–10

Jadhav P, Muhammad N, Bhuyar P, Krishnan S, Abd Razak AS, Zularisam AW, Nasrullah M (2021) A review on the impact of conductive nanoparticles (CNPs) in anaerobic digestion: applications and limitations. Environ Technol Innov 23:101526

Kessler A, Hedberg J, Blomberg E, Odnevall I (2022) Reactive oxygen species formed by metal and metal oxide nanoparticles in physiological media—a review of reactions of importance to nanotoxicity and proposal for categorization. Nanomaterials 12(11):1922

Khan SS, Mukherjee A, Chandrasekaran N (2011) Impact of exopolysaccharides on the stability of silver nanoparticles in water. Water Res 45(16):5184–5190

Lee J, Wu J, Deng Y, Wang J, Wang C, Wang J, Chang C, Dong Y, Williams P, Zhang LH (2013) A cell-cell communication signal integrates quorum sensing and stress response. Nat Chem Biol 9(5):339–343

Li L, Liu Q, Wang YX, Zhao HQ, He CS, Yang HY, Gong L, Mu Y, Yu HQ (2016a) Facilitated biological reduction of nitroaromatic compounds by reduced graphene oxide and the role of its surface characteristics. Sci Rep 6(1):30082

Li L, Hu J, Shi X, Fan M, Luo J, Wei X (2016b) Nanoscale zero-valent metals: a review of synthesis, characterization, and applications to environmental remediation. Environ Sci Pollut Res 23:17880–17900

Lin R, Cheng J, Zhang J, Zhou J, Cen K, Murphy JD (2017) Boosting biomethane yield and production rate with graphene: the potential of direct interspecies electron transfer in anaerobic digestion. Biores Technol 239:345–352

Lombi E, Donner E, Tavakkoli E, Turney TW, Naidu R, Miller BW, Scheckel KG (2012) Fate of zinc oxide nanoparticles during anaerobic digestion of wastewater and post-treatment processing of sewage sludge. Environ Sci Technol 46(16):9089–9096

Lu JS, Chang JS, Lee DJ (2020) Adding carbon-based materials on anaerobic digestion performance: a mini-review. Biores Technol 300:122696

Luna-delRisco M, Orupõld K, Dubourguier HC (2011) Particle-size effect of CuO and ZnO on biogas and methane production during anaerobic digestion. J Hazard Mater 189(1–2):603–608

Mu H, Chen Y (2011) Long-term effect of ZnO nanoparticles on waste activated sludge anaerobic digestion. Water Res 45(17):5612–5620

Mu H, Su Y, Zhang Y, Qiu Q, Zhu L, Zhao C (2022) Reduction of ZnO nanoparticles toxicity to methanogenic wastewater treatment by the presence of Fe_3O_4 nanoparticles: focusing on nanomaterial interaction and direct interspecies electron transfer. Biochem Eng J 180:108365

Mu H, Zhao Y, Zhao C (2023) Alleviation of ZnO nanoparticles-induced methanogenic inhibition by granular activated carbon. Environ Technol 44(28):4352–4362

Nguyen D, Visvanathan C, Jacob P, Jegatheesan V (2015) Effects of nano cerium (IV) oxide and zinc oxide particles on biogas production. Int Biodeterior Biodegrad 102:165–171

Qi L, Liu X, Miao Y, Chatzisymeon E, Yang P, Lu H, Pang L (2021) Response of cattle manure anaerobic digestion to zinc oxide nanoparticles: methane production, microbial community, and functions. J Environ Chem Eng 9(6):106704

Sun C, Lee JS, Zhang M (2008) Magnetic nanoparticles in MR imaging and drug delivery. Adv Drug Deliv Rev 60(11):1252–1265

Tou F, Yang Y, Feng J, Niu Z, Pan H, Qin Y, Guo X, Meng X, Liu M, Hochella MF (2017) Environmental risk implications of metals in sludges from waste water treatment plants: the

Zheng X, Wu L, Chen Y, Su Y, Wan R, Liu K, Huang H (2015) Effects of titanium dioxide and zinc oxide nanoparticles on methane production from anaerobic co-digestion of primary and excess sludge. J Environ Sci Health Part A 50(9):913–921

Zheng L, Zhang Z, Tian L, Zhang L, Cheng S, Li Z, Cang D (2019) Mechanistic investigation of toxicological change in ZnO and TiO_2 multi-nanomaterial systems during anaerobic digestion and the microorganism response. Biochem Eng J 147:62–71 discovery of vast stores of metal-containing nanoparticles. Environ Sci Technol 51(9):4831–4840

Ünşar EK, Çığgın AS, Erdem AY, Perendeci NA (2016) Long and short term impacts of CuO, Ag and CeO_2 nanoparticles on anaerobic digestion of municipal waste activated sludge. Environ Sci Process Impacts 18(2):277–288

Wang D, Chen Y (2016) Critical review of the influences of nanoparticles on biological wastewater treatment and sludge digestion. Crit Rev Biotechnol 36(5):816–828

Wang T, Zhang D, Dai L, Chen Y, Dai X (2016) Effects of metal nanoparticles on methane production from waste-activated sludge and microorganism community shift in anaerobic granular sludge. Sci Rep 6(1):25857

Wang D, Liu X, Zeng G, Zhao J, Liu Y, Wang Q, Chen F, Li X, Yang Q (2018) Understanding the impact of cationic polyacrylamide on anaerobic digestion of waste activated sludge. Water Res 130:281–290

Wang S, Chen L, Yang H, Liu Z (2021) Influence of zinc oxide nanoparticles on anaerobic digestion of waste activated sludge and microbial communities. RSC Adv 11(10):5580–5589

Wang C, Wei W, Zhang YT, Ni BJ (2022a) Evaluating the role of biochar in mitigating the inhibition of polyethylene nanoplastics on anaerobic granular sludge. Water Res 221:118855

Wang J, Ma D, Feng K, Lou Y, Zhou H, Liu B, Xie G, Ren N, Xing D (2022b) Polystyrene nanoplastics shape microbiome and functional metabolism in anaerobic digestion. Water Res 219:118606

Wang C, Wei W, Zhang YT, Chen X, Ni BJ (2023) Hydrochar alleviated the inhibitory effects of polyvinyl chloride microplastics and nanoplastics on anaerobic granular sludge for wastewater treatment. Chem Eng J 452:139302

Wang D, Pan Q, Yang J, Gong S, Liu X, Fu Y (2024) Effects of mixtures of engineered nanoparticles and cocontaminants on anaerobic digestion. Environ Sci Technol 58(6):2598–2614

Wei W, Huang QS, Sun J, Dai X, Ni BJ (2019a) Revealing the mechanisms of polyethylene microplastics affecting anaerobic digestion of waste activated sludge. Environ Sci Technol 53(16):9604–9613

Wei W, Huang QS, Sun J, Wang JY, Wu SL, Ni BJ (2019b) Polyvinyl chloride microplastics affect methane production from the anaerobic digestion of waste activated sludge through leaching toxic bisphenol-A. Environ Sci Technol 53(5):2509–2517

Wei W, Hao Q, Chen Z, Bao T, Ni BJ (2020) Polystyrene nanoplastics reshape the anaerobic granular sludge for recovering methane from wastewater. Water Res 182:116041

Wu Y, Wu J, Shen Q, Zheng X, Chen Y (2021) Anaerobic fermentation metabolism of *Moorella thermoacetica* inhibited by copper nanoparticles: comprehensive analyses of transcriptional response and enzyme activity. Water Res 197:117081

Yang Y, Guo J, Hu Z (2013) Impact of nano zero valent iron (NZVI) on methanogenic activity and population dynamics in anaerobic digestion. Water Res 47(17):6790–6800

Yang Y, Zhang Y, Li Z, Zhao Z, Quan X, Zhao Z (2017) Adding granular activated carbon into anaerobic sludge digestion to promote methane production and sludge decomposition. J Clean Prod 149:1101–1108

Zhang L, He X, Zhang Z, Cang D, Nwe KA, Zheng L, Li Z, Cheng S (2017) Evaluating the influences of ZnO engineering nanomaterials on VFA accumulation in sludge anaerobic digestion. Biochem Eng J 125:206–211

Zhang L, Zhang Z, He X, Zheng L, Cheng S, Li Z (2019) Diminished inhibitory impact of ZnO nanoparticles on anaerobic fermentation by the presence of TiO_2 nanoparticles: phenomenon and mechanism. Sci Total Environ 647:313–322

Zhang J, Zhao M, Li C, Miao H, Huang Z, Dai X, Ruan W (2020) Evaluation the impact of polystyrene micro and nanoplastics on the methane generation by anaerobic digestion. Ecotoxicol Environ Saf 205:111095

Zhang B, Tang X, Fan C, Hao W, Zhao Y, Zeng Y (2021) Cationic polyacrylamide alleviated the inhibitory impact of ZnO nanoparticles on anaerobic digestion of waste activated sludge through reducing reactive oxygen species induced. Water Res 205:117651

1 Introduction

Antibiotics are wildly used in human medicine, veterinary practices, and agriculture. Most antibiotics could not be fully metabolized by humans and animals, and residual antibiotics create a selective pressure for the emergence of antibiotic resistance bacteria carrying antibiotic resistance genes (ARGs). Wastewater treatment plants are important hotspot of ARGs because they receive various sources of wastewater containing unmetabolized antibiotics, antibiotic resistance bacteria, and ARGs from human and animal urines and feces. A previous study revealed the presence of 19 ARG types and 384 ARG subtypes in sewage sludge through metagenomic analysis (Yin et al. 2019). Qualitative polymerase chain reaction (qPCR) results indicated that the abundances of tetracycline resistance genes, such as *tet*(Q) and *tet*(G), in activated sludge reached ~10^{13} gene copies/g solids (Auerbach et al. 2007). Thus, wastewater treatment plants provide a suitable environment for the development and dissemination of antibiotic resistance. The presence of antibiotics along with biocides and heavy metals can intensify the selective pressure on the proliferation of antibiotic resistance bacteria and facilitate the horizontal gene transfer of ARGs from antibiotic resistance bacteria to other microorganisms during wastewater and sewage sludge treatment (Leiva et al. 2021). Inadequate treatment of wastewater or sewage sludge may lead to the introduction of ARGs into receiving environments through wastewater discharges or the reuse and land application of sewage sludge.

Anaerobic digestion stands as a reliable method for stabilizing organic wastes including sewage sludge. In addition to biogas production, the resultant digested organic wastes can be repurposed as beneficial biofertilizers for agricultural land application. While anaerobic digestion was reported to provide benefit for the ARG removal from sewage sludge, the efficacy of anaerobic digestion in eliminating ARGs remains a contentious topic because ARG reduction is regulated by various factors. ARGs present in the digested sludge can be acquired by bacteria in natural environment via horizontal gene transfer, including conjugation, transformation, and transduction, during land application (Martínez 2008). This process facilitates the transmission of ARGs to the natural environment. Therefore, it is imperative to investigate the impact of anaerobic digestion on changes in ARG abundance to minimize transmission risks associated with digestate land application.

The elimination of ARGs during anaerobic digestion hinges on various factors, including the ARG forms (intracellular and extracellular ARGs), operational conditions (e.g., solids retention time (SRT), temperature, and pH), pretreatment methods (thermal hydrolysis pretreatment, alkali pretreatment, microwave pretreatment, and ultrasonic pretreatment), and conductive additives (Fig. 1). Operational conditions during sludge digestion play pivotal roles in shaping microbial communities, constraining bacterial hosts of ARGs, curbing horizontal transfer of ARGs, and facilitating DNA hydrolysis.

Fate of Antibiotic Resistance Genes in Sewage Sludge and Its Mitigation During Anaerobic Digestion

Renxin Zhao, Haowei Li, Liping Wang, and Yuepeng Sun

Abstract Sewage sludge is a significant reservoir of antibiotic resistance genes (ARGs) in wastewater treatment plants, posing risks when applied to agricultural soils. Anaerobic digestion shows promise in reducing ARGs in sewage sludge, but its efficacy varies. Factors like forms of ARGs, temperature, pH, solids retention time (SRT), and pretreatment methods influence ARG reduction. Thermophilic anaerobic digestion generally outperforms mesophilic anaerobic digestion in ARG removal. The effect of SRT on ARG removal depends on operational factors, temperature, pretreatment method, and ARG subtypes. While pretreatment methods benefit ARG reduction, ARG rebound during subsequent anaerobic digestion is commonly observed. It is necessary to deeply reveal the effects of anaerobic digestion on ARG fate under different operational conditions and the underlying mechanisms, which will be essential for developing strategies to minimize the environmental impact of ARG.

Keywords Anaerobic digestion · Antibiotic resistance genes · Pretreatment · Temperature

R. Zhao · L. Wang · Y. Sun (✉)
School of Ecology and Environment, Inner Mongolia University, Hohhot 010021, Inner Mongolia, China
e-mail: 111989064@imu.edu.cn

Inner Mongolia Key Laboratory of Environmental Pollution Control & Waste Resource Reuse, Inner Mongolia University, Hohhot 010021, Inner Mongolia, China

H. Li
School of Energy and Environmental Engineering, Hebei University of Engineering, Handan 056000, Hebei, China

G. Wu (ed.), *Anaerobic Digestion*, Green Energy and Technology,
https://doi.org/10.1007/978-3-031-69378-6_10

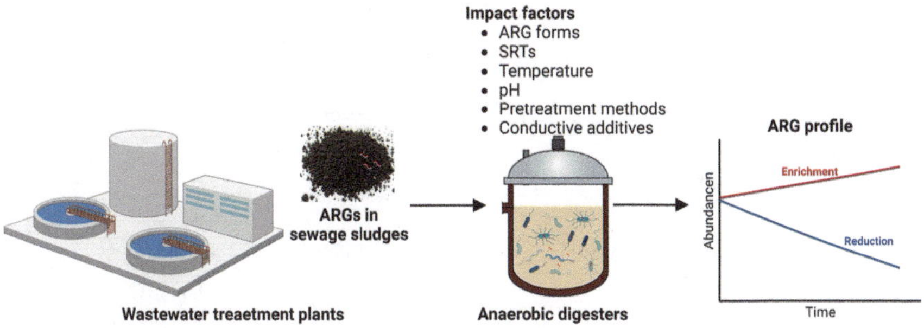

Fig. 1 Factors impacting ARG removal from sewage sludge via anaerobic digestion

2 ARG Forms and Testing Methods

ARGs are presented as intracellular (Fig. 2a) and extracellular ARGs (e.g., plasmid-carrying or free ARGs, and ARGs in bacteriophage and extracellular vesicles) (Fig. 2b–d), and both have potential to participate in the dissemination of antibiotic resistance through horizontal gene transfer. The absolute abundances of intracellular and extracellular ARGs (*bla_{TEM}*, *bla_{SHV}*, *erm*(B), *sul*1, *sul*2, *tet*(A), *tet*(O), *tet*(W) and *tet*(X)) in sewage sludge ranged from 1.0×10^5 to 1.4×10^{13} and from 7.3×10^3 to 3.4×10^{11} copies/g dry weight (Dong et al. 2019; Zou et al. 2020). The percentages of extracellular ARG abundances to the total ARG abundances showed a wide distribution ranging from 1.5 to 86.5% in waste activated sludge (Zou et al. 2020). Intracellular and extracellular ARGs respond differently during anaerobic digestion. For example, mesophilic anaerobic digesters significantly reduce gene copies of total ARGs in intracellular DNA but not in extracellular DNA (Cai et al. 2022a). Haffiez et al. (2023) suggested that anaerobic digesters with thermal hydrolysis pretreatments could effectively reduce the abundances of extracellular *erm, bla, tet,* and *sul*, but increased the abundances of intracellular *bla, erm,* and *intI*1, in waste activated sludges. The difference in the reduction between intracellular and extracellular ARGs might be attributed to their different degradation mechanisms. The degradation of intracellular ARGs might be related to the decay of their bacterial hosts, while extracellular ARGs could be cleaved by extracellular DNases in digested sludge. In addition, extracellular ARGs can also protect themselves through the adsorption to sludge or cell debris (Yuan et al. 2022). Investigating ARGs in the forms of intracellular and extracellular ARGs remains challenging due to the limited methods available for their extraction. The precipitation method with the addition of trimethylammonium bromide is generally used for the extraction of extracellular ARGs from sewage sludge (Zhang et al. 2013; Zou et al. 2020).

ARGs are not only present in bacterial fractions but also encapsulated within bacteriophage or extracellular vesicles in sewage sludge (Calero-Cáceres et al. 2014; Liu et al. 2024). Using a metagenomic approach, Liu et al. (2024) identified ARGs in phage communities from a full-scale anaerobic digester treating sludge.

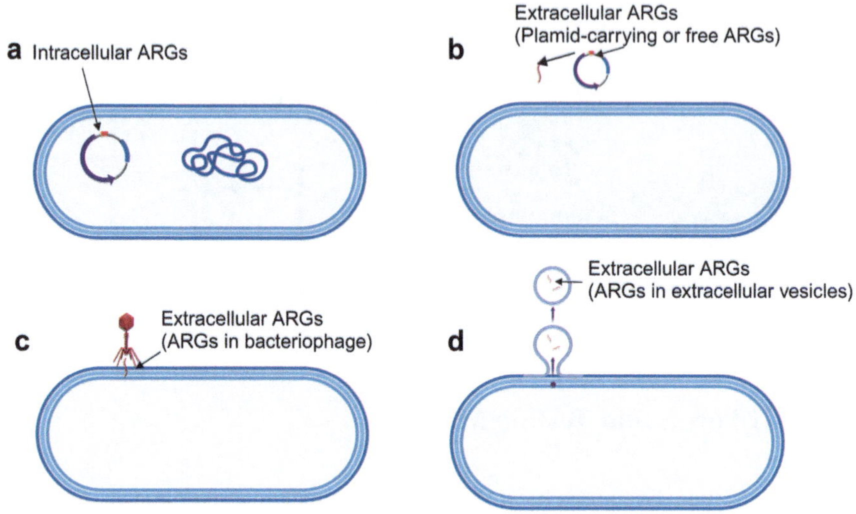

Fig. 2 ARG are present as intracellular (**a**), and extracellular ARGs including plasmid-carrying or free ARGs (**b**), and ARGs in bacteriophage (**c**) and extracellular vesicles (**d**)

ARGs including vancomycin type (*van*U, *van*S, and *van*R), macrolide-lincosamide-streptogramin type [*erm*(A) and *erm*(T)], and tetracycline type (*tet*(M)) were identified on the phage contigs (Liu et al. 2024). The abundance of ARGs, including bla_{TEM}, $bla_{CTX\text{-}M}$, *qnr*A, *qnr*S, and *sul*1, was estimated to be ~10^6 gene copies per gram sludge in the sludge phage fraction (Calero-Cáceres et al. 2014). The information regarding the fate of sludge ARGs in phage fractions during anaerobic digestion is very limited in the literature. One study demonstrated that only 0.57% of viruses were found to carry ARGs during anaerobic digestion (Zhang et al. 2024). These viral communities played a greater role in reducing ARGs than the proliferation, indicating the value of phage-based control of ARGs during anaerobic digestion (Zhang et al. 2024). Another study revealed an enrichment effect of anaerobic digestion on ARGs carried by phage (Xu et al. 2022a). Both thermophilic and mesophilic anaerobic digestion increased the total absolute abundance of phage ARGs [i.e., *erm*(X), *erm*(F), bla_{CTX}, bla_{VIM}, *tet*(W), and *tet*(G)] in cow manure by 47.60% and 59.22%, respectively (Xu et al. 2022a). Although ARGs in bacterial fractions have been extensively studied, research on ARGs within bacteriophage fractions is relatively limited partially because of the great challenge in distinguishing phage particles from vesicle-like particles.

Techniques including qPCR, metagenomics, and high-throughput PCR are used for analyzing ARGs, providing data on ARG copy numbers and ARG diversity. The qPCR relies on the primer design based on the known sequence of ARGs, enabling the measurement of absolute abundance (gene copies per gram or liter) and relative abundance (in relation to 16S rRNA gene copies). However, its application is limited by the complex primer design process and the restricted number of ARGs detected. High-throughput PCR overcomes this limitation by allowing the simultaneous quantification of hundreds of ARGs in a single run (Wang et al. 2014). In addition, metagenomics targets whole-community DNA, enabling the detection of both known and unknown ARGs in various environments. This method generates short reads that offer limited information about ARG sequences. However, with bioinformatics pipelines, these short reads can be assembled into longer overlapping DNA contigs. Thus, the reconstructed partial or complete genomes can then be used for ARG annotation against databases including ARDB (Liu and Pop 2009), CARD (Alcock et al. 2023), and SARG (Yin et al. 2018). Metagenomics can provide information about the relative abundance and diversity of ARGs. Overall, these methods offer complementary strengths, with quantitative PCR providing precise quantification, high-throughput PCR allowing broad detection, and metagenomics offering comprehensive insights into ARG diversity and abundance.

3 Factors Affecting the Fate of Sludge ARGs During Anaerobic Digestion

3.1 SRT

Dynamics of microbial communities in anaerobic digesters play a crucial role in the elimination of ARGs, with SRT emerging as a pivotal operational parameter shaping the microbial communities. Some studies underscore the significance of SRT in influencing the fate of ARGs during anaerobic digestion (Ma et al. 2011; Sun et al. 2019; Zhang et al. 2019). Ma et al. (2011) applied qPCR methods to investigate ARG removal in mesophilic co-digestion of primary and secondary sludge under SRTs of 10 and 20 days. An extended SRT of 20 days in mesophilic anaerobic digestion yielded higher removal rates of *tet*(G), *sul*1, and *sul*2 compared to a shorter SRT of ten days (Ma et al. 2011). More ARG removal at a long SRT can be attributed to the reduced presence of bacterial hosts carrying ARGs under oligotrophic conditions developed during extended residence times (Sun et al. 2019).

The influence of SRT on ARG removal is contingent upon various factors such as operational configuration, temperature, pretreatment methods, and targeted ARG subtypes. For instance, Zhang et al. (2019) observed that shortening the SRT from 20 to 15 days led to a decrease in the absolute abundances of *erm*(B), *erm*(F), *mef* A, *sul*2, and *tet*(X) in a two-stage mesophilic digester with microwave pretreatment. In addition, mesophilic anaerobic digestion needs longer SRTs (40 days) to remove *erm*(F), *sul*1, *sul*2, *tet*(W), *tet*(X), *intI*1, and *intI*2, while thermophilic anaerobic digestion is more effective at shorter SRTs (20 days) (Mortezaei et al. 2023). Furthermore, Sun et al. (2022) demonstrated that a shorter SRT of 10 days outperformed longer SRT of 12.5 or 15 days in reducing *sul*1, *tet*(Q), and *intI*1 during mesophilic anaerobic digestion with thermal hydrolysis pretreatment.

The assessment of ARG abundance variation at different SRTs necessitates the consideration of ARG subtypes, temperature, and pretreatment methods. Mesophilic digestion typically requires longer SRTs for effective ARG removal, whereas thermophilic digesters or those employing pretreatments may achieve satisfactory removal at shorter SRTs.

3.2 *Temperature*

Temperature can influence microbial communities carrying ARGs and consequently the ARG removal (Diehl and LaPara 2010; Ghosh et al. 2009; Ma et al. 2011). Thermophilic anaerobic digestion generally performed better than mesophilic anaerobic digestion in ARG removal from sewage sludge (Diehl and LaPara 2010; Mortezaei et al. 2023; Sun et al. 2019; Tian et al. 2016; Zou et al. 2020). With qPCR methods, Zou et al. (2020) observed that the reduction rates of extracellular *bla, sul,* and *tet* in waste activated sludge during anaerobic digestion were obviously higher at 55°C (from 52.7% to 96.6%) compared to that at 35°C (from 22.8% to 93.9%). Similarly, Diehl and Lapara (2010) indicated that digesters operating at 55°C achieved significant reduction in *tet*(X), *tet*(A), *tet*(L), *tet*(O), and *tet*(W) in wastewater biosolids compared to other operating temperatures (22°C, 37°C, and 46°C).

The mechanism regarding the different reduction characteristics of ARGs under different temperature conditions is still under investigation. Possible explanations (Fig. 3) for the different ARGs reduction between thermophilic and mesophilic conditions include: (1) thermophilic conditions limit the range of the ARG hosts (Ma et al. 2011); (2) thermophilic conditions had a smaller portion of plasmids, insertion sequences and integrons than that of mesophilic conditions, suggesting the lower horizontal transfer potential of ARGs that mediated by mobile genetic elements (Tian et al. 2016); and (3) thermophilic and mesophilic conditions shape different microbial communities, and microbial communities have been shown to be correlated with the ARG composition (Guo et al. 2017; Sun et al. 2019).

High temperatures do not always guarantee better ARG removal. Zhang et al. (2015) found that there was no significant difference in the total abundance and diversity of ARGs in sewage sludge between thermophilic and mesophilic conditions. However, specific ARG subtypes such as *aad*A, *mac*B, and *sul*1 were enriched during thermophilic anaerobic digestion, while *sul*1 and *tet*(M) were enriched during mesophilic anaerobic digestion (Zhang et al. 2015). Taken together, thermophilic anaerobic digestion remains efficient for eliminating ARGs, although the mechanism underlying ARG reduction and enrichment remains unclear.

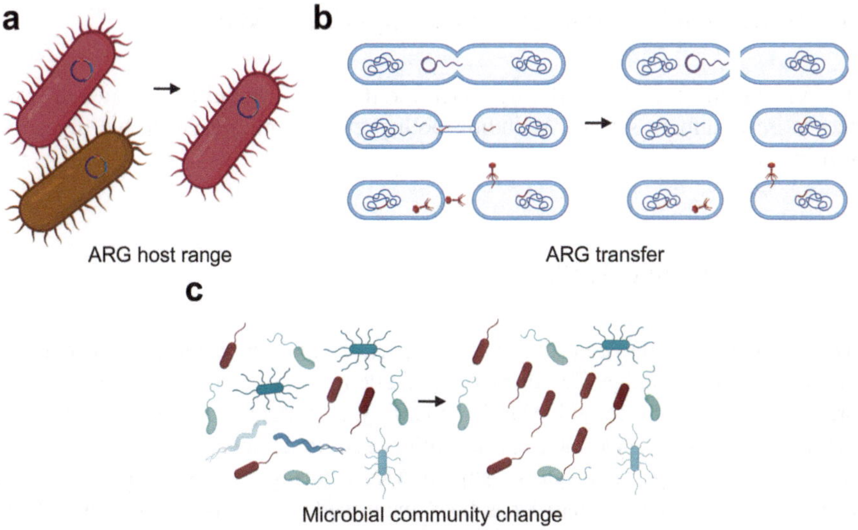

Fig. 3 Potential mechanisms of ARG reduction during anaerobic digestion. The operational conditions such as high temperature could limit the range of ARG host (**a**), lower the horizontal transfer potential of ARGs (**b**), and shape different microbial communities (**c**)

3.3 pH

pH is a crucial factor influencing the performance of anaerobic digestion and the reduction of ARGs in sewage sludge. The ARG level in sludge tends to decrease more significantly under alkaline pH conditions compared to acidic pH conditions (Chen et al. 2019; Huang et al. 2016, 2017; Wu et al. 2016). For instance, alkaline fermentation at pH of 10 resulted in a greater reduction in *sul* and *tet* genes compared to neutral or acidic pH levels (Chen et al. 2019; Huang et al. 2017). Employing qPCR methods, Huang et al. (2016) demonstrated that pH of 10 led to a removal of 0.70–1.31 log units of *tet*(O), *tet*(Q), *tet*(C), and *tet*(X), whereas pH of 4 resulted in 0.58–1.18 log units increase during mesophilic anaerobic digestion. Mechanistic exploration has indicated that alkaline conditions restrict the transfer of ARGs carried on plasmid DNA, extracellular DNA, and phage DNA (Huang et al. 2016, 2017). Additionally, alkaline pH conditions shape microbial communities by limiting the proliferation of sulfonamide resistance bacteria (Chen et al. 2019; Wu et al. 2016). Thus, adjusting pH represents an alternative approach for achieving efficient ARG removal from sewage sludge during anaerobic digestion.

4 Treatment Technologies for Mitigating ARGs

The management of ARGs during anaerobic digestion should be considered from the entire treatment stream, including both pretreatment and anaerobic digestion processes (Fig. 4).

4.1 Pretreatment Methods

The efficiency of anaerobic digestion is generally limited by the poor biodegradability of sewage sludge. To address this challenge, various pretreatment methods are employed to enhance sludge biodegradability. Importantly, coupled with pretreatment technologies, anaerobic digestion offers additional benefits for ARG removal from sewage sludge. The fate of sludge ARGs during anaerobic digestion with different pretreatment methods has been examined, including thermal hydrolysis (Ma et al. 2011; Sun et al. 2022; Wang et al. 2019; Zou et al. 2023), alkali treatment (Wang et al. 2019; Zou et al. 2023), microwave (Tong et al. 2016; Zhang et al. 2016, 2017), and ultrasonic pretreatment methods (Mortezaei et al. 2024; Wang et al. 2019).

Thermal hydrolysis pretreatment has attracted significant attention because it has been well investigated in lab-scale studies and has been applied to the polit-scale (Sun et al. 2022) and full-scale (Redhead et al. 2020; Tong et al. 2019) facilities. Thermal hydrolysis pretreatment is designed to accelerate the rate-limiting step of hydrolysis of organic solid wastes by applying high temperature (120–180°C) and high pressures (1 to 9 bars), prior to anaerobic digestion (Luo et al. 2022). Both high temperature and pressure can not only sterilize the sludge, destroy cell walls, and release readily degradable components, but also promote the DNA hydrolytic

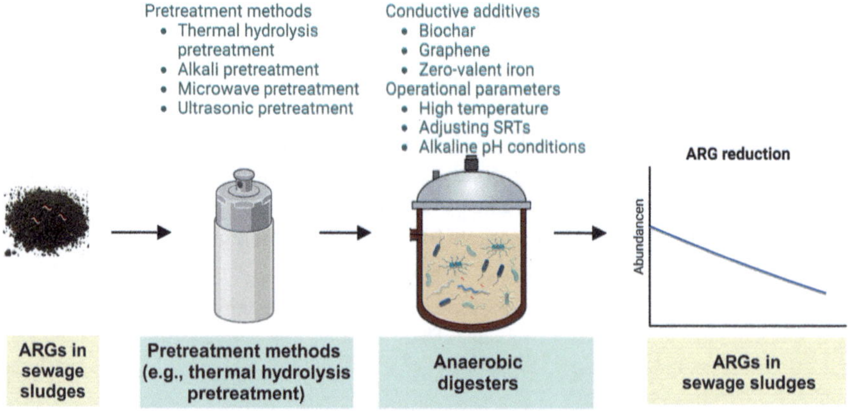

Fig. 4 Potential treatment technologies for mitigating ARGs from sewage sludge

destruction (Ma et al. 2011). Thermal hydrolysis as an anaerobic digestion pretreatment method can benefit the reduction of ARGs (Ma et al. 2011; Pei et al. 2016; Sun et al. 2019, 2022; Tong et al. 2019). Ma et al. (2011) revealed that lab-scale thermal hydrolysis pretreatment (150°C and 4.80 bar for 30 min) resulted in a remarkable reduction of *erm, sul,* and *tet* genes in sewage sludge with the range of reduction from 1.59 to 2.60 log units. Pei et al. (2016) reported that thermal hydrolysis pretreatment (170°C and 8.0 bars for 30 min) can effectively reduce *tet* genes in pharmaceutical waste sludge and municipal waste sludge, with reduction of 2.01 to 3.79 log units. Thermal hydrolysis pretreatment performed better than alkaline and enzymatic pretreatments in decreasing absolute abundances of both intracellular and extracellular ARGs (1.3–5.1 log units reduction), including bla_{SHV}, bla_{TEM}, *sul*1, *sul*2, *tet*(A), *tet*(O), and *tet*(X) (Zou et al. 2023). However, ARG rebound was observed during subsequent anaerobic digestion following thermal hydrolysis pretreatment as revealed using both qPCR (Ma et al. 2011; Sun et al. 2022; Tong et al. 2019) and metagenomics (Sun et al. 2019) techniques. This rebound effect might be mainly due to the proliferation of the bacterial hosts carrying ARG, which respond to the selection pressures imposed by the digester operating conditions (Ma et al. 2011), and the horizontal gene transfer of ARGs that survive during thermal hydrolysis pretreatment (Cai et al. 2022b). Thermal hydrolysis pretreatment significantly reduced the absolute abundance of total ARGs in raw sewage sludge but increased extracellular ARGs by 0.3–1.4 log units under practical conditions (140°C for 30 min) (Cai et al. 2022b). The released extracellular ARGs during thermal hydrolysis pretreatment could be transferred to the microbes in the digesters via transformation (Cai et al. 2022b). Compared to the rebound effect of both intracellular and extracellular *tet, sul,* and *bla* genes during anaerobic digestion followed thermal hydrolysis pretreatment, the rebound effect was not observed for alkaline and enzymatic pretreatment methods (Zou et al. 2023). Manipulating operating conditions of thermal hydrolysis pretreatment might be useful in mitigating the dissemination of particular ARG subtypes during anaerobic digestion (Haffiez et al. 2022; Sun et al. 2022). Sludge ARGs such as *sul*1, *sul*2, *tet*(X), and *intI*1 showed significantly higher removal during anaerobic digestion with thermal hydrolysis pretreatment at 160°C compared to 135°C (Sun et al. 2022). Thus, thermal hydrolysis pretreatment plays a positive role in reducing ARGs but the ARG rebound effect under various operational conditions may limit its application.

Microwave pretreatment facilitated the reduction of specific ARG subtypes in sewage sludge (Tong et al. 2016; Zhang et al. 2016, 2017). Tong et al. (2016) showed that the total relative abundance of the ARGs (e.g., bla_{SHV}, *tet*(C), *tet*(M) *tet*(O), and *tet*(X)) was reduced by 42.7% with the microwave-acid pretreatment. Unfortunately, most of targeted ARGs tended to enrich during the subsequent mesophilic anaerobic digestion (Tong et al. 2016). Zhang et al. (2019) suggested that microwave pretreatment increased both the absolute and relative abundances of bla_{TEM}, *erm*(F), *sul*1, *sul*2, *tet*(M), and *tet*(X) in waste activated sludge during anaerobic digestion with SRTs of both 15 and 20 days. Tong et al. (2016) hypothesized that the rebound effect observed during anaerobic digestion could be attributed to the persistence of residual ARGs during microwave pretreatment. The residual ARGs might be easily captured by microbes in subsequent digesters through horizontal gene transfer (Tong et al. 2016).

Alkaline pretreatment at pH > 12 resulted in a considerable reduction of ARGs by 0.62–1.36 log units in spectinomycin mycelial residues, attributed to the hydrolysis of phosphodiester bond of DNA in strong alkaline solution (Song et al. 2020). Huang et al. (2017) revealed that alkaline fermentation at pH > 12 reduced the gene copies of *sul*1, *sul*2, *tet*(C), *tet*(O), *tet*(Q), and *tet*(X) by 0.87 to 1.04 log units in waste activated sludge. Zou et al. (2023) demonstrated that alkaline pretreatment at pH > 11 reduced the total absolute abundances of ARGs including bla_{TEM}, *sul*1, *sul*2, *tet*(A), *tet*(Q), and *tet*(X) by 1.0 log unit from sewage sludge. Remarkably, intracellular and extracellular ARGs respond differently during alkaline pretreatment. The alkaline pretreatment at pH > 12 achieved the maximum removal for intracellular *tet*(X) by 1.7 log units, while alkaline pretreatment at pH > 11 achieved the maximum removal of extracellular bla_{TEM} by 1.4 log units (Zou et al. 2023).

Ultrasonic pretreatment at 30–60 min under 240 W and 40 kHz significantly reduces the relative abundance of *sul*1, *sul*2, *tet*(A), *tet*(O), and *tet*(W) in sewage sludge, with the removal efficiency of 34.8–41.7% (Mortezaei et al. 2024). Whether ultrasonic pretreatment resulted in ARG rebound during subsequent anaerobic digestion is still being debated. Ultrasonic pretreatment followed by anaerobic digestion resulted in greater ARG reduction compared to the control anaerobic digestion system without pretreatment. Most of targeted ARGs can be further reduced during subsequent mesophilic anaerobic digestion, with an overall removal efficiency ranged from 43.8% to 80.3% (Mortezaei et al. 2024). Li et al. (2023b) argued that ultrasonic pretreatment (0.2 W/mL power density for 30 min) significantly increased the abundance of bacteria carrying ARGs such as tetracycline-resistance bacteria and the conjugation transfer frequency in sewage sludge during mesophilic anaerobic digestion. Although ultrasonic pretreatment (20 kHz, 2.5 W/mL, 30 min) reduced the relative abundances of bla_{TEM}, *erm*(B), and *intI*1 by 67.6%, 21.7%, and 56.8%, respectively, in waste activated sludge, most of them could be rebounded when incubated at 30°C for 10 days (Wang et al. 2021a).

4.2 Other Methods

In addition to manipulating the operational parameters of anaerobic digesters, the addition of conductive materials including biochar (Sun et al. 2018; Xu et al. 2022b), graphene (Wang et al. 2021b), and zero-valent iron (Li et al. 2023a; Xiang et al. 2019) has been reported to be effective in mitigating ARGs during anaerobic digestion of sewage sludge. For instance, Xu et al. (2022b) showed that biochar generated from pig manure significantly reduced the total abundance of ARGs by 34.9% in the liquid phase during anaerobic co-digestion of pig manure and sewage sludge. Similarly, bamboo biochar at the concentrations of 20 g/L reduced the relative abundances of *tet*(C), *tet*(G), *tet*(W), and *tet*(X) by 71.6% during anaerobic digestion of cattle farm wastewater compared to the control without the addition of biochar (Sun et al.

2018). Graphene and graphene oxide enhance ARG removal. For example, graphene showed the greatest elimination effects of *bla*$_{OXA-1}$, *erm*(B), *erm*(F), *tet*(Q), and *tet*(X), with removal efficiency of 57.75%–96.11%; while graphene oxide had the most obvious removal effect on *sul*1, *sul*2, *tet*(M), *tet*(O), and *tet*(W), with removal efficiency of 75.66%–90.50%, during anaerobic co-digestion of sewage sludge and food waste (Wang et al. 2021b). In addition, zero-valent iron has been shown to enhance ARG removal during anaerobic digestion. For instance, the addition of zero-valent iron at 100 mg/g total solids during thermophilic anaerobic digestion of waste activated sludge reduced the abundance of 15 ARG types and halved the number of high-risk ARGs. High-risk ARGs here represent ARGs with the genome sequences that contain mobile genetic elements and are associated with human pathogenic species (Li et al. 2023a). Gao et al. (2017) indicated that absolute abundances of *tet* genes including *tet*(A), *tet*(C), *tet*(G), *tet*(M), *tet*(O), *tet*(W), and *tet*(X) were reduced by 1.44–3.91 log units with the addition of zero-valent iron, significantly higher than the reduction of 1.56–2.81 log units in the control without the addition of zero-valent iron, during thermophilic anaerobic co-digestion of waste sludge and kitchen waste (Gao et al. 2017). The addition of conductive materials serves as an effective strategy for attenuating ARGs during anaerobic digestion of sewage sludge. However, their impact on the stability and performance of the digestion system, such as methane production and sludge reduction, must be considered. Additionally, the environmental and ecological risks of applying digested sludge with these additives to agricultural soils need to be thoroughly evaluated.

5 Future Perspectives

The differentiation between intracellular and extracellular ARGs offers an opportunity to accurately quantify the ARGs and provides insights into the effects of anaerobic digestion on ARGs in sludge. Thus, this understanding can further inform the development of digesters aimed at controlling the transmission of ARGs. Additionally, understanding the occurrence and fate of phage associated ARGs during anaerobic digestion will enhance our comprehension of the role that phage communities play in the ARG transmission. This knowledge is crucial for improving ARG risk assessments related to the land application of sewage sludge. However, current studies on ARGs in the phage fraction face a significant challenge in distinguishing phage particles from vesicle-like particles. Consequently, results obtained from both metagenomics and qPCR methods often conflate the fate of ARGs in phage with that in vesicle-like particles. This limitation underscores the need for improved methodologies that can accurately differentiate phage particles from vesicle-like particles. Furthermore, the underlying mechanism regarding the ARG reduction under various operational conditions needs to be investigated. Whether anaerobic digestion lowers the risk of ARGs (e.g., ARGs carried by the mobile genetic elements and associated with human pathogens) warrants further study using new risk assessment models, which will be helpful for the ARG risk assessment. Moreover, while pretreatments have been applied to mitigate the ARGs from sewage sludge, the rebound effect in the subsequent anaerobic digestion warrants further investigation. Although the addition of conductive additives enhances the ARG removal, future research could investigate those combined effects on both the performance of anaerobic digesters in eliminating ARGs and bioenergy production.

6 Take-Home Messages

(1) ARGs are present in both intracellular and extracellular forms, as well as within bacteriophage fractions. The distinction between intracellular and extracellular ARGs is significant, as they may respond differently to anaerobic digestion. Investigating the distribution and dynamics of both forms of ARGs is essential for understanding how anaerobic digestion affects ARGs and for developing strategies to mitigate their spreading.
(2) Thermophilic anaerobic digestion typically exhibits superior performance compared to mesophilic anaerobic digestion in the removal of ARG from sewage sludge.
(3) The influence of SRT on ARG removal from sewage sludge is contingent upon various factors such as operational configuration, temperature, pretreatment methods, and targeted ARG subtypes.
(4) Alkaline conditions are more effective than acidic conditions in removing ARGs from sewage sludge.
(5) Pretreatment methods, including thermal hydrolysis pretreatment, alkaline pretreatment, microwave pretreatment, and ultrasonic pretreatment, showed positive effects on the reduction of sludge ARGs. However, rebound effects of tested ARGs are commonly observed during subsequent anaerobic digestion, and current studies have not yet provided sufficient evidence to fully uncover the underlying mechanism.

References

Alcock BP, Huynh W, Chalil R, Smith KW, Raphenya AR, Wlodarski MA, Edalatmand A, Petkau A, Syed SA, Tsang KK (2023) CARD 2023: expanded curation, support for machine learning, and resistome prediction at the comprehensive antibiotic resistance database. Nucleic Acids Res 51(D1):D690–D699

Auerbach EA, Seyfried EE, McMahon KD (2007) Tetracycline resistance genes in activated sludge wastewater treatment plants. Water Res 41(5):1143–1151

Cai C, Huang X, Dai X (2022a) Differential variations of intracellular and extracellular antibiotic resistance genes between treatment units in centralized sewage sludge treatment plants. Water Res 222:118893

Cai C, Hui X, Yang W, Hua Y, Liu H, Dai X (2022b) Implications for mitigation of antibiotic resistance: differential response of intracellular and extracellular antibiotic resistance genes to sludge fermentation coupled with thermal hydrolysis. Water Res 209:117876

Calero-Cáceres W, Melgarejo A, Colomer-Lluch M, Stoll C, Lucena F, Jofre J, Muniesa M (2014) Sludge as a potential important source of antibiotic resistance genes in both the bacterial and bacteriophage fractions. Environ Sci Technol 48(13):7602–7611

Chen Y, Huang H, Zheng X (2019) Fate of sulfonamide resistance genes during sludge anaerobic fermentation: roles of sludge components and fermentation pHs. Biores Technol 289:121636

Diehl DL, LaPara TM (2010) Effect of temperature on the fate of genes encoding tetracycline resistance and the integrase of class 1 integrons within anaerobic and aerobic digesters treating municipal wastewater solids. Environ Sci Technol 44(23):9128–9133

Dong P, Wang H, Fang T, Wang Y, Ye Q (2019) Assessment of extracellular antibiotic resistance genes (eARGs) in typical environmental samples and the transforming ability of eARG. Environ Int 125:90–96

Gao P, Gu C, Wei X, Li X, Chen H, Jia H, Liu Z, Xue G, Ma C (2017) The role of zero valent iron on the fate of tetracycline resistance genes and class 1 integrons during thermophilic anaerobic co-digestion of waste sludge and kitchen waste. Water Res 111:92–99

Ghosh S, Ramsden SJ, LaPara TM (2009) The role of anaerobic digestion in controlling the release of tetracycline resistance genes and class 1 integrons from municipal wastewater treatment plants. Appl Microbiol Biotechnol 84:791–796

Guo J, Li J, Chen H, Bond PL, Yuan Z (2017) Metagenomic analysis reveals wastewater treatment plants as hotspots of antibiotic resistance genes and mobile genetic elements. Water Res 123:468–478

Haffiez N, Azizi SMM, Zakaria BS, Dhar BR (2022) Propagation of antibiotic resistance genes during anaerobic digestion of thermally hydrolyzed sludge and their correlation with extracellular polymeric substances. Sci Rep 12(1):6749

Haffiez N, Zakaria BS, Azizi SMM, Dhar BR (2023) Fate of intracellular, extracellular polymeric substances-associated, and cell-free antibiotic resistance genes in anaerobic digestion of thermally hydrolyzed sludge. Sci Total Environ 855:158847

Huang H, Chen Y, Zheng X, Su Y, Wan R, Yang S (2016) Distribution of tetracycline resistance genes in anaerobic treatment of waste sludge: the role of pH in regulating tetracycline resistant bacteria and horizontal gene transfer. Biores Technol 218:1284–1289

Huang H, Zheng X, Chen Y, Liu H, Wan R, Su Y (2017) Alkaline fermentation of waste sludge causes a significant reduction of antibiotic resistance genes in anaerobic reactors. Sci Total Environ 580:380–387

Leiva AM, Piña B, Vidal G (2021) Antibiotic resistance dissemination in wastewater treatment plants: a challenge for the reuse of treated wastewater in agriculture. Rev Environ Sci Bio/technol 20:1043–1072

Li W, Pang L, Chatzisymeon E, Yang P (2023a) Effects of micron-scale zero valent iron on behaviors of antibiotic resistance genes and pathogens in thermophilic anaerobic digestion of waste activated sludge. Biores Technol 376:128895

Li X, Chen G, Liu L, Wang G (2023b) Anaerobic sludge digestion elevates dissemination risks of bacterial antibiotic resistance in effluent supernatant. J Environ Manage 338:117854

Liu W, Xu C, Li T, Ren Z, Hao S, Chen Z, Huang X, Wen X (2024) Temporal dynamics and contribution of phage community to the prevalence of antibiotic resistance genes in a full-scale sludge anaerobic digestion plant. Environ Sci Technol 58(14):6296–6304

Liu B, Pop M (2009) ARDB—antibiotic resistance genes database. Nucleic Acids Res 37(suppl_1):D443–D447

Luo H, Sun Y, Taylor M, Nguyen C, Strawn M, Broderick T, Wang ZW (2022) Impacts of aluminum- and iron-based coagulants on municipal sludge anaerobic digestibility, dewaterability, and odor emission. Water Environ Res 94(1):e1684

Ma Y, Wilson CA, Novak JT, Riffat R, Aynur S, Murthy S, Pruden A (2011) Effect of various sludge digestion conditions on sulfonamide, macrolide, and tetracycline resistance genes and class I integrons. Environ Sci Technol 45(18):7855–7861

Martínez JL (2008) Antibiotics and antibiotic resistance genes in natural environments. Science 321(5887):365–367

Mortezaei Y, Williams MR, Demirer GN (2023) Effect of temperature and solids retention time on the removal of antibiotic resistance genes during anaerobic digestion of sludge. Bioresour Technol Rep 21:101377

Mortezaei Y, Williams MR, Demirer GN (2024) The fate of antibiotic resistance genes during anaerobic digestion of sewage sludge with ultrasonic pretreatment. Environ Sci Pollut Res 31(4):5513–5525

Pei J, Yao H, Wang H, Ren J, Yu X (2016) Comparison of ozone and thermal hydrolysis combined with anaerobic digestion for municipal and pharmaceutical waste sludge with tetracycline resistance genes. Water Res 99:122–128

Redhead S, Nieuwland J, Esteves S, Lee DH, Kim DW, Mathias J, Cha CJ, Toleman M, Dinsdale R, Guwy A (2020) Fate of antibiotic resistant *E. coli* and antibiotic resistance genes during full scale conventional and advanced anaerobic digestion of sewage sludge. PLoS One 15(12):e0237283

Song S, Jiang M, Yao J, Liu H, Dai X, Wang G (2020) Alkaline-thermal pretreatment of spectinomycin mycelial residues: insights on anaerobic biodegradability and the fate of antibiotic resistance genes. Chemosphere 261:127821

Sun W, Gu J, Wang X, Qian X, Tuo X (2018) Impacts of biochar on the environmental risk of antibiotic resistance genes and mobile genetic elements during anaerobic digestion of cattle farm wastewater. Biores Technol 256:342–349

Sun C, Li W, Chen Z, Qin W, Wen X (2019) Responses of antibiotics, antibiotic resistance genes, and mobile genetic elements in sewage sludge to thermal hydrolysis pre-treatment and various anaerobic digestion conditions. Environ Int 133:105156

Sun Y, Luo H, Iboleon R, Wang Z (2022) Fate of antibiotic resistance genes and class 1 integrons during sludge treatment using pilot-scale anaerobic digestion with thermal hydrolysis pretreatment. Biores Technol 364:128043

Tian Z, Zhang Y, Yu B, Yang M (2016) Changes of resistome, mobilome and potential hosts of antibiotic resistance genes during the transformation of anaerobic digestion from mesophilic to thermophilic. Water Res 98:261–269

Tong J, Liu J, Zheng X, Zhang J, Ni X, Chen M, Wei Y (2016) Fate of antibiotic resistance bacteria and genes during enhanced anaerobic digestion of sewage sludge by microwave pretreatment. Biores Technol 217:37–43

Tong J, Fang P, Zhang J, Wei Y, Su Y, Zhang Y (2019) Microbial community evolution and fate of antibiotic resistance genes during sludge treatment in two full-scale anaerobic digestion plants with thermal hydrolysis pretreatment. Biores Technol 288:121575

Wang FH, Qiao M, Su JQ, Chen Z, Zhou X, Zhu YG (2014) High throughput profiling of antibiotic resistance genes in urban park soils with reclaimed water irrigation. Environ Sci Technol 48(16):9079–9085

Wang M, Li R, Zhao Q (2019) Distribution and removal of antibiotic resistance genes during anaerobic sludge digestion with alkaline, thermal hydrolysis and ultrasonic pretreatments. Front Environ Sci Eng 13:1–10

Wang M, Chen H, Liu S, Xiao L (2021a) Removal of pathogen and antibiotic resistance genes from waste activated sludge by different pre-treatment approaches. Sci Total Environ 763:143014

Wang P, Zheng Y, Lin P, Li J, Dong H, Yu H, Qi L, Ren L (2021b) Effects of graphite, graphene, and graphene oxide on the anaerobic co-digestion of sewage sludge and food waste: attention to methane production and the fate of antibiotic resistance genes. Biores Technol 339:125585

Wu Y, Cui E, Zuo Y, Cheng W, Rensing C, Chen H (2016) Influence of two-phase anaerobic digestion on fate of selected antibiotic resistance genes and class I integrons in municipal wastewater sludge. Biores Technol 211:414–421

Xiang Y, Yang Z, Zhang Y, Xu R, Zheng Y, Hu J, Li X, Jia M, Xiong W, Cao J (2019) Influence of nanoscale zero-valent iron and magnetite nanoparticles on anaerobic digestion performance and macrolide, aminoglycoside, β-lactam resistance genes reduction. Biores Technol 294:122139

Xu L, Gu J, Wang X, Song Z, Jiang H, Li N, Lei L, Xie J, Hu T, Ding Q (2022a) Risk of horizontal transfer of intracellular, extracellular, and bacteriophage antibiotic resistance genes during anaerobic digestion of cow manure. Biores Technol 351:127007

Xu S, Duan Y, Zou S, Liu H, Luo L, Wong JW (2022b) Evaluations of biochar amendment on anaerobic co-digestion of pig manure and sewage sludge: waste-to-methane conversion, microbial community, and antibiotic resistance genes. Biores Technol 346:126400

Yin X, Deng Y, Ma L, Wang Y, Chan LY, Zhang T (2019) Exploration of the antibiotic resistome in a wastewater treatment plant by a nine-year longitudinal metagenomic study. Environ Int 133:105270

Yin X, Jiang XT, Chai B, Li L, Yang Y, Cole JR, Tiedje JM, Zhang T (2018) ARGs-OAP v2.0 with an expanded SARG database and Hidden Markov Models for enhancement characterization and quantification of antibiotic resistance genes in environmental metagenomes. Bioinformatics 34(13):2263–2270

Yuan Q, Yu P, Cheng Y, Zuo P, Xu Y, Cui Y, Luo Y, Alvarez PJ (2022) Chlorination (but not UV disinfection) generates cell debris that increases extracellular antibiotic resistance gene transfer via proximal adsorption to recipients and upregulated transformation genes. Environ Sci Technol 56(23):17166–17176

Zhang Y, Snow DD, Parker D, Zhou Z, Li X (2013) Intracellular and extracellular antimicrobial resistance genes in the sludge of livestock waste management structures. Environ Sci Technol 47(18):10206–10213

Zhang T, Yang Y, Pruden A (2015) Effect of temperature on removal of antibiotic resistance genes by anaerobic digestion of activated sludge revealed by metagenomic approach. Appl Microbiol Biotechnol 99:7771–7779

Zhang J, Chen M, Sui Q, Wang R, Tong J, Wei Y (2016) Fate of antibiotic resistance genes and its drivers during anaerobic co-digestion of food waste and sewage sludge based on microwave pretreatment. Biores Technol 217:28–36

Zhang J, Liu J, Wang Y, Yu D, Sui Q, Wang R, Chen M, Tong J, Wei Y (2017) Profiles and drivers of antibiotic resistance genes distribution in one-stage and two-stage sludge anaerobic digestion based on microwave-H_2O_2 pretreatment. Biores Technol 241:573–581

Zhang J, Liu J, Lu T, Shen P, Zhong H, Tong J, Wei Y (2019) Fate of antibiotic resistance genes during anaerobic digestion of sewage sludge: Role of solids retention times in different configurations. Biores Technol 274:488–495

Zhang J, Lu T, Song Y, Rocha UNd, Liu J, Nikolausz M, Wei Y, Richnow HH (2024) Viral communities contribute more to the lysis of antibiotic-resistant bacteria than the transduction of antibiotic resistance genes in anaerobic digestion revealed by metagenomics. Environ Sci Technol 58(5):2346–2359

Zou Y, Tu W, Wang H, Fang T (2020) Anaerobic digestion reduces extracellular antibiotic resistance genes in waste activated sludge: the effects of temperature and degradation mechanisms. Environ Int 143:105980

Zou Y, Tu W, Wu M, Liu X, Wang H (2023) Fates of intracellular and extracellular antibiotic resistance genes during sludge anaerobic digestion with different pretreatments. Chem Eng J 454:140356

Kinetic Modelling of Typical Biological Reactions During Anaerobic Digestion

Guangxue Wu, Huanhuan Chang, and Bang Du

Abstract In parallel with novel process developments and experimental research in anaerobic digestion, considerable effort has been put into kinetic modelling. The use of mathematical models results in a better understanding of the system performance and process dynamics, including microbial growth and substrate degradation. Typically, various kinetic models, such as the Monod model, are employed to describe microbial growth dynamics. Substrate degradation within anaerobic digestion systems relies on the cooperation among multiple steps, with hydrolysis or methanogenesis often serving as the rate-limiting step. Additionally, the effluent substrate concentration (or the biogas production) can be estimated through models without detailed steps, such as the Chen and Hashimoto model. Ultimately, the accuracy and reliability of kinetic models depend on the goodness of fit to experimental data, ensuring robust evaluation of system performance based on the established kinetic models.

Keywords Biogas yield · Kinetic modelling · Microbial growth · Substrate degradation

1 Introduction

The practical application of anaerobic digestion depends not only on the acclimation of functional microorganisms, but also on their reaction kinetics. Microbial kinetics control the required reaction duration for anaerobic digestion, which can be described by diverse types of kinetic models. This means that kinetic model can provide valuable information for the reaction rate and conversion potential of biodegradation of organic carbon, which can further reveal the bottlenecks limiting digestibility of feedstock and methane production (Karki et al. 2022). In addition, kinetic model can provide key parameters for designing anaerobic digestion systems such as the

G. Wu (✉) · H. Chang · B. Du
Civil Engineering, School of Engineering, College of Science and Engineering, University of Galway, Galway H91 TK33, Ireland
e-mail: guangxue.wu@universityofgalway.ie

G. Wu (ed.), *Anaerobic Digestion*, Green Energy and Technology,
https://doi.org/10.1007/978-3-031-69378-6_11

reaction rate, the maximum methane production, and the lag phase, which are crucial for the analysis of system performance such as the inhibition and system efficiency. Finally, through the kinetic model, the correlation between system performance and kinetic characteristics can be obtained, which can reveal the kinetic mechanisms of system instability (Li et al. 2018).

Optimization of kinetic models of anaerobic digestion is an iterative process, which requires comprehensive understanding of potential kinetic mechanisms. For example, different modelling parameters can provide a good insight into the bioreaction mechanism of anaerobic digestion (Hassan et al. 2017). These fundamental understanding of the reaction mechanism and kinetic parameters can benefit practical applications such as the design of anaerobic digesters, the prediction of biogas yield, and the design of hydraulic and sludge retention times (Wang et al. 2021). On the other hand, mechanistic studies lead to in-depth understanding of complex transport phenomena, microbial biochemical kinetics, and stochiometric relationships associated with anaerobic digestion, which can further advance the kinetic modelling (Yu et al. 2013).

2 Microbial Growth in Anaerobic Digestion Systems

2.1 Microbial Growth Phases and Their Corresponding Reaction Rates

Generally, microbial growth under batch conditions can be divided into lag phase, exponential phase, stationary phase, and decay phase (Fig. 1). The concept proposed by Angelidaki et al. (1983) is introduced, where the microbial growth phases and their corresponding reaction rates are presented. The lag phase (also known as the adaptation phase) represents the duration before biogas production, which can also indicate the duration for microorganisms to acclimatize to the environment (such as the quality of the substrate, the concentration of biomass, and the concentration of inhibitors) (Njoya et al. 2021). Therefore, the lag phase is determined by both the microorganisms' physiology status and the environmental conditions. In the acceleration phase, microbial growth rate increases, while this phase can be neglected under most cases and is not included in modelling; microbial growth mainly occurs during the exponential phase, where microbial growth rate is constant (Kythreotou et al. 2014).

The lag time (λ) is a key parameter for kinetic modelling, especially for methane production. From the technology application aspect, the anaerobic digestion duration is also an important design parameter. Therefore, the effective digestion time (T_{ef}) has been proposed, which can be calculated by subtracting the lag time (λ) from the time taken to achieve 80% or 90% of the maximum cumulative methane production (T_{80} or T_{90}) (Li et al. 2018). Generally, a shorter T_{ef} with a longer λ indicates a shorter anaerobic digestion period and irreversible process inhibition, a longer T_{ef} with

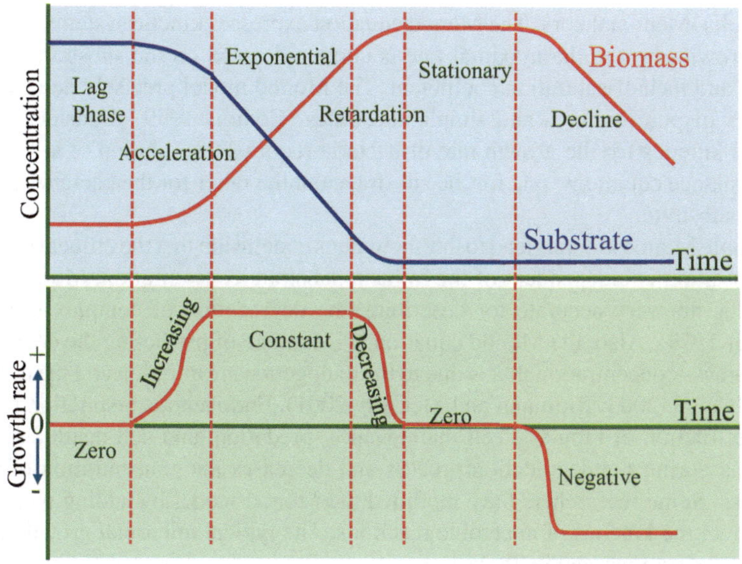

Fig. 1 Phases of microbial growth and the respective growth rate. Revised from Angelidaki et al. (1983)

a longer λ indicates longer periods of anaerobic digestion and reversible process inhibition, and a shorter T_{ef} with a shorter λ could reveal a high biogas production rate and a short anaerobic digestion period (Mao et al. 2017).

Figure 1 mainly describes scenarios from feed-batch reactors, where dynamic substrate and biomass concentrations occur. The reactor configuration and operational mode can also affect microbial kinetics, such as the plug-flow reactor system and the system with biofilm growth should be different from the continuously stirred reactor system (Angelidaki et al. 1983). Therefore, kinetic parameters for describing microbial growth should be measured separately for discontinuous and continuous feeding processes (Kythreotou et al. 2014). However, the underlying principles from one reactor configuration or operational condition can be applied to other reactor configurations and operational conditions if revised properly.

2.2 *Kinetic Models for Microbial Growth*

The basis for modelling the kinetics of bacterial growth was derived by Michaelis and Menten (1913). The model describes the enzyme activity depending on the substrate concentration. This dependency can be transferred to bacterial growth because microbial growth is also an autocatalytic reaction. Monod (1949) identified the non-linear relationship between the specific growth rate and the limiting substrate concentration while studying the bacterial cultures growth and its parallelism to the

Michaelis–Menten theory. The Monod equation expresses kinetic systems involving cells growing where the maximal rate is corrected based on the substrate concentration and its half-saturation coefficient. The Monod model precisely describes the kinetics of pure cultures and simple substrates (Contois 1959). Consideration of Monod kinetics for the growth rate of bacteria requires the solution of at least two mass balance equations, one for the substrate and the other for the bacteria growing on the substrate.

Simple Monod kinetics lead to the anomalous conclusion that the effluent substrate concentration is independent of the influent substrate concentration. Moreover, the model is not very accurate for describing the degradation of complex substrates (Pfeffer 1974). Also, the Monod equation is incapable of predicting the decrease of the biomass concentration that is due to the endogenous respiration and the cell lysis (Gavala et al. 2003; Rittmann and McCarty 2001). Endogenous respiration (i.e. the self-destruction of biomass, cell maintenance, predation, and cell death and lysis) happens during the degradation process and decreases the concentration of active biomass. Some researchers have modified the Monod model by adding parameters that affect the kinetics of anaerobic reactions. The typical microbial growth kinetic models are summarized in Table 1.

Moser (1958) upgraded the Monod model with a parameter m (usually $m > 1$) to integrate effects of adaption of microorganisms to stationary processes by mutation. The Moser model describes microbial growth which differs from the exponential characteristic. For $m = 1$, the expression of the Moser model aligns with that of the Monod model. The Contois equation (Contois 1959) involves not only the substrate dependency, but also the cell concentration to calculate the specific growth rate. The Contois equation considers the effects of both limiting substrate and biomass concentration on microbial growth (Wang and Li 2014). The major difference is that the effluent substrate concentration is related to the influent substrate concentration in the Contois equation whilst it is totally independent of the influent substrate concentration in the Monod equation (Hu et al. 2002). Furthermore, when existing more slowly biodegradable organic carbons, their degradation requires the production of extracellular enzymes by microorganisms for hydrolysis, which requires a long duration. In such cases, the Contois equation may be more appropriate to describe the kinetics of this situation (Hu et al. 2002; Noike et al. 1985). The Contois model yields good results both for discontinuous and continuous processes, but its capability to model dynamic processes is strongly limited (Beba and Atalay 1986).

McCarty (1965) modified the Monod equation considering the endogenous respiration and the cell lysis. To account for the effect of loss of active biomass on the net growth rate, a microorganism decay rate has been defined. Generally, the decay coefficient (b) is around 5% of the maximum specific growth rate, while the decay coefficient for the methanogens is about 1% of their maximum specific growth rate, which can be ignored when modelling and simulations of anaerobic processes (Angelidaki et al. 1983; Gavala et al. 2003; Maleki et al. 2018). For the conversion of acetic acid to biogas by *Methanosarcina barkeri*, around 95% of the acetic acid is converted to biogas, 3% to cell material, and 2% to energy supply (Webb 1963).

Table 1 Kinetic models applied for microbial growth in anaerobic digestion systems

Model	Microbial growth kinetics	Remarks	References
Monod	$\mu = \mu_{\max}\frac{S}{K_s+S}$	• Accurate for pure cultures and simple substrates • Not accurate for degradation of complex substrates and heterogeneous cultures	Monod (1949)
Moser	$\mu = \mu_{\max}\cdot\frac{S^m}{S^m+K_s}$	• $m > 1$, integrates effects of adoption of microorganisms to stationary processes by mutation • $m = 1$, Monod model	Moser (1958)
Contois	$\mu = \frac{\mu_{\max}S}{K_X X+S}$	Involves both substrate and microorganism concentrations	Contois (1959)
McCarty	$\mu = \frac{\mu_{\max}S}{K_s+S} - b$	Microorganism decay rate has been defined	McCarty (1965)
Powell	$\mu = \mu_{\max}\cdot\frac{(K+L+S)}{2L}\left(1-\sqrt{1-\frac{4LS}{(K+L+S)^2}}\right)$	• L is for diffusion & permeability of the substrate through the cell wall • K relates to enzyme activity	Powell (1967)
Grau	$\mu = \frac{\mu_{\max}S}{S_0} - b$	Consists of a second order chemical reaction kinetics combined with the Monod model	Grau et al. (1975)
Chen & Hashimoto	$\mu = \frac{\mu_{\max}S}{KS_0+(1-K)S} - b$	Cell concentration is related to initial substrate concentration and substrate concentration (S/S_0)	Chen and Hashimoto (1978)

(continued)

Table 1 (continued)

Model	Microbial growth kinetics	Remarks	References
Bergter	$\mu = \mu_{\max} \frac{S}{K_s+S}\left[1 - \exp(-t/T)\right]$	• Considered effect of deceleration during the lag phase • t: time • T: lag time	Bergter (1983)
Modified Stover-Kincannon	$\frac{dS}{dt} = \frac{U_{\max}\frac{QS_0}{V}}{K_B+\frac{QS_0}{V}}$	Developed for anaerobic filters	Stover and Kincannon (1982), Yu et al. (1998)

Revised from Maleki et al. (2018)

μ is the specific growth rate; $\mu_{\max}$ is the maximum specific growth rate; S_0 and S are the substrate concentrations in the influent and the studied time, respectively; K_S is the half-saturation coefficient; m is the Moser index; X is the microorganism concentration; b is the specific microorganism decay rate; K_X is Contois kinetic constant; K is the Chen and Hashimoto kinetic constant; Q is the influent flow rate; V is the volume of the reactor; $U_{\max}$ is the constant of maximum consumption; and K_B is the constant of saturation value.

Powell (1967) included diffusion and permeation of substrate in addition to reaction kinetics through the cell wall with two additional parameters K and L. K describes the kinetics of growth due to enzyme activity, and the parameter L describes the diffusion and permeability. Grau et al. (1975) developed a linear multi-component substrate model to describe and predict substrate concentration, which consists of a second order chemical reaction kinetics combined with the Monod model. Chen and Hashimoto (1978) modified a model from Contois, to include the cell concentration via the relation between the observed substrate concentration and the initial substrate concentration. The Chen and Hashimoto's model predicts the total substrate utilization rate as a function of residence time for wastes with known refractory content. The Chen and Hashimoto model has been recognized as a reliable tool in predictions regarding digestion of wastes with a substantial total solids content (Fongsatitkul et al. 2012). Moreover, the Chen and Hashimoto model has been satisfactorily used for both the continuous and batch anaerobic digestion processes (Chen and Hashimoto 1980; Kafle et al. 2013; Kafle and Kim 2012). The modified Monod model by Bergter (1983) considers deceleration during the lag phase by introducing an exponential part of the mathematical model. Stover and Kincannon (1982) modified the Monod model for rotating biological contactors. According to this model, both the hydraulic loading rate ($\frac{Q}{V}$) and the influent substrate concentration (S_o) affect substrate removal rate and removal efficiency. Therefore, the total organic loading rate ($\frac{QS_o}{V}$) is applied to replace the substrate concentration in the equation.

2.3 The Goodness of Fit of Kinetic Models

In kinetic modelling of anaerobic digestion, various metrics beyond the correlation coefficient (R^2) have been applied to assess the performance of kinetic models. These includes indexes such as the difference between the measured (B_{measured}) and predicted ($B_{\text{predicted}}$) results [Diff.(%)], the residual sum of squares (RSS), the root mean square error (RMSE), the Bayesian information criterion (BIC), as well as the Akaike's information criterion (AIC). These metrics offer a more comprehensive evaluation of model accuracy and reliability (Karki et al. 2022; Zhen et al. 2016). The applied goodness of fit of kinetic models is summarized in Table 2.

The Theil's Inequality Coefficient (TIC) allows judging whether there is a considerable difference between measured and predicated results. A value of the TIC < 0.3 indicated a good agreement between experimental and predicated data (Wang et al. 2015). RMSE is interpreted as the standard deviation between the predicted and measured values with a lower RMSE indicating a better fit. AIC is an alternative method used for comparing models and determining which model is more likely to be correct using the information theory. The AIC value can be positive or negative, which varies using different units for the express data, and the smallest absolute value of AIC is the most likely correct one (Abdelwahab et al. 2023).

The adoption of these error functions can affect the model comparison. For example, Karki et al. (2022) observed that regarding the same feedstock, different

Table 2 Error functions applied for comparison of different kinetic models used for anaerobic digestion systems

Error functions	Equations	Remarks	References
Theil's Inequality Coefficient (TIC)	$\text{TIC} = \frac{\sqrt{\sum_t (B_{\text{measured}} - B_{\text{predicted}})^2}}{\sqrt{\sum_t B_{\text{measured}}^2} + \sqrt{\sum_t B_{\text{predicted}}^2}}$	TIC < 0.3 indicating a good agreement	Zhou (1993)
The residual sum of squares (RSS)	$\text{RSS} = \sum_{i-1}^{N} (B_{\text{predicted}} - B_{\text{measured}})^2$	A lower RSS indicating a better fit	Karki et al. (2022)
The root mean square error (RMSE)	$\text{RMSE} = \sqrt{\frac{\text{RSS}}{N}} = \sqrt{\frac{\sum_{i-1}^{N} (B_{\text{predicted}} - B_{\text{measured}})^2}{N}}$	A lower RMSE indicating a better fit	Wang et al. (2011)
Akaike's information criterion (AIC)	$\text{AIC} = \begin{cases} N \ln\left(\frac{\text{RSS}+2n}{N}\right), \text{ when } \frac{N}{n} \geq 40 \\ N \ln\left(\frac{\text{RSS}}{N}\right) + 2n + \frac{2n(n+1)}{N-n-1}, \text{ when } \frac{N}{n} < 40 \end{cases}$	A smaller absolute value of AIC indicating a better fit	Motulsky and Christopoulos (2003)
The Bayesian information criterion (BIC)	$\text{BIC} = N \ln\left(\frac{\text{RSS}}{N}\right) + n \ln(N)$	A lower BIC indicating a better fit	Nguyen et al. (2019)

N is number of data points; n is the number of model parameters best-fitting models can be obtained, in which the modified Gompertz model with the second order equation is the best based on R^2 and RMSE, and the Cone model is the best based on AIC. Generally, R^2 and AIC are more useful for comparison of models, while R^2 and RMSE can be adopted for overall comparison (Karki et al. 2022).

3 Kinetics for Limiting Stages of Anaerobic Digestion

From a biochemical point of view, anaerobic digestion may be described as four subsequent steps: hydrolysis, acidogenesis, acetogenesis, and methanogenesis (Appels et al. 2008; Lyberatos and Skiadas 1999). However, from a process-engineering point of view, the biogas production process can be divided into two steps: an acidification step, comprising both hydrolysis and acidogenesis, and a methane production step, comprising both acetogenesis and methanogenesis (Karakashev et al. 2006; Shin and Song 1995). The simplistic assumption of a two-step process does not account for the complexity of biochemical and biological interactions at work in anaerobic digestion systems. Nevertheless, this basic differentiation is often applied to describe the overall balance of the biological system. Following basic chemistry kinetics, these two steps may appear as two consecutive first-order reactions: the first step of acidification generates intermediate products, i.e. volatile fatty acids (VFAs), which are then converted into the final product, i.e. biogas, in the second step of methane production (Hobson and Wheatley 1993; Shin and Song 1995). Generally, the limiting step of anaerobic digestion is the hydrolysis stage for solid waste and organic matters with a high proportion of recalcitrant components, while the methanogenesis process is the limiting step for organic matters with high biodegradability (Pavlostathis and Giraldo-Gomez 1991).

3.1 Kinetic Models for Hydrolysis

Particulate organic material must be hydrolyzed to solutes capable of being actively or passively transported across cell membranes before they can be microbially metabolized. There are two mechanisms to describe the hydrolytic process (Batstone et al. 2002): (i) the microbes secrete into the bulk liquid enzymes that adsorb onto a particle or react with a soluble substrate (Jain et al. 1992), and (ii) the microbes adhere to a particle and consume soluble products released by reactions that were catalyzed by enzymes that they locally produced (Vavilin et al. 1996) (Table 3).

The prevalent model for simulating hydrolysis typically employs a first-order rate equation in terms of the concentration of degradable organic material (Vavilin et al. 2008). Vavilin et al. (2008) summarizes a wide range of values of the first-order rate coefficient obtained from composite and simple organic materials, including carbohydrates, lipids, and proteins sources. The wide range of values could be due to different experimental conditions, different hydrolytic biomass to substrate ratios,

Table 3 Kinetic models for hydrolysis during anaerobic digestion

Model	Hydrolysis kinetics	Remarks	References
First order	$\frac{dS}{dt} = -kS$	k is the first-order coefficient	Vavilin et al. (2008)
Michaelis–Menten	$\frac{dS}{dt} = k_h E \frac{S}{K_m+S} = V_m \frac{S}{K_m+S}$	E is the enzyme concentration; $V_m = k_h E$ is the maximum hydrolysis rate; k_h is the maximum hydrolysis rate constant; and K_m is the half-saturation rate coefficient	Batstone et al. (2002)
Non-degradability in the first-order kinetics	$\frac{dS}{dt} = -k(S - \beta S_0)$	β is the non-degradable fraction of the substrate	Vavilin et al. (2008)
Surface-related hydrolysis kinetics	$\rho = \hat{\rho}_m \frac{\beta X}{1+\beta X} \frac{S}{K_S+S}$	ρ and $\hat{\rho}_m$ are the current and maximum hydrolysis rates, respectively; β is the equilibrium constant	Vavilin et al. (1996)

and the lumped effect of disintegration and hydrolysis. First-order kinetics can only be applied when the rate-limiting factor is the surface of the particulate substrate, and bioavailability or biodegradability related phenomena do not interfere (Sanders et al. 2003). While hydrolysis of a soluble substrate can be described by Michaelis–Menten kinetics (Batstone et al. 2002).

The biodegradability and hence the biogas production potential of a complex substrate depends on the content of biodegradable carbohydrates (including cellulose, hemicellulose, and lignin fractions), proteins, and lipids (Angelidaki and Sanders 2004). The complete biodegradability of a given substrate can be determined through the long-term batch anaerobic digestion study (Kayhanian 1995), but the reaction period should be specified in these cases. Equation 1 shows the applicability of the first-order kinetics corrected by the non-degradable fraction of a complex substrate (Vavilin et al. 2008):

$$\frac{dS}{dt} = -k(S - \beta S_0) \quad (1)$$

where S_0 is the initial substrate concentration, and β is the non-degradable fraction of the substrate. This latter parameter can greatly fluctuate depending on the organic substrate (Kayhanian et al. 1991). Kayhanian and Tchobanoglous (1992) reported that the biodegradable fractions of various substrates, expressed in the unit of volatile solids (VS), are 82% (food waste), 72% (yard waste), 82% (office paper), 67% (mixed paper), and 22% (newsprint). Especially, municipal solid waste should be divided into easily degradable and recalcitrant fractions to model its degradation (Vavilin et al. 2006).

A first-order function may be most appropriate for complex, heterogeneous substrates, while other hydrolysis functions may be more appropriate for single homogeneous substrates (Eastman and Ferguson 1981). McCarty and Mosey (1991) proposed that hydrolysis could be considered as a microbial process, and besides first-order kinetics, they suggested a "pseudo-Monod" equation with a high saturation constant of 5000–10,000 mg/L. Vavilin et al. (1996) developed the following rate function of hydrolysis for particles with different shapes:

$$r = kS_0^{1-D}S^D \tag{2}$$

where S and S_0 are the current and initial substrate concentrations, respectively; and D is the degree index that equals 2/3, 1/2, and 0 for spherical, cylinder, and plate-form particles, respectively. The last case is equivalent to zero-order kinetics. The hydrolytic constant is a function of the ratio between the characteristic sizes of bacteria and particles hydrolyzed (Vavilin et al. 1996):

$$k = 6r_m \frac{\rho_X}{\rho_S} \frac{\delta}{d_S} \quad \text{(spherical particles)} \tag{3}$$

$$k = 4r_m \frac{\rho_X}{\rho_S} \frac{\delta}{d_S} \quad \text{(cylinder particles)} \tag{4}$$

where r_{m} is the maximum specific hydrolysis rate; ρ_{X} and ρ_{S} are the bacterial and particle densities, respectively; δ denotes the depth of the bacterial layer and d_{S} is the current diameter of particles. Thus, for different particle shapes, a rate constant value is inversely proportional to the characteristic size of the particle.

The surface-related hydrolysis kinetics model that considers the colonization of waste particles by hydrolytic bacteria has been developed by Vavilin et al. (1996):

$$\rho = \hat{\rho}_m \frac{\beta X}{1 + \beta X} \frac{S}{K_S + S} \tag{5}$$

where ρ and $\hat{\rho}_m$ are the current and maximum hydrolysis rates, respectively; S is the volatile solid waste concentration; X is the concentration of hydrolytic (acidogenic) biomass; and β is the equilibrium constant equal to the ratio between the adsorption and desorption rate constants in the Langmuir function.

3.2 *Kinetic Models for Methanogenesis*

Kythreotou et al. (2014) reported that the kinetic model of biogas production (i.e. product-formation model) was the most important among the reaction-kinetic models. Therefore, understanding the kinetics of biogas and biomethane production from substrates is crucial for designing and evaluating anaerobic digesters or reactors. Among the numerous reaction-kinetic models utilized for describing methane production, the most common kinetic model expressions are presented in Table 4.

Table 4 Kinetic models applied for methanogenesis in anaerobic digestion systems

Model	Methane production kinetics	Remarks	References
First-order exponential model	$M(t) = M_0\left(1 - e^{(-k_h t)}\right)$	k_h : Methane production rate constant (first-order disintegration rate constant) (1/day)	Pavlostathis and Giraldo-Gomez (1991)
Modified first-order exponential model	$M(t) = M_0\left(1 - e^{(-k_h(t-\lambda))}\right)$	The lag phase was included	Mochizuki and Hattori (1987)
Modified Gompertz model	$M(t) = M_0 \exp(-\exp(\frac{R_{max}e(\lambda-t)}{M_0} + 1)$	Allows the determination of important parameters of anaerobic digestion	Zwietering et al. (1990)
Logistic model	$M(t) = \frac{M_0}{1+\exp\left(\frac{4R_{max}(\lambda-t)}{M_0}+2\right)}$	Fits the global shape of the biogas production kinetics	Zwietering et al. (1990)
Cone model	$M(t) = \frac{M_0}{1+(k_C t)^{-Fs}}$	k_C: Cone kinetic constant (1/day); Fs: shape factor (dimensionless, value affected by object's shape)	Lima et al. (2018)
Transference model	$M(t) = M_0\left(1 - \exp\left(-\left(\frac{R_{max}(t-\lambda)}{M_0}\right)\right)\right)$	Mainly suitable for easily biodegradable substrate	Donoso-Bravo et al. (2010)
Monod model	$M(t) = M_0\left(\frac{kt}{1+kt}\right)$	/	
Modified Monod model	$M(t) = M_0\left(\frac{kt}{1+kt}\right)[1 - \exp(-\frac{t}{\lambda})]$	t for reaction time andλ for lag time	Bergter (1972)
Modified Gompertz with second order model	$M(t) = M_{01}\left(\frac{t}{k_{01}+t}\right) + M_{02} \cdot e\left\{-e^{\frac{R_{max}}{M_{02}}}(\lambda - t) + 1\right\}$	M_{01} and M_{02}: specific methane yields at the first and second stages (mL/g VSS), k_{01}: first-stage half-saturation time (days)	Karki et al. (2022)

$M(t)$: cumulative methane yield (mL/g VSS_{added}) at a digestion time t days; M_0: methane potential of the substrate (mL/g VSS_{added}); k_h: methane production rate constant or the first-order disintegration rate constant (1/d); t: time (d); R_{max}: maximum methane production rate (mL/g VSS_{added}/d); λ: the lag phase (d)

The first-order kinetic model assumes that the rate-limiting step of anaerobic digestion is hydrolysis (Pavlostathis and Giraldo-Gomez 1991). However, this model does not allow the estimation of neither the specific methane production rate nor the lag phase. The maximum biogas yield and the apparent hydrolysis rate coefficient (k_h) were obtained using the first-order equation in the case of anaerobic degradation of solid waste (such as waste-activated sludge) (Batstone et al. 2009). The modified Gompertz model is an empirical model that allows the determination of important parameters of anaerobic digestion such as the lag phase, the specific methane production rate, and the maximum cumulative methane production (Zwietering et al. 1990). The modified Gompertz is also used when inhibition is observed in the anaerobic digestion process assuming methane production reflects microbial growth (Mao et al. 2019).

The Logistic model fits the global shape of the biogas production kinetics: an initial exponential increase and a final stabilization at a maximal production level. This model assumes that the rate of gas production is proportional to the amount of gas already produced, the maximum production rate, and the maximum capacity of biogas production. This model has been used for anaerobic fermentation, as well as, for estimating the methane production in landfill leachate (Mawson et al. 1991; Pommier et al. 2007). Transference model is derived from the first-order kinetic model where the kinetic constant is substituted for the ratio of the maximum methane production rate to the methane production potential of substrate (Zahan et al. 2018). The Transference model is used mainly for control purposes, conceptualizing any process as a system receiving inputs and generating outputs (Donoso-Bravo et al. 2010). Transference model is used where there is no lag phase in biogas production and it is mainly suitable for easily biodegradable substrate (Zhao et al. 2016).

The Cone model is an empirical model that was developed to evaluate the methane production from specific substrate in the presence of ruminal microorganisms (Lima et al. 2018). The Cone model estimates the behaviour of methane production by the shape factor (Fs), which indicates the presence or absence of the lag phase (Groot et al. 1996). As the Monod equation does not have the lag phase, which cannot be applied accurately for modelling methane production. The modified Monod model by Bergter (1972) considers deceleration during the lag phase by introducing an exponential part of the mathematical model, with t for time and λ for the lag time. Mochizuki and Hattori (1987) also introduced the lag time into the first-order kinetics model.

The Logistic model and the modified Gompertz model can be applied to predict the methane production and evaluate the bio-kinetic parameters (Panigrahi et al. 2020). Usually, the modified Gompertz model and the Logistic model are better for describing phases of lag, log, and stationary during anaerobic digestion of complex organic matter due to their sigmoidal shape of the curve, while the Transference model is better for anaerobic digestion of easily biodegradable substrate without the lag phase in biogas production (Panigrahi et al. 2020). While for feedstock with both readily degradable and recalcitrant components, methane production with a two-stage model would be better for modelling, which can be achieved by modifying the Gompertz model with the incorporation of another equation (Karki et al. 2022).

The modified Gompertz model with the second order equation was modified to fit the methane production dynamics in anaerobic co-digestion processes, which typically showing two stages of methane production (Masih-Das and Tao 2018). The kinetic model can be affected by both the types of feedstocks and the operational systems (such as mono- or co-digestion). For example, during anaerobic digestion of coffee pulp and cattle manure, the Cone model is better for mono-digestion, while the modified Gompertz model with the second order equation is more suitable for co-digestion (Karki et al. 2022).

In addition to the cumulative biogas production, the Gaussian equation (Eq. 6) can be used to model the biogas production rate, assuming that biogas production rates follow the normal distribution (Ghatak and Mahanta 2014).

$$y = a_1 \cdot \exp\left[-0.5\left(\frac{T - T_0}{a_2}\right)^2\right] \tag{6}$$

where y = biogas production rate (mL/L/d) at time T; T = time needed for digestion (d); a_1 (mL/L/d) and a_2 (d) are the constants; T_0 = time where the maximum biogas production rate took place (d).

3.3 *Integration of Hydrolysis and Methanogenesis Processes*

The successful operation of anaerobic digestion systems depends on the cooperation among all steps rather than only on one limiting step (hydrolysis or methanogenesis). Therefore, the comparison between the hydrolysis and methanogenesis constants is of paramount importance for system optimization. As mentioned above, the first-order kinetic model is usually applied for the hydrolysis step, through which the hydrolysis rate constant (k_h) can be calculated. While the modified Gompertz model is generally applied to the methanogenesis step, where the maximum methane production rate (R_m, mL CH_4/(L·d) or mL CH_4/(g VSS·d)) is calculated. For comparison, the constant for methane production can be revised, such as adjusting R_m to R_m' to set the same unit of k_h (Li et al. 2018), which can be achieved through $R_m' = R_m$/(Theoretical g COD/g VS *350 mL CH_4/g COD). The k_h/R_m' ratio indicates the ratio between the hydrolysis rate constant to the methanogenesis rate constant, with the k_h/R_m' ratio above 1 indicating a higher hydrolysis reaction rate, while below 1 showing the higher methanogenesis rate. Also, this ratio can indicate the production and consumption of VFAs (Li et al. 2018).

4 Kinetics for Anaerobic Digestion Systems Without Considering Detailed Steps

For practical engineering applications, if only the effluent substrate concentration is considered for system monitor and assessment, the concentration can be predicted by different models (Table 5). The main difference for these models is that the Monod model is independent of the influent substrate concentration, while the others such as the Contois model are dependent on the influent substrate concentration (Hu et al. 2002).

To predict the effluent substrate concentration, specific kinetic models require kinetic parameters derived from daily experimental data under steady-state conditions, such as the influent substrate concentration S_0, the effluent substrate concentration S, and the solids retention time (SRT) (Hu et al. 2002; Njoya et al. 2021). Furthermore, the organic loading rate (OLR) is a key applied parameter in anaerobic reactor

Table 5 Kinetic models applied for anaerobic digestion systems predicting the effluent substrate concentration

Model	Substrate utilization kinetics	Remarks	References
Monod	$S = \frac{K_s(1+b\theta_c)}{\theta_c(\mu_{\max}-b)-1}$ S is independent of S_0	K_s: half-velocity coefficient or half-saturation constant* b: specific microorganism decay rate $\mu_{\max}$: the maximum specific growth rate θ_c: solids retention time	Monod (1949)
Simple first-order kinetic model	$S = \frac{S_0}{k\theta_c+1}$ S is a function of S_0	k: the first-order reaction rate constant	Linke (2006)
Contois	$S = \frac{S_0BY}{BY+\theta_c(\mu_{\max}-b)-1}$ S is a function of S_0	B: kinetic parameter Y: growth yield coefficient	Contois (1959)
Grau	$S = \frac{S_0(1+b\theta_c)}{\mu_{\max}\theta_c}$ S is a function of S_0	/	Grau et al. (1975)
Chen & Hashimoto	$S = \frac{S_0K}{\mu_{\max}\theta_c+K-1}$ S is a function of S_0	K: the Chen and Hashimoto kinetic constant (dimension less)	Chen and Hashimoto (1980)

Revised from Maleki et al. (2018)
S is the effluent substrate concentration for a reactor at steady state; and S_0 represents the initial substrate concentration
*Substrate concentration at 50% of maximum specific growth rate

operation, which can be expressed with SRT and S_0 as OLR = S_0/SRT. In general, the use of kinetic models for predicting substrate consumption or biogas production, without considering detailed steps, should consider factors such as substrate type, reactor type/operational mode, and the fitness of model. The application of typical models is discussed below as an example.

4.1 The Chen and Hashimoto Model

The Chen and Hashimoto model has been satisfactorily used for both the continuous and batch anaerobic digestion processes, and has been recognized as a reliable tool in predictions regarding digestion of wastes with a substantial solids content (Chen and Hashimoto 1980; Fongsatitkul et al. 2012; Kafle et al. 2013; Kafle and Kim 2012). When adopting the Chen and Hashimoto model (Chen and Hashimoto 1980; Kafle and Kim 2012), the kinetic parameters K and μ_{max} can be obtained through the linearization of Eq. (7), where plotting S_0/S against θc can yield a straight line with the intercept of (1–1/K) and the slope equivalent to μ_{max}/K.

$$\frac{S_0}{S} = \frac{\mu_{max}\theta_c}{K} + \left(1 - \frac{1}{K}\right) \tag{7}$$

The correlation between substrate concentration S and biogas yield y at time t is expressed as Eq. (8). y_m represents the maximum biogas yield, which can be obtained from a simple batch test.

$$y = y_m \frac{S_0 - S}{S_0} \tag{8}$$

4.2 The Modified Chen and Hashimoto Model

The modified Chen and Hashimoto model (Andriamanohiarisoamanana et al. 2020) can be written as Eq. (9). The biogas yield y can be expressed as the function of retention time θ_c or OLR.

$$y = y_m\left(1 - \frac{K}{\mu_{max}\theta_c + K - 1}\right) = y_m \frac{\mu_{max}S_0 - \text{OLR}}{\mu_{max}S_0 + (K-1)\text{OLR}} \tag{9}$$

Similar to substrate consumption, the kinetic parameters K and μ_{max} of the modified Chen and Hashimoto model can be obtained by plotting $y/(y_m - y)$ against θc (Eq. 10).

$$\theta_c = \frac{K}{\mu_{\max}} \frac{y}{y_m - y} + \frac{1}{\mu_{\max}} \tag{10}$$

Equation 10 can calculate the critical retention time. θ_c is determined as a function of $y/(y_m - y)$ with $K/\mu_{\max}$ and $1/\mu_{max}$ of the slope and intercept, respectively. The intercept ($1/\mu_{\max}$) of Eq. (10) equals the critical retention time when biomass washout occurs. For example, the critical retention time of a mesophilic anaerobic system treating swine manure is 9.5 d when the μ_{max} is obtained as 1.05 1/d (Kafle and Kim 2012). A shorter critical retention time indicates better reactor performance.

Additionally, thc targeted biogas proportion ($p = y/y_m$) can be achieved through adjusting θ_c, as expressed with Eq. (11).

$$p = 1 - \frac{K}{\mu_{\max}\theta_c + K - 1} \tag{11}$$

For instance, when $\mu_{\max}$ in an anaerobic system is assumed as 0.1 1/d, an experience curve of p for different values of θ_c and K can be obtained for engineering purpose (Fig. 2). θc increases with an increase in K. The required θ_c is 55 days to achieve 90% of y_m ($p = 0.90$) with K of 0.5.

Another application involves calculating the effective biogas production period (Kafle and Kim 2012). The time for different percentages of biogas production can be determined from the modified Chen and Hashimoto model. The correlation between retention time and p is derived from Eq. (12).

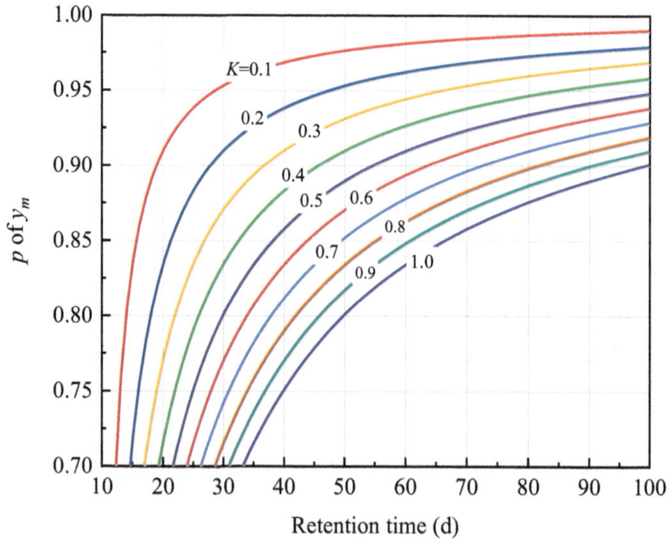

Fig. 2 An experience curve of proportion p for different values of retention time and the Chen and Hashimoto kinetic constant K with $\mu_{\max}$ of 0.1 1/d

$$\theta_c = \frac{K}{\mu_{\max}} \frac{p}{1-p} + \frac{1}{\mu_{\max}} \tag{12}$$

For instance, T_{80} represents the retention time at 80% of total biogas production (p = 80%). The effective biogas production time (T_{Ef}) is derived by subtracting the lag time from T_{80} ($T_{(80)\mathrm{Ef}} = T_{80} - \lambda$).

5 Future Perspectives

With wide applications of mathematical models in anaerobic digestion, the requirement of reliability of model results is becoming more and more important. Uncertainty analysis of the model, especially the reliability analysis of the simulation results, is particularly necessary. Moreover, expanding the kinetic parameter across various substrates, operational conditions, and models is essential for applying suitable models in practical applications. Establishing such a parameter platform can provide a significant reference for other projects, facilitates problem tracing, and ensures stable system operation.

Enhancing the comprehensiveness of the anaerobic digestion model involves accounting for various limiting factors. pH, mass transfer, and ammonia are considered as typical inhibiting factors in anaerobic digestion modelling (Batstone et al. 2002). Moreover, thermodynamic constraint is frequently neglected in kinetic models, yet it serves as a strong driving force included in the rate law (Jin and Bethke 2007). Kinetic simulations may proceed even when biological reactions could be thermodynamically unfavourable (Jin and Bethke 2007). Therefore, thermodynamic factors should be included in the kinetic model to guarantee a more precise depiction of anaerobic digestion processes.

Integration of kinetic modelling into the digital twin framework enables comprehensive supervision and early warning throughout the entire anaerobic digestion process. This integration can enhance decision-making capabilities and has the potential to significantly reduce investment costs while efficiently processing waste/wastewater and facilitating resource recovery. Therefore, advanced techniques should be incorporated into the kinetic models for better simulation and predication.

6 Take-Home Messages

(1) Microbial growth curve can be described with different kinetic models to better understand the different growth stages.

(2) The degradation of substrates in anaerobic digestion systems relies on the cooperative interactions among multiple steps, with hydrolysis and methanogenesis often acting as the rate-limiting processes. Considering the kinetic characteristics of each step and integrating these steps are crucial for developing a reliable kinetic model.
(3) In practical engineering applications, when assessing system performance based solely on the effluent substrate concentration or biogas yield, simplified models like the Chen and Hashimoto model can be used to estimate these parameters without requiring detailed stepwise calculations.
(4) It is essential to evaluate and compare the goodness of fit of various kinetic models to select the most suitable model for specific reactor types, substrates, and operational parameters, ensuring accurate predictions and reliable simulations.

References

Abdelwahab TA, Mohanty MK, Sahoo PK, Behera D, Fo dah AE (2023) Cobalt nanoparticles to enhance anaerobic digestion of cow dung: focusing on kinetic models for biogas yield and effluent utilization. Biomass Conv Biorefinery 13:11657–11669

Andriamanohiarisoamanana FJ, Ihara I, Yoshida G, Umetsu K (2020) Kinetic study of oxytetracycline and chlortetracycline inhibition in the anaerobic digestion of dairy manure. Biores Technol 315:123810

Angelidaki I, Sanders W (2004) Assessment of the anaerobic biodegradability of macropollutants. Rev Environ Sci Bio/technol 3:117–129

Angelidaki I, Ellegaard L, Ahring BK (1983) A mathematical model for dynamic simulation of anaerobic digestion of complex substrates: focusing on ammonia inhibition. Biotechnol Bioeng 42(2):159–166

Appels L, Baeyens J, Degrève J, Dewil R (2008) Principles and potential of the anaerobic digestion of waste-activated sludge. Prog Energy Combust Sci 34(6):755–781

Batstone DJ, Keller J, Angelidaki I, Kalyuzhnyi SV, Pavlostathis SG, Rozzi A, Vavilin VA (2002) The IWA anaerobic digestion model no 1 (ADM1). Water Sci Technol 45(10):65–73

Batstone DJ, Tait S, Starrenburg D (2009) Estimation of hydrolysis parameters in full-scale anerobic digesters. Biotechnol Bioeng 102(5):1513–1520

Beba A, Atalay FS (1986) Mathematical models for methane production in batch fermenters. Biomass 11(3):173–184

Bergter F (1972) Wachstum von mikroorganismen: Experimente und Modelle. VEB Gustav Fischer Verlag, Jena

Bergter F (1983) Wachstum von mikroorganismen: Experimente und Modelle, 2nd edn. VEB Gustav Fischer Verlag, Jena

Chen YR, Hashimoto AG (1978) Kinetics of methane fermentation. In: Symposium on biotechnology and bioengineering symposium, vol 8, pp 269–282

Chen YR, Hashimoto AG (1980) Substrate utilization kinetic model for biological treatment process. Biotechnol Bioeng 22(10):2081–2095

Contois DE (1959) Kinetics of bacterial growth: Relationship between population density and specific growth rate of continuous cultures. Microbiology 21(1):40–50

Donoso-Bravo A, Pérez-Elvira SI, Fdz-Polanco F (2010) Application of simplified models for anaerobic biodegradability tests. Evaluation of pre-treatment processes. Chem Eng J 160(2):607–614

Eastman JA, Ferguson JF (1981) Solubilization of particulate organic carbon during the acid phase of anaerobic digestion. J Water Pollut Control Fed 53(3):352–366

Fongsatitkul P, Elefsiniotis P, Wareham DG (2012) Two-phase anaerobic digestion of the organic fraction of municipal solid waste: estimation of methane production. Waste Manage Res 30(7):720–726

Gavala HN, Angelidaki I, Ahring BK (2003) Kinetics and modeling of anaerobic digestion process. Adv Biochem Eng Biotechnol 81:57–93

Ghatak MD, Mahanta P (2014) Comparison of kinetic models for biogas production rate from saw dust. Int J Res Eng Technol 3(7):248–254

Grau P, Dohanyos M, Chudoba J (1975) Kinetics of multicomponent substrate removal by activated sludge. Water Res 9(7):637–642

Groot JC, Cone JW, Williams BA, Debersaques FM, Lantinga EA (1996) Multiphasic analysis of gas production kinetics for in vitro fermentation of ruminant feeds. Anim Feed Sci Technol 64(1):77–89

Hassan M, Umar M, Ding W, Mehryar E, Zhao C (2017) Methane enhancement through co-digestion of chicken manure and oxidative cleaved wheat straw: stability performance and kinetic modeling perspectives. Energy 141:2314–2320

Hobson PN, Wheatley AD (1993) Anaerobic digestion: modern theory and practice. Elsevier Applied Science, London, pp 7–72

Hu WC, Thayanithy K, Forster CF (2002) A kinetic study of the anaerobic digestion of ice-cream wastewater. Process Biochem 37(9):965–971

Jain S, Lala AK, Bhatia SK, Kudchadker AP (1992) Modelling of hydrolysis controlled anaerobic digestion. J Chem Technol Biotechnol 53(4):337–344

Jin Q, Bethke CM (2007) The thermodynamics and kinetics of microbial metabolism. Am J Sci 307(4):643–677

Kafle GK, Kim SH (2012) Kinetic study of the anaerobic digestion of swine manure at mesophilic temperature: a lab scale batch operation. J Biosyst Eng 37(4):233–244

Kafle GK, Kim SH, Sung KI (2013) Ensiling of fish industry waste for biogas production: a lab scale evaluation of biochemical methane potential (BMP) and kinetics. Biores Technol 127:326–336

Karakashev D, Batstone DJ, Trably E, Angelidaki I (2006) Acetate oxidation is the dominant methanogenic pathway from acetate in the absence of *Methanosaetaceae*. Appl Environ Microbiol 72(7):5138–5141

Karki R, Chuenchart W, Surendra KC, Sung S, Raskin L, Khanal SK (2022) Anaerobic co-digestion of various organic wastes: kinetic modeling and synergistic impact evaluation. Biores Technol 343:126063

Kayhanian M (1995) Biodegradability of the organic fraction of municipal solid waste in a high-solids anaerobic digester. Waste Manage Res 13(2):123–136

Kayhanian M, Tchobanoglous G (1992) Computation of C/N ratios for various organic fractions. Biocycle 33(5):58–60

Kayhanian M, Lindenauer K, Hardy S, Tchobanoglous G (1991) Two-stage process combines anaerobic and aerobic methods. Biocycle 32(3):48–53

Kythreotou N, Florides G, Tassou SA (2014) A review of simple to scientific models for anaerobic digestion. Renew Energy 71:701–714

Li L, He Q, Zhao X, Wu D, Wang X, Peng X (2018) Anaerobic digestion of food waste: correlation of kinetic parameters with operational conditions and process performance. Biochem Eng J 130:1–9

Lima DRS, Adarme OFH, Baêta BEL, Gurgel LVA, de Aquino SF (2018) Influence of different thermal pretreatments and inoculum selection on the biomethanation of sugarcane bagasse by solid-state anaerobic digestion: a kinetic analysis. Ind Crops Prod 111:684–693

Linke B (2006) Kinetic study of thermophilic anaerobic digestion of solid wastes from potato processing. Biomass Bioenerg 30(10):892–896

Lyberatos G, Skiadas IV (1999) Modelling of anaerobic digestion: a review. Global NEST J 1(2):63–76

Maleki E, Bokhary A, Liao BQ (2018) A review of anaerobic digestion bio-kinetics. Rev Environ Sci Bio/technol 17:691–705

Mao C, Wang X, Xi J, Feng Y, Ren G (2017) Linkage of kinetic parameters with process parameters and operational conditions during anaerobic digestion. Energy 135:352–360

Mao C, Xi J, Feng Y, Wang X, Ren G (2019) Biogas production and synergistic correlations of systematic parameters during batch anaerobic digestion of corn straw. Renew Energy 132:1271–1279

Masih-Das J, Tao W (2018) Anaerobic co-digestion of food waste with liquid dairy manure or manure digestate: co-substrate limitation and inhibition. J Environ Manage 223:917–924

Mawson AJ, Earle RL, Larsen VF (1991) Degradation of acetic and propionic acids in the methane fermentation. Water Res 25(12):1549–1554

McCarty PL (1965) Thermodynamics of biological synthesis and growth. Air Water Pollut 9(10):621–639

McCarty PL, Mosey FE (1991) Modelling of anaerobic digestion processes (a discussion of concepts). Water Sci Technol 24(8):17–33

Michaelis L, Menten ML (1913) Die Kinetik der Invertinwirkung. Biochem Z 49:333–369

Mochizuki M, Hattori T (1987) Kinetic study of growth throughout the lag phase and the exponential phase of *Escherichia coli*. FEMS Microbiol Ecol 3(5):291–296

Monod J (1949) The growth of bacterial cultures. Annu Rev Microbiol 3:371–394

Moser H (1958) The dynamics of bacterial populations maintained in the chemostat. Carnegie Institution of Washington, Washington DC, USA

Motulsky HJ, Christopoulos A (2003) Fitting models to biological data using linear and nonlinear regression: a practical guide to curve fitting. GraphPad Software Inc., San Diego CA

Nguyen DD, Jeon BH, Jeung JH, Rene ER, Banu JR, Ravindran B, Chang SW (2019) Thermophilic anaerobic digestion of model organic wastes: evaluation of biomethane production and multiple kinetic models analysis. Biores Technol 280:269–276

Njoya M, Basitere M, Ntwampe SK, Lim JW (2021) Performance evaluation and kinetic modeling of down-flow high-rate anaerobic bioreactors for poultry slaughterhouse wastewater treatment. Environ Sci Pollut Res 28:9529–9541

Noike T, Endo G, Chang JE, Yaguchi JI, Matsumoto JI (1985) Characteristics of carbohydrate degradation and the rate-limiting step in anaerobic digestion. Biotechnol Bioeng 27(10):1482–1489

Panigrahi S, Sharma HB, Dubey BK (2020) Anaerobic co-digestion of food waste with pretreated yard waste: a comparative study of methane production, kinetic modeling and energy balance. J Clean Prod 243:118480

Pavlostathis SG, Giraldo-Gomez E (1991) Kinetics of anaerobic treatment: a critical review. Crit Rev Environ Sci Technol 21(5–6):411–490

Pfeffer JT (1974) Temperature effects on anaerobic fermentation of domestic refuse. Biotechnol Bioeng 16(6):771–787

Pommier S, Chenu D, Quintard M, Lefebvre X (2007) A logistic model for the prediction of the influence of water on the solid waste methanization in landfills. Biotechnol Bioeng 97(3):473–482

Powell EO (1967) The growth rate of microorganisms as a function of substrate concentration. In: Microbial physiology and continuous culture, 3rd international symposium, Salisbury, HMSO, London, pp 34–56

Rittmann BE, McCarty PL (2001) Environmental biotechnology: principles and applications. McGraw-Hill, New York

Sanders WTM, Veeken AHM, Zeeman G, Van Lier JB (2003) Analysis and optimisation of the anaerobic digestion of the organic fraction of municipal solid waste. In: Mata-Alvarez J (ed)

Biomethanization of the organic fraction of municipal solid waste. IWA Publishing, London, UK, pp 63–88

Shin HS, Song YC (1995) A model for evaluation of anaerobic degradation characteristics of organic waste: Focusing on kinetics, rate-limiting step. Environ Technol 16:775–784

Stover EL, Kincannon DF (1982) Rotating biological contactor scale-up and design. In: Proceedings of the 1st international conference on fixed film biological processes, Kings Island, Ohio, USA, pp 1667–1687

Vavilin VA, Rytov SV, Lokshina LY (1996) A description of hydrolysis kinetics in anaerobic degradation of particulate organic matter. Biores Technol 56(2–3):229–237

Vavilin VA, Jonsson S, Ejlertsson J, Svensson BH (2006) Modelling MSW decomposition under landfill conditions considering hydrolytic and methanogenic inhibition. Biodegradation 17:389–402

Vavilin VA, Fernandez B, Palatsi J, Flotats X (2008) Hydrolysis kinetics in anaerobic degradation of particulate organic material: an overview. Waste Manage 28(6):939–951

Wang ZW, Li Y (2014) A theoretical derivation of the Contois equation for kinetic modeling of the microbial degradation of insoluble substrates. Biochem Eng J 82:134–138

Wang M, Tang SX, Tan ZL (2011) Modeling *in vitro* gas production kinetics: derivation of logistic-exponential (LE) equations and comparison of models. Anim Feed Sci Technol 165(3–4):137–150

Wang R, Li Y, Wang W, Chen Y, Vanrolleghem PA (2015) Effect of high orthophosphate concentration on mesophilic anaerobic sludge digestion and its modeling. Chem Eng J 260:791–800

Wang K, Yun S, Xing T, Li B, Abbas Y, Liu X (2021) Binary and ternary trace elements to enhance anaerobic digestion of cattle manure: focusing on kinetic models for biogas production and digestate utilization. Biores Technol 323:124571

Webb JL (1963) Enzyme and metabolic inhibitors. Academic Press, New York and London

Yu H, Wilson F, Tay JH (1998) Kinetic analysis of an anaerobic filter treating soybean wastewater. Water Res 32(11):3341–3352

Yu L, Wensel PC, Ma J, Chen S (2013) Mathematical modeling in anaerobic digestion (AD). J Biorem Biodegrad S4:003

Zahan Z, Othman MZ, Muster TH (2018) Anaerobic digestion/co-digestion kinetic potentials of different agro-industrial wastes: a comparative batch study for C/N optimisation. Waste Manage 71:663–674

Zhao C, Yan H, Liu Y, Huang Y, Zhang R, Chen C, Liu G (2016) Bio-energy conversion performance, biodegradability, and kinetic analysis of different fruit residues during discontinuous anaerobic digestion. Waste Manage 52:295–301

Zhen G, Lu X, Kobayashi T, Kumar G, Xu K (2016) Anaerobic co-digestion on improving methane production from mixed microalgae (*Scenedesmus* sp., *Chlorella* sp.) and food waste: Kinetic modeling and synergistic impact evaluation. Chem Eng J 299:332–341

Zhou X (1993) A new method with high confidence for validation of computer simulation models of flight systems. J Syst Eng Electron 4(4):43–52

Zwietering MH, Jongenburger I, Rombouts FM, Van't Riet KJAEM (1990) Modeling of the bacterial growth curve. Appl Environ Microbiol 56(6):1875–1881

Modeling with the Anaerobic Digestion Model No. 1 for Enhanced Understanding and Application of Anaerobic Treatment Processes

Rongrong Mo, Wenjie Guo, Damien Batstone, Jacek Makinia, and Yongmei Li

Abstract Mathematical model is a valuable tool for prediction and recovery of resources and energy through anaerobic digestion of organic wastes. The International Water Association (IWA) Anaerobic Digestion Model No. 1 (ADM1) is the most commonly used structured anaerobic digestion model. However, as substrates become more complex and our deeper understanding of the anaerobic digestion mechanism, both systematic and specific modifications have been applied to ADM1. The modified models have provided a diverse range of applications besides anaerobic digestion processes, such as fermentation and biogas upgrading processes. A brief introduction to the original structure of ADM1 was presented, followed by extensions to extracellular and intracellular biochemical processes as well as physicochemical processes of the model. Then the introduction and modifications to kinetics (including inhibition functions) and stoichiometry are proposed. Furthermore, the applications of the model to a variety of anaerobic digestion scenarios are presented. Finally, the need for improvements in simulating anaerobic digestion and biogas upgrading processes, as well as the lack of full-scale application to other substrates besides sludge are highlighted. Future research perspectives are provided for the model development based on the in-depth understanding of the anaerobic treatment processes of different organic wastes and the recovery of valuable resources.

R. Mo · W. Guo · Y. Li (✉)
State Key Laboratory of Pollution Control and Resource Reuse, College of Environmental Science and Engineering, Tongji University, Shanghai 200092, China
e-mail: liyongmei@tongji.edu.cn

Y. Li
Shanghai Institute of Pollution Control and Ecological Security, Shanghai 200092, China

D. Batstone
Australian Centre for Water and Environmental Biotechnology, The University of Queensland, Brisbane, QLD 4072, Australia
e-mail: d.batstone@uq.edu.au

J. Makinia
Faculty of Civil and Environmental Engineering, Gdansk University of Technology, Narutowicza Street 11/12, 80-233 Gdansk, Poland
e-mail: jmakinia@pg.gda.pl

G. Wu (ed.), *Anaerobic Digestion*, Green Energy and Technology,
https://doi.org/10.1007/978-3-031-69378-6_12

Keywords Anaerobic digestion model no. 1 (ADM1) · Model elements · Model modification · Practical application · Resources recovery

1 Introduction

Mathematical modeling is a valuable approach for predicting biogas production and optimizing the anaerobic digestion system. Many mathematical models, including unstructured (not considering intermediates) and structured (considering intermediates) models, have been developed to monitor and predict anaerobic digestion processes (Batstone et al. 2002). Unstructured models generally consider hydrolysis or methanogenesis to be the rate limiting step (i.e., single step), and include first order (Dennehy et al. 2016), Monod (Lokshina et al. 2001), Contois (Karim et al. 2007), and modified Gompertz (Pan et al. 2019) models. These models have been widely used because of their simple structure and limited number of parameters. However, due to the consideration of only one limiting step, these models are unable to predict some major phenomena inducing system failure (such as acidification) and important outputs (such as pH, volatile fatty acids (VFAs) concentrations, and methane content).

The International Water Association (IWA) Anaerobic Digestion Model No. 1 (ADM1) is a commonly used structured model, which was developed by an IWA task group (Batstone et al. 2002). ADM1 is based on the "four-stage theory of anaerobic digestion" (i.e., hydrolysis, acidogenesis, acetogenesis and methanogenesis) (Zeikus et al. 1985), which can simulate dynamics of substrates, intermediates, final products, and major microbial clades in anaerobic digestion. Batstone et al. (2006a) further proposed extensions to ADM1 in the early years (2002–2005), including the model extensions, such as sulfate reduction and simple phosphorus balances. Later, Batstone et al. (2015) provided a further improvement of ADM1 in plant-wide modeling, identifying developments in the linked phosphorus-sulfur-iron processes across the cycle, and proposing the concept of the removal of composite particulates (X_C). Up to date, ADM1 is still a powerful tool for simulating and predicting anaerobic digestion systems.

2 The Structure of ADM1

ADM1 is a structured model describing the biochemical and physicochemical processes of anaerobic digestion, and it was originally proposed focusing on sewage sludge. The new discovery on anaerobic digestion has refreshed the understanding, which requires extension or revision of ADM1. In spite of the wider application of ADM1, in some cases, key mechanisms are not included in the model (Kerroum et al. 2010). This requires the addition of new components (state variables) and processes (reactions), representing a structural modification to the model.

2.1 The Original ADM1 Structure

The original model is described in considerable detail in the relevant scientific report (Batstone et al. 2002). Components, processes, stoichiometric coefficients, kinetic expressions, and the relevant parameters are the major elements of ADM1, and all these elements are involved in a Petersen matrix to quantitatively describe the anaerobic digestion process (Batstone et al. 2002).

ADM1 is a mechanistic biochemical model, and intensive state variables are utilized, mainly comprising of soluble components and particulate components. Particulate components include substrate particulates and microbial biomass, and the latter was separated into several functional communities. Both biochemical and chemical processes act as source/sink elements in the general mass balance. Biochemical processes in ADM1 include five steps: disintegration, hydrolysis, acidogenesis, acetogenesis and methanogenesis. An overview of the components and bioconversion processes in ADM1 is presented in Fig. 1. ADM1 originally proposed a two-step solubilization process, with composite particulates being "disintegrated" to particulate proteins, carbohydrates, lipids, and inert substances, and the degradable disintegration products then being "hydrolyzed" to amino acids, sugars, and long-chain fatty acids (LCFAs), respectively. In the following acidogenesis process, sugars and amino acids are fermented into VFAs, hydrogen and carbon dioxide by different degraders, and LCFAs are oxidized to produce acetate and hydrogen. Furthermore, propionic acid, butyric acid and valeric acid are anaerobically converted into acetate, carbon dioxide and hydrogen. The last step of methanogenesis includes aceticlastic methanogenesis where methane is produced by cleavage of acetate, and hydrogenotrophic methanogenesis, in which carbon dioxide is converted by hydrogen-utilizing methanogenic archaea into methane. Biomass production is implicit in substrate uptake, which is linked by a yield coefficient. Decay processes of microbial species are also considered in the model, where the decayed biomass returns to the fraction of complex organics and undergoes disintegration and hydrolysis.

Particulate components are those that cannot pass through microbial cellular walls and degrade extracellularly, including fractions involved in disintegration and hydrolysis (composites-X_C, proteins-X_{pr}, carbohydrates-X_{ch}, lipids-X_{li}, inert substances-X_I) and microbial species (sugar degraders-X_{su}, amino acid degraders-X_{aa}, LCFAs degraders-X_{fa}, butyrate and valerate degraders-X_{c4}, propionate degraders-X_{pro}, acetate degraders-X_{ac} and hydrogen degraders-X_{h2}). Soluble components are capable of directly passing through cell walls, and their degradation is mediated by organisms intracellularly. Soluble components include the monomers of complex polymers (sugars-S_{su}, amino acids-S_{aa}, LCFAs-S_{fa}), VFAs (propionate-S_{pro}, butyrate-S_{bu}, valerate-S_{va} and acetate-S_{ac}), hydrogen (S_{h2}), and methane (S_{ch4}).

In addition to the organic species, inorganic carbon (S_{IC}) and nitrogenous species (S_{IN}) are included as acid–base active compounds and used as elemental balance closures. Chemical oxygen demand (COD) is used as the principal unit for organic

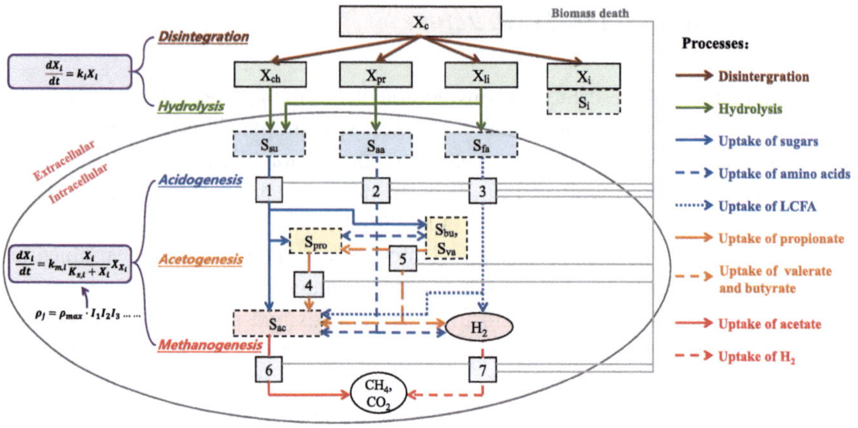

Fig. 1 Conceptual model for ADM1 (Mo et al. 2023): (1) sugar degraders (X_{su}), (2) amino acid degraders (X_{aa}), (3) LCFAs degraders (X_{fa}), (4) propionate degraders (X_{pro}), (5) butyrate and valerate degraders (X_{c4}), (6) acetate degraders (X_{ac}) and (7) hydrogen degraders (X_{h2})

compounds, while the zero-COD nitrogenous species and inorganic carbon species are described in terms of their molar concentrations.

The physicochemical processes in ADM1 involve six acid/base equilibria (acetic acid/acetate, propionic acid/propionate, butyric acid/butyrate, valeric acid/valerate, dissolved carbon dioxide/bicarbonate, and ammonium/ammonia) in association with pH solved by a charge balance approach and three gas–liquid transfer processes for CH_4, CO_2, and H_2.

Acid–base reactions relating to physicochemical reactions in the liquid phase can be implemented as differential–algebraic equations (DAE) or differential equations (DE). If implemented as a DAE system, the acid/base pairs are normally lumped as a combined dynamic state variable, and the concentration of individual acid or base is calculated from the following acid–base equilibria equation:

$$S_{i^-} - \frac{K_a S_i}{K_a + S_{H^+}} = 0 \quad (1)$$

where K_a and S_{H^+} denote the acid–base equilibrium constant and the variable for hydrogen ion concentration, respectively.

pH is determined by the proton concentration, which is determined directly by solution of the implicit algebraic as shown in Eq. 1. A charge balance is used to complete the implicit set of equations. If the liquid phase equations are implemented as DE, acid–base transfer would be described by an additional dynamic rate equation shown as:

$$\rho_{A|B,Hi} = k_{A|B}\left(S_{\text{liq},H} S_{\text{liq},i} - K_{a,Hi} S_{\text{liq},Hi}\right) \quad (2)$$

where $k_{A|B}$ is the rate coefficient for the base to acid reaction, $S_{\text{liq},H}$, $S_{\text{liq},i}$ and $S_{\text{liq},Hi}$ are concentrations of hydrogen ion, acid ion and acid, respectively, and $K_{a,Hi}$ is the acid association and dissociation constant.

During dynamic anaerobic digestion, gaseous components (CH_4, CO_2, H_2) become supersaturated in the liquid phase and transfer into the gas phase (Zhang et al. 2015), which is described by the gas transfer rate equation shown as follows:

$$\rho_{T,j} = k_L a_j \left(S_{\text{liq},j} - K_{H,j} P_{\text{gas},j}\right) \tag{3}$$

where $S_{\text{liq},j}$ is the dynamic state variable for soluble gas, $P_{\text{gas},j}$ is the pressure of gas, $k_L a_j$ is the gas–liquid transfer coefficient, and $K_{H,j}$ is the Henry's law coefficient.

2.2 *Extracellular Disintegration and Hydrolysis in ADM1 and Their Extensions*

Disintegration and hydrolysis are usually rate limiting steps in the case of a slowly degradable particulate substrate (Ramirez et al. 2009). An accurate description of these processes can improve the predictive ability of the model.

Disintegration describes the conversion of complex solid mixture, X_C, into three biodegradable organic components (X_{pr}, X_{li} and X_{ch}) and inert components (S_i and X_i). To simplify the model structure, the original ADM1 assumed that X_C was homogeneous and subject to a single first-order disintegration coefficient. In the case when the complex input substrates are not homogeneous, different decomposition efficiencies can be assumed. For instance, Yasui et al. (2008) found that degradable organics in the primary sludge contained three distinct kinetically differentiable fractions ($X_{\text{settle-I}}$, $X_{\text{settle-II}}$ and $X_{\text{settle-III}}$) and developed a modified ADM1 structure including separate degradation processes for the three identified solid fractions. The division of X_C into readily and slowly biodegradable solid fractions (X_{CR} and X_{CS}), as shown in Fig. 2a (pathway 1), was proposed based on the different biodegradability obtained from interpreting the methane production curve. The concept was applied to several studies, resulting in reasonable predictions (Garcia-Gen et al. 2015; Montecchio et al. 2019). When simulating co-digestion of more than one input substrates, the co-substrates were incorporated by using separate particulate biodegradable components $X_{C\_\text{substrate}}$ with different kinetic processes (Donoso-Bravo et al. 2020). For instance, when simulating the co-digestion of waste activated sludge (WAS) and food waste (FW), Montecchio et al. (2019) divided X_C into two terms related to WAS (X_{C_WAS}) and FW (X_{C_FW}). Finally, with only hydrolysis as the rate-limiting of solubilization process, the most common approach is to characterize input solids as a mix of proteins, carbohydrates, lipids and inert; this approach can not only avoid imposing the second-order hydrolysis kinetics, but also allow varying carbohydrate/protein/lipid/inert ratios in the inputs (Batstone et al. 2015).

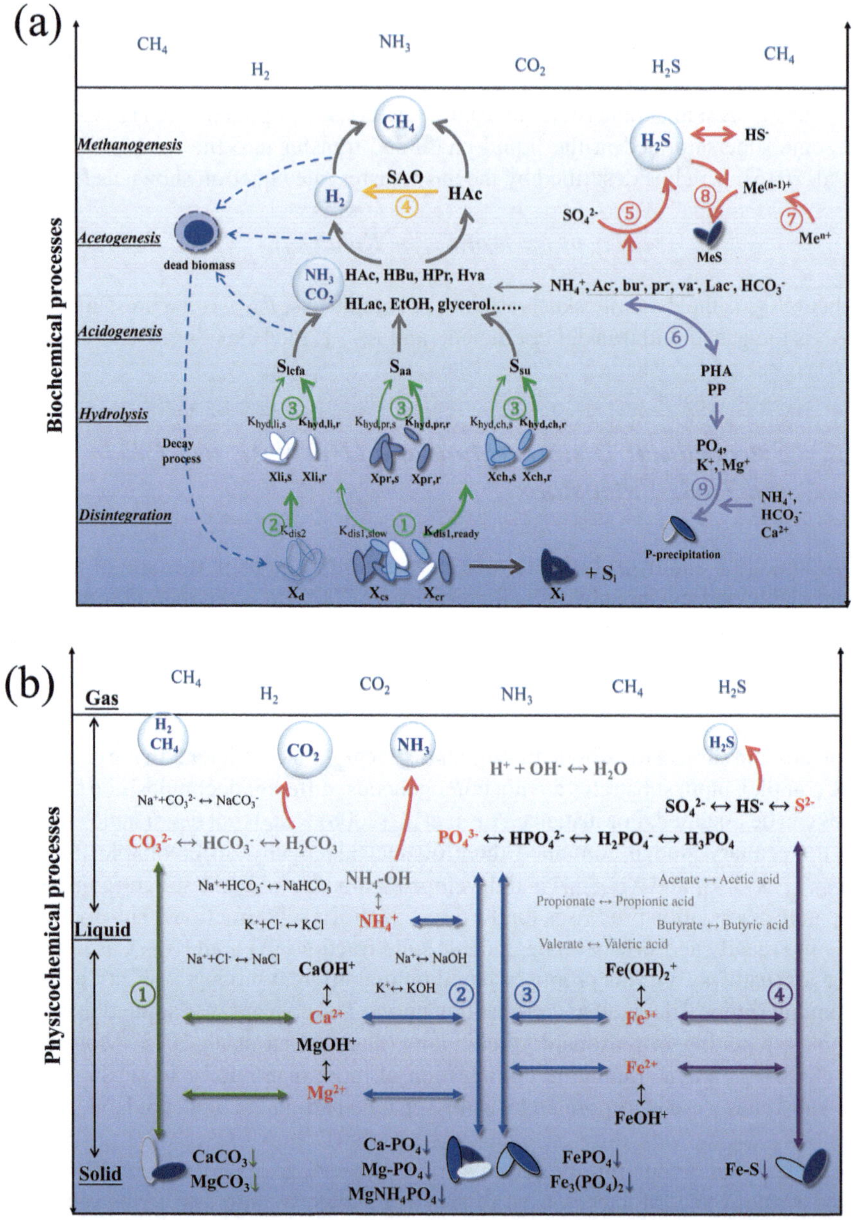

Fig. 2 Conceptual model of **biochemical processes (a)**, including detailed disintegration and hydrolysis processes (green line); conversions of S, Fe and P (red line and purple line) (Batstone et al. 2015); syntrophic acetate oxidation (yellow line) and original processes (black line), and **physicochemical processes (b)** (Batstone and Flores-Alsina 2022), including $CO_3{}^{2-}$ precipitation (dark green line), $PO_4{}^{3-}$ precipitation (navy blue line), and S^{2-} precipitation (deep purple line), for the modified ADM1

Dead biomass of all species of degrading organisms is recycled to the fraction X_C and undergoes the same disintegration process. However, since disintegration occurs as an initial step of the anaerobic process scheme and decay is continuously performed in a recycling approach throughout the retention time, these two processes are not coupled and have different nitrogen contents. To solve this problem, a modified model ADM1xp was developed, in which the products of the decay process were not completely related to the fraction X_C, but to the X_{pr}, X_{li}, X_{ch} and the new inert fraction (X_p) (Hinken et al. 2014). X_p represents inert endogenous products, which was originally proposed in the Activated Sludge Model No. 1 (ASM1) (Henze et al. 1986), and allows separate tracking of input inert and endogenous inert. This modification has been widely accepted, and Satpathy et al. (2016) found that the ADM1xp could significantly reduce the difference between measured and simulated biogas productions from 48 to 1%. To separate the kinetics of biomass decay and degradation of input substrates, a new component, dead biomass (X_d) with different decomposition kinetics, was introduced to ADM1 to further subdivide the disintegration process (pathway 2 in Fig. 2a) (Garcia-Gen et al. 2013; Montecchio et al. 2019).

Biodegradable disintegration products in the original ADM1 are presented as lumped variables (X_{ch}, X_{pr} and X_{li}), and they are also assumed to be homogeneous, with only one hydrolysis pathway for each of them (Batstone et al. 2002). With the diversification of digestion substrates, especially when X_C is excluded as an input, the primary particulate substrates may be further fractionated. It was proposed to separate each biodegradable disintegration product into two fractions, i.e., readily biodegradable fractions and slowly biodegradable fractions, and both fractions had different hydrolysis rates (pathway 3 in Fig. 2a) (Bai et al. 2015; Garcia-Gen et al. 2015). Bareha et al. (2019) divided the particulate organic matter into protein and non-protein hydrolysable fractions. They were then separately split into four different bioaccessible fractions by a modified "extracellular polymeric substances" fractionation method, which was based on successive alkaline extractions ending with strong acid extraction. When simulating the dynamic anaerobic digestion process of lignocellulosic waste, different degradation pathways can be distinguished between cellulose and other biodegradable components. Therefore, the lignocellulosic substrate could be initially disintegrated into three fractions: readily hydrolysable (carbohydrates and proteins), slowly hydrolysable (degradable cellulose) and inert (undegradable cellulose) (Li et al. 2021a; Shi et al. 2014).

2.3 *Intracellular Processes and Components in ADM1 and Their Extensions*

Acidogenesis, acetogenesis and methanogenesis are intracellular biochemical processes. Due to the complexity of these processes, the original ADM1 excludes many intermediate components to maintain a simple structure. Consideration of additional intermediates is a key point in model extension. Some important extensions are introduced as follows and given in Table 1.

2.3.1 Lactate and Ethanol

Lactate and ethanol are excluded from ADM1 due to their low production in most anaerobic digestion systems. However, in the anaerobic digestion systems fed with rapidly degradable carbohydrate-rich substrates or in fermentation processes, lactate and ethanol are important intermediates or by-products (Soda et al. 2011). For instance, the accumulation of lactate (but not ethanol) and other organic acids has a strong effect on pH (Hinken et al. 2014). Therefore, it is desirable to include both lactate and ethanol in ADM1 under these conditions. An uptake of lactate/ethanol and decay of their degraders were added to the model in terms of four new variables: soluble components (lactate-S_{la}, ethanol-S_{et}) and their degraders (lactate degraders-X_{la}, ethanol degraders-X_{et}). S_{la} and S_{et} are converted from sugar and then decomposed into VFAs and hydrogen with the growth of X_{la} and X_{et} (Soda et al. 2011). This extension has widely been used in anaerobic digestion systems with high carbohydrate substrates, such as organic waste (Soda et al. 2011), silage (Waszkielis et al. 2022), food waste (Parra-Orobio et al. 2020; Satpathy et al. 2016), starch wastewater (Hinken et al. 2014), and sugarcane vinasse (Couto et al. 2022). Garcia-Gen et al. (2014) proposed a new approach presenting the fermentation components, such as lactate and ethanol, derived from sugar in terms of equivalent glucose fermentation by a generic group of fermenters (X_{fer}) instead of additional microbial groups.

2.3.2 Syntrophic Acetate Oxidation

It has been reported that the aceticlastic methanogenesis is likely replaced by a non-aceticlastic methanogenesis, known as syntrophic acetate oxidation (SAO), under high ammonia (NH_4^+) concentrations (Wilson et al. 2012) or high temperature (Wett et al. 2014). To simulate these conditions, a modified ADM1, accounting for the SAO pathway (pathway 4 in Fig. 2a) and SAO bacteria (X_{SAOB}) (Yang et al. 2020), was implemented in anaerobic digestion systems with high temperature or high solid content, which are prone to NH_4^+ inhibition (Montecchio et al. 2017b; Rivera-Salvador et al. 2014; Wett et al. 2014). For instance, Capson-Tojo et al. (2021) modified the model with SAO as acetate-consuming pathway and threshold inhibition function for free ammonia inhibition and obtained a better simulation with the correlation coefficient of methane production increased from 0.610 to 0.938.

Some extensions are also proposed under specific conditions. For instance, Fezzani and Cheikh (2009) extended ADM1 with the degradation of phenolic compounds and achieved accurate simulation for anaerobic digestion of olive mill wastes. Uhlenhut et al. (2018) divided the original uptake of propionate into two more detailed pathways with three new genera of propionate-oxidizing bacteria (X_{POB})

Table 1 Detailed information of the extensions to the ADM1 model for intercellular processes and components

	Processes added	Components added	Transformations	Applicable situation/ substrate	References
Fermentation	Uptake of lactate; decay of lactate degraders	Lactate—S_{la}; lactate degraders—X_{la}	Sugar $\xrightarrow{X_{su}}$ Lactate Lactate $\xrightarrow{X_{la}}$ Propionate + Acetate	Fermentation process; carbohydrate-rich substrates	Parra-Orobio et al. (2020), Satpathy et al. (2016), Thamsiriroj et al. (2012), Waszkielis et al. (2022)
	Uptake of ethanol; decay of ethanol degraders	Ethanol—S_{et}; ethanol degraders—X_{et}	Sugar $\xrightarrow{X_{su}}$ Ethanol Ethanol $\xrightarrow{X_{et}}$ Acetate	Fermentation process; carbohydrate-rich substrates	Antonopoulou et al. (2012b), Soda et al. (2011)
Phosphorus, iron and sulfur transformations	Uptake of H_2 with sulfate; decay of sulfate-reducing bacteria	Soluble sulfate—S_{SO4}; Sulfide—S_{IS}; Sulfate-reducing bacteria—X_{SRB_H2}	H_2 Soluble sulfate $\xrightarrow{X_{SRB\_H2}}$ Sulfide	Sulfate-rich substrates	Batstone (2006)
	Uptake of VFAs with sulfate; Uptake of H_2 with sulfate; decay of sulfate-reducing bacteria	Soluble sulfate—S_{SO4}; Sulfide—S_{IS}; sulfate-reducing bacteria—X_{SRB_H2}, X_{SRB_ac}, X_{SRB_pro}, X_{SRB_c4}	VFA, H_2 Soluble sulfate $\xrightarrow{X_{SRB\_H2}, X_{SRB\_ac}, X_{SRB\_pro}, X_{SRB\_c4}}$ Sulfide	Sulfate-rich substrates	Barrera et al. (2015), Fedorovich et al. (2003), Flores-Alsina et al. (2016), Pokorna-Krayzelova et al. (2017)
	Uptake of H_2S (sulfide oxidation); decay of sulfide-oxidizing bacteria	Oxygen—S_{O2}; sulfide-oxidizing bacteria—X_{SOB}	O_2 $H_2S \xrightarrow{X_{SOB}} S$	Biogas desulfurization	Pokorna-Krayzelova et al. (2017)
	Storage of VFAs to form polyhydroxyalkanoates; Decay of phosphorus-accumulating organisms; Lysis of polyhydroxyalkanoates and polyphosphates	Phosphorus-accumulating organisms—X_{PAO}; polyhydroxyalkanoates—X_{PHA}; polyphosphates—X_{PP}	VFAs $\xrightarrow{X_{PAO}}$ Polyhydroxyalkanoates Polyhydroxyalkanoates → VFAs	Enhanced biological phosphorus removal (EBPR) sludge; plant-wide simulation	Flores-Alsina et al. (2016), Wang et al. (2016)

(continued)

Table 1 (continued)

	Processes added	Components added	Transformations	Applicable situation/ substrate	References
	Conversion of iron	Ferric iron—$S_{Fe}{}^{3+}$; Ferrous iron—$S_{Fe}{}^{2+}$	Ferric iron $\xrightarrow{H_2,\text{sulfides}}$ Ferrous iron	Plant-wide simulation	Flores-Alsina et al. (2016)
	Corrosion of ZVI; degradation of LBPs	Zero valent iron—$S_{Fe}{}^{0}$; Ferrous iron—$S_{Fe}{}^{2+}$; less biodegradable pollutants—S_{LBPs}; metabolic products of LBPs—$S_{(CH)}$	$Fe^0 + 2H^+ \rightarrow Fe^{2+} + H_2$ LBPs + H_2 → less toxic products	ZVI-based anaerobic systems	*X*iao et al. (2013)
Syntrophic acetate oxidation	Syntrophic acetate oxidation; decay of SAO bacteria	SAO bacteria—X_{SAOB}	Acetate $\xrightarrow{X_{SAOB}}$ $H_2 + CO_2$	Under high ammonia concentration or high temperature	Montecchio et al. (2017b), Rivera-Salvador et al. (2014), Wett et al. (2014)

and achieved a better fit between simulation and realistic scenarios. The difference between measured and simulated values for mean methane content reduced from 13.4 to 0.6% after that modification.

2.3.3 Phosphorus (P), Iron (Fe) and Sulfur (S) Transformations

Only carbon (C) and nitrogen (N) transformations were considered in ADM1, while other elemental transformations were excluded due to limited precedent work at the time of publication. However, it is crucial to include P, S and Fe transformations in the model under some conditions. For sulfate-rich substrates, the sulfate transformations can be described in two different ways (Table 1). First, soluble sulfate (S_{SO4}) is reduced to sulfide (S_{IS}) by one group of sulfate-reducing bacteria (X_{SRB_H2}) using hydrogen as the electron donor (Batstone 2006). Secondly, a widely used approach has assumed that S_{SO4} is also reduced to sulfides (S_{IS}) with multiple electron donors (S_{bu}, S_{pro}, S_{ac} and S_{h2}) by relevant bacteria species (X_{SRB_c4}, X_{SRB_pro}, X_{SRB_ac} and X_{SRB_H2}) (pathway 5 in Fig. 2a) (Barrera et al. 2015; Flores-Alsina et al. 2016; Pokorna-Krayzelova et al. 2017). To optimize the microaeration process for biogas desulfurization, Pokorna-Krayzelova et al. (2017) extended ADM1 with one additional process of sulfide oxidation to elemental sulfur by sulfide-oxidizing bacteria, X_{SOB}. Ferric iron ($S_{Fe}{}^{3+}$) is converted to ferrous iron ($S_{Fe}{}^{2+}$) by utilizing hydrogen and/or sulfides (S_{IS}) as electron donors (pathway 7 in Fig. 2a) (Flores-Alsina et al. 2016).

P-related components include polyphosphate-accumulating organisms (X_{PAO}), polyhydroxyalkanoates (X_{PHA}) and polyphosphates (X_{PP}). Their incorporation brings seven new processes: uptakes of S_{va}, S_{bu}, S_{pro} and S_{ac} to form X_{PHA} as well as lysis of X_{PAO}, X_{PP} and X_{PHA} (pathway 6 in Fig. 2a) (Flores-Alsina et al. 2016; Wang et al. 2016). The extension of P transformations and its close link to the S and Fe cycles are necessary for plant-wide P simulation, especially considering P recovery.

In addition, electron release from zero-valent iron (ZVI) corrosion, H_2 formation from ZVI corrosion and transformation of less biodegradable pollutants (LBPs), were integrated into ADM1. The modified model could provide more precise strategies for the design, development, and application of ZVI-based anaerobic systems, especially for treating various LBPs-containing wastewaters (Xiao et al. 2013).

2.4 *Physicochemical Processes in ADM1 and Their Extensions*

The physicochemical model applied originally to ADM1 was the most advanced physicochemical model for IWA standardized models at the time, in that it could provide pH as an output, but has been since replaced by more advanced models which consider ion activity, ion pairing and solids precipitation in much more detail

(Solon et al. 2017). Physicochemical extensions have been extensive since the publication of ADM1 has been proposed for plant wide (and extra-plant) use via the IWA Physicochemical Model No. 1 (Batstone and Flores-Alsina 2022) as shown in Fig. 2b.

Liquid–gas exchange processes for H_2, CH_4 and CO_2 were incorporated in the original ADM1 as two-film (liquid film-controlled) rate expressions. Due to the substrate complexity or changing conditions, it is necessary to consider comprehensive liquid–gas exchanges when new gas components, such as H_2S (Flores-Alsina et al. 2016) and NH_3 (especially above 10 g N/L and under thermophilic conditions) (Zhang et al. 2015), are added to ADM1.

Furthermore, to better describe the acid–base reactions and their effects on the anaerobic digestion processes, more inorganic components and acid/base equilibria should be incorporated into the model. For example, the equilibrium among four inorganic phosphorus components ($PO_4^{3-}/HPO_4^{2-}/H_2PO_4^-/H_3PO_4$) was introduced to the model, and a charge balance equation used to calculate the pH value was refined (Zhang et al. 2015).

In addition, ion pairing and ion activity have important impacts on physicochemical processes, and have been extended in ADM1 (Batstone and Flores-Alsina 2022). Some new components, such as S_{Na+}, S_{K+}, S_{Cl-}, and ion pairing processes, such as $NaCO_3^- \leftrightarrow Na^+ + CO_3^{2-}$, were considered in the model. Ion strength (as ion activity correction) was corrected by multiplying each ion concentration with an activity coefficient (γ) (Batstone and Flores-Alsina 2022; Solon et al. 2015). Solon et al. (2015) demonstrated that ion strength correction led to significant differences in predicting the process performance, and activity corrections were more important than ion pairing effects for pH prediction.

Precipitation is the major liquid–solid process, but it was not involved in the original ADM1 due to a wide range of precipitating ions and precipitate types. However, liquid–solid reactions are very important in systems with high levels of cations, which can affect both other physicochemical processes and biochemical processes. For instance, struvite ($MgNH_4PO_4$) formation can affect the pH values in anaerobic digestion systems (Britton et al. 2005). To improve the model ability to describe the formation rate of precipitates, Zhang et al. (2015) considered calcium ions (Ca^{2+}) and magnesium ions (Mg^{2+}) as two major M^{m+} ions in the model, which could cause five major types of precipitates: $CaCO_3$, $MgCO_3$, $MgNH_4PO_4$, $MgHPO_4$, and $Ca_3(PO_4)_2$. Comprehensive P precipitation processes, including the formation of $Ca_3(PO_4)_2$, $Ca_5(PO_4)_3(OH)$, $Ca_8H_2(PO_4)_6$, $MgNH_4PO_4$, $MgHPO_4$ and $KMgPO_4$, were extended to ADM1 in later studies to simulate P behavior under the P-rich influent (pathway 9 in Fig. 2a, and pathway 1 and 2 in Fig. 2b) (Flores-Alsina et al. 2016; Wang et al. 2015). Due to strong interactions between S and Fe, Flores-Alsina et al. (2016) and Puyol et al. (2017) extended the precipitation processes required for the P, S and Fe cycles (formation of FeS, and $FePO_4/Fe_3(PO_4)_2$, pathway 8 in Fig. 2a, and pathway 3 and 4 in Fig. 2b) by a multiple mineral precipitation model. Subsequently, to consider the effects of trace elements on physicochemical processes of anaerobic digestion, the reversible precipitation processes, involving Fe, Ni, Co and major anionic components, were added to the model (Maharaj et al. 2019).

Precipitation kinetics is generally based on the saturation index (SI) that is used to indicate the degree of supersaturation. Musvoto et al. (2000) indicated that the precipitation rate only depends on the SI. To refine the expression of precipitation kinetics, Kazadi Mbamba et al. (2015) included the effect of precipitate concentration, and Lizarralde et al. (2015) considered the delay of nucleation and the influence of crystal seeding and total suspended solids (TSS) concentration.

Suitable expressions can be introduced into ADM1 for the extension of precipitation processes. For instance, based on the fundamental relationship in the crystallization process, the following equation was introduced to describe mechanisms of solids precipitation involving Ca, Mg (Zhang et al. 2015), Fe, Co and Ni (Maharaj et al. 2019):

$$r_{M_{v+}A_{v-}} = k'_{r,M_{v+}A_{v-}}\left[\left(\left[M^{m+}\right]^{v+}\left[A^{a-}\right]^{v-}\right)^{\frac{1}{v}} - k'_{sp,M_{v+}A_{v-}}{}^{\frac{1}{v}}\right]^{n} \quad (4)$$

where $\left[M^{m+}\right]$ and $\left[A^{a-}\right]$ denote the concentrations of cations and anions, respectively; $k'_{r,M_{v+}A_{v-}}$ is the precipitation rate constant; $k'_{sp,M_{v+}A_{v-}}$ is the solubility product; v^{+} and v^{-} are the total number of cationic and anionic charges, respectively, and $v = v^{+} + v^{-}$; and n is a constant for crystallization of sparingly soluble salts.

3 Model Kinetics

3.1 Kinetic Expression

Substrate conversion rates are expressed as several process kinetics. The disintegration and hydrolysis of complex organic matters which are extracellular processes are described as first-order kinetics (Eq. 5) based on the substrate concentration. Decay of biomass is also presented by first-order kinetics. This function is an empirical expression that is assumed to reflect the cumulative effects of extracellular processes. While Monod-type kinetic expressions, as shown in Eq. 6, are used as the basis for all intracellular biochemical processes (acidogenesis, acetogenesis and methanogenesis).

$$r_{X_i} = -k_i X_i \quad (5)$$

$$r_{S_i} = -k_{m,i}\frac{S_i}{K_{s,i}+S_i}X_i \quad (6)$$

where X_i and S_i are the concentrations of components expressed as dynamic state variables, k_i is the first-order rate constant, and $k_{m,i}$ and $K_{s,i}$ are the specific maximum uptake rate and half saturation constant, respectively.

In the original ADM1, extracellular processes (disintegration and hydrolysis) were assumed to be first-order kinetics, which were independent of microbial activity and proportional to substrate concentration. However, hydrolysis is an enzymatic process where hydrolytic bacteria cover the surface of solids and produce enzymes (Mottet et al. 2013). Vavilin et al. (2008) assessed different hydrolysis models and found that the Contois kinetics provide a better fit at high organic loading rates. The Contois model incorporates a nonparameterized microbial concentration term, and hence describes microbe, as well as substrate limited conditions. Ramirez et al. (2009) developed a modified ADM1 using the Contois kinetics instead of the original first-order kinetics to account for the growth of hydrolytic bacteria. The Contois model was found to be more robust in describing a broad range of experiments with minimal changes to the kinetic coefficients (Mairet et al. 2011). As a result, it has been widely used in simulating high solid content substrates, such as sludge and cattle slurry (Bai et al. 2015; Bareha et al. 2019; Mottet et al. 2013; Normak et al. 2015). In addition, Liotta et al. (2015) proposed a linear function that dynamically expresses the kinetic constants of three specific processes (disintegration, uptakes of acetate and propionate) to account for the effect of the total solids (TS) content on dry anaerobic digestion.

For intracellular processes, ADM1 used substrate uptake Monod kinetics to describe the biochemical reaction rate (Batstone et al. 2002). The main modification has adopted the microbial growth-based (Siegrist et al. 2002) rather than the substrate uptake-based kinetics. However, mathematically, both kinetics have no difference.

3.2 Inhibition Function

Intracellular conversion processes can be inhibited by inappropriate pH (Latif et al. 2017), insufficient nutrients or the accumulation of intermediate products, such as free ammonia (Astals et al. 2018), molecular hydrogen (Cazier et al. 2019), etc. The effects of these inhibitions are quantified as inhibition functions (I) and expressed in kinetics. For multiple inhibitors, ADM1 utilizes a multiplicative inhibition function as follows:

$$\rho_j = \rho_{\max} \cdot I_1 I_2 I_3 \ldots \tag{7}$$

where $\rho_{\max}$ is the Monod-type kinetic equation without inhibitions and I_i is the inhibition function.

The kinetics of disintegration, hydrolysis and decay of biomass processes were related only to kinetic constants and substrate concentrations, without considering any inhibitors in the original ADM1. However, it has been suggested that inhibitors of hydrolysis should be considered due to microbial inactivation and enzyme denaturation under adverse conditions. For example, Normak et al. (2015) introduced hydrolysis kinetics containing VFAs inhibition (I_{TVFA}) to simulate the start-up process when anaerobic digestion of cow manure. Li et al. (2020) considered the inhibition of hydrolysis caused by high solids content (I_{TS}) and used a non-competitive inhibition function to reflect inhibition of hydrolysis. Furthermore, various inhibitors can increase biomass decay, which is completely neglected in the original ADM1. Sun et al. (2020) developed a low-pH inhibition switching function I_{pH} (Table 2) to describe increased decay at low pH.

For intracellular processes, pH inhibition was implemented using empirical equations, while other inhibitions were presented as non-competitive functions (Batstone et al. 2002). Non-competitive functions have widely been used in literature, but sometimes they cannot describe inhibition processes accurately. Non-competitive function was only influenced by the concentration of inhibitors and the influence of substrate concentration was ignored. Bai et al. (2017) used the Monod function to simulate the inhibition of free ammonia on VFAs generation during anaerobic fermentation of sludge with a high solids content. The determination coefficient of VFAs increased to over 0.95 compared to less than 0.8 when a non-competitive function was used. The Hill function, proposed by Ramirez et al. (2009), has also been used to describe pH inhibition (Barrera et al. 2015; Wang et al. 2015). However, these kinetic inhibition functions have limitations, including allowing the reaction to proceed even under unfavorable thermodynamic conditions. Specifically, the uptake of propionate and butyrate is fundamentally impacted by concentration of products (i.e., H_2 and acetate) due to positive Gibbs free energy at elevated concentrations. To address this, the thermodynamic inhibition function (I_{th}, shown in Table 2) of H_2 was introduced into the kinetic expression of the uptake of propionate and butyrate (Shi et al. 2016). However, within biofilms, the non-competitive function rather than the thermodynamic inhibition function has a great impact on simulation accuracy, which could be due to that other processes (such as microbial interactions and substrate diffusion) rather than the thermodynamic are controlling factors (Batstone et al. 2006b).

In the original ADM1, the effect of acid inhibition was mainly hidden in the empirical pH inhibition function. However, low pH inhibition is partly due to the concentration of free acids, and it is reasonable to include this separately. The non-competitive inhibitory term I_{VFA} was added into ADM1 to represent the inhibitory effect of high concentration of VFAs on methanogenesis (Fezzani and Cheikh 2009; Thamsiriroj et al. 2012). Considering the interaction of VFAs, Li et al. (2019) added a non-competitive inhibition function (I_{ac}) to include acetic acid inhibition on the uptake of propionate and butyrate processes.

ADM1 includes several commonly used inhibition functions, but specific inhibition functions have been utilized, as summarized in Table 2. Trace elements, such as Fe, Ni, and Co, have explicitly been considered by introducing an inhibition term (${I_{Me}}^{2+}$), as they are essential constituents in enzyme systems (Maharaj et al. 2019). LCFAs inhibition (I_{lcfa}) has also been included in ADM1, as acetogenesis and methanogenesis are highly sensitive to LCFAs and can be inhibited at high LCFAs concentrations. Palatsi et al. (2010) proposed a new inhibition function which replaced K_{Ilcfa} with $K'_{Ilcfa}X_{lcfa}/S_{lcfa}$ (Table 2) to consider a relationship between LCFAs inhibitory substrate concentration and the specific biomass content. In addition, Arnell et al. (2016) used the Gaussian function (threshold function that

Table 2 Inhibition terms added to the ADM1 kinetic expressions

Inhibition item	Process	Inhibition term	References
I_{TS}	Hydrolysis	$I_{TS} = \frac{1}{1+S_{TS}/K_{TS}}$	Li et al. (2020)
I_{pH}	Decay of X_{ac}; decay of X_{h2}	Specific decay rate $= \frac{b}{IpH}$ $I_{pH} = \begin{cases} \exp\left(-3\left(\frac{pHUL-pH}{pHUL-pHLL}\right)^n\right) & \text{if pH} < \text{pHUL} \\ 1 & \text{if pH} \geq \text{pHUL} \end{cases}$	Sun et al. (2020)
I_{NH3}	Uptake of acetate	Hill function $I_{NH_3} = b\left(1 - \frac{S^{\alpha}_{NH_3}}{KS^{\alpha}_{NH_3} + S^{\alpha}_{NH_3,lim}}\right)$	Ramirez et al. (2009)
	Uptake of acetate	Threshold inhibition function	Astals et al. (2018)
I_{VFA}	Uptake of acetate	$I_{TVFA} = \frac{1}{1+S_{TVFA}/K_{I,TVFA}}$	Boubaker and Ridha (2008), Fezzani and Cheikh (2009)
	Hydrolysis	$I_{VFA} = \frac{1}{1+\left(\frac{M_{vfa}}{IM_{vfa}}\right)^n}$	Normak et al. (2015)
I_{ac}	Uptake of propionate; uptake of butyrate	$I_{ac} = \frac{1}{1+S_{ac}/K_{I,ac}}$	Li et al. (2019)
I_{LCFA}	Hydrolysis; uptake of sugars; uptake of LCFAs; uptake of VFAs	$I_{lcfa} = \frac{1}{1+S_{lcfa}/K_{I,lcfa}}$	Ramirez et al. (2009)
	Uptake of LCFAs; uptake of acetate; uptake of H_2	$I_{lcfa} = \frac{1}{1+S^2_{lcfa}/(K'_{I,lcfa} \cdot X_{I,lcfa})}$	Palatsi et al. (2010)
	Uptake of acetate	Gaussian function $I_{lcfa} =$ $\begin{cases} e^{-2.77259\left(\frac{S_{fa}-K_{I,fa,low}}{K_{I,fa,high}-K_{I,fa,low}}\right)^2} & \text{for } S_{fa} > K_{I,fa,low} \\ 1 & \text{for } S_{fa} \leq K_{I,fa,low} \end{cases}$	Arnell et al. (2016)

(continued)

Table 2 (continued)

Inhibition item	Process	Inhibition term	References
I_{lac}	Uptake of acetate	$I_{lac} = \frac{1}{1+S_{lac}/K_{I,lac}}$	Thamsiriroj et al. (2012)
$I_{Na}{}^{+}$	Uptake of acetate	$I_{Na^+} = \frac{1}{1+S_{Na^+}/K_{I,Na^+}}$	Hierholtzer and Akunna (2012)
$I_{cations}$	Uptake of acetate	$I_{cations} = \frac{1}{1+\frac{S_{Na^+}}{K_{I,Na^+}}+\frac{S_{Mg^{2+}}}{K_{I,Mg^{2+}}}+\frac{S_{Ca^{2+}}}{K_{I,Ca^{2+}}}+\frac{S_{K^+}}{K_{I,K^+}}}$	Hierholtzer and Akunna (2014), Song et al. (2018)
$I_{Me}{}^{2+}$	All biochemical processes	$I_{Me^{2+}} = \frac{a_1 S_{Me^{2+}}+a_2}{\left(S_{Me^{2+}}\right)^2+b_1 S_{Me^{2+}}+b_2}$	Maharaj et al. (2018)
I_{H2S}	Acetogenesis and methanogenesis	$I_{H_2S} = \frac{1}{1+S_{H_2S}/K_{I,H_2S}}$	Barrera et al. (2015)
	Acetogenesis and methanogenesis	$I_{H_2S} = \begin{cases} \left(1-\frac{S_{H_2S}}{K_{I,H_2S}}\right)^2 & \text{if } S_{H_2S} < K_{I,H_2S} \\ 10^{-6} & \text{if } S_{H_2S} \geq K_{I,H_2S} \end{cases}$	Pokorna-Krayzelova et al. (2017)
I_{PO4}	Uptake of VFAs; Uptake of LCFAs	$I_{PO_4} = \frac{1}{1+\frac{K_{S,PO_4}}{S_{PO_4}}+\left(\frac{S_{PO_4}}{K_{I,PO_4}}\right)^2}$	Wang et al. (2015)
I_{phenol}	Uptake of acetate	$I_{phenol} = \frac{1}{1+S_{phenol}^2/K_{I,phenol}}$	Fezzani and Cheikh (2009)
I_{th}	Uptake of VFAs (H_2 inhibition)	$I_{th} = \begin{cases} 1-\frac{\Delta G_R}{R_{th}T} & \text{if } \Delta G_R < 0 \\ 0 & \text{if } \Delta G_R \geq 0 \end{cases}$	Patón and Rodríguez (2017)

has been used for pH inhibition in ADM1) to describe LCFAs inhibition on the uptake of acetate, which explicitly allowed for determining the onset of the inhibition.

Sulfide is also inhibitory to acetogenesis and methanogenesis, and the extension of sulfide inhibition (I_{H2S}) is needed when modeling anaerobic digestion systems with sulfur-rich substrates (e.g., cane-molasses vinasse) to predict the failure of anaerobic digestion (especially methanogenesis) (Barrera et al. 2015). Non-competition inhibition function and a staged inhibition function were used for sulfide inhibitions (Barrera et al. 2015; Pokorna-Krayzelova et al. 2017). Considering the influence of orthophosphate on the biological processes during anaerobic digestion, Wang et al. (2015) modified ADM1 by including generalized Haldane equations for orthophosphate inhibition (I_{PO4}) into the kinetics for uptake of VFAs and LCFAs. In addition, ADM1 has also been extended with other inhibition for the acetogenesis and methanogenesis, such as lactate inhibition (Hinken et al. 2014; Thamsiriroj et al. 2012), phenol inhibition (Fezzani and Cheikh 2009), cation inhibition (Song et al. 2018), and pharmaceuticals inhibition (Fountoulakis et al. 2008).

3.3 Kinetic Parameters

Kinetic parameters involved in biochemical and physicochemical rate equations are numerical descriptions of the component conversion rate. These parameters would vary with conditions and substrates, and the modification of parameters is one of the key points of model adjustment. Therefore, for accurate model simulation under different conditions, it is necessary to calibrate the parameters using more convenient, rapid, and effective methods.

Kinetic parameters are generally calibrated using the method described in Fig. 3. Many kinetic parameters are involved in the model, but only a limited number can be changed in most situations. Those for specific cases can be selected based on previous research, commonly related to the rate-limiting step, which is normally hydrolysis for particulate substrates (Rivera-Salvador et al. 2014) or mathematical algorithms (sensitivity analysis) (Yu et al. 2012). The parameters are estimated by minimizing the error between the experimental and simulated data. The "trial and error" method is a simple manual calibration that is suitable for situations where a few parameters need to be calibrated or the simulated results are similar to the experimental data. However, for cases involving many complex parameters, iterative methods (optimization algorithms), e.g., nonlinear least-square (Montecchio et al. 2017b), genetic algorithm (Koutrouli et al. 2009), differential evolution algorithms (Rivera-Salvador et al. 2014), implemented simplex method (Hinken et al. 2014), and particle swarm optimization (Bai et al. 2015, 2017), are more efficient and rapid. Any iterative or single-step optimization technique should also include statistical analysis of parameter uncertainty, including joint confidence intervals (Batstone et al. 2009). Experimental methods can also be used to determine kinetic parameters, as demonstrated by Uhlenhut et al. (2018), who used Lineweaver–Burk diagrams to estimate the parameters for uptake of propionate ($k_{m,pro}$ and $K_{s,pro}$). Recently, Ge et al. (2023) used machine learning to predict the kinetic parameters, and enhanced the performance of ADM1. This indicates that machine learning methods can be combined with the conventional mechanism models to improve simulation accuracy.

Parameters of hydrolysis and methanogenesis are the most frequently modified parameters. Hydrolysis is considered the rate-limiting step in anaerobic digestion, and the rate of hydrolysis varies considerably with the substrate. For example, hemicellulose and amorphous cellulose are slower to be hydrolyzed than common carbohydrate and protein substrates (Li et al. 2021a). Therefore, the hydrolysis parameters (k_{dis} or k_{hyd}) are more often modified. Since methane production is commonly used as a simulation indicator (Montecchio et al. 2019), kinetic parameters for the uptake of acetate and H_2 (k_{m_ac} and K_{s_ac}, k_{m_h2} and K_{s_h2}) which can directly affect methane production, are also frequently modified.

In some specific simulation situations, kinetic parameters of physicochemical processes are also important. For instance, when dry anaerobic digestion of cardboard, in addition to the decrease in the first-order hydrolysis rate (k_{hyd_ch}), a reduced

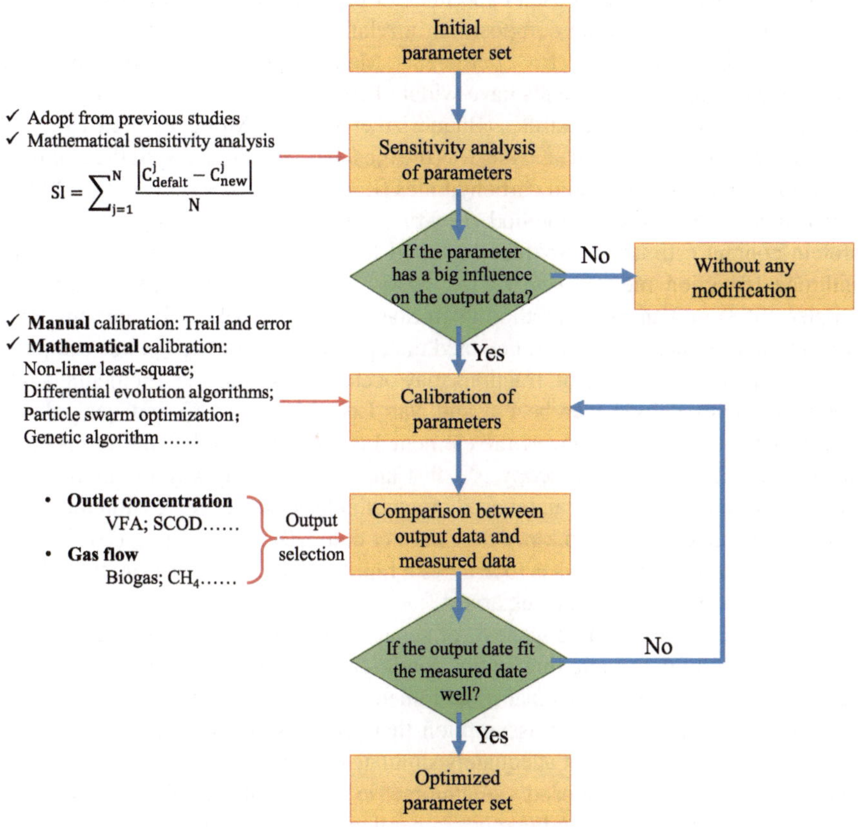

Fig. 3 Modeling methodology used for kinetic parameter calibration

gas transfer resulted in a significant decrease in the liquid/gas mass transfer coefficient, k_T, from its default value of 200 d^{-1} for wet anaerobic digestion of activated sludge to 0.5 d^{-1} (Abbassi-Guendouz et al. 2012). Apart from kinetic coefficients (gas transfer, solids precipitation), most parameters in the physicochemical model are fundamentally fixed by free energy and are not contestable.

4 Model Stoichiometry

Stoichiometric coefficients describe the stoichiometric conversion relationships among components in the anaerobic digestion system, and they contain the yield parameters of the products to the substrates (f) and the yield parameters of the organisms to the substrates (Y).

Stoichiometric coefficients of disintegration ($f_{\mathrm{ch,xc}}$, $f_{\mathrm{pr,xc}}$, $f_{\mathrm{li,xc}}$, $f_{\mathrm{si,xc}}$ and $f_{\mathrm{xi,xc}}$) represent biodegradable and non-biodegradable fractions of composite solids (X_{C}). Many researchers have proposed experimental protocols or methodologies for the characterization of substrate composition in relation to ADM1 input variables and have modified the stoichiometric coefficients. Methane production curves obtained from bio-methane potential tests have widely been used to estimate biodegradable and inert fractions of the substrates (Baquerizo et al. 2021; Mottet et al. 2013).

Direct physical–chemical analysis was the most basic method to split the complex solids into proteins, lipids, and carbohydrates (Li et al. 2021b; Ramirez et al. 2009), which include the Lowry's method (Lowry et al. 1951) or Kjeldahl nitrogen for protein content estimation, Soxhlet's method for lipid content determination, and anthrone reduction method (Dreywood 1946), Weender analysis (Naumann and Bassler 1993) or Van Soest method (Van Soest and Wine 1967) for carbohydrate content measurements. It should be noted that, particularly for high-solids samples, loss or denaturation of organic fractions may occur during analysis. Elemental analysis was proposed by Kleerebezem and Van Loosdrecht (2006) and generalized by Zaher et al. (2009) to convert the elemental composition and COD into ADM1 inputs. Near infrared spectroscopy is a fast and cost-efficient way to characterize solid wastes content, and it can be used in a partial least square regression model to estimate methane production kinetics (Charnier et al. 2017). Guo et al. (2023) used combined instrumental analyses to achieve a rapid and accurate fractionation of the primary organic matters in the substrates.

The stoichiometry for acidogenesis of sugars can be altered as they may depend on reactor conditions, including redox conditions and pH (Shi et al. 2019). The parameters involved in the original stoichiometry for the acidogenesis ($f_{\mathrm{ac,su}}$, $f_{\mathrm{pro,su}}$, $f_{\mathrm{bu,su}}$, $f_{\mathrm{h2,su}}$) are based on the assumption that products are produced in constant proportions, but they may not adequately simulate experimental results. In response, Rodríguez et al. (2006) proposed a stoichiometric system with variable acidogenesis products controlled by thermodynamics rather than kinetics. To account for electron-sink capacity (incorporating pH), which highly affects acidogenesis, some other variable stoichiometric systems have been developed and added to ADM1. These systems depend on the concentration of undissociated acids (Penumathsa et al. 2008), hydrogen partial pressure (p_{H2}) (Thamsiriroj et al. 2012), and NADH/NAD$^+$ (Shi et al. 2019).

The implementation of the variable stoichiometry system expands ADM1 applicability to fermentative systems and reduces the parameter numbers to be estimated (Shi et al. 2019). This development is particularly important because the stoichiometric parameters for acidogenesis of sugars are subject to reactor conditions.

5 Applications of the ADM1

ADM1 was developed as a generalized model, but focused on sewage sludge (Batstone and Keller 2003; Batstone et al. 2004). With the modification of the model, the model can be more widely used in different anaerobic digestion systems with complex substrates, such as food waste, agriculture waste, or other co-substrates with sewage sludge. So far, the modified ADM1 has been used to simulate the lab-scale or full-scale anaerobic digestion processes in reactors including continuously stirred tank reactor (CSTR), up-flow anaerobic sludge blanket reactor (UASB), anaerobic baffled reactors (ABR), two-phase anaerobic digestion, temperature-phase anaerobic digestion, anaerobic membrane bioreactor (AnMBR), and good simulation results on various output indexes were obtained, as given in Table 3.

The well-fitted ADM1 model can be applied to simulate anaerobic digestion processes of urban, agricultural, and industrial wastes (Table 3). For urban application, ADM1 has been used for simulation and design of anaerobic digestion systems of urban organic wastes, such as food and garden wastes (Curry and Pillay 2012; Fitamo et al. 2016), municipal wastewater (Ozgun 2019), gray water, black water, feces (Elmitwalli et al. 2006, 2011), and primary or waste activated sludges from water resource recovery facilities (WRRFs) (Baquerizo et al. 2021). ADM1 is also useful in agricultural applications, such as anaerobic digestion processes of grass silage (Koch et al. 2010; Wichern et al. 2009), sweet sorghum extract (Antonopoulou et al. 2012a), and manure (Li et al. 2020). The industrial applications of ADM1 are much more extensive (Table 3) and more complex substrates are involved, such as various kinds of industrial wastewater (Hinken et al. 2014), olive mill solid waste (Boubaker and Ridha 2008), and vinasse from sugarcane ethanol distillation (Barrera et al. 2015; Couto et al. 2022).

ADM1 can be used as a management tool for the design and operation of the engineered anaerobic digestion processes in practice (Fig. 4). The anaerobic digestion start-up process is one of the most important issues when a new biogas plant is built or restarted (Normak et al. 2015). This can be unstable or fail due to poor inoculum (Pandey et al. 2011) or high organic loading rate (OLR) (Zhao et al. 2010). The extended ADM1 can better predict the start-up process, such as the activity of microorganisms (Zhao et al. 2010) and dynamics of VFAs (Shi et al. 2016), and provide a start-up procedure, including dynamic control (Batstone et al. 2010). Normak et al. (2015) simulated the uninoculated start-up of the anaerobic digestion

Table 3 Application of the ADM1 to different anaerobic digestion scenarios

	Substrates	Conditions	Scale and reactor	Simulation items	Simulation results and analysis	References
Urban applications	PS + WAS	Part of a WTTP Mesophilic	Full-scale CSTR	Biogas, CH_4% TS, VS	The differences between the ADM1 results are all within 2%	Shang et al. (2005)
	Sewage, grey water, black water and feces	Part of urban sanitation Low and high temperature (15–35 °C)	Full-scale; UASB + UASB septic-tank	CH_4 flow, CH_4% COD, COD removal	Determination of a suitable design for each system	Elmitwalli et al. (2006)
	OFMSW +	Mesophilic	Full-scale	Biogas flow, CH_4%, CO_2% pH, COD, TVFA, IN, IC	Acceptable simulating results	Derbal et al. (2009)
	Domestic wastewater	Mesophilic	Pilot-scale UASB	Biogas flow pH, TCOD, SCOD	With average experimental values within 10% of the simulated results	Lohani et al. (2016)
	Municipal wastewater	Mesophilic	Full-scale	CH_4 flow COD, alkalinity, pH	Acceptable simulating results	Ozgun (2019)
	Urban wastewater	Part of a WRRF Mesophilic	Full-scale	Biogas flow, CH_4% COD, TSS, VSS, N, P	Acceptable simulating results	Baquerizo et al. (2021)
	Food waste	Continuous mode Mesophilic	Pilot-scale Two-stage (CSTR + ADR)	Biogas flow, VFAs, pH	In CSTR: R^2 (pH)>0.85, R^2(biogas) = 0.42 In ADR: R^2 (pH) = 0.81, R^2 (biogas) = 0.96, R^2 (VFAs) = 0.97	Yu et al. (2012)
		Semi-continuous mode Thermophilic	Lab-scale	CH_4 flow pH, TCOD, acetate, NH_4^+	Average absolute percent relative error based on methane optimization: 13.7% of CH_4, 9.4% of pH, 58.2% of NH_4^+, 110.2% of ammonia, 70.16% of TCOD	Atallah et al. (2014)

(continued)

Table 3 (continued)

	Substrates	Conditions	Scale and reactor	Simulation items	Simulation results and analysis	References
	Fruit and vegetable wastes	Semi-continuous mode Mesophilic	Lab-scale CSTR	Biogas flow pH, VS	Acceptable simulating results	Garcia-Gen et al. (2015)
	Food and garden waste	Thermophilic	Lab-scale CSTR	CH_4 flow NH_4^+, VFAs	Simulate the optimal feedstock composition for CH_4 production	Fitamo et al. (2016)
	Food waste + WAS	Semi-continuous mode Mesophilic	Lab-scale CSTR	CH_4 flow pH, TAN, COD	Nash Sutcliffe efficiency coefficient = 0.76–0.82. Correlation coefficient = 0.9–0.96. Root mean squared error = 0.09–0.12	Montecchio et al. (2019)
Industrial applications	Corn processing wastewaters	Mesophilic	Full-scale EGSBR	CH_4 flow pH, COD	The mean absolute relative error values are 18%, 10%, and 1%, respectively, for methane production, COD and pH	Ersahin et al. (2007)
	Olive mill wastewater and solid waste	Semi-continuous mode Mesophilic	Lab-scale Tubular digesters	Gas flow, CH_4%, CO_2% pH, TVFA	Acceptable simulating results; the reactor failure at high OLR and short HRT was predicted by model	Boubaker and Ridha (2008)
	Evaporator condensate from pulp mill	Semi-continuous mode Mesophilic	Lab-scale 5L glass contact CSTR	CH_4 flow, CH_4% SCOD, COD removal, pH, VFA	Acceptable simulating results	Silva et al. (2009)
	Opium alkaloid effluents	Continuous mode Mesophilic	Lab-scale UASB	CH_4 and CO_2 flow pH, COD	Acceptable simulating results except biogas flows in the latter periods	Dereli et al. (2010)
	Starch wastewater	Continuous mode Mesophilic	Lab-scale UASB	CH_4 flow Lactate, VFAs, COD	Acceptable simulating results	Hinken et al. (2014)
	Vinasse	Mesophilic	Lab-scale 3.5 L UASB reactor	Biogas and H_2S flow COD, pH, SO_4^{2-}, VFAs	Acceptable predictions; predict failure of AD when the sulfate loading rate increased	Barrera et al. (2015)

(continued)

Table 3 (continued)

	Substrates	Conditions	Scale and reactor	Simulation items	Simulation results and analysis	References
		Part of biogas plant Mesophilic	Full-scale covered in-ground anaerobic reactor	Biogas flow, $H_2S\%$, $CH_4\%$	The values of relative absolute error are under 20%	Silva Neto et al. (2019)
		Continuous mode Thermophilic	Lab-scale	H_2 flow Lactate, VFAs	The stability of the Markov Chain was greater than 96%; modeling combined with microbial analysis	Couto et al. (2022)
	Confectionery effluents	Mesophilic	Full-scale EGSBR	CH_4 flow COD, TVFA, pH	MRAE for COD, CH_4 flow, TVFA and pH were 22%, 16%, 29% and 1%	Dereli (2019)
	Industrial wastewater	Mesophilic	Lab-scale AnMBR	CH_4 flow Inhibitions by free ammonia and cations	Long SRT values cause high CH_4 rate for MBR; inhibitions of free ammonia and cations to osmosis membrane	Song et al. (2018)
		Mesophilic	Full-scale Internal circulation reactor	CH_4, CO_2 and H_2S flow pH, TVFA, COD, SO_4^{2-}, IN (NH_x), IP ($H_xPO_4^{3-x}$)	Average deviations < 15%	Feldman et al. (2018)
Agricultural applications	Sweet sorghum extract	Batch mode Mesophilic	Lab-scale 1350 mL flasks	H_2, flow pH, VFAs	Acceptable simulating results except overestimated pH	Antonopoulou et al. (2012b)
		Continuous Mesophilic	Pilot-scale Two-stage CSTR	CH_4 and biogas flow pH	The deviation values for CH_4: 1.9% (15d HRT) and 1.1% (10d HRT); the deviation of pH and $CH_4\%$ < 5%	Antonopoulou et al. (2012a)

(continued)

Table 3 (continued)

	Substrates	Conditions	Scale and reactor	Simulation items	Simulation results and analysis	References
	Grass silage	Semi-continuous mode Mesophilic	Lab-scale 35L fermenters	Gas flow and composition, H_2 pH, TS, VFAs	Final square error sum = 28.3	Wichern et al. (2009)
		Mesophilic	Lab-scale Two-loop reactor	Biogas flow, CH_4%, CO_2%, TN, NH_4^+, VFAs	Acceptable simulating results (75% coverage) except CH_4% and CO_2%	Koch et al. (2010)
	Vegetable crop residues	Batch mode Mesophilic	Lab-scale 500 ml flasks	CH_4	R^2 (CH_4) = 0.982	Li et al. (2021b)
	Dairy manure	Semi-continuous mode Mesophilic	Lab-scale bench-scale digester	Biogas flow, CH_4%, CO_2% COD reduction, NH_4^+, VFAs	CH_4%, CO_2%, TVFA and COD reduction fit well; biogas and NH_4^+ were overpredicted	Page et al. (2008)
		Mesophilic	Full-scale plug-flow digesters	Biogas flow, CH_4%, CO_2% COD reduction, NH_4^+, VFAs	Biogas, biogas composition and COD fit well; VFAs and NH_4^+ were overpredicted	Page et al. (2008)
	Pig manure	Semi-continuous mode Mesophilic	Lab-scale	CH_4 flow SCOD, NH_4^+, PO_4^{3-}	R^2 (methane) = 0.92; R^2 (SCOD) = 0.852; R^2 (NH_4^+-N) = 0.728; R^2 (PO_4^{3-}) = 0.685	Li et al. (2020)
	Dairy manure + mushroom substrate	Semi-continuous mode Mesophilic	Lab-scale CSTR	CH_4 flow pH	Acceptable simulating results	Shi et al. (2014)

PS primary sludge; *OFMSW* organic fraction of municipal solid waste; *TPAD* temperature-phased anaerobic digestion; *EGSBR* expanded granular sludge bed reactor; *ADR* advective–diffusive reactor of cattle slurry using the modified ADM1 with an inhibition function and identified an inhibitory threshold value of VFAs to hydrolysis of 9.87 g/L during the start-up period. During stable operation, the model can be used to adjust anaerobic digestion condition and realize the maximum recovery of energy and other by-products with the minimum manual effort or ancillary measurements. OLR and hydraulic retention time (HRT) are significant operational parameters in the anaerobic digestion process. Based on simulation using a modified model, Mendes et al. (2015) concluded that the optimum HRT of the mesophilic anaerobic sludge digestion system was in the range of 6–9 days. This application was supported by Montecchio et al. (2017a), who proposed that the maximization of the methane and energy production for one unit of digester volume in thermophilic biogas plants fed with WAS was achieved at HRT range of 10–12 days. Some researchers used ADM1 to predict the collapse threshold of the anaerobic digester (Boubaker and Ridha 2008; Koutrouli et al. 2009). When multiple substrates are co-digested, the model can be used to determine the optimal ratio between different substrates to maximize the biogas production (Fitamo et al. 2016; Zaher et al. 2009). For instance, Zhou et al. (2019) simulated the anaerobic co-digestion (AcoD) of WAS and maize with different mixing ratios, and demonstrated the optimal ratio was 2:1. Zhou et al. (2020) developed a rule-based proportional-integral-derivative controller for the feeding control of the AcoD process, which were combined with ADM1 to keep methane production stable at the setpoint, resisting the disturbances and maximizing methane production.

The development of plant-wide modeling is driven by the fact that the sludge originates from the wastewater treatment and that the nature of the sludge is closely related to the wastewater treatment process. In addition, return streams from digestion are a significant load on the main treatment process. To compare the performance of different control strategies in a unified framework, the Benchmark Simulation Model No. 2 (BSM2) was presented as a plant-wide model, which contains both ASM series and ADM1 (Gernaey et al. 2006, 2014; Vreck et al. 2006). With the emphasis on P recovery, to simulate the P recovery in wastewater treatment processes, Batstone et al. (2015) and Flores-Alsina et al. (2016) proposed a plant-wide phosphorus model (including S and Fe) based on BSM2. Plant-wide modeling of wastewater treatment has been a key focus, particularly in the development of commercial software (Batstone et al. 2015).

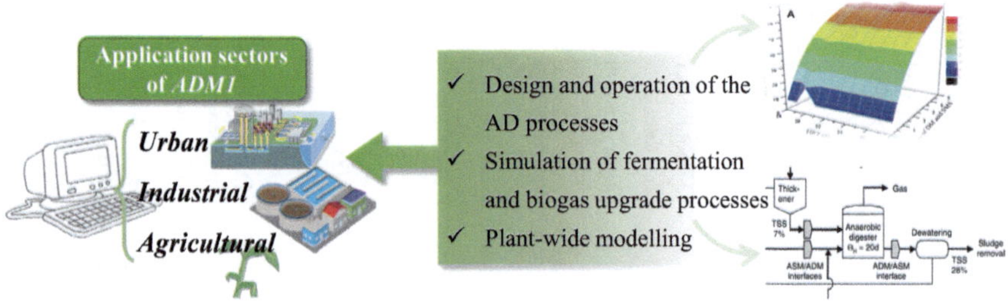

Fig. 4 Practical application of the ADM1

The application scope of the modified ADM1 has significantly been expanded. Beyond its conventional use in the anaerobic digestion process, the modified models can offer valuable insights into process optimization, substrate utilization, and product formation of the fermentation processes. Various kinds of valuable by-products can be produced during anaerobic fermentation, such as lactate, VFAs, and H_2, and the modified ADM1 can be used to predict the production of those compounds and regulate optimal fermentation conditions (Shi et al. 2016).

In addition, biogas upgrading processes are crucial for enhancing the quality and potential utilization of biogas. ADM1 has been extended and adapted to facilitate the simulation of biogas upgrading by biogas desulfurization. In a study of Oliveros-Muñoz et al. (2021), the modified ADM1-S/O was used to optimize microaeration conditions, revealing that the optimal initiation time for microaeration was day 11, whereas the dissolved oxygen concentration in the digester was 1.936×10^{-4} ppm. However, ADM1 application in biogas upgrading processes remains relatively limited. In the future, the model could potentially serve as a useful tool for exploring additional biogas upgrading processes, such as CO_2 removal and CH_4 enrichment.

6 Future Perspectives

ADM1 is the most widely used structural anaerobic digestion model and is a powerful tool to evaluate the performance of anaerobic digesters. It has been modified based on different purposes. However, ADM1 still has some limitations. Firstly, although previous modifications of the model have broadened its scope (Table 3), there are still some unsatisfactory or even neglected areas. For example, during the co-digestion of WAS with beverage wastewater, although the biogas production was predicted well, the VFAs and total ammonia nitrogen (TAN) underperformed (Donoso-Bravo et al. 2020), and few studies have simulated the change of amino acids, sugars and LCFAs. In many cases, this is because the underlying mechanisms have not been fully investigated. Secondly, high VFAs or ammonia can inhibit hydrolysis, but there is very little literature describing this in an ADM1 context. Finally, most ADM1 simulations of other substrates (such as food waste and agricultural waste) besides sludge are still at the lab-scale level, and verification of the model in full-scale plants is lacking. Additionally, to improve biogas production, these substrates have been co-digested with sludge, causing changes in feed properties. However, model simulation studies on co-digestion have also been performed on the lab-scale.

To address the abovementioned limitations and promote the application of ADM1, a more comprehensive consideration of the intermediate production and uptake processes is needed to improve the poor fitting of the current model to the intermediate products. Additionally, the inhibition functions of hydrolysis need further development to make the model work in both normal and extreme anaerobic digestion conditions. As the model is extended, the number of parameters involved is gradually increasing, which requires a fast and accurate methodology for parameter estimation and calibration. Moreover, some novel biogas upgrading processes, such as CO_2 removal and CH_4 upgradation, are still promising applications of the modified ADM1 in future research. In addition to biogas, ADM1 can be further modified and aid in understanding and optimizing the production of valuable intermediates such as ethanol, lactic acid, and VFAs. In addition, chain elongation is a potential process to produce valuable biochemicals and biofuels, such as caproate and caprylate, from organic wastes. However, ADM1 has not been applied in this area. Because of the open structure and common nomenclature of ADM1, the model may be valuable in other fields as well.

In the long term, with the development of modern instrumental analytical technology, especially molecular technique, a deeper understanding of the mechanism of the anaerobic fermentation and digestion processes will be achieved. It is necessary to continuously improve ADM1 based on this deeper understanding, so that it can simulate these anaerobic processes more accurately.

7 Take-Home Messages

(1) ADM1 was originally proposed focusing on sewage sludge. To serve varying purposes and with deeper understanding of anaerobic digestion, the structure of the model, including extracellular and intracellular biochemical processes as well as physicochemical processes, have been widely modified.
(2) Effects of microbes have been considered in hydrolysis by replacing first-order kinetics with Contois kinetics. However, a more comprehensive consideration of high VFA or ammonia inhibitions to hydrolysis is needed in the future.
(3) Compared to constant stoichiometric coefficients, variable stoichiometric systems based on redox conditions and pH offer improved description of the acidogenesis of sugar.
(4) The parameters are calibrated to minimize the error between the experimental and simulated data by the "trial and error" or iterative methods (optimization algorithms). Parameters of hydrolysis and methanogenesis are the most frequently modified parameters for anaerobic digestion processes.
(5) ADM1 has a wide range of applications, including urban, agricultural, and industrial applications. In addition to anaerobic digestion processes, the modified model can also be used for fermentation and biogas upgrading processes. However, the applications of the model to substrates other than sludge in full-scale are still limited.
(6) It is highlighted that ADM1 can be further developed based on the in-depth understanding of the anaerobic treatment processes of organic wastes and the recovery of valuable products.

References

Abbassi-Guendouz A, Brockmann D, Trably E, Dumas C, Delgenes JP, Steyer JP, Escudie R (2012) Total solids content drives high solid anaerobic digestion via mass transfer limitation. Biores Technol 111:55–61

Antonopoulou G, Gavala HN, Skiadas IV, Lyberatos G (2012a) ADM1-based modeling of methane production from acidified sweet sorghum extract in a two stage process. Biores Technol 106:10–19

Antonopoulou G, Gavala HN, Skiadas IV, Lyberatos G (2012b) Modeling of fermentative hydrogen production from sweet sorghum extract based on modified ADM1. Int J Hydrogen Energy 37(1):191–208

Arnell M, Astals S, Amand L, Batstone DJ, Jensen PD, Jeppsson U (2016) Modelling anaerobic co-digestion in benchmark simulation model no. 2: parameter estimation, substrate characterisation and plant-wide integration. Water Res 98:138–146

Astals S, Peces M, Batstone DJ, Jensen PD, Tait S (2018) Characterising and modelling free ammonia and ammonium inhibition in anaerobic systems. Water Res 143:127–135

Atallah NM, El-Fadel M, Ghanimeh S, Saikaly P, Abou-Najm M (2014) Performance optimization and validation of ADM1 simulations under anaerobic thermophilic conditions. Biores Technol 174:243–255

Bai J, Liu H, Yin B, Ma H (2015) Modeling of enhanced VFAs production from waste activated sludge by modified ADM1 with improved particle swarm optimization for parameters estimation. Biochem Eng J 103:22–31

Bai J, Liu H, Yin B, Ma H, Chen X (2017) Modified ADM1 for modeling free ammonia inhibition in anaerobic acidogenic fermentation with high-solid sludge. J Environ Sci 52:58–65

Baquerizo G, Fiat J, Buffiere P, Girault R, Gillot S (2021) Modelling the dynamic long-term performance of a full-scale digester treating sludge from an urban WRRF using an extended version of ADM1. Chem Eng J 423:128870

Bareha Y, Girault R, Guezel S, Chaker J, Tremier A (2019) Modeling the fate of organic nitrogen during anaerobic digestion: development of a bioaccessibility based ADM1. Water Res 154:298–315

Barrera EL, Spanjers H, Solon K, Amerlinck Y, Nopens I, Dewulf J (2015) Modeling the anaerobic digestion of cane-molasses vinasse: extension of the anaerobic digestion model no. 1 (ADM1) with sulfate reduction for a very high strength and sulfate rich wastewater. Water Res 71:42–54

Batstone DJ, Flores-Alsina X (2022) Generalised physicochemical model (PCM) for wastewater processes. IWA Publishing

Batstone DJ (2006) Mathematical modelling of anaerobic reactors treating domestic wastewater: rational criteria for model use. Rev Environ Sci Biotechnol 5(1):57–71

Batstone DJ, Keller J (2003) Industrial applications of the IWA anaerobic digestion model No. 1 (ADM1). Water Sci Technol 47(12):199–206

Batstone DJ, Balthes C, Barr K (2010) Model assisted startup of anaerobic digesters fed with thermally hydrolysed activated sludge. Water Sci Technol 62(7):1661–1666

Batstone DJ, Keller J, Angelidaki I, Kalyuzhnyi SV, Pavlostathis SG, Rozzi A, Sanders WT, Siegrist H, Vavilin VA (2002) The IWA anaerobic digestion model no 1 (ADM1). Water Sci Technol 45(10):65–73

Batstone DJ, Keller J, Blackall LL (2004) The influence of substrate kinetics on the microbial community structure in granular anaerobic biomass. Water Res 38(6):1390–1404

Batstone DJ, Keller J, Steyer JP (2006a) A review of ADM1 extensions, applications, and analysis: 2002–2005. Water Sci Technol 54(4):1–10

Batstone DJ, Picioreanu C, van Loosdrecht MCM (2006b) Multidimensional modelling to investigate interspecies hydrogen transfer in anaerobic biofilms. Water Res 40(16):3099–3108

Batstone DJ, Puyol D, Flores-Alsina X, Rodríguez J (2015) Mathematical modelling of anaerobic digestion processes: applications and future needs. Rev Environ Sci Biotechnol 14(4):595–613

Batstone DJ, Tait S, Starrenburg D (2009) Estimation of hydrolysis parameters in full-scale anerobic digesters. Biotechnol Bioeng 102(5):1513–1520

Boubaker F, Ridha BC (2008) Modelling of the mesophilic anaerobic co-digestion of olive mill wastewater with olive mill solid waste using anaerobic digestion model No. 1 (ADM1). Bioresour Technol 99(14):6565–6577

Britton A, Koch FA, Mavinic DS, Adnan A, Oldham WK, Udala B (2005) Pilot-scale struvite recovery from anaerobic digester supernatant at an enhanced biological phosphorus removal wastewater treatment plant. J Environ Eng Sci 4(4):265–277

Capson-Tojo G, Astals S, Robles A (2021) Considering syntrophic acetate oxidation and ionic strength improves the performance of models for food waste anaerobic digestion. Biores Technol 341:125802

Cazier EA, Trably E, Steyer JP, Escudie R (2019) Reversibility of hydrolysis inhibition at high hydrogen partial pressure in dry anaerobic digestion processes fed with wheat straw and inoculated with anaerobic granular sludge. Waste Manage 85:498–505

Charnier C, Latrille E, Jimenez J, Torrijos M, Sousbie P, Miroux J, Steyer JP (2017) Fast ADM1 implementation for the optimization of feeding strategy using near infrared spectroscopy. Water Res 122:27–35

Couto PT, Eng F, Bovio-Winkler P, Cavalcante WA, Etchebehere C, Fuentes L, Nopens I, Zaiat M, Ribeiro R (2022) Modeling of hydrogen and organic acid production using different concentrations of sugarcane vinasse under thermophilic conditions and a link with microbial community 16S rRNA gene sequencing data. J Clean Prod 370:133437

Curry N, Pillay P (2012) Biogas prediction and design of a food waste to energy system for the urban environment. Renew Energy 41:200–209

Dennehy C, Lawlor PG, Croize T, Jiang Y, Morrison L, Gardiner GE, Zhan X (2016) Synergism and effect of high initial volatile fatty acid concentrations during food waste and pig manure anaerobic co-digestion. Waste Manage 56:173–180

Derbal K, Bencheikh-Lehocine M, Cecchi F, Meniai AH, Pavan P (2009) Application of the IWA ADM1 model to simulate anaerobic co-digestion of organic waste with waste activated sludge in mesophilic condition. Biores Technol 100(4):1539–1543

Dereli RK (2019) Modeling long-term performance of full-scale anaerobic expanded granular sludge bed reactor treating confectionery industry wastewater. Environ Sci Pollut Res 26(24):25037–25045

Dereli RK, Ersahin ME, Ozgun H, Ozturk I, Aydin AF (2010) Applicability of anaerobic digestion model no. 1 (ADM1) for a specific industrial wastewater: opium alkaloid effluents. Chem Eng Jo 165(1):89–94

Donoso-Bravo A, Olivares D, Lesty Y, Bossche HV (2020) Exploitation of the ADM1 in a XXI century wastewater resource recovery facility (WRRF): the case of codigestion and thermal hydrolysis. Water Res 175:115654

Dreywood (1946) Qualitative test for carbohydrate material. Industr Eng Chem Analy Ed 18:499–504

Elmitwalli T, Feng Y, Behrendt J, Otterpohl R (2006) Anaerobic digestion potential for ecological and decentralised sanitation in urban areas. Water Sci Technol 53(9):45–54

Elmitwalli T, Zeeman G, Otterpohl R (2011) Modelling anaerobic digestion of concentrated black water and faecal matter in accumulation system. Water Sci Technol 63(9):2039–2045

Ersahin ME, Insel G, Dereli RK, Ozturk I, Kinaci C (2007) Model based evaluation for the anaerobic treatment of corn processing wastewaters. CLEAN—Soil, Air, Water 35(6):576–581

Fedorovich V, Lens P, Kalyuzhnyi S (2003) Extension of enaerobic digestion model no. 1 with processes of sulfate reduction. Appl Biochem Biotechnol 109(1):33–45

Feldman H, Flores-Alsina X, Kjellberg K, Jeppsson U, Batstone DJ, Gernaey KV (2018) Model-based analysis and optimization of a full-scale industrial high-rate anaerobic bioreactor. Biotechnol Bioeng 115(11):2726–2739

Fezzani B, Cheikh RB (2009) Extension of the anaerobic digestion model No. 1 (ADM1) to include phenolic compounds biodegradation processes for the simulation of anaerobic co-digestion of olive mill wastes at thermophilic temperature. J Hazard Mater 162(2–3):1563–1570

Fitamo T, Boldrin A, Dorini G, Boe K, Angelidaki I, Scheutz C (2016) Optimising the anaerobic co-digestion of urban organic waste using dynamic bioconversion mathematical modelling. Water Res 106:283–294

Flores-Alsina X, Solon K, Kazadi Mbamba C, Tait S, Gernaey KV, Jeppsson U, Batstone DJ (2016) Modelling phosphorus (P), sulfur (S) and iron (Fe) interactions for dynamic simulations of anaerobic digestion processes. Water Res 95:370–382

Fountoulakis MS, Stamatelatou K, Lyberatos G (2008) The effect of pharmaceuticals on the kinetics of methanogenesis and acetogenesis. Biores Technol 99(15):7083–7090

Garcia-Gen S, Lema JM, Rodriguez J (2013) Generalised modelling approach for anaerobic co-digestion of fermentable substrates. Biores Technol 147:525–533

Garcia-Gen S, Rodriguez J, Lema JM (2014) Optimisation of substrate blends in anaerobic co-digestion using adaptive linear programming. Biores Technol 173:159–167

Garcia-Gen S, Sousbie P, Rangaraj G, Lema JM, Rodriguez J, Steyer JP, Torrijos M (2015) Kinetic modelling of anaerobic hydrolysis of solid wastes, including disintegration processes. Waste Manage 35:96–104

Ge Y, Tao J, Wang Z, Chen C, Mu L, Ruan H, Rodríguez Yon Y, Su H, Yan B, Chen G (2023) Modification of anaerobic digestion model No.1 with Machine learning models towards applicable and accurate simulation of biomass anaerobic digestion. Chem Eng J 454:140369

Gernaey K, Jeppsson U, Vanrolleghem P, Copp J (2014) Benchmarking of control strategies for wastewater treatment plants. IWA Scientific and Technical Report No. 23, IWA Publishing

Gernaey KV, Rosen C, Jeppsson U (2006) WWTP dynamic disturbance modelling–an essential module for long-term benchmarking development. Water Sci Technol 53(4–5):225–234

Guo W, Li D, Zhang Z, Mo R, Peng Y, Li Y (2023) A novel approach for the fractionation of organic components and microbial degraders in ADM1 and model validation based on the methanogenic potential. Water Res 236:119945

Henze M, Grady L, Gujer W, Marais GVR, Matsuo T (1986) Activated sludge model no.1. IWAPRC, London

Hierholtzer A, Akunna JC (2012) Modelling sodium inhibition on the anaerobic digestion process. Water Sci Technol 66(7):1565–1573

Hierholtzer A, Akunna JC (2014) Modelling start-up performance of anaerobic digestion of saline-rich macro-algae. Water Sci Technol 69(10):2059–2065

Hinken L, Huber M, Weichgrebe D, Rosenwinkel KH (2014) Modified ADM1 for modelling an UASB reactor laboratory plant treating starch wastewater and synthetic substrate load tests. Water Res 64:82–93

Karim K, Klasson KT, Drescher SR, Ridenour W, Borole AP, Al-Dahhan MH (2007) Mesophilic digestion kinetics of manure slurry. Appl Biochem Biotechnol 142(3):231–242

Kazadi Mbamba C, Batstone DJ, Flores-Alsina X, Tait S (2015) A generalised chemical precipitation modelling approach in wastewater treatment applied to calcite. Water Res 68:342–353

Kerroum D, Mossaab BL, Abdessalam HM (2010) Use of ADM1 model to simulate the anaerobic digestion process used for sludge waste treatment in thermophilic conditions. Turk J Eng Environ Sci 34(2):121–129

Kleerebezem R, Van Loosdrecht MC (2006) Waste characterization for implementation in ADM1. Water Sci Technol 54(4):167–174

Koch K, Lübken M, Gehring T, Wichern M, Horn H (2010) Biogas from grass silage—Measurements and modeling with ADM1. Biores Technol 101(21):8158–8165

Koutrouli EC, Kalfas H, Gavala HN, Skiadas IV, Stamatelatou K, Lyberatos G (2009) Hydrogen and methane production through two-stage mesophilic anaerobic digestion of olive pulp. Biores Technol 100(15):3718–3723

Latif MA, Mehta CM, Batstone DJ (2017) Influence of low pH on continuous anaerobic digestion of waste activated sludge. Water Res 113:42–49

Li H, Chen Z, Fu D, Wang Y, Zheng Y, Li Q (2020) Improved ADM1 for modelling C, N, P fates in anaerobic digestion process of pig manure and optimization approaches to biogas production. Renew Energy 146:2330–2336

Li P, Liu D, Pei Z, Zhao L, Shi F, Yao Z, Li W, Sun Y, Wang S, Yu Q, Chen L, Liu J (2021a) Evaluation of lignin inhibition in anaerobic digestion from the perspective of reducing the hydrolysis rate of holocellulose. Biores Technol 333:125204

Li P, Pei Z, Liu D, Shi F, Wang S, Li W, Sun Y, Liu J, Gao Y, Yu Q (2021b) Application of anaerobic digestion model no. 1 for modeling anaerobic digestion of vegetable crop residues: Fractionation of crystalline cellulose. J Clean Prod 285:124865

Li X, Yang Z, Liu G, Ma Z, Wang W (2019) Modified anaerobic digestion model No.1 (ADM1) for modeling anaerobic digestion process at different ammonium concentrations. Water Environ Res 91(8):700–714

Liotta F, Chatellier P, Esposito G, Fabbricino M, Frunzo L, van Hullebusch ED, Lens PN, Pirozzi F (2015) Modified anaerobic digestion model no.1 for dry and semi-dry anaerobic digestion of solid organic waste. Environ Technol 36(5–8):870–880

Lizarralde I, Fernandez-Arevalo T, Brouckaert C, Vanrolleghem P, Ikumi DS, Ekama GA, Ayesa E, Grau P (2015) A new general methodology for incorporating physico-chemical transformations into multi-phase wastewater treatment process models. Water Res 74:239–256

Lohani SP, Wang S, Lackner S, Horn H, Khanal SN, Bakke R (2016) ADM1 modeling of UASB treating domestic wastewater in Nepal. Renew Energy 95:263–268

Lokshina LY, Vavilin VA, Kettunen RH, Rintala JA, Holliger C, Nozhevnikova AN (2001) Evaluation of kinetic coefficients using integrated monod and haldane models for low-temperature acetoclastic methanogenesis. Water Res 35(12):2913–2922

Lowry OH, Rosebrough NJ, Farr AL, Randall RJ (1951) Protein measurement with the Folin phenol reagent. J Biol Chem 193(1):265–275

Maharaj BC, Mattei MR, Frunzo L, Hullebusch EDV, Esposito G (2019) ADM1 based mathematical model of trace element complexation in anaerobic digestion processes. Biores Technol 276:253–259

Maharaj BC, Mattei MR, Frunzo L, van Hullebusch ED, Esposito G (2018) ADM1 based mathematical model of trace element precipitation/dissolution in anaerobic digestion processes. Biores Technol 267:666–676

Mairet F, Bernard O, Ras M, Lardon L, Steyer JP (2011) Modeling anaerobic digestion of microalgae using ADM1. Biores Technol 102(13):6823–6829

Mendes C, Esquerre K, Matos Queiroz L (2015) Application of anaerobic digestion model no. 1 for simulating anaerobic mesophilic sludge digestion. Waste Manage 35:89–95

Mo R, Guo W, Batstone D, Makinia J, Li Y (2023) Modifications to the anaerobic digestion model no. 1 (ADM1) for enhanced understanding and application of the anaerobic treatment processes—a comprehensive review. Water Res 244:120504

Montecchio D, Astals S, Di Castro V, Gallipoli A, Gianico A, Pagliaccia P, Piemonte V, Rossetti S, Tonanzi B, Braguglia CM (2019) Anaerobic co-digestion of food waste and waste activated sludge: ADM1 modelling and microbial analysis to gain insights into the two substrates' synergistic effects. Waste Manage 97:27–37

Montecchio D, Braguglia CM, Gallipoli A, Gianico A (2017a) A model-based tool for reactor configuration of thermophilic biogas plants fed with waste activated sludge. Renew Energy 113:411–419

Montecchio D, Esposito G, Gagliano MC, Gallipoli A, Gianico A, Braguglia CM (2017b) Syntrophic acetate oxidation during the two-phase anaerobic digestion of waste activated sludge: microbial population, Gibbs free energy and kinetic modelling. Int Biodeterior Biodegrad 125:177–188

Mottet A, Ramirez I, Carrère H, Déléris S, Vedrenne F, Jimenez J, Steyer JP (2013) New fractionation for a better bioaccessibility description of particulate organic matter in a modified ADM1 model. Chem Eng J 228:871–881

Musvoto EV, Wentzel MC, Loewenthal RE, Ekama GA (2000) Integrated chemical–physical processes modelling—I. Development of a kinetic-based model for mixed weak acid/base systems. Water Res 34(6):1857–1867

Naumann C, Bassler R (1993) Die Chemische Untersuchung von Futtermitteln. VDLUFA-Verlag, Darmstadt

Normak A, Suurpere J, Suitso I, Jõgi E, Kokin E, Pitk P (2015) Improving ADM1 model to simulate anaerobic digestion start-up with inhibition phase based on cattle slurry. Biomass Bioenergy 80:260–266

Oliveros-Muñoz JM, Martínez-Villalba JA, Jiménez-Islas H, Luna-Porres MY, Escamilla-Alvarado C, Ríos-Fránquez FJ (2021) Luus-Jaakola method and ADM1 based optimization of hydrogen sulfide in anaerobic digestion of cow manure. Biochem Eng J 171:108012

Ozgun H (2019) Anaerobic digestion model no. 1 (ADM1) for mathematical modeling of full-scale sludge digester performance in a municipal wastewater treatment plant. Biodegradation 30(1):27–36

Page DI, Hickey KL, Narula R, Main AL, Grimberg SJ (2005) Modeling anaerobic digestion of dairy manure using the IWA anaerobic digestion model no. 1 (ADM1). Water Sci Technol 58(3):689–695

Palatsi J, Illa J, Prenafeta-Boldu FX, Laureni M, Fernandez B, Angelidaki I, Flotats X (2010) Long-chain fatty acids inhibition and adaptation process in anaerobic thermophilic digestion: batch tests, microbial community structure and mathematical modelling. Biores Technol 101(7):2243–2251

Pan Y, Zhi Z, Zhen G, Lu X, Bakonyi P, Li YY, Zhao Y, Rajesh BJ (2019) Synergistic effect and biodegradation kinetics of sewage sludge and food waste mesophilic anaerobic co-digestion and the underlying stimulation mechanisms. Fuel 253:40–49

Pandey PK, Ndegwa PM, Soupir ML, Alldredge JR, Pitts MJ (2011) Efficacies of inocula on the startup of anaerobic reactors treating dairy manure under stirred and unstirred conditions. Biomass Bioenergy 35(7):2705–2720

Parra-Orobio BA, Donoso-Bravo A, Torres-Lozada P (2020) Energy balance and carbon dioxide emissions comparison through modified anaerobic digestion model No 1 for single-stage and two-stage anaerobic digestion of food waste. Biomass Bioenergy 142:105814

Patón M, Rodríguez J (2017) Dynamic thermodynamic simulation of ADM1 validates the hydrogen inhibition approach and suggests an unfeasible butyrate degradation pathway. In: Mannina G (ed) Frontiers in wastewater treatment and modelling. Springer Cham, pp 260–265

Penumathsa BK, Premier GC, Kyazze G, Dinsdale R, Guwy AJ, Esteves S, Rodriguez J (2008) ADM1 can be applied to continuous bio-hydrogen production using a variable stoichiometry approach. Water Res 42(16):4379–4385

Pokorna-Krayzelova L, Mampaey KE, Vannecke TPW, Bartacek J, Jenicek P, Volcke EIP (2017) Model-based optimization of microaeration for biogas desulfurization in UASB reactors. Biochem Eng J 125:171–179

Puyol D, Flores-Alsina X, Segura Y, Molina R, Jerez S, Gernaey KV, Melero JA, Martinez F (2017) ZVI addition in continuous anaerobic digestion systems dramatically decreases P recovery potential: dynamic modelling. In: Mannina G (ed) Frontiers in wastewater treatment and modelling. Springer Cham, pp 211–217

Ramirez I, Mottet A, Carrere H, Deleris S, Vedrenne F, Steyer JP (2009) Modified ADM1 disintegration/hydrolysis structures for modeling batch thermophilic anaerobic digestion of thermally pretreated waste activated sludge. Water Res 43(14):3479–3492

Rivera-Salvador V, Lopez-Cruz IL, Espinosa-Solares T, Aranda-Barradas JS, Huber DH, Sharma D, Toledo JU (2014) Application of Anaerobic Digestion Model No. 1 to describe the syntrophic acetate oxidation of poultry litter in thermophilic anaerobic digestion. Bioresour Technol 167:495–502

Rodríguez J, Lema JM, van Loosdrecht MCM, Kleerebezem R (2006) Variable stoichiometry with thermodynamic control in ADM1. Water Sci Technol 54(4):101–110

Satpathy P, Biernacki P, Cypionka H, Steinigeweg S (2016) Modelling anaerobic digestion in an industrial biogas digester: application of lactate-including ADM1 model (Part II). J Environ Sci Health A 51(14):1226–1232

Shang Y, Johnson BR, Sieger R (2005) Application of the IWA anaerobic digestion model (ADM1) for simulating full-scale anaerobic sewage sludge digestion. Water Sci Technol 52(1–2):487–492

Shi E, Li J, Leu SY, Antwi P (2016) Modeling the dynamic volatile fatty acids profiles with pH and hydraulic retention time in an anaerobic baffled reactor during the startup period. Biores Technol 222:49–58

Shi E, Li J, Zhang M (2019) Application of IWA Anaerobic Digestion Model No. 1 to simulate butyric acid, propionic acid, mixed acid, and ethanol type fermentative systems using a variable acidogenic stoichiometric approach. Water Res 161:242–250

Shi X, Yuan X, Wang Y, Zeng S, Qiu Y, Guo R, Wang L (2014) Modeling of the methane production and pH value during the anaerobic co-digestion of dairy manure and spent mushroom substrate. Chem Eng J 244:258–263

Siegrist H, Vogt D, Garcia-Heras JL, Gujer W (2002) Mathematical model for meso- and thermophilic anaerobic sewage sludge digestion. Environ Sci Technol 36(5):1113–1123

Silva F, Nadais H, Prates A, Arroja L, Capela I (2009) Modelling of anaerobic treatment of evaporator condensate (EC) from a sulphite pulp mill using the IWA anaerobic digestion model no. 1 (ADM1). Chem Eng J 148(2–3):319–326

Silva Neto JVD, Elaiuy MLC, Nour EAA (2019) ADM1 approach to the performance optimisation and biogas H_2S prediction of a large-scale anaerobic reactor fed on sugarcane vinasse. Water Sci Technol 80(9):1774–1786

Soda S, Wada K, Okuda M, Ike M (2011) Application of modified ADM1 to long-term experiments for methane/hydrogen production from model organic waste. Water Pract Technol 6(1):wpt2011009

Solon K, Flores-Alsina X, Kazadi Mbamba C, Ikumi D, Volcke EIP, Vaneeckhaute C, Ekama G, Vanrolleghem PA, Batstone DJ, Gernaey KV, Jeppsson U (2017) Plant-wide modelling of phosphorus transformations in wastewater treatment systems: impacts of control and operational strategies. Water Res 113:97–110

Solon K, Flores-Alsina X, Mbamba CK, Volcke EI, Tait S, Batstone D, Gernaey KV, Jeppsson U (2015) Effects of ionic strength and ion pairing on (plant-wide) modelling of anaerobic digestion. Water Res 70:235–245

Song YC, Kim M, Shon H, Jegatheesan V, Kim S (2018) Modeling methane production in anaerobic forward osmosis bioreactor using a modified anaerobic digestion model No. 1. Bioresour Technol 264:211–218

Sun M, Liu B, Yanagawa K, Ha NT, Goel R, Terashima M, Yasui H (2020) Effects of low pH conditions on decay of methanogenic biomass. Water Res 179:115883

Thamsiriroj T, Nizami AS, Murphy JD (2012) Why does mono-digestion of grass silage fail in long term operation? Appl Energy 95:64–76

Uhlenhut F, Schlüter K, Gallert C (2018) Wet biowaste digestion: ADM1 model improvement by implementation of known genera and activity of propionate oxidizing bacteria. Water Res 129:384–393

Van Soest PJ, Wine RH (1967) Use of detergents in the analysis of fibrous feeds IV. Determination of plant cell-wall constituents. Assoc Off Analy Chem 28:50–59

Vavilin VA, Fernandez B, Palatsi J, Flotats X (2008) Hydrolysis kinetics in anaerobic degradation of particulate organic material: an overview. Waste Manage 28(6):939–951

Vreck D, Gernaey KV, Rosen C, Jeppsson U (2006) Benchmark simulation Model no 2 in Matlab-simulink: towards plant-wide WWTP control strategy evaluation. Water Sci Technol 54(8):65–72

Wang R, Li Y, Chen W, Zou J, Chen Y (2016) Phosphate release involving PAOs activity during anaerobic fermentation of EBPR sludge and the extension of ADM1. Chem Eng J 287:436–447

Wang R, Li Y, Wang W, Chen Y, Vanrolleghem PA (2015) Effect of high orthophosphate concentration on mesophilic anaerobic sludge digestion and its modeling. Chem Eng J 260:791–800

Waszkielis K, Białobrzewski I, Bułkowska K (2022) Application of anaerobic digestion model No. 1 for simulating fermentation of maize silage, pig manure, cattle manure and digestate in the full-scale biogas plant. Fuel 317:123491

Wett B, Takacs I, Batstone D, Wilson C, Murthy S (2014) Anaerobic model for high-solids or high-temperature digestion - additional pathway of acetate oxidation. Water Sci Technol 69(8):1634–1640

Wichern M, Gehring T, Fischer K, Andrade D, Lübken M, Koch K, Gronauer A, Horn H (2009) Monofermentation of grass silage under mesophilic conditions: measurements and mathematical modeling with ADM 1. Biores Technol 100(4):1675–1681

Wilson CA, Novak J, Takacs I, Wett B, Murthy S (2012) The kinetics of process dependent ammonia inhibition of methanogenesis from acetic acid. Water Res 46(19):6247–6256

Xiao X, Sheng GP, Mu Y, Yu HQ (2013) A modeling approach to describe ZVI-based anaerobic system. Water Res 47(16):6007–6013

Yang Z, Sun H, Zhao Q, Kurbonova M, Zhang R, Liu G, Wang W (2020) Long-term evaluation of bioaugmentation to alleviate ammonia inhibition during anaerobic digestion: Process monitoring, microbial community response, and methanogenic pathway modeling. Chem Eng J 399:125765

Yasui H, Goel R, Li YY, Noike T (2008) Modified ADM1 structure for modelling municipal primary sludge hydrolysis. Water Res 42(1–2):249–259

Yu L, Zhao Q, Ma J, Frear C, Chen S (2012) Experimental and modeling study of a two-stage pilot scale high solid anaerobic digester system. Biores Technol 124:8–17

Zaher U, Li R, Jeppsson U, Steyer JP, Chen S (2009) GISCOD: general integrated solid waste co-digestion model. Water Res 43(10):2717–2727

Zeikus JG, Kerby R, Krzycki JA (1985) Single-carbon chemistry of acetogenic and methanogenic bacteria. Science 227(4691):1167–1173

Zhang Y, Piccard S, Zhou W (2015) Improved ADM1 model for anaerobic digestion process considering physico-chemical reactions. Biores Technol 196:279–289

Zhao BH, Mu Y, Dong F, Ni BJ, Zhao JB, Sheng GP, Yu HQ, Li YY, Harada H (2010) Dynamic modeling the anaerobic reactor startup process. Ind Eng Chem Res 49(16):7193–7200

Zhou H, Lv S, Ying Z, Wang Y, Liu J, Liu W (2019) Characteristics of two-phase mesophilic anaerobic digestion of co-substrates consisting of waste activated sludge and corn silage based on modified ADM1. Waste Manage 91:168–178

Zhou H, Ying Z, Cao Z, Liu Z, Zhang Z, Liu W (2020) Feeding control of anaerobic co-digestion of waste activated sludge and corn silage performed by rule-based PID control with ADM1. Waste Manage 103:22–31

Bioreactor Principles for Anaerobic Digestion

Huijuan Sun, Lei Zhang, and Yang Liu

Abstract Anaerobic digestion contributes significantly to meeting sustainability goals across environmental, economic, and social dimensions. It aligns with the principles of the circular economy by minimizing waste discharge and enabling the recovery of bioenergy and nutrients. The anaerobic digestion process typically takes place within anaerobic bioreactors. The bioreactor design and operation largely rely on the key parameters like substrate composition, pH, carbon/nitrogen ratio, hydraulic retention time, organic loading rate, solids retention time, and temperature. Particulate hydrolysis is commonly identified as a limiting step, especially for high-solid content feedstocks. Various strategies, including mixing, biomass immobilization, pre-treatments, and the addition of conductive materials to facilitate direct interspecies electron transfer, are employed to bioreactors to enhance anaerobic digestion efficiency. Emerging research directions involve enhancing the microbiome and digestibility of feedstocks through abiotic stress, addressing emerging pollutants, exploring novel anaerobic digester configurations, and more. Innovations in the design of anaerobic bioreactor will focus on creating value-added products and enhancing biogas purification methods, heralding a new era of efficiency and environmental stewardship in bioenergy production. These advancements promise to broaden the utility of biogas as a renewable energy source and achieve higher yields of economically valuable bioproducts, significantly contributing to global sustainability efforts.

Keywords Anaerobic bioreactors · Circular economy · Feedstock · Organic loading rate · Temperature

H. Sun · Y. Liu (✉)
Department of Civil and Environmental Engineering, University of Alberta, Edmonton, AB, Canada
e-mail: Dr.Liu@qut.edu.au

H. Sun
e-mail: huijuan3@ualberta.ca

L. Zhang · Y. Liu
School of Civil and Environmental Engineering, Queensland University of Technology, Brisbane, QLD, Australia
e-mail: l236.zhang@qut.edu.au

G. Wu (ed.), *Anaerobic Digestion*, Green Energy and Technology,
https://doi.org/10.1007/978-3-031-69378-6_13

1 Introduction

The anaerobic digestion process is a reliable and practical technology for recycling and reclaiming organic waste. From socioeconomic and environmental perspectives, anaerobic digestion emerges as the optimal waste management solution due to its capacity to generate renewable energy and value-added products like biofertilizer (Aworanti et al. 2023; Gao 2020; Sun 2023; Zhang et al. 2024a). This technology can be categorized according to several factors: the total solid (TS) content of feedstocks, the feeding mode, the operational temperature, the number of operational stages, and the organic loading rate (OLR). The TS content of the feedstocks determines whether the anaerobic digestion process is dry (15–40% TS) or wet (0.5–15% TS). The feeding mode determines whether the process operates in a batch mode or a continuous mode. The operational temperature distinguishes between mesophilic conditions (20–40 °C) and thermophilic conditions (50–65 °C). By considering the physical separation of the four digestion steps (i.e., hydrolysis, acidogenesis, acetogenesis, and methanogenesis) in various reactors, the anaerobic digestion systems include single-stage, two-stage, three-stage, and even four-stage setups. Different growth factors (e.g., temperature, pH, and nutrients) and operational conditions (e.g., retention time and OLR) are controlled in different stages to facilitate one or several digestion steps (Van et al. 2020). Anaerobic digestion technology can also be differentiated into low-rate and high-rate bioreactors based on factors such as OLR or biodegradation rate. Each category has its advantages and disadvantages, necessitating a comprehensive analysis before selecting a specific anaerobic digestion process for treating a particular type of waste or wastewater.

2 Types of Bioreactors and Assessment of Their Design and Operation

2.1 Anaerobic Bioreactors

The anaerobic digestion process typically takes place within an enclosed vessel known as a digester or an anaerobic bioreactor. A typical anaerobic bioreactor comprises a digestion chamber, a biogas collection dome, an inlet for feed, outlets for treated wastewater and produced biogas, as well as an outlet for waste sludge (Abdelgadir et al. 2014). The configuration of an anaerobic bioreactor is one of the most important factors that influence the efficacy of an anaerobic digestion process. Figure 1 illustrates the diagram of corresponding anaerobic reactors, including plug flow reactor (PFR), anaerobic membrane bioreactor (AnMBR), upflow anaerobic sludge blanket (UASB) reactor, continuously stirred tank reactor (CSTR), fluidized bed reactor (FBR), fixed-bed reactor, expanded granular sludge bed (EGSB) reactor, and anaerobic sequencing batch reactor (ASBR). Table 1 outlines the characteristics, advantages, disadvantages, and applicable scenarios of widely used types of anaerobic reactors.

The PFR or plug flow digester is a straightforward continuous reactor usually lacking mixing and suspended biomass retention. It comprises an elongated tubular digester available in various sizes and orientations, either vertical or horizontal. It is usually operated at 35–55 °C and has a long hydraulic retention time (HRT) (e.g., 15–20 d). The design is particularly well-suited for handling feedstocks with high TS contents, typically ranging from 11 to 14% such as cattle manure (Aworanti et al. 2023; Rocha-Meneses et al. 2022).

The AnMBR represents a cutting-edge wastewater treatment technology that integrates membrane filtration within the anaerobic digestion process. The membrane serves to retain colloidal and suspended solids, facilitating complete separation between solids and liquid. There are three primary configurations based on the membrane's placement. The first variant involves an anaerobic reactor with an external membrane unit, where the mixed liquor is pressurized and transferred from the reactor to the membrane unit. The second type, known as the submerged AnMBR,

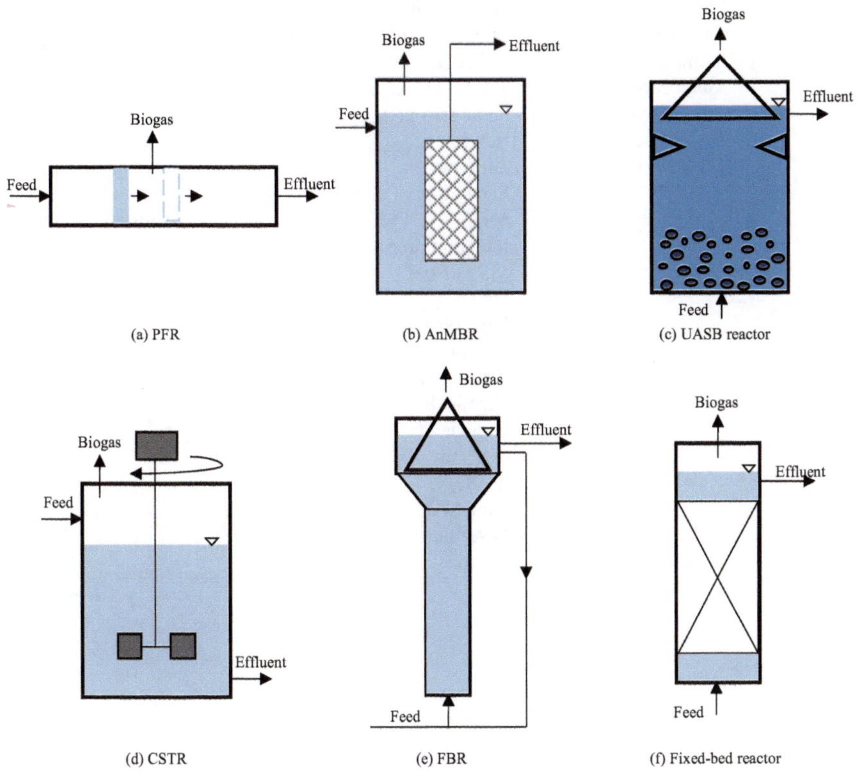

Fig. 1 Schematic of different types of anaerobic digestion reactors and processes

Table 1 Characteristics, advantages, disadvantages, and applicable scenarios of different types of anaerobic reactors

Anaerobic reactor	Characteristics	Advantages	Disadvantages	Applicable scenarios
PFR	• The reaction mixture moves at the same speed along the flow direction • Fluid is mixed in the radial direction but not in the axial direction • A PFR can be placed horizontally or vertically	• Simple configuration and low operation cost • Rapid heat transfer due to a high surface area-to-volume ratio • Minimized side reactions due to radial mixing • Inexpensive and portable	• Long hydraulic retention time (HRT) (20–50 d) • Inhomogeneous mass distribution and low mass transfer due to the lack of mixing • Floating and thermal stratification if substrate is below 10% TS	• TS content of 11–14% • Semisolid flow of raw materials such as sludge, vegetable waste, and cow dung • Production of biodiesel and other biofuels
AnMBR	• Membrane acts as a total barrier for biomass retention • Membrane can be submerged directly in the reactor or placed externally	• High quality of effluent that is free of solids and pathogens and rich in nutrients • Separation of solids retention time (SRT) from HRT • High operation stability • Effective in chemical oxygen demand (COD) reduction • Favor nutrient recovery	• Organic fouling in the membrane • Flux reduction at the membrane outlet • Ineffective at a low temperature	• High-strength wastewater • Low-strength wastewater • Industrial wastewater containing inhibitory or toxic substances

(continued)

Table 1 (continued)

Anaerobic reactor	Characteristics	Advantages	Disadvantages	Applicable scenarios
UASB reactor	• It is a biomethane-producing reactor • A blanket of granular sludge forms at the lower part of the reactor • A three-phase separator can separate biogas, treated effluent, and sludge mixture • The typical upflow velocity is 0.5–1.0 m/h	• Retention of biomass • Decouple SRT from HRT • Effective in producing biogas • Can handle organic shock loads effectively	• Post-treatment requirement • Generation of reduced sulfur compounds • Long start-up period • Ineffective at a low temperature	• High-strength wastewater • Industrial wastewater from breweries and beverage, distilleries and fermentation, food, pulp and paper, and textile industries
CSTR	• Content is mixed continuously to maintain the solids in suspension and to form a homogenous mixture	• Simplicity in configuration, easy functioning, and efficient uniform stirring • Uniform microbial cultures, composition, and temperature • Low resistance in mass transfer • Scalability	• Long HRT (20–50 d) • Low levels of cell inventory • High energy requirement due to mixing • "Short-circuit" • An external settler to retain biomass	• TS content > 5% • High solid feedstock such as organic fraction of municipal solid waste (OFMSW) • Biohydrogen production
FBR	• Microorganisms suspend on inert particles such as sand and activated carbon • Biomass is maintained in a fluidized state by the force of incoming waste stream • It is a biofilm treatment method	• Good temperature control • Enhanced heat and mass transfer • High sludge activity • No channeling or clogging	• Long start-up period • Difficult to control biolayer thickness • High energy consumption • Process instability due to shock loads and inhibitory substances	• Bio-oil production • High-strength and high solid content wastewater such as municipal sludge and corn ethanol thin stillage

(continued)

Table 1 (continued)

Anaerobic reactor	Characteristics	Advantages	Disadvantages	Applicable scenarios
Fixed-bed reactor	• The reactor contains a fixed bed of packing media such as bioball, natural zeolite, and bioring • The microorganisms attach the media and grow as a biofilm • Feed can be fed continuously or intermittently	• Efficient in removing high concentrations of substrates • Efficient in removing soluble COD • Less sensitive to shock loading • Cost-effective in operation and maintenance	• Clogging • Limited mixing • Difficulty in scale up	• Wastewater with high pollution loads • Agricultural wastewater • Landfill leachate • Industrial wastewater • Municipal wastewater
EGSB	• It is the family of UASB reactor • The upflow velocity is higher (4–10 m/h) due to a high recycle ratio • The upflow velocity can be achieved by a high height-to-width ratio (4–5) or the effluent recirculation	• Intensified hydraulic mixing • Reduced dead zone inside the reactor • Improved contact between substrate and sludge • Accommodate very high OLRs	• Particulate hydrolysis is limited • More energy input	• Soluble wastewater • High-strength wastewater • Low-strength wastewater • Acidified wastewater at a psychrophilic condition • Wastewater containing lipids and inhibitory compounds
ASBR	• It is similar to CSTR, but operated in a batch mode • Mixing is required • It provides biomass and liquid separation within the reactor • An external filter is not required	• No short circuit • High efficiency for both COD removal and gas production • No primary and secondary settling • Flexible control	• Poor settling and retention of solids • Inefficient if overloaded	• Rapidly degradable compounds • Slowly degradable compounds • Protein-rich wastewater

features a membrane unit immersed directly within the reactor. The third configuration entails a membrane unit immersed in an external anaerobic chamber filled with biomass (Kanafin et al. 2021).

The UASB reactor relies on upflow velocity of the substrate to be treated, which is introduced from the bottom of the reactor. The substrate undergoes treatment as it passes through a suspended sludge blanket, comprised of microbial granules that agglomerate and accumulate into dense pellets at the lower section of the reactor. The UASB reactor is characterized by its unique three-phase separator, facilitating the separation of biomass from biogas and treated effluent. The UASB reactor is widely recognized as the most popular anaerobic granular sludge system treating a diverse range of high-strength industrial wastewater (Show et al. 2020).

The CSTR operates continuously, with a steady feed introduced into the reactor while a product stream is concurrently withdrawn. Various stirring methods, such as mechanical stirrers with impellers, hydraulic stirrers with liquid circulation, or pneumatic stirrers with bubble diffusers, are employed to ensure a uniform mixture within the CSTR. The CSTR is commonly integrated into multiple-stage anaerobic digestion systems, with nearly 90% of the first-stage reactors being CSTRs (Holl et al. 2022).

The FBR is characterized by its elevated upflow velocity, typically ranging from 10 to 25 m/h, and a substantial biomass concentration resulting from the adherence of biofilm onto inert particles. The FBR is a prominent technology due to its high resistance to disturbances, improved heat and mass transfer, broad particle size distribution, and uniform mixing (Rocha-Meneses et al. 2022).

Within the fixed-bed reactor, microorganisms thrive on carrier materials boasting substantial specific surface areas, which may include plastics, metals, mineral inert substances, or wood. This reactor can function in either a downflow or upflow configuration. Unlike many digesters that require heating, the fixed-bed reactor achieves high treatment efficacy even at ambient effluent temperatures (Holl et al. 2022).

The EGSB reactor is a modification of the UASB reactor. Like the UASB reactor, granules are cultivated within the EGSB reactor. However, unlike the UASB reactor, the EGSB reactor features a significantly higher upflow velocity, facilitated by an elevated effluent recirculation rate. Consequently, the EGSB reactor exhibits greater resilience to toxic compounds due to dilution and enhanced mass transfer, enabling higher loading rates (Show et al. 2020).

The ASBR is similar to the CSTR, but the ASBR is operated in batch mode. In contrast to continuous processes, the ASBR eliminates the need for an external filter by integrating biomass separation from liquid within one reactor. Consequently, the ASBR provides extended retention times for biomass, thereby enhancing treatment efficiency (Holl et al. 2022).

Figure 2 illustrates peer-reviewed publications categorized by various types of anaerobic reactors from 2019 to 2023, sourced from the Clarivate Web of Science™ database. The application of high-rate bioreactors, particularly AnMBRs and UASB reactors, is predominant. Nevertheless, low-rate bioreactors also hold invaluable advantages due to their reliability and economic viability across different scales of operation (Abbasi et al. 2012).

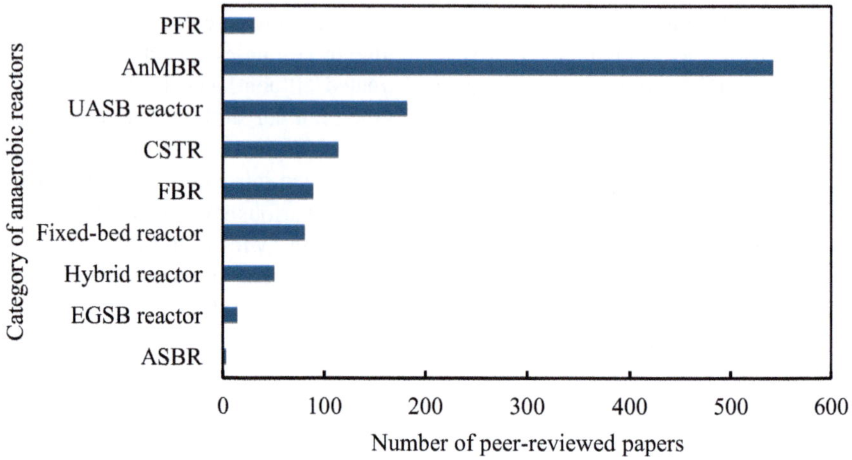

Fig. 2 Publication of peer-reviewed papers in recent five years from 2019 to 2023. Clarivate Web of Science™ database was used for the paper search. The keywords used included a specific type of anaerobic reactor and "Anaerobic digestion"

2.2 Technology Based on Feedstock

Biogas is one main product of anaerobic digestion. It mainly comprises CH_4 (55–70%) and CO_2 (30–45%). Additionally, biogas contains other gases such as water vapor, nitrogen gas, oxygen, hydrogen sulfide (H_2S), ammonia, hydrocarbons, and siloxanes (Awe et al. 2017). The type of feedstock is one of the most important factors that impact anaerobic digestion process efficiency. The feedstocks that can be utilized for anaerobic digestion include agricultural wastes or residues (e.g., rice straw, corn stover, wheat straw, and manure), forest residues (e.g., branches, leaves, and bark), municipal solid wastes [e.g., organic fraction of municipal solid waste (OFMSW), sewage sludge, and food waste], industrial wastes/wastewater (e.g., dairy wastewater, food/beverage processing wastewater, pulp and paper processing wastewater, and textile wastewater), and aquatic biomass (Aworanti et al. 2023; Rocha-Meneses et al. 2022). Feedstocks comprising readily biodegradable carbohydrates, proteins, and lipids can serve as individual feedstocks (Bhajani and Pai 2022). For instance, human excreta, food leftovers, kitchen residue, and cow dung have high biodegradable content (i.e., high volatile to total solid ratio), making them good feedstock sources for biogas generation (Gao et al. 2019; Osei-Owusu et al. 2023; Zhang et al. 2021a, b, 2024b). However, relying solely on a single feedstock (i.e., mono-digestion) often fails to yield sufficient biogas due to challenges such as the recalcitrant nature of lignocellulosic materials, limited microbial diversity, imbalanced carbon/nitrogen (C/N) ratios, and the influence of operating conditions (Gao et al. 2020; Ibro et al. 2022). Furthermore, specific feedstocks containing lignin, such as wood products, paper, and straw, can impede digestion due to the slow hydrolysis rate associated with lignin (Ibro et al. 2022).

To overcome the limitations associated with biogas production from a single feedstock, co-digestion offers an economical solution. Co-digestion involves combining multiple feedstocks in a single digester to enhance methane yield and ensure process stability. These systems offer several advantages, including the dilution of toxic compounds, enhanced buffering capacity, improved nutrient balance, synergistic effects among microorganisms, and accelerated biodegradation through biostimulation (Kumar et al. 2024). By implementing co-digestion strategies (e.g., combining sewage sludge with food waste, blackwater with food waste, food waste with hydrothermal liquid digestate, Napier grass with food waste, cow manure with food waste, food waste with algae and cow dung, cheese whey with septage, and leather waste with wheat straw), methane yield gains ranging from 13% to 176% compared to mono-digestion can be achieved (Chow et al. 2020). However, despite the benefits of co-digestion, adverse effects stemming from incompatible feedstock mixing ratios can lead to issues such as overloading, acidification, and system failure (Chow et al. 2020).

The feasibility of co-digestion is influenced by key factors such as the mixing ratio of co-substrates, pH levels, alkalinity, and nutrient balance (e.g., C/N ratio).

Different ratios of co-substrates significantly influence the anaerobic digestion efficiency. For instance, it is advised to include 20–50 vol% of organic fraction of OFMSW in the co-substrates. Agricultural waste should range from 5 to 50 vol%. Maintaining a ratio of 5–60 vol% for fat, oil, and grease (FOG) is preferable. Additionally, crude glycerol is recommended to constitute 0.5–1.2 vol% of the co-substrates (Chow et al. 2020).

Hydrolysis and acidogenesis require a pH range of 5.5–6.5, while methanogenesis requires a pH range of 6.5–8.2 (Kanafin et al. 2021). Either a low (e.g., $pH < 3$) or a high pH level (e.g., $pH > 12$) can have detrimental effects on anaerobic digestion functioning. The production of volatile fatty acids (VFAs) during acidogenesis can lead to a decline in pH, posing challenges for methanogens. Nevertheless, the hydrolysis of proteins (represented by $C_xH_yNO_z$) generates hydroxide ions (OH^-), which help counteract pH decreases, as shown from Eqs. (1–3) (Sun 2023; Zhang et al. 2021a).

$$2C_xH_yNO_z + (4x - 2z - 4)\,H_2O \rightarrow 2CH_3COO^- + 2H^+ + 2NH_3 + (2x - 4)\,CO_2 + (4x + y - 2z - 11)\,H_2 \quad (1)$$

$$CO_2 + H_2O \leftrightarrow HCO_3^- + H^+ \quad (2)$$

$$NH_3 + H_2O \leftrightarrow NH_4^+ + OH^- \quad (3)$$

The operating temperature exerts an influence on the pH of wastewater within an anaerobic digestion system. Lower temperatures promote increased solubility of gaseous compounds such as H_2S and CO_2, potentially inducing alterations in pH levels. Consequently, the maintenance of pH within the anaerobic digestion system relies on careful management of operating temperature, alkalinity, and organic composition, all of which constitute crucial determinants in pH regulation.

The C/N ratio represents the relationship between the quantities of carbon and nitrogen found in organic matter. This ratio determines the nutritional balance necessary for the growth of microbes. It can be calculated using Eq. (4). The C/N ratio has been recognized to be one of the most important factors affecting the total ammonia nitrogen/free ammonia nitrogen and the methane production (Sun et al. 2023). A high C/N ratio indicates a rapid utilization of nitrogen by microorganisms, which can limit microbial growth. As a result, the microorganisms may face a nitrogen shortage, hindering efficient digestion, slowing down the process, and reducing biogas production. Additionally, excess carbon may lead to the accumulation of VFAs, causing process imbalances and potential acidification. Conversely, a low C/N ratio can trigger ammonia inhibition and raise pH levels above 8.5, which is toxic to methanogens. In severe scenarios, ammonia accumulation can cause process failure. Typically, an optimum C/N ratio of 15–35 is usually recommended for anaerobic digestion, as microorganisms utilize carbon 25 to 30 times faster than nitrogen (Morozova et al. 2020). However, the C/N ratio itself cannot determine the performance of anaerobic digestion process; other factors like carbon source type also exert influence (Choi et al. 2020). A broader C/N ratio spectrum of 2.3–152 has been applied for anaerobic digestion (Xu et al. 2018). Nevertheless, the C/N ratio remains an essential reference index in anaerobic digestion.

$$C/N = \frac{W_1 \times C_1 + W_2 \times C_2 + \ldots + W_n \times C_n}{W_1 \times N_1 + W_2 \times N_2 + \ldots + W_n \times N_n} \tag{4}$$

where W_1, W_2, and W_n are the weights of different carbon sources in the wastewater being treated in kg. C_1, C_2, and C_n are the percentages of organic carbon contents in the carbon sources. N_1, N_2, and N_n are the percentages of nitrogen contents in the carbon sources.

The suitability of a particular substrate or mixture for anaerobic digestion can be assessed through the biochemical methane potential (BMP) test. The BMP is a crucial metric for assessing the ultimate biomethane yield of a substrate, expressed as the ratio of the produced methane chemical oxygen demand (COD) to the substrate COD (g CH_4-COD/g substrate COD). If the targeted product of an anaerobic digestion process is biomethane, the BMP test is commonly employed to determine the theoretical maximum biomethane yield achievable from the substrate to be used. However, the absence of a standardized BMP testing methodology limits the comparability of BMP results across different studies (Filer et al. 2019). Widely employed BMP test protocols can refer to Hansen et al. (2004), Angelidaki et al. (2009), Möller and Müller (2012), the Association of German Engineers standard procedure VD1 4630 (2016), and Holliger et al. (2020).

2.3 Anaerobic Digestion Based on the Loading Rate

Anaerobic bioreactors can be categorized into low-rate and high-rate bioreactors based on factors such as the biodegradation rate or the OLR. The OLR is a measurement of the capacity of a bioreactor. It is defined as the quantity of organic matter that a bioreactor can process per unit volume per day, consequently governing the volumetric rate of biogas production within the bioreactor. The OLR can be expressed as Eq. (5).

$$\begin{aligned} \text{OLR} &= \frac{\text{COD concentration} \times \text{Flow rate of substrate entering the bioreactor}}{\text{Bioreactor volume}} \\ &= \frac{\text{COD} \times Q}{V} \\ &= \frac{\text{COD}}{\text{HRT}} \end{aligned} \tag{5}$$

where OLR is the organic loading rate in kg COD/m^3/d, COD is the chemical oxygen demand of substrate in kg/m^3, V is the bioreactor volume in m^3, Q is the flow rate of substrate that enters the bioreactor in m^3/d, and HRT is the hydraulic retention time in d. The HRT is a measure of the average time that the substrate remains in a bioreactor. The HRT determines the working volume of the bioreactor, thereby directly impacting the economic viability of the anaerobic digestion process. It can be calculated using Eq. (6).

$$\begin{aligned} \text{HRT} &= \frac{\text{Bioreactor volume}}{\text{Flow rate of substrate entering the bioreactor}} \\ &= \frac{V}{Q} \end{aligned} \tag{6}$$

Although there are no rigid definitions for the classifications of low-rate and high-rate bioreactors, low-rate ones typically manage substrates with low OLRs (0.5–1.6 kg VS/m^3/d) (Rittmann and McCarty 2012) and necessitate prolonged HRTs to achieve satisfactory substrate degradation. Low-rate anaerobic reactors are relatively simple technologies that use waste such as manure to produce biogas. The typical characteristics of these types of reactors include long HRT (30–60 d), no heating requirement, and no stirring capacity. There is always a stratification inside a low-rate bioreactor with four zones including a scum layer, a liquid layer, a layer of digesting solids, and a layer of digested solids (Van et al. 2020). Commonly used low-rate bioreactors include PFRs, septic tanks, anaerobic ponds, and covered anaerobic lagoons (Nkuna et al. 2022; Rocha-Meneses et al. 2022). A low loading rate usually corresponds to a long HRT. It fosters increased contact time between microorganisms and substrate, facilitating biomass accumulation within the reactor. Yet, this low loading rate often translates to diminished treatment capacity, low COD removal efficiency, and large reactor volume.

The advancement of high-rate bioreactors represents a significant achievement in enhancing anaerobic wastewater treatment facilities (van Lier et al. 2008). High-rate bioreactors are typically employed in the treatment of high-strength wastewater with elevated OLRs (1.6–4.8 kg VS/m^3/d) (Rittmann and McCarty 2012) and short HRTs due to their ability to sustain high biomass concentrations. This category of reactors usually requires smaller reactor volumes when compared to low-rate ones. Examples of high-rate bioreactors include CSTRs, UASB reactors, FBRs, EGSB reactors, AnMBRs, and innovative hybrid reactors (Abdel-Fattah et al. 2022).

The determination of an optimum OLR is influenced by factors such as HRT, solids retention time (SRT), temperature, and characteristics of the wastewater being treated. In anaerobic digestion, a higher OLR is usually preferred because it reduces the reactor size, lowers the energy requirement, and finally reduces the investment cost (Kumar et al. 2024). A high loading rate associated with a reduced HRT could potentially mediate the development of a hydrogen-forming microbial consortium, known for generating VFAs and H_2 (Lago et al. 2023). The greater VFA availability and more efficient hydrogenotrophic (i.e., H_2 and CO_2 are utilized) pathway for methanogens lead to enhanced process efficiency. However, an excessively high OLR can lead to decreased bioreactor performance and stability due to proliferation of hydrolytic and acidogenic bacteria, intermediatory VFA accumulation, pH reduction, inhibition of methanogens, and ultimately, disruption of microbial community structure (Musa et al. 2018). Temperature regulates the highest achievable OLR for stable operation. A thermophilic condition typically allows for a higher endurable OLR due to accelerated hydrolysis rates and a reduced HRT, particularly in the treatment of high-solid content wastewater (Zhang et al. 2021a). Additionally, increasing the substrate concentration elevates the OLR, thereby subsequently followed by the increasing biogas production. But an excessively high substrate concentration may lead to inhibitory effects and eventually compromise the effluent quality. Under mesophilic conditions, an AnMBR fed with sewage and food waste can achieve an optimal OLR of 16 kg COD/m^3/d (Cheng et al. 2024). Similarly, under thermophilic conditions, a UASB reactor fed with blackwater can achieve an optimal OLR of 12.9 kg COD/m^3/d, while under mesophilic conditions, it can achieve 16 kg COD/m^3/d (Zhang et al. 2021a, b). Additionally, a UASB reactor fed with food waste and blackwater can achieve an optimal OLR of 10 kg COD/m^3/d under mesophilic conditions (Gao et al. 2020).

The SRT is a measure of the average time that the solid fraction can retain within a bioreactor. It is an important parameter in delineating the timeframe available for substrate degradation and microbial proliferation (Vanwonterghem et al. 2015). Within an anaerobic bioreactor, the SRT can be estimated by Eq. (7) (van Lier et al. 2008).

$$\mathrm{SRT} = \frac{\mathrm{Inventory}}{\mathrm{Wastage}} = \frac{V \times X_r}{Q_e \times X_e + Q_w \times X_w} \tag{7}$$

where SRT is the solids retention time in d, X_r is the concentration of viable biomass in the bioreactor in kg/m^3, Q_e is the effluent flow rate in m^3/d, X_e is the concentration of viable biomass in the effluent in kg/m^3, Q_w is the waste sludge flow rate in m^3/d, and X_w is the concentration of viable biomass in the waste sludge that is discharged from the bioreactor in kg/m^3.

In reactors without biomass recirculation systems like CSTRs, the SRT is equal to the HRT. However, in reactors such as AnMBRs, UASB reactors, fixed-bed reactors, and ASBRs, the SRT is decoupled from the HRT (Lago et al. 2023). For instance, an AnMBR can operate with a long SRT of 150 d, while its HRT can vary from 2 h to 20 d (Skouteris et al. 2012). In this scenario, the SRT is independent of the OLR. However, the SRT plays an important role in the hydrolysis of carbohydrates, proteins, and lipids. The most substantial portion of organic matter hydrolysis occurs primarily within the initial 15 d at 25 °C and 10 d at 35 °C (Mahmoud et al. 2004). Process imbalance was found with a 9-d SRT, characterized by issues such as foaming, VFA accumulation, and increased alkalinity during the treatment of dewatered-sewage sludge using CSTRs under both thermophilic and mesophilic conditions (Nges and Liu 2010). Shortening the SRT is favorable for the VFA production while unfavorable for the methanogenesis, making SRT regulation an effective strategy for fermentation-focused processes. However, the specific type of VFA produced is heavily influenced by temperature (Vanwonterghem et al. 2015).

2.4 Anaerobic Digestion Systems Based on Temperature

The effectiveness of bioreactors largely relies on the metabolic functions of microorganisms cultivated within the bioreactors. Temperature emerges as a particularly crucial factor among the various influences on microbial growth and enzyme activities (Kim et al. 2003). Temperature regulates microbiological processes by imposing effects on the nature of metabolism, the nutritional requirements, the biomass composition, and the reaction rates (Lin et al. 2016). Microorganisms are active across four temperature ranges including psychrophilic, mesophilic, thermophilic, and hyperthermophilic, and nevertheless, anaerobic bioreactors are usually operated at psychrophilic (below 20 °C), mesophilic (typically 35–40 °C), and thermophilic (typically 55–60 °C) conditions.

Psychrophilic anaerobic digestion offers distinct advantages over mesophilic and thermophilic anaerobic digestion processes, primarily due to its cost efficiency stemming from low-temperature operation. This approach opens avenues for extending anaerobic digestion technology to cold climate regions and provides an integrated solution for agricultural waste management, particularly beneficial for farmers utilizing ambient temperature digesters (Akindolire et al. 2022). In regions like Canada, Europe, Russia, and northern US, where the ambient temperature remains consistently below 20 °C year-round, the climate conditions favor psychrophilic anaerobic digestion processes, reducing energy input requirements (Tiwari et al. 2021). Psychrophilic anaerobic digestion presents a stable and robust method for treating ammonia-rich waste such as animal manure, ensuring efficient digestion without ammonia toxicity concerns (Akindolire et al. 2022). However, the low temperature can impede vital cellular processes, potentially rendering cell biology inactive and leading to reduced biogas yields. Additionally, cold climates may alter the physical and chemical properties of wastewater, decreasing substrate availability to microbes (Tiwari et al. 2021). The slower rate of psychrophilic conditions imposes limitations on the minimum retention time of microorganisms to prevent biomass washout in continuous systems. Typically, the start-up of psychrophilic anaerobic digestion requires a longer period to allow for the retention of sufficient slow-growing microorganisms (Saady and Massé 2015). To address these challenges, reducing the OLR proves beneficial for methane production under psychrophilic conditions. Furthermore, modifications in reactor configurations and the use of hybrid reactors have shown promise in enhancing methane generation (Tiwari et al. 2021).

Both thermophilic and mesophilic conditions are frequently preferred for anaerobic digestion operations. Mesophilic conditions foster the growth of robust microbial communities capable of adapting to environmental fluctuations, thereby ensuring stable performance. Mesophilic anaerobic digestion exhibits a reduced risk of ammonium and VFA inhibition, along with high COD and TS removal rates. Additionally, mesophilic reactors are simpler to maintain, entail lower investment costs, and lead to operational cost savings compared to thermophilic anaerobic digestion (Ahmad Mir et al. 2019). However, nutrient imbalances may arise (Rodríguez-Jiménez et al. 2022). Considering the elevated energy demands associated with a thermophilic condition, a mesophilic condition is predominantly favored for effective biogas production.

Thermophilic conditions offer numerous advantages including accelerated hydrolysis, higher OLR, increased methane production, effective pathogen elimination, and reduced viscosity of mixed liquor (Wu et al. 2020). Digestion rates are notably enhanced under thermophilic conditions, up to 2.25 times faster compared to mesophilic conditions (Kasinski 2020). Thermophilic anaerobic digestion is particularly well-suited for treating high-solid content wastewater such as concentrated blackwater, animal manure, and municipal solid waste. However, stability concerns can arise in thermophilic reactors due to the sensitivity of methanogens to toxic substances like ammonia and VFAs, as well as environmental fluctuations (Arsova 2010). To enhance process stability in thermophilic anaerobic digestion, various strategies can be employed, including the development of resilient thermophilic microbiomes, co-digestion of compatible feedstocks, utilization of microbial electrochemical technologies, and incorporation of conductive materials (Ryue et al. 2020).

The composition of methanogenic archaeal communities varies according to temperature conditions. A significant shift occurs from mesophilic to thermophilic conditions, characterized by a decrease in aceticlastic *Methanosarcina* and *Methanothrix* species and an increase in thermophilic *Methanocella* species, which are hydrogenotrophic (Conrad 2023). Under thermophilic conditions, hydrogenotrophic methanogens predominate, leading to a reduction in the diversity of the archaeal community (Dong et al. 2018). Conversely, under low-temperature conditions, the diversity of methanogenic archaea appears to increase, with enhanced homoacetogenic activities observed. Aceticlastic genera are particularly abundant in these conditions (Conrad 2023). Moreover, the hydrogenotrophic pathway has been shown to become the predominant methanogenic pathway following prolonged operation of psychrophilic anaerobic digestion (McKeown et al. 2009).

3 Operating Principles and Strategies to Improve System Performance

3.1 Performance Comparison Between Different Bioreactors

The primary objective of operating anaerobic bioreactors is the complete removal of the feed COD. The COD removal efficiency is one important measure for assessing the efficacy of an anaerobic digestion process. Despite the potential for a significantly reduced COD concentration in the bioreactor effluent compared to the influent, the "removed" COD is not always converted into the desired product, such as biomethane, but instead accumulates in the reactor sludge due to the insufficient hydrolysis of particulate organic matter (Sun et al. 2022). Theoretically, the feed COD can be allocated to three main fates: (1) conversion to methane or other targeted products; (2) discharge in effluent; and (3) accumulation in sludge (including waste sludge and reactor sludge), collectively encompassing the entire feed COD. However, discrepancies often arise between the feed COD and the detected COD in the three fates due to occasional sludge washout, uncertainties in metabolic processes, and analytical errors. Consequently, an unknown conversion factor is introduced to reconcile the imbalance between the feed COD and its detectable conversions. The COD mass balance test is typically conducted to evaluate the fates of feed COD. In the case where methane is the desired product, the COD mass balance equation is expressed as shown in Eq. (8).

$$\begin{aligned}\mathrm{COD_{inf}}Q\Delta t = & \sum_{i=1}^{i=\Delta t} \mathrm{COD_{methane}} V_i + \mathrm{COD_{eff}} Q\Delta t \\ & + \mathrm{COD_{dis}} V_{\mathrm{dis}} + \left(\mathrm{COD_{slu,\Delta t}} V_{\Delta t} - \mathrm{COD_{slu,0}} V_0\right) \\ & + \mathrm{COD_{unkown}} \end{aligned} \quad (8)$$

where Δt is the duration of COD mass balance test in d. $\mathrm{COD_{inf}}$, $\mathrm{COD_{eff}}$, and $\mathrm{COD_{dis}}$ are the CODs of influent, effluent, and discharged sludge in g/L. $\mathrm{COD_{slu,0}}$ is the COD of sludge in the reactor at $t = 0$, and $\mathrm{COD}_{slu,\Delta t}$ is the COD of sludge in the reactor at $t = \Delta t$. $\mathrm{COD_{methane}}$ is the COD of produced methane in g/L. Q is the volumetric flow rate in L/d. V_i is the methane volume in L at $t = i$. V_{dis} is the volume of discharged waste sludge in L. $V_{\Delta t}$ and V_0 are the volumes of sludge in the reactor in L at $t = \Delta t$ and $t = 0$, respectively. $\mathrm{COD_{unkown}}$ is the introduced compensation to meet the COD mass balance.

Table 2 encapsulates representative studies on anaerobic bioreactors conducted over the past five years, offering exemplary insights for bioreactor design. The extent of COD removal varies depending on factors such as feed composition, reactor configuration, and operational parameters like temperature and OLR. Temperature exerts a direct influence on the physicochemical properties of all constituents within the bioreactor, as well as the thermodynamics and kinetics of all reactions (Sathiyah et al. 2023). The mesophilic condition is commonly favored for achieving optimal COD removal and ensuring stable operation. The OLR of an anaerobic bioreactor has a direct impact on the feed COD removal. To mitigate the risk of anaerobic digestion process failure, bioreactors are typically initiated with a relatively low OLR. Following the establishment of stable operation, the OLR is incrementally increased until an optimum level is reached, resulting in the highest achievable COD removal efficiency. However, the continuous elevation of OLR may lead to the decreased COD removal due to the VFA accumulation.

Biogas produced from anaerobic bioreactors is another important objective. The theoretical maximum CH_4 production at a standard temperature and pressure condition (1 atm, 0 °C) is 350 mL/g COD, however, the quantity and quality of biogas are primarily influenced by the composition of wastewater, pH level, and operating condition including temperature, OLR, HRT, C/N ratio, and mixing (Singh et al. 2020). Basically, the bioreactors operated at a mesophilic or thermophilic condition, a higher OLR, and a C/N ratio of 15–35 have higher biogas yield. In addition, the use of conductive materials and nanoparticles in reactors also contributes to biogas production (Dompara et al. 2023).

Managing and disposing of waste sludge generated from biological processes poses a significant challenge for wastewater treatment plants, constituting a substantial portion of their operating cost, ranging from 25 to 65% (Pérez-Elvira et al. 2006). Mitigating sludge production is a key strategy to alleviate the burden of sludge post-treatment. Compared to aerobic systems, anaerobic processes typically yield lower amounts of sludge. For example, a sewage treatment plant equipped with anaerobic digesters demonstrated a 31% reduction in sewage sludge production (Kim and Phae 2022). In terms of reactor operation, the SRT stands out as a critical operational parameter linked to sludge discharge and effluent quality. Extending the SRT duration aids in reducing sludge production; however, an excessively prolonged SRT may result in the release of recalcitrant compounds, hindering acidification and potentially causing biomass washout, thereby diminishing effluent quality (Chen et al. 2018).

3.2 Mixing

In anaerobic digesters such as CSTRs and moving bed biofilm reactors (MBBRs), mixing is commonly implemented to ensure homogeneity in the fluid mixture and provide an even environment for anaerobes (Singh et al. 2020; Swain et al. 2022). Adequate mixing can provide efficient utilization of digester volume, prevent stratification, foam, scum, and crust formation, avoid pH and temperature gradients, maintain close contact between microorganisms, enzymes, and substrate, and facilitate heat and mass transfer (Singh et al. 2021). However, optimizing mixing conditions is

Table 2 Representative studies on anaerobic bioreactors in the last five years

Bioreactor	Wastewater	Temperature	pH	HRT	Optimal OLR	SRT	Removal	Production	References
Anaerobic membrane bioreactor (AnMBR) (15 L)	Sewage sludge and food waste	37 and 55 °C	7.12–7.68	15 d	16 kg COD/ m^3/d		95.7–99.5% COD	295–303 L CH_4/kg COD added	Cheng et al. (2024)
AnMBR (5 L)	Slaughterhouse wastewater	Room temperature (15–30 °C)	7.4	24 h	1.1 kg COD/ m^3/d		90% COD	290 NL CH_4/kg COD removed	Viet et al. (2023)
Moving bed AnMBR with polyethylene glycol granules (8 L)	Synthetic low-strength wastewater	20 °C	7.0–7.2	8 h	1.4 kg COD/ m^3/d	–	> 90% COD	280 L CH_4/kg COD removed	Tran et al. (2023)
Anaerobic bioreactor with fixed-structured bed (5.5 L)	Landfill leachate and glycerin	30 °C	7.4	35.2 h	7.1 kg COD/ m^3/d	–	89.9% COD	300 L CH_4/kg COD removed	de Castro et al. (2021)
AnMBR (6.5 L)	Pulp and paper mill primary sludge	50 °C	7.2	8 d	2.5 kg COD/ m^3/d	32.5 d		189 L biogas/kg MLSS added	Bokhary et al. (2021)
Covered in-ground anaerobic reactor (5 m^3)	Palm oil mill effluent	30 °C	4.65–4.98	44 d	1.4 kg COD/ m^3/d	–	93.0% COD	120 L CH_4/kg COD removed	Haryanto et al. (2021)
AnMBR (3.31 m^3)	Beet molasses	–	5.5	–	10 kg COD/ m^3/d	–	92% COD	370 L CH_4/kg COD removed	Turker and Dereli (2021)

(continued)

Table 2 (continued)

Bioreactor	Wastewater	Temperature	pH	HRT	Optimal OLR	SRT	Removal	Production	References
AnMBR (40 L)	Municipal wastewater	14–26 °C	7.2	0.5 d	0.9 kg COD/ m^3/d	50 d	78% COD	256 L biogas/ kg COD removed	Plevri et al. (2021)
UASB reactor (2 L)	Blackwater	52 °C	6.6	2.5 d	12.9 kg COD/ m^3/d	–	81.9% COD	2.4 N L CH_4/L/ d	Zhang et al. (2021a)
UASB reactor (1 L)	Blackwater	35 °C		0.25 d	16 kg COD/ m^3/d	–	75.6% COD	8.4 g CH_4-COD/L/d	Zhang et al. (2021b)
UASB reactor (3.5 L)	Food waste and blackwater	35 °C	8.0–8.5	2.6 d	10 kg COD/ m^3/d	–	85% COD	2.42 L CH_4/L/d	Gao et al. (2020)
Anaerobic dynamic membrane bioreactor (3.6 L)	Domestic wastewater	20–25 °C	–	1–8 h	0.82–6.8 kg COD/ m^3/d	–	60.4–77.3% COD	50–120 L CH_4/ kg COD removed	Yang et al. (2020)
Fixed-bed anaerobic reactor seeded with granular sludge (10 L)	Artificial wastewater consisting of molasses, cat food, and tap water	4 °C	6.8–7.2	6 d	4.33 kg COD/m^3/d	–	59.8% COD	206 L biogas/ kg COD removed	Zhao et al. (2020)
Semi-continuous CSTR (2 L)	Municipal solid waste	35 and 55 °C	6.49–7.73	22 d	9 kg VS/m^3/d	–	–	222 L biogas/ VS added (35 °C) and 239 L biogas/VS added (55 °C)	Jiang et al. (2020)

challenging. Generally, continuous mixing is a waste of energy and does not improve anaerobic digestion performance. Alternatively, intermittent mixing can be optimized to improve anaerobic digestion efficiency (Kariyama et al. 2018). However, the capital, operating, and maintenance cost associated with the mixing equipment is substantial.

There are typically three mixing methods, namely gas recirculation, liquid recirculation, and mechanical (impeller) mixing (Kariyama et al. 2018). Particularly, effluent recirculation, commonly employed in UASB reactors, is used to improve both mixing and upflow velocity within the reactor. Mixing can create favorable hydrodynamic conditions, facilitating heat and mass transfer in the reactor. Additionally, mixing induced by effluent recirculation aids in diluting feedwater, thereby mitigating toxicity and reducing VFA accumulation. Furthermore, effluent recirculation may extend the SRT of a UASB reactor by retaining biomass within the reactor (López-López et al. 2015; Lukitawesa et al. 2018; Zamanzadeh et al. 2016). However, it is challenging to determine definitively whether effluent recirculation should be applied to a UASB reactor operated under specific scenarios. The benefits to bioreactor performance from effluent recirculation are contingent upon bioreactor design (e.g., single-stage or two-stage), substrate characteristics (e.g., solid content, pollutant concentration, buffer capacity, and viscosity), operating conditions (e.g., temperature, OLR, and continuous or intermittent operation mode), and the ratio between recirculation volume and influent volume (Sun et al. 2022). For substrates with high solid content, effluent recirculation might reduce methane production due to inefficient hydrolysis of particulate organic matter (Sun et al. 2022; Zhang et al. 2020).

3.3 *Biomass Immobilization*

Immobilization is a novel technology that confines biomass materials (e.g., microorganisms and enzymes) within a defined spatial extent, restraining their free movement while preserving their activity (Salter and Kell 1991). Immobilizing biomass can enhance environmental adaptability, increase the potential for recovery, and diminish the likelihood of secondary pollution compared to non-immobilized biomass materials (Liu et al. 2022). In comparison to suspended biomass systems, immobilized biomass systems demonstrate greater efficiency in removing heavy metals, pesticides, polycyclic aromatic hydrocarbons, and trace organics (Girijan and Kumar 2019). Immobilization can be accomplished through three distinct methods: physical method, chemical method, and physicochemical method.

In the physical approach, biomass materials are immobilized via adsorption or entrapment. Adsorption involves the binding of biomass materials onto carrier materials' surfaces or within their matrices, utilizing mechanisms such as hydrogen bonding, hydrophobic interactions, electron affinity, and van der Waals forces (Ren et al. 2020). Meanwhile, entrapment entails the confinement of biomass materials within carriers, such as natural gels or synthetic polymers, forming small particles to prevent leakage and minimize external influences (Liu et al. 2022). Although physical methods maintain the natural structures of biomass materials, they may encounter challenges such as weak affinity in adsorption and limited mass transfer in entrapment.

Chemical immobilization, on the other hand, involves covalent binding or cross-linking. Covalent binding establishes bonds between chemical groups on carrier surfaces and nucleophilic groups on biomass materials (Datta et al. 2020). While covalent binding enhances the affinity between carriers and biomass materials, chemical agents utilized in the process may impair cell integrity and reduce microbial metabolic activity (Liu et al. 2022). Cross-linking involves the binding of microbial cells or enzymes with one another using cross-linking agents like glutaraldehyde, diacetamide, hexanediamine, maleic anhydride, and isocyanate derivatives (Liu et al. 2022).

The physicochemical method, known as the biofilm method, centers on forming biofilms on carriers through surface adsorption and internal fixation. The core mechanism of the biofilm method is selecting appropriate carriers conducive to biofilm growth, ensuring high affinity for biomass materials and efficient contaminant removal. While traditional carriers like activated carbon, clay, chitosan, agar, silica, and polymeric materials have been utilized, novel carriers with large surface areas and high loading capacities, such as magnetic materials, mesoporous materials, and metal–organic frameworks, are garnering increasing attention (Liu et al. 2022).

3.4 Pre-treatments

Anaerobic digestion has found extensive use in treating various waste streams such as agricultural residues, food wastes, waste activated sludge, and high-strength wastewater like blackwater. However, a notable challenge lies in the rate-limiting step, i.e., hydrolysis. To address this challenge, pre-treatment methods have been employed, including chemical addition, mechanical approaches, and thermal treatment. Chemical addition involves the use of various agents such as acidic compounds (e.g., sulfuric acid and hydrochloric acid), alkaline substances (e.g., sodium hydroxide and calcium hydroxide), and oxidizing agents (e.g., hydrogen peroxide and calcium hypochlorite).

Pre-treatment of waste activated sludge with calcium hypochlorite has been demonstrated to enhance its solubilization, biodegradability, and ability to fix phosphorus (Yu et al. 2021a). Ultrasonication has been utilized for pre-treating lignocellulosic waste and food waste. The vibration and subsequent bursting of microbubbles generated during ultrasonication impart energy that alters the structure of lignocellulose, thereby improving the accessibility of organic materials to microorganisms (Zhang et al. 2019). Thermal treatment has been extensively employed for pre-treating waste activated sludge. Heat disrupts chemical bonds in cell membranes, leading to the release of intracellular components from biomass (Zhang et al. 2019).

However, most pre-treatment methods incur high operating costs and require significant energy inputs, thus there is still a demand for pre-treatment technologies that are both energy- and cost-effective.

3.5 *Direct Interspecies Electron Transfer (DIET)*

Incorporating conductive materials into anaerobic digestion processes facilitates direct interspecies electron transfer (DIET), thereby enhancing the conversion of organics into methane through syntrophic interactions. In this mode, electrons are transferred directly from electron donors to acceptors, involving fewer biological enzymatic reactions compared to interspecies hydrogen transfer (IHT) (Chen et al. 2022). Research has indicated that the electron transfer rate via DIET is 8.57 times higher than that of IHT (Storck et al. 2015), leading to a shorter lag phase, reduced start-up time, increased biomethane production, and enhanced system stability (Chen et al. 2022).

The DIET mechanism for methane generation is depicted in the following reaction (Eq. 9). The effectiveness of DIET hinges on electrical connections between cells through either biological conduit (e.g., conductive pili and outer surface cytochromes) or abiotic conductive pathways (e.g., granular activated carbon and biochar) (Zhang and Liu 2023). However, the precise biological pathways involved in DIET remain unclear. Moreover, the operational complexity of reactors limits the practical utilization of conductive materials in engineered anaerobic digestion systems.

$$CO_2 + 8e^- + 8H^+ \rightarrow CH_4 + 2H_2O \tag{9}$$

3.6 *Others*

Bioaugmentation holds promise for enhancing the stability and efficiency of anaerobic digestion processes, but its economic feasibility warrants careful consideration. Assessing cost-effectiveness involves evaluating factors such as the balance between energy input and output (improved methane yield), digester volume, expenses related to microbial cultures, and the reduction in waste volume and environmental contamination (Roopnarain et al. 2021).

Introduction of small amounts of oxygen, known as micro-aeration, into anaerobic digestion systems has been demonstrated to enhance the hydrolysis rates of particulate organic matter (Fu et al. 2016). Additionally, oxygen addition can mitigate VFA accumulation and reduce H_2S levels (Roopnarain et al. 2021). This strategy has been applied in the treatment of various recalcitrant feedstocks such as corn straw, primary sludge, blackwater, and food waste. Low doses of oxygen can stimulate the activity of hydrolytic and acidogenic bacteria. However, excessive oxygen may promote the growth of aerobic microorganisms and inhibit methanogenic activity (Guo et al. 2021; Yu et al. 2020).

Combining microbial electrolysis cells (MEC) with anaerobic digestion has shown promise in boosting biomethane recovery from diverse high-strength organics. This integrated approach demonstrates improved process stability compared to anaerobic digestion alone by reducing the accumulation of short-chain VFAs at high OLRs. Additionally, it helps alleviate inhibitory effects caused by toxic and refractory compounds like phenolic compounds and salts (Huang et al. 2022, 2023). Further exploration is needed to understand the underlying mechanisms governing microbial interactions and scalability of this combined MEC and anaerobic digestion system (Huang et al. 2022).

4 Techno-Economic and Environmental Analysis

4.1 Cost-Benefit Analysis of Anaerobic Bioreactors

The economic viability of operating an anaerobic digestion system hinges on several factors including scale, location, feedstock type, targeted products, and local regulations. Table 3 presents essential considerations for conducting a cost–benefit analysis of a biogas-generating anaerobic digestion system.

The capital cost primarily encompasses the fabrication, transportation, and installation of equipment (e.g., digesters, feed and gas storage tanks, pumps, and blenders) utilized in the anaerobic digestion system. Expenses related to system installation and testing personnel should also be factored into the capital cost. The operating cost is associated with the routine functioning of the anaerobic digestion system, including

Table 3 Possible expenditure and income of an anaerobic digestion system

Expenditure		Income	
Capital cost	Equipment fabrication	Biogas utilization	Generated electricity
	Equipment installation		Generated heat
	Transportation	Digestate utilization	Biofertilizer
	Manpower	Gate fee	Disposal
Operating cost	Energy consumption (heat, electricity)		
	Water consumption		
	Chemical use		
	Salaries of operators		
Maintenance cost	Routine maintenance		
	Waste disposal		

energy consumption (heat and electricity), water usage, and any chemical additives. Operator salaries represent another significant operating expense. The maintenance cost arises from the regular maintenance of the anaerobic digestion system and the disposal of waste sludge.

Revenue sources primarily include proceeds from biogas and digestate sales, as well as gate fees for waste disposal. Biogas produced can be utilized in a combined heat and power (CHP) engine to generate electricity and heat, with consideration given to the conversion efficiency of the CHP engine (Chowdhury 2021). Digestate contains digested organic matter and nutrients such as nitrogen (N), phosphorus (P), and potassium (K). It can serve as a biofertilizer to substitute chemical fertilizer or be used as a raw material for producing chemical fertilizers (e.g., slow-release struvite fertilizer) through nutrient recovery processes.

4.2 Environmental Impact Assessment and Sustainability Aspects

The contribution of anaerobic digestion to sustainable development encompasses the environmental, economic, and social dimensions. Table 4 presents the principal environmental impacts of anaerobic digestion. Anaerobic digestion mitigates the potential for water pollution and offers a sustainable approach to organic waste management, requiring minimal land usage (reserved solely for anaerobic digestion plants) compared to traditional incineration or landfill methods. Biogas derived from anaerobic digestion serves as a renewable energy source, enhancing energy security, reducing dependence on fossil fuels, and curbing greenhouse gas emissions resulting from non-renewable energy sources (Sun 2023). Furthermore, the digestate produced serves as a valuable fertilizer resource for agriculture, thereby decreasing reliance on chemical fertilizers. However, anaerobic digestion also poses potential adverse effects such as human toxicity and ecotoxicity, soil acidification, the formation of particulate matter and photochemical oxidants, and ozone depletion.

From an economic standpoint, the operation of anaerobic digestion systems creates both direct and indirect employment opportunities, encompassing roles in routine operations, waste management, transportation, and within the supply chain and support services. The utilization of biogas and biofertilizer also contributes to job creation (Piadeh et al. 2024). Revenue is generated through the sale of biogas and biofertilizer. Moreover, the adoption of biogas improves living standards in low-income areas by offering a reliable and sustainable energy source (Martinez-Sanchez et al. 2017). Similarly, the use of digestate supports sustainable agricultural practices and bolsters food security (Silverman et al. 2020).

From a social perspective, the global expansion of biogas generation and its utilization presents a clean energy alternative. In remote communities, biogas usage mitigates the adverse health effects associated with traditional firewood use and contributes to forest conservation efforts, thereby combating climate change (Piadeh

Table 4 Environmental impact assessment of anaerobic digestion (Piadeh et al. 2024; Slorach et al. 2019)

Environmental perspective	Impact assessment
Water-related impact	Enhancing safety and affordability of water resources by proper waste management
	Enhancing water quality
	Helping protect areas such as forests, rivers, aquifers and lakes through using collected organic waste
Waste management	Using less land than burning or landfilling
Global warming potential	Reducing greenhouse gas emission by utilizing renewable biogas
	Causing possible N_2O emission from digestate application
Depletion of fossil and metal resources	Reducing fossil fuel consumption in heat and energy production by utilizing renewable energy
	Consuming metal by CHP engine
Human toxicity and ecotoxicity	Coming from transportation of chemicals and metals
Eutrophication	Helping reduce water pollution
	Stabilizing and recovering nutrients such as N and P
Acidification	Causing gaseous emission of ammonia from digestate application on the land
Particulate matter	Forming particulate matter due to ammonia emission from the digestate application on the land
Photochemical oxidants	Producing nitrous oxides and volatile organic compounds from the combustion of fuel in transport vehicles and of biogas in CHP engine
Ozone depletion	Generating chlorofluorocarbons during transportation of waste and digestate
Land occupation	Agricultural or urban land occupation by anaerobic digestion plants

et al. 2024). Furthermore, treating organic waste in anaerobic digesters, rather than resorting to burning or landfilling, enhances environmental hygiene and curbs the spread of infectious pathogens (Sohoo et al. 2021). The proliferation of job opportunities and reduced conflicts related to energy contribute to an overall improvement in quality of life, education, and social well-being (Gibellato et al. 2023).

4.3 Integration with Circular Economy Principles

In broad terms, the circular economy operates on two fundamental principles: (1) to minimize waste generation and pollution; and (2) to extend the lifespan of materials through reuse, recovery, and recycling practices (Tawfik et al. 2022). Anaerobic digestion aligns with these principles, facilitating the transition toward next-generation waste and wastewater management systems that are essential for a sustainable economy (Diamantis et al. 2021). Anaerobic digestion stands out as one of the most economically viable technologies for treating biowaste and wastewater. Advances in technology have enabled anaerobic bioreactors, particularly high-rate ones, to effectively treat contaminants and recalcitrant substances, boasting high treatment capacities. With or without pre-treatment, anaerobic digestion enables the recovery of energy from raw feedstock and nutrients from both the raw feedstock and digestate. Moreover, digestate serves as an excellent fertilizer or organic amendment for agricultural purposes.

Traditionally, biogas has been the primary bioenergy product recovered from anaerobic digestion of sewage sludge in wastewater treatment plants. However, other valuable products such as nutrients, polyhydroxyalkanoates, cellulose fiber, VFAs, H_2, heat, and metals can also be extracted from anaerobic digestion processes involving various types of feedstocks or co-digestion of different materials (Owusu-Agyeman et al. 2023). For example, struvite, a slow-release fertilizer, can be produced through chemical precipitation from raw blackwater (Sun 2023) as well as from anaerobic blackwater digestate (Yee et al. 2019). Similarly, valuable products like calcium phosphate can be obtained in the form of anaerobic granules alongside biogas yields (Zhang et al. 2021a, b). Nonetheless, large-scale production of these value-added products necessitates the implementation of innovative technologies.

5 Future Perspectives

5.1 Addressing Challenges in Implementation and Scalability

Anaerobic digestion has a rich history in waste management, offering a sustainable remedy to conventional waste management issues like odor control, greenhouse gas mitigation, and economic returns (Adami et al. 2020; Uddin and Wright 2021). Biogas and its derivative chemical products are increasingly competitive with other renewable energy sources (Uddin and Wright 2021), with public incentives further driving the adoption of sustainable waste management practices, including anaerobic digestion systems.

However, one challenge lies in digesting potential feedstocks with diverse physical and chemical properties. Factors such as low biodegradability, inhibitory effects, and toxicity to digesters pose constraints on the adaptation of potential feedstocks to anaerobic digestion systems. To overcome these hurdles, advanced technologies have been developed, and the standardization of plug-in models for various feedstocks could enhance the implementation of anaerobic digestion systems (Bhajani and Pal 2022). Despite these advancements, economic viability remains a significant bottleneck for the widespread implementation and scaling up of anaerobic digestion systems (Uddin and Wright 2021). Government support, especially in developing countries, may prove pivotal in determining the success of anaerobic digestion system implementation (Issahaku et al. 2024).

5.2 Emerging Trends and Innovations in Anaerobic Digestion

An emerging avenue for future research aligning with global trends is the concept of 'waste-to-energy', which primarily involves the production of biogas through anaerobic digestion of sewage sludge, the exploration of bioenergy alternatives beyond biomethane from anaerobic waste and wastewater treatment, and the direct or indirect utilization of digestate for further nutrient recovery (Gao 2020; Sun 2023; Zhang et al. 2024a). While biogas remains as the primary targeted product derived from anaerobic digestion processes, the complexity of organic sources poses challenges. Commercializing biogas production encounters difficulties in maintaining optimal operating conditions and upgrading biogas to elevate biomethane content (Bhajani and Pal 2022). Enhancing biogas production and quality in bioreactors through methods like DIET with additives such as activated carbon, biochar, and phenazine has garnered increasing attention (Yu et al. 2021b; Zhang et al. 2022a, b).

As technology advances and our understanding of the anaerobic digestion process deepens, new strategies are emerging to address inhibition factors in anaerobic digestion, thereby enhancing its efficiency. One such trend involves boosting the microbiome and digestibility of feedstocks through abiotic stress factors to promote biomethanation (Thakur et al. 2023). These stress factors encompass a range of elements, including but not limited to salinity, micro-aeration, micronutrients (e.g., Cu, Co, Ni, Fe, Mg, and Ca), vitamins, antibiotics (e.g., macrolides, quinolones, sulfonamides, and tetracyclines), and organic substances (i.e., carbon polymers) (Thakur et al. 2023). While these stress factors may enhance the anaerobic digestion process through various mechanisms at appropriate concentrations, excessive doses could potentially suppress specific stages or pathways. Further research is required to gain a better understanding of the effects of combining multiple stressors on specific feedstocks.

Various emerging pollutants, including chemical compounds found in personal care products, pharmaceuticals, hormones, pesticides, and household and industrial chemicals, have been detected in treated wastewater at trace levels. The presence of these pollutants in water bodies poses risks to both aquatic ecosystems and human health (Tawfik et al. 2022). Moreover, these pollutants can accumulate in biosolids during the primary and secondary treatment processes of wastewater and sewage sludge, thereby hindering the anaerobic digestion process if anaerobic digestion is used for wastewater treatment or post-treatment of sewage sludge. Therefore, it is essential to remove these pollutants from wastewater and sewage sludge. To address this challenge, various approaches have been explored, including the use of hybrid anaerobic bioreactors (e.g., anaerobic fluidized membrane bioreactor), bioaugmentation methods, and pre-treatment technologies such as electro-oxidation, ultrasonication, thermal hydrolysis, and ozonation, to mitigate the inhibition caused by these emerging pollutants on anaerobic digestion. However, there is still limited understanding of the persistent transformation products of these pollutants, highlighting the need for further research in this area (Chyoshi et al. 2022; Tawfik et al. 2022).

5.3 Future Prospects and Research Directions

The potential of anaerobic digestion is not fully realized due to limited understanding of microbial dynamics and the need for optimized digester designs. Addressing these challenges through advanced bioreactor technology is crucial for enhancing stability and efficiency, managing variable waste inputs, and integrating with other waste treatment technologies. Ongoing research and development aimed at innovating bioreactor designs are essential to expand the efficiency and applicability of anaerobic digestion systems globally, enabling them to handle a broader range of substrates and produce higher yields of biogas and valuable products.

Novel bioreactor designs can be utilized to enhance the conversion of diverse waste types into valuable biofuels, including hydrogen, biogas, and biohythane. These systems currently implement a two-stage reactor process, where hydrogen and biogas are produced in each stage to boost both productivity and energy recovery. Further refinements in bioreactor technology involve optimizing biochemical processes and design improvements. These advancements optimize critical parameters such as pH, temperature, and retention time, which enhance the quality of biohythane and the stability of the process. Such optimizations not only shorten fermentation times but also allow flexible adjustments of hydrogen to methane ratios, making the production process more efficient and economically viable. Besides, further design improvements can focus on simplifying the process, exploring highly active microbial community, and constructing environment conditions to facilitate biohythane production. These improvements reduce the capital cost and decrease the maintenance and operation difficulty.

The potential of novel bioreactor designs to boost the production of value-added products, such as medium-chain fatty acids and alcohols (ethanol) from biomass waste, is substantial. These advanced systems use specific microbial consortia or microbial community with mix culture to increase both the yield and efficiency of bioproducts. Utilizing anaerobic digestion digestate as a substrate in biorefining processes not only reduces costs but also enhances saccharification and fermentation efficiency. Process integration, for instance gasification process and anaerobic digestion, can resolve the slow hydrolysis and stability issue for the various organic solid waste. Metagenomic and metatranscriptomic analyses can be utilized to provide deeper insights into microbial interactions and pathways, further enhance production capabilities. Ongoing research and optimization for anaerobic digestion are crucial for achieving industrial scalability and economic viability, indicating a promising future for biotechnology in resource recovery and green chemistry. Therefore, the application of innovative bioreactor designs in anaerobic digestion processes substantially increases biomass conversion versatility and efficiency, supporting the sustainable production of diverse bio-based products and contributing to environmental sustainability goals.

The production of biogas through anaerobic digestion involves the generation of methane and carbon dioxide, along with impurities like H_2S, ammonia, and water vapor. These contaminants, especially CO_2, reduce the biogas's calorific value, complicating its use as a transportation fuel or for grid injection. Current biogas purification methods—such as water scrubbing, pressure swing adsorption, and membrane separation—are criticized for being energy-intensive, costly, and chemically demanding, posing scalability issues for smaller installations. Future research in biogas purification should focus on innovative bioreactor technologies to overcome these challenges. Potential directions include integrating anaerobic digestion with bioelectrochemical systems to enhance CO_2 capture and conversion efficiency, thus converting impurities directly into valuable products like methane or hydrogen. Exploring biological purification methods using specialized microbial consortia could offer sustainable alternatives to chemical methods. Advances in bioreactor design are also crucial, involving better control systems and tailored microbial environments to optimize impurity transformation. Additionally, coupling biogas purification with renewable energy sources and promoting pilot projects could further reduce the energy demands of these processes and help scale innovative solutions. For instance, using solar energy to drive photobioreactors for CO_2 sequestration and conversion into biofuels. This integrated approach could significantly advance the sustainability and efficiency of biogas purification, aligning with global energy sustainability goals.

6 Take-Home Messages

(1) Anaerobic digestion contributes to the circular economy by minimizing waste discharge and pollution generation through efficient contaminant removal and enabling bioenergy and nutrient recovery.
(2) The successful design and operation of an anaerobic digestion system relies on the factors including feedstock characteristics, OLR, HRT, SRT, and operating temperature.
(3) Hydrolysis is often the primary limiting step in anaerobic digestion of high-solid content feedstocks like waste activated sludge, crop residuals, blackwater, food waste, and animal manure. Various strategies have been devised to address this limitation and enhance anaerobic digestion efficiency, including increasing

mixing intensity, immobilizing biomass, employing pre-treatment methods such as chemical addition, mechanical techniques, and thermal treatment, bioaugmentation, micro-aeration, and promoting DIET through the addition of conductive materials like granular activated carbon and biochar.

(4) Future research directions for anaerobic bioreactors include enhancing the digestibility of feedstocks through abiotic stress, removing/emulating emerging pollutants, exploring novel anaerobic digester configurations, developing technologies for "waste-to-energy", and utilizing AI for modeling and optimizing anaerobic digestion processes.

References

Abbasi T, Tauseef SM, Abbasi SA (2012) Low-rate and high-rate anaerobic reactors/digesters/fermenters. In: Abbasi T, Tauseef SM, Abbasi SA (eds) Biogas energy. Springer, New York, NY, pp 35–39

Abdel-Fattah HK, El-Seddik MM, Galal MM, Mostafa NG (2022) A comparative study for hybrid UASB reactor performance using polyethylene media and luffa sponge as biofilm support. J Environ Health Sci Eng 20(2):749–756

Abdelgadir A, Chen X, Liu J, Xie X, Zhang J, Zhang K, Wang H, Liu N (2014) Characteristics, process parameters, and inner components of anaerobic bioreactors. BioMed Res Int, 841573

Adami L, Schiavon M, Torretta V, Costa L, Rada EC (2020) Evaluation of conventional and alternative anaerobic digestion technologies for applications to small and rural communities. Waste Manage 118:79–89

Ahmad Mir M, Hussain A, Verma C (2019) Design considerations and operational performance of anaerobic digester: a review. Cogent Eng 3(1):1181696

Angelidaki I, Alves M, Bolzonella D, Borzacconi L, Campos JL, Guwy AJ, Kalyuzhnyi S, Jenicek P, van Lier JB (2009) Defining the biomethane potential (BMP) of solid organic wastes and energy crops: a proposed protocol for batch assays. Water Sci Tech 59(5):927–934

Akindolire MA, Rama H, Roopnarain A (2022) Psychrophilic anaerobic digestion: a critical evaluation of microorganisms and enzymes to drive the process. Renew Sustain Energy Rev 161:112394

Arsova L (2010) Anaerobic digestion of food waste: current status, problems and an alternative production. Columbia University, Berlin

Awe OW, Zhao Y, Nzihou A, Minh DP, Lyczko N (2017) A review of biogas utilisation, purification and upgrading technologies. Waste Biomass Valoriz 8:267–283

Aworanti OA, Agbede OO, Agarry SE, Ajani AO, Ogunkunle O, Laseinde OT, Rahman SMA, Fattah IMR (2023) Decoding anaerobic digestion: a holistic analysis of biomass waste technology, process kinetics, and operational variables. Energies 16(8):3378

Bhajani SS, Pal SL (2022) Review: factors affecting biogas production. Int J Res Appl Sci Eng Tech 10:79–88

Bokhary A, Leitch M, Liao BQ (2021) Effect of organic loading rate on the biological performance of the thermophilic anaerobic membrane bioreactor treating pulp and paper primary sludge. Clean Technol Environ Policy 23:2669–2681

Chen Y, Xiao K, Jiang X, Shen N, Zeng RJ, Zhou Y (2018) Long solid retention time (SRT) has minor role in promoting methane production in a 65°C single-stage anaerobic sludge digester. Biores Technol 247:724–729

Chen L, Fang W, Chang J, Liang J, Zhang P, Zhang G (2022) Improvement of direct interspecies electron transfer via adding conductive materials in anaerobic digestion: mechanisms, performances, and challenges. Front Microbiol 13:860749
Cheng H, Qin H, Li Y, Zuo G, Liu J, Li Y (2024) Comparative study of high-performance mesophilic and thermophilic anaerobic membrane bioreactors in the co-digestion of sewage sludge and food waste: methanogenic performance and energy recovery potential. Sci Total Environ 912:169518
Choi Y, Ryu J, Lee SR (2020) Influence of carbon type and carbon to nitrogen ratio on the biochemical methane potential, pH, and ammonia nitrogen in anaerobic digestion. J Animal Sci Tech 62:74–83
Chow WL, Chong S, Lim JW, Chan YJ, Chong MF, Tiong TJ, Chin JK, Pan G-T (2020) Anaerobic co-digestion of wastewater sludge: a review of potential co-substrates and operating factors for improved methane yield. Processes 8(1):39
Chowdhury TH (2021) Technical-economical analysis of anaerobic digestion process to produce clean energy. Energy Rep 7:247–253
Chyoshi B, Coelho LHG, García J, Subtil EL (2022) Fate and removal of emerging contaminants in anaerobic fluidized membrane bioreactor filled with thermoplastic gel as biofilm support. Chemosphere 300:1234557
Conrad R (2023) Complexity of temperature dependence in methanogenic microbial environments. Front Microbiol 14:1232946
Datta S, Veena R, Samuel MS, Selvarajan E (2020) Immobilization of laccases and applications for the detection and remediation of pollutants: a review. Environ Chem Lett 19:521–538
de Castro TM, Arantes EJ, de Mendonça Costa MSS, Gotardo JT, Passig FH, de Carvalho KQ, Gomes SD (2021) Anaerobic co-digestion of industrial waste landfill leachate and glycerin in a continuous anaerobic bioreactor with a fixed-structured bed (ABFSB): effects of volumetric organic loading rate and alkaline supplementation. Renew Energy 164:1436–1446
Diamantis V, Eftaxias A, Stamatelatou K, Noutsopoulos C, Vlachokostas C, Aivasidis A (2021) Bioenergy in the era of circular economy: anaerobic digestion technological solutions to produce biogas from lipid-rich wastes. Renew Energy 168:438–447
Dompara I, Maragkaki A, Papastefanakis N, Floraki C, Vernardou D, Manios T (2023) Effects of different materials on biogas production during anaerobic digestion of food waste. Sustainability 15(7):5698
Dong N, Bu F, Zhou Q, Khanal SK, Xie L (2018) Performance and microbial community of hydrogenotrophic methanogenesis under thermophilic and extreme-thermophilic conditions. Biores Technol 266:454–462
Filer J, Ding HH, Chang S (2019) Biochemical methane potential (BMP) assay method for anaerobic digestion research. Water 11(5):921
Fu S, Wang F, Shi X, Guo G (2016) Impacts of microaeration on the anaerobic digestion of corn straw and the microbial community structure. Chem Eng J 287:523–528
Gao M (2020) Anaerobic treatment of source-diverted blackwater-maximizing biomethane recovery. University of Alberta, Canada
Gao M, Zhang L, Florentino AP, Liu Y (2019) Performance of anaerobic treatment of blackwater collected from different toilet flushing systems: can we achieve both energy recovery and water conservation? J Hazard Mater 365:44–52
Gao M, Zhang L, Liu Y (2020) High-loading food waste and blackwater anaerobic co-digestion: maximizing bioenergy recovery. Chem Eng J 394:124911
Gibellato S, Ballestra LV, Fiano F, Graziano D, Gregori GL (2023) The impact of education on the energy trilemma index: a sustainable innovativeness perspective for resilient energy systems. Appl Energy 330(B):120352
Girijan S, Kumar M (2019) Immobilized biomass systems: an approach for trace organics removal from wastewater and environmental remediation. Current Opinion Environ Sci Heal 12:18–29
Guo B, Yu N, Weissbrodt DG, Liu Y (2021) Effects of micro-aeration on microbial niches and antimicrobial resistances in blackwater anaerobic digesters. Water Res 196:117035

Hansen TL, Schmidt JE, Angelidaki I, Marca E, la Cour JJ, Mosbæk H, Christensen TH (2004) Method for determination of methane potentials of solid organic waste. Waste Manage 24(4):393–400
Haryanto A, Shintawati, Hasanudin U (2021) Anaerobic treatment for palm oil mill effluent using covered in-the ground anaerobic reactor (CIGAR). J Eng Tech Sci 53(6):210609
Holl E, Steinbrenner J, Merkle W, Krümpel J, Lansing S, Baier U, Oechsner H, Lemmer A (2022) Two-stage anaerobic digestion: state of technology and perspective roles in future energy systems. Biores Technol 360:127633
Holliger C, Fruteau de Laclos H, Hafner SD, Koch K, Weinrich S, Astals S, Alves M, Andrade D, Angelidaki I, Appels L, Azman S, Bagnoud A, Baier U, Fernandez YB, Bartacek J, Battista F, Bolzonella D, Bougrier C, Braguglia C, Buffi P, Wierinck I (2020) Requirements for measurement and validation of biochemical methane potential (BMP). Standard BMP methods document 100, version 1.8. Available online: https://www.dbfz.de/en/BMP (accessed on February 1, 2024)
Huang Q, Liu Y, Dhar BR (2022) A critical review of microbial electrolysis cells coupled with anaerobic digester for enhanced biomethane recovery from high-strength feedstocks. Crit Rev Environ Sci Technol 52(1):50–89
Huang Q, Liu Y, Dhar BR (2023) Boosting resilience of microbial electrolysis cell-assisted anaerobic digestion of blackwater with granular activated carbon amendment. Biores Technol 381:129136
Ibro MK, Ancha VR, Lemma DB (2022) Impacts of anaerobic co-digestion on different influencing parameters: a critical review. Sustainability 14(15):9387
Issahaku M, Derkyi NSA, Kemausuor F (2024) A systematic review of the design considerations for the operation and maintenance of small-scale biogas digesters. Heliyon 10(1):e24019
Jiang J, He S, Kang X, Sun Y, Yuan Z, Xing T, Guo Y, Li L (2020) Effect of organic loading rate and temperature on the anaerobic digestion of municipal solid waste: process performance and energy recovery. Front Energy Res 8:89
Kanafin YN, kanafina D, Malamis S, Katsou E, Inglezakis VJ, Poulopoulos SG, Arkhangelsky E (2021) Anaerobic membrane bioreactors for municipal wastewater treatment: a literature review. Membranes 11(12):967
Kariyama ID, Zhai X, Wu B (2018) Influence of mixing on anaerobic digestion efficiency in stirred tank digesters: a review. Water Res 143:503–517
Kasinski S (2020) Mesophilic and thermophilic anaerobic digestion of organic fraction separated during mechanical heat treatment of municipal waste. Appl Sci 10(7):2412
Kim D, Phae C (2022) Analysis of the effect of sludge reduction and energy production from introducing anaerobic digest plant into a sewage treatment plant. Water 14(11):1821
Kim M, Gomec CY, Ahn Y, Speece RE (2003) Hydrolysis and acidogenesis of particulate organic material in mesophilic and thermophilic anaerobic digestion. Environ Technol 24(9):1183–1190
Kumar DJP, Mishra RK, Chinnam S, Binnal P, Dwivedi N (2024) A comprehensive study on anaerobic digestion of organic solid waste: a review on configurations, operating parameters, techno-economic analysis and current trends. Biotechnology Notes 5:33–39
Lago A, Greses S, Aboudi K, Moreno I, González-Fernández C (2023) Effect of decoupling hydraulic and solid retention times on carbohydrate-rich residue valorization into carboxylic acids. Sci Rep 13:20590
Lin Q, He G, Rui J, Fang X, Tao Y, Li J, Li X (2016) Microorganism-regulated mechanisms of temperature effects on the performance of anaerobic digestion. Microb Cell Fact 15:96
Liu D, Yang X, Zhang L, Tang Y, He H, Liang M, Tu Z, Zhu H (2022) Immobilization of biomass materials for removal of refractory organic pollutants from wastewater. Int J Environ Res Public Health 19(21):13830
López-López A, León-Becerril E, Rosales-Contreras ME, Villegas-García E (2015) Influence of alkalinity and VFAs on the performance of an UASB reactor with recirculation for the treatment of Tequila vinasses. Environ Technol 36(19):2468–2476
Lukitawesa, Wikandari R, Millati R, Taherzadeh MJ, Niklasson C (2018) Effect of effluent recirculation on biogas production using two-stage anaerobic digestion of citrus waste. Molecules 23(12):3380

Mahmoud N, Zeeman G, Gijzen H, Lettinga G (2004) Anaerobic stabilisation and conversion of biopolymers in primary sludge-effect of temperature and sludge retention time. Water Res 38(4):983–991

Martinez-Sanchez V, Levis JW, Damgaard A, DeCarolis JF, Barlaz MA, Astrup TF (2017) Evaluation of externality costs in life-cycle optimization of municipal solid waste management systems. Environ Sci Technol 51(6):3119–3127

McKeown RM, Scully C, Enright A-M, Chinalia FA, Lee C, Mahony T, Collins G, O'Flaherty V (2009) Psychrophilic methanogenic community development during long-term cultivation of anaerobic granular biofilms. ISME J 3:1231–1242

Möller K, Müller T (2012) Effects of anaerobic digestion on digestate nutrient availability and crop growth: a review. Eng Life Sci 12:242–257

Morozova I, Nikulina N, Oechsner H, Krümpel J, Lemmer A (2020) Effects of increasing nitrogen content on process stability and reactor performance in anaerobic digestion. Energies 13(5):1139

Musa MA, Idrus S, Man HC, Daud NNN (2018) Wastewater treatment and biogas recovery using anaerobic membrane bioreactors (AnMBRs): strategies and achievements. Energies 11:1675

Nges IA, Liu J (2010) Effects of solid retention time on anerobic digestion of dewatered-sewage sludge in mesophilic and thermophilic conditions. Renew Energy 35(10):2200–2206

Nkuna R, Roopnarain A, Rashama C, Adeleke R (2022) Insights into organic loading rates of anaerobic digestion for biogas production: a review. Crit Rev Biotechnol 42(4):487–507

Osei-Owusu BA, Baidoo MF, Arthur R, Oduro-Kwarteng S (2023) Physico-chemical characterization of selected feedstocks as co-substrates for household biogas generation in Ghana. Int J Sustain Eng 16(1):117–128

Owusu-Agyeman I, Plaza E, Elginöz N, Atasoy M, Khatami K, Perez-Zabaleta M, Cabrera-Rodríguez C, Yesil H, Tugtas AE, Calli B, Cetecioglu Z (2023) Conceptual system for sustainable and next-generation wastewater resource recovery facilities. Sci Total Environ 885:163758

Pérez-Elvira SI, Nieto Diez P, Fdz-Polanco F (2006) Sludge minimisation technologies. Rev Environ Sci Bio/Tech 5:375–398

Piadeh F, Offie I, Behzadian K, Rizzuto JP, Bywater A, Córdoba-Pachón J-R, Walker M (2024) A critical review for the impact of anaerobic digestion on the sustainable development goals. J Environ Manage 349:119458

Plevri A, Mamais D, Noutsopoulos C (2021) Anaerobic MBR technology for treating municipal wastewater at ambient temperatures. Chemosphere 275:129961

Ren D, Wang Z, Jiang S, Yu H, Zhang S, Zhang X (2020) Recent environmental applications and development prospects for immobilized laccase: a review. Biotech Genetic Eng Rev 36:81–131

Rittmann BE, McCarty PL (2012) Environmental biotechnology: principles and applications. McGraw-Hill Education, New York

Rocha-Meneses L, Zannerni R, Inayat A, Abdallah M, Shanableh A, Ghenai C, Kamil M, Kikas T (2022) Current progress in anaerobic digestion reactors and parameters optimization. Biomass Conv Biorefinery. https://doi.org/10.1007/s13399-021-02224-z

Rodríguez-Jiménez LM, Pérez-Vidal A, Torres-Lozada P (2022) Research trends and strategies for the improvement of anaerobic digestion of food waste in psychrophilic temperatures conditions. Heliyon 8(10):e11174

Roopnarain A, Rama H, Ndaba B, Bello-Akinosho M, Bamuza-Pemu E, Adeleke R (2021) Unravelling the anaerobic digestion 'black box': biotechnological approaches for process optimization. Renew Sustain Energy Rev 152:111717

Ryue J, Lin L, Kakar FL, Elbeshbishy E, Al-Mamun A, Dhar BR (2020) A critical review of conventional and emerging methods for improving process stability in the thermophilic anaerobic digestion. Energy Sustain Dev 54:72–84

Saady NMC, Massé DI (2015) A start-up psychrophilic anaerobic sequence batch reactor digesting a 35% total solids feed of dairy manure and wheat straw. AMB Express 5:55

Salter GJ, Kell DB (1991) New materials and technology for cell immobilization. Curr Opin Biotechnol 2:385–389

Sathiyah D, Ngema L, Tetteh EK, Chollom MN, Rathilal S (2023) Effect of operational parameters on anaerobic digestion of municipal and sugar industry wastewater. Int J Energy Production Manage 8(3):161–167

Show K-Y, Yan Y, Yao H, Guo H, Li T, Show D-Y, Chang J-S, Lee D-J (2020) Anaerobic granulation: a review of granulation hypotheses, bioreactor designs and emerging green applications. Biores Technol 300:122751

Silverman RE, Flores RJ, Brouwer J (2020) Energy and economic assessment of distributed renewable gas and electricity generation in a small disadvantaged urban community. Appl Energy 280:115974

Singh B, Szamosi Z, Siménfalvi Z (2020) Impact of mixing intensity and duration on biogas production in an anaerobic digester: a review. Crit Rev Biotechnol 40(4):508–521

Singh B, Kovács KL, Bagi Z, Nyári J, Szepesi GL, Petrik M, Siménfalvi Z, Szamosi Z (2021) Enhancing efficiency of anaerobic digestion by optimization of mixing regimes using helical ribbon impeller. Fermentation 7(4):251

Skouteris G, Hermosilla D, López P, Negro C, Blanco Á (2012) Anaerobic membrane bioreactors for wastewater treatment: a review. Chem Eng J 198:138–148

Slorach PC, Jeswani HK, Cuéllar-Franca R, Azapagic A (2019) Environmental sustainability of anaerobic digestion of household food waste. J Environ Manage 236:798–814

Sohoo I, Ritzkowski M, Heerenklage J, Kuchta K (2021) Biochemical methane potential assessment of municipal solid waste generated in Asian cities: a case study of Karachi, Pakistan. Renew Sustain Energy Rev 135:110175

Storck T, Virdis B, Batstone DJ (2015) Modelling extracellular limitations for mediated versus direct interspecies electron transfer. ISME J 10:621–631

Sun H (2023) Towards more efficient and economical phosphorus and bioenergy recovery from source-diverted blackwater. University of Alberta, Canada

Sun H, Yu N, Mou A, Yang X, Liu Y (2022) Effluent recirculation weakens the hydrolysis of high-solid content feeds in upflow anaerobic sludge blanket reactors. J Environ Chem Eng 10(3):107913

Sun H, Yang X, Zhang L, Liu Y (2023) The role of protein contents in promoting wastewater phosphorus and bioenergy recovery during anaerobic digestion. Biomass Bioenerg 169:106694

Swain G, Maurya KL, Sonwani RK, Singh RS, Jaiswal RP, Rai BN (2022) Effect of mixing intensity on biodegradation of phenol in a moving bed biofilm reactor: process optimization and external mass transfer study. Biores Technol 351:126921

Tawfik A, Mohsen M, Ismail S, Alhajeri N, Osman AI, Rooney DW (2022) Methods to alleviate the inhibition of sludge anaerobic digestion by emerging contaminants: a review. Environ Chem Lett 20:3811–3836

Thakur N, Sharma M, Alghamdi H, Zheng Y, Xue W, Jeon B-H, Salama El-S, Li X (2023) A recent trend in anaerobic digestion (AD): enhancement of microbiome and digestibility of feedstocks via abiotic stress factors for biomethanation. Chem Eng J 472:145047

Tiwari BR, Rouissi T, Brar SK, Surampalli RY (2021) Critical insights into psychrophilic anaerobic digestion: novel strategies for improving biogas production. Waste Manage 131:513–526

Tran HS, Viet NTT, Heaven S, Banks CJ (2023) Moving bed anaerobic membrane bioreactor for low-strength wastewater treatment. Case Stud Chem Environ Eng 8:100391

Turker M, Dereli RK (2021) Long term performance of a pilot scale anaerobic membrane bioreactor treating beet molasses based industrial wastewater. J Environ Manage 278(Pt 1):111403

Uddin MdM, Wright MM (2021) Anaerobic digestion fundamentals, challenges, and technological advances. Phys Sci Rev 8(9):2819–2837

Van DP, Fujiwara T, Tho BL, Toan PPS, Minh GH (2020) A review of anaerobic digestion systems for biodegradable waste: configurations, operating parameters, and current trends. Environ Eng Res 25(1):1–17

Van Lier JB, Mahmoud N, Zeeman G. Anaerobic wastewater treatment. In: Henze M, van Loosdrecht MCM, Ekama GA, Brdjanovic D (eds) (2008) Biological wastewater treatment: principles, modelling and design. IWA Publishing, London, UK

Vanwonterghem I, Jensen PD, Rabaey K, Tyson GW (2015) Temperature and solids retention time control microbial population dynamics and volatile fatty acid production in replicated anaerobic digesters. Sci Rep 5:8496

VDI 4630 (2016). Fermentation of organic materials-characterization of the substrate, sampling, collection of material data, fermentation tests. VDI, Germany

Viet NTT, Vu DC, Duong TH (2023) Effect of hydraulic retention time on performance of anaerobic membrane bioreactor treating slaughterhouse wastewater. Environ Res 233:116522

Wu Z, Lin Z, Sun Z, Gou M, Xia Z, Tang Y (2020) A comparative study of mesophilic and thermophilic anaerobic digestion of municipal sludge with high-solids content: reactor performance and microbial community. Biores Technol 302:122851

Xu R, Zhang K, Liu P, Khan A, Xiong J, Tian F, Li X (2018) A critical review on the interaction of substrate nutrient balance and microbial community structure and function in anaerobic co-digestion. Biores Technol 247:1119–1127

Yang Y, Zang Y, Hu Y, Wang XC, Ngo HH (2020) Upflow anaerobic dynamic membrane bioreactor (AnDMBR) for wastewater treatment at room temperature and short HRTs: process characteristics and practical applicability. Chem Eng J 383:123186

Yee RA, Alessi DS, Ashbolt NJ, Hao W, Konhauser K, Liu Y (2019) Nutrient recovery from source-diverted blackwater: optimization for enhanced phosphorus recovery and reduced co-precipitation. J Clean Prod 235:417–425

Yu N, Guo B, Zhang Y, Zhang L, Zhou Y, Liu Y (2020) Different micro-aeration rates facilitate production of different end-products from source-diverted blackwater. Water Res 177:115783

Yu N, Sun H, Mou A, Liu Y (2021a) Calcium hypochlorite enhances the digestibility of and the phosphorus recovery from waste activated sludge. Biores Technol 340:125658

Yu N, Guo B, Zhang Y, Zhang L, Zhou Y, Liu Y (2021b) Self-fluidized GAC-amended UASB reactor for enhanced methane production. Chem Eng J 420:127652

Zamanzadeh M, Hagen LH, Svensson K, Linjordet R, Horn SJ (2016) Anaerobic digestion of food waste–effect of recirculation and temperature on performance and microbiology. Water Res 96:246–254

Zhang L, Xu M, Liu Y (2019) Pretreatment for anaerobic blackwater treatment: ultrasonication and thermal hydrolysis. J Environ Eng Sci 14(1):32–36

Zhang L, Guo B, Mou A, Li R, Liu Y (2020) Blackwater biomethane recovery using a thermophilic upflow anaerobic sludge blanker reactor: impacts of effluent recirculation on reactor performance. J Environ Manage 274:111157

Zhang L, Mou A, Guo B, Sun H, Anwar MN, Liu Y (2021a) Simultaneous phosphorus recovery in energy generation reactor (SPRING): high rate thermophilic blackwater treatment. Resour Conserv Recycl 164:105163

Zhang L, Mou A, Sun H, Zhang Y, Zhou Y, Liu Y (2021b) Calcium phosphate granules formation: key to high rate of mesophilic UASB treatment of toilet wastewater. Sci Total Environ 773:144972

Zhang Y, Guo B, Dang H, Zhang L, Sun H, Yu N, Tang Y, Liu Y (2022a) Roles of granular activated carbon (GAC) and operational factors on active microbiome development in anaerobic reactors. Biores Technol 343:126104

Zhang Y, Zhang L, Yu N, Guo B, Liu Y (2022b) Enhancing the resistance to H_2S toxicity during anaerobic digestion of low-strength wastewater through granular activated carbon (GAC) addition. J Hazard Mater 430:128473

Zhang Y, Liu Y (2023) Harnessing carbonaceous materials' multifaceted roles for enhanced anaerobic digestion performance. Chem Eng J 477:146931

Zhang L, Yuan Y, Li C, Zhang Y, Sun H, Xu R, Liu Y (2024a) Biomineralization of phosphorus during anaerobic treatment of distillery wastewaters. Sci Total Environ 925:171431

Zhang L, Yuan Y, Zhang Y, Liu Y (2024b) Switching from wet to dry anaerobic digestion of food waste with different dilution times under no mechanical mixing condition. Chemosphere 352:141299

Zhao H, Yan F, Li X, Piao R, Wang W, Cui Z (2020) Impact of organic loading rate on performance and methanogenic microbial communities of a fixed-bed anaerobic reactor at 4 °C. Water 12(9):2586

Anaerobic Membrane Bioreactor for Wastewater Treatment

Gaojun Wang, Bo Zhang, and Rong Chen

Abstract As an emerging wastewater treatment technology, anaerobic membrane bioreactor (AnMBR) shows advantages in excellent efficiency for pollutant removal and energy recovery. The promotion of AnMBR is determined by the requirement, the advancement of fundamental principles, and the development of novel configurations. The materials of membrane greatly influence AnMBR performance. Both in situ and ex situ controlling strategies have been developed depending on the type of membrane fouling. Following, the progress and application cases of AnMBR in treating low-strength municipal wastewater (pilot scale) and high-strength industrial wastewater (full scale) are comprehensively introduced in terms of advantages and bottlenecks. For municipal wastewater treatment, the major bottlenecks toward full-scale applications underline in the trade-off between low energy recovery potential and energy consumption for stable operation, as well as carbon emissions from dissolved methane at low temperatures. For industrial wastewater treatment, membrane fouling when treating wastewater with high suspended solids is challenging due to the sustained operation and maintenance cost. The new research trends in AnMBR include but are not limited to emerging contaminants removal and advanced strategies for mitigating membrane fouling. The development of AnMBR can significantly contribute to low-carbon wastewater treatment practices.

Keywords AnMBR · Emerging contaminants · Energy recovery · Industrial wastewater · Membrane fouling · Municipal wastewater

G. Wang · B. Zhang · R. Chen (✉)
Key Lab of Environmental Engineering, Xi'an University of Architecture and Technology, Xi'an 710055, Shaanxi, China
e-mail: chenrong@xauat.edu.cn

G. Wang
e-mail: gjwang@xauat.edu.cn

G. Wu (ed.), *Anaerobic Digestion*, Green Energy and Technology,
https://doi.org/10.1007/978-3-031-69378-6_14

1 Introduction

With the acceleration of global economic development and urbanization, the energy and water crises have become crucial issues challenging sustainable development. Wastewater treatment has transited toward energy and resource recovery from wastewater (Diaz-Elsayed et al. 2019). Anaerobic membrane bioreactor (AnMBR) has been received intensive attention as an emerging wastewater treatment technology, due to its low energy consumption, high-quality effluent, stable energy recovery efficiency, and small footprint (Hu et al. 2022b). The integration of anaerobic digestion with membrane separation was first proposed by H. E. Grethlein in 1978 (Grethlein 1978). The combined septic tank with a semi-permeable membrane module removed biochemical oxygen demand (BOD) by 85–95% and even achieved complete removals of turbidity and pathogens. Since then, AnMBR has gone through flourishing development from lab to practical applications.

Compared with conventional wastewater treatment technologies, AnMBR not only can guarantee a high treatment efficiency, but also can achieve renewable energy (in the form of CH_4) recovery from organic pollutants, potentially reducing the consumption of fossil fuels for wastewater treatment. Currently, AnMBR has been successfully applied to treat different types of industrial wastewater, such as those derived from food manufacturing, livestock farming, and pharmaceutical industry (An et al. 2023). Meanwhile, in 2011, Prof. Perry McCarty proposed that AnMBR could be an important option for seeking energy recovery from municipal wastewater (McCarty et al. 2011). Since then, numerous attempts have been conducted to exploring the feasibility of treating municipal wastewater with AnMBR (Aslam et al. 2022). Although many bottlenecks still exist, the pilot-scale AnMBR has been established to treat municipal wastewater with acceptable energy balance.

2 Principle and Characteristics of AnMBR

2.1 Principle of AnMBR

AnMBR is a combined technology of anaerobic digestion and membrane filtration, with the incorporation of both biological degradation and physical separation. In AnMBR, a high biomass concentration can be achieved through membrane retention, thereby efficiently separating hydraulic retention time (HRT) from sludge retention time (SRT). The high biomass concentration contributes to excellent chemical oxygen demand (COD) removal with the removal percentage higher than 95%. At the same time, the high biomass concentration makes high-efficient methane yield under low HRT or high loading rate conditions feasible. Furthermore, AnMBR can reduce footprint requirement for building infrastructures. Besides retaining biomass, membrane can extend the residence time of refractory substances, e.g., pharmaceutical and personal care products (PPCP) and antibiotics, making their high-efficient removal possible (Ji et al. 2020a).

Meanwhile, AnMBR has some technical bottlenecks limiting its applications. The critical issue is the high cost of system configuration and the maintenance of membrane modules. During AnMBR operation, membrane fouling is unavoidably aggravated due to foulants deposited on the membrane surface (De Vela 2021). This would lead to a decrease in the membrane flux. In situ and ex situ cleaning operation and even replacing membrane are necessary to recover the production capacity of membrane (Hu et al. 2020; Nabi et al. 2023). All these increase construction and operation costs of AnMBR systems.

2.2 AnMBR Configurations

Three types of AnMBR configurations can be classified based on the location of the membrane module, i.e., external AnMBR, submerged AnMBR, and external submerged AnMBR (Zhen et al. 2019) (Fig. 1). In an external AnMBR, the membrane module is separated from the anaerobic digestion unit, where a pump is necessary to feed the anaerobic digestion effluent to the membrane module for separation. Usually, the pressure-driven crossflow is applied to generate the surface shearing force for achieving the in situ cleaning and mitigating membrane fouling. Consequently, there is no need to disrupt the anaerobic digestion unit during in situ membrane cleaning. However, without mechanistic driving force, the application of only crossflow makes the membrane flux of this configuration relatively low. To mitigate the deposited suspended flocs on the membrane surface, a high shear force should be maintained, causing a high energy consumption.

For the submerged AnMBR configuration, both the anaerobic digestion unit and the membrane module are within the same reactor. A pump is used to generate the driving force of negative pressure in the inner side of membrane, thereby pulling the water from outside to inside of membrane. No sludge recirculation is required in the submerged AnMBR, which can save energy consumption. Alternatively, in situ membrane cleaning can be achieved by recirculating the generated biogas, which can also maintain stable membrane flux. Chemical cleaning (e.g., alkaline and acids cleaning) may disturb microbial activity, and is generally applied as the ex situ cleaning. However, the ex situ chemical cleaning may disturb the anaerobic condition within the bioreactor and cause the inhibition of methanogens.

External submerged AnMBR is an integration of the above two types of configurations and uses biogas recirculation for in situ membrane cleaning. The extensive membrane cleaning can be easily achieved without disturbing the main anaerobic digestion unit. This overcomes the limitations of weak driving force of external AnMBR and submerged AnMBR. Therefore, the external submerged AnMBR has become the popular configuration for practical applications.

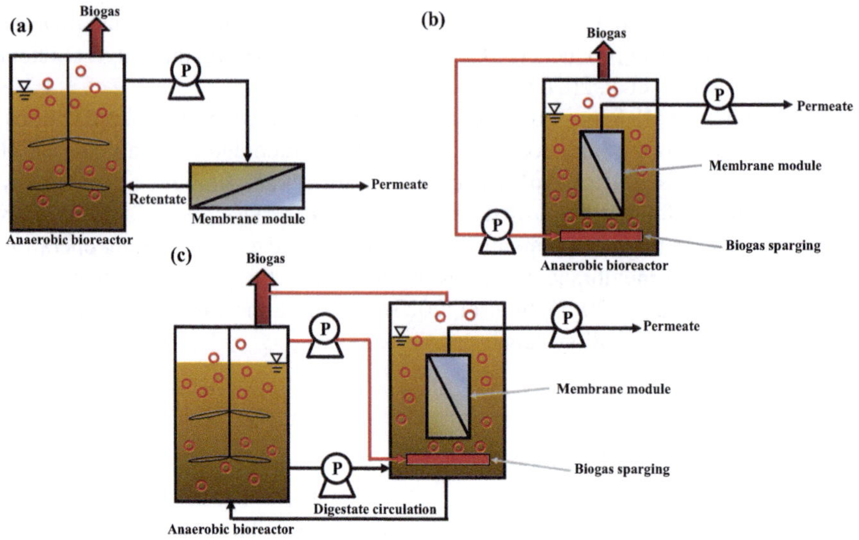

Fig. 1 The three types of configurations of AnMBR, i.e., **a** external AnMBR, **b** submerged AnMBR, and **c** external submerged AnMBR

2.3 Characteristics of Membrane Used in AnMBR

2.3.1 Membrane Material

Ceramic and polymers are two common compounds for membrane manufacturing. The hydrophilic ceramic membrane has a long lifespan, strong mechanistic strength, and excellent filtration capacity. Moreover, the inorganic ceramic possesses stable chemical features, which makes ceramic membrane potentially outcompeting polymer membrane when facing alkali/acids cleaning in AnMBR. Ceramic membrane is used mostly in the early studies of AnMBR (Jeong et al. 2018; Liu et al. 2019b). Ghyoot and Verstraete (1997) filtrated the digestate of primary sludge using a ceramic membrane, and achieved a short term flux as high as 200–250 L/m^2/h (LMH). Currently, extensive studies of ceramic AnMBR have been conducted for treating municipal wastewater at both lab- and pilot-scale, and achieved a COD removal efficiency of approximately 90% (Yue et al. 2015). Wu et al. (2023b) used a ceramic AnMBR for treating municipal solid waste leachate, and 95% COD removal efficiency was achieved under the flux of 6.2 LMH, suggesting the feasibility of ceramic AnMBR for treating high-strength wastewater.

The polymeric materials for membrane manufacturing of AnMBR include polyvinylidene difuoride (PVDF), polyethersulfone (PES), polytetrafluoroethylene (PTFE), polysulfone (PSF), and polyethylene (PE). Among these, the PVDF membrane is the most popular one, due to its advantages of low cost, excellent mechanical strength, and robust resistance to corrosive chemicals (Mahmood et al. 2022). Polymeric membrane based AnMBR can be used to treat municipal wastewater, food wastewater, livestock wastewater, and chemical industry wastewater.

Filtration property and membrane fouling are different between ceramic membrane and polymer membrane in AnMBR. Jeong et al. (2018) showed that the ceramic membrane had higher pollutant removal efficiency than the polymer membrane due to its better retention capacity of biomass. Meanwhile, the super-hydrophilic property of ceramic membrane could resist the accumulation of hydrophobic microbial products on membrane surface, resulting in a lower fouling rate than polymer membrane (Jeong et al. 2018). Liu et al. (2019b) distinguished irreversible fouling substances on ceramic membrane and polymer membrane and found that more organics involved in polymer membrane fouling while inorganic fouling occurred in ceramic membrane. Therefore, different cleaning strategies based on membrane materials should be adopted.

2.3.2 Membrane Pore Size and Module Type

Microfiltration (MF) membrane and ultrafiltration (UF) membrane are commonly used in AnMBR, with pore size ranging from 0.1–10 μm to 0.002–0.1 μm, respectively. This pore size range can guarantee the efficiency of AnMBR to retain most microbes and suspended substances. The membrane pore size determines the intrinsic resistance of membrane and the potential of suspended substances deposited in membrane pore. MF-AnMBR is the most-commonly used system treating various types of wastewaters due to its high membrane flux and low energy consumption for generating driving force. Ji et al. (2020b) compared performances of two MF-AnMBR with different pore sizes, and MF of 0.4 μm pore size showed better filtration property than the MF of 0.05 μm pore size. Compared with MF, AnMBR implemented by UF possesses better performance in the removal of dissolved organic substances and the generation of high-quality water. However, the high cost and energy consumption hinder UF in large-scale AnMBR application.

2.4 Membrane Fouling in AnMBR

2.4.1 Mechanism of Membrane Fouling

The unavoidable membrane fouling is the major challenge for AnMBR applications (Charfi et al. 2012). The mechanism of membrane fouling involves the complex temporal deposition of different types of substances (e.g., suspended and colloidal particles, organic and inorganic compounds, pollutants, and microbial products) on membrane surface or in membrane inner pore. Generally, membrane fouling development in AnMBR can be classified into three phases (Wang et al. 2022b). A rapid increase in transmembrane pressure (TMP) occurs initially after operating a new

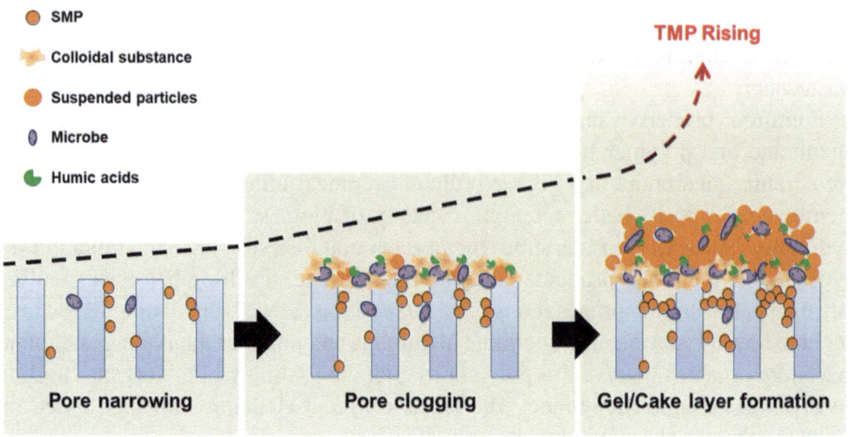

Fig. 2 The process of membrane fouling in AnMBR (Wang et al. 2022b)

membrane, followed by a slow rate but long-term duration of TMP increase. When the TMP rises approximately to 30 kPa, TMP jumping may occur, causing a rapid membrane flux decline. Consistently with the occurrence of three phases chronologically, the fouling behaviors of pore narrowing, pore clogging, and gel/cake fouling layer deposition occur (Fig. 2).

During filtration in AnMBR, micro-molecular foulants with smaller size than the average membrane pore size likely accumulate in the inner side of membrane pore, inducing pore narrowing. Typically, soluble microbial products (SMP) are considered important compounds responsible for pore narrowing (Xiong et al. 2016), and SMP comprise of hydrophilic proteins, polysaccharides, lipids and humic-like substances. For instance, the addition of SMP compounds aggravated the attachment of the foulants in membrane pore space, suggesting that SMP govern pore narrowing (Meng et al. 2017). SMP are classified to utilization-associated products (UAP) and biomass-associated products (BAP). UAP are smaller molecules, which can be utilized for microbial growth. In contrast, BAP are mainly derived from endogenous respiration, contributing more to pore narrowing.

Pore clogging is mainly caused by macro-molecular SMP, extracellular polymeric substances (EPS), colloidal particles, and cell debris. Regarding SMP, both UAP and BAP likely contribute to pore clogging. BAP with higher molecules easily occupy the narrow pore space directly, occurring in the early stage of pore clogging (Ni et al. 2011). UAP with low molecules tend to attach with the deposited foulants in the pore space, leading to the intermediate pore clogging (Bolton et al. 2006). Moreover, the colloidal polysaccharides, proteins, and humic-like substances in EPS could cause membrane pore clogging during the initial fouling stage by depositing in the pore space, while their interactions with foulants cause intermediate pore clogging (Zhou et al. 2019). Meanwhile, the interaction of metal–organic complexation could trigger inorganic compounds bridging on organic foulants (Xin et al. 2015).

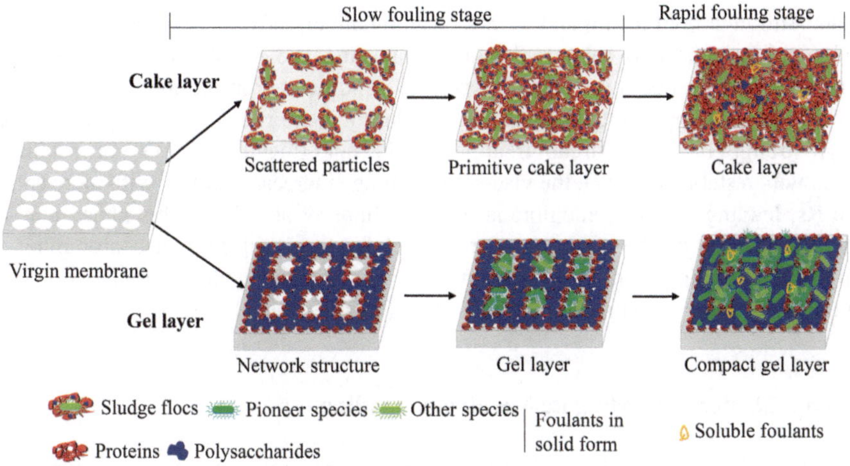

Fig. 3 The proposed mechanism of cake and gel layers formation (Lei et al. 2021b)

Following pore narrowing and clogging, a fouling layer covers the membrane surface, which inhibits more foulants deposited on pore space. During fouling layer formation, the foulants start to accumulate on the fouling layer, and the attractive force of foulants–foulants rather than foulants–membrane governs the fouling rate. Usually, fouling layer contributes to most of the fouling resistance. In situ cleaning strategy of recirculating biogas flow near membrane surface in submerged AnMBR and hydrodynamic crossflow of external AnMBR could effectively reverse layer fouling. Layer fouling is classified into gel layer and cake layer, respectively (Lei et al. 2021b) (Fig. 3). The gel layer is mainly formed from sticky substances, e.g., polysaccharides and proteins of EPS, macro-molecular SMP, colloidal pollutants in wastewater and inorganic compounds for chelation. The cake layer is commonly much thicker than the gel layer, consisting of suspended particles in wastewater, floc biomass, and metal–organic complexation substances. For gel/cake layer formation, the covalent complexation is a crucial factor to enhance the hydrophobicity of organic foulants.

2.4.2 Factors Influencing Membrane Fouling

The sludge characteristics in AnMBR, e.g., volatile solids (VS) content and viscosity, influence membrane fouling and flux (De Vela 2021), which are determined by system parameters of SRT, HRT, and organic loading rates (OLR). Usually, an excessive long SRT with a high suspended solids concentration can aggravate membrane fouling (Mahmoud and Liao 2017). Thus, a trade-off between a long SRT to guarantee treatment efficiency and a short SRT to avoid high membrane fouling potential should be evaluated. Generally, a short HRT means a high OLR, which requires a high membrane flux, potentially causing an increased rate of TMP or membrane fouling rate. Although both in situ and ex situ cleaning are effective to prolong the lifespan of membrane, how to balance the membrane flux and membrane fouling rate is crucial for stable operation under short HRT conditions. Furthermore, the viscosity of sludge is an important characteristic regulating membrane fouling, especially for high-strength bio-waste stream treatment (Lutze and Engelhart 2021). For instance, even with a stable VS level, the viscosity of sludge increased rapidly with increasing OLRs, leading to severe membrane fouling (Jiang et al. 2023). This phenomenon is likely caused by the secretion of microbial products with high viscosity under a high OLR. Using polymeric conditioners to regulate sludge characteristic could be a feasible strategy to control membrane flux (Odriozola et al. 2021).

2.4.3 Strategies to Mitigate Membrane Fouling

In situ cleaning and ex situ cleaning strategies are generally used to recover the flux of a fouled membrane (Shahid et al. 2020). In situ cleaning strategies include biogas sparging in submerged AnMBR and hydrodynamic scouring in external AnMBR. For the biogas sparging strategy, the generated biogas is recirculated in the submerged AnMBR, forming a crossflow liquid–gas phase to induce in situ shear on membrane surface. The benefits of biogas sparging are cake layer removal and permeate flux maintenance, whereas the efficiency is closely related to the sparging duration and flux. Nonetheless, the trade-off between system performance and energy consumption of biogas sparging should be evaluated when considering engineering-scale applications (Wang et al. 2018b).

The water cleaning strategy includes crossflow flushing membrane surface and backwashing flux vertically cross membrane space. The crossflow mitigates the accumulation of foulants on membrane surface, which efficiently controls the cake layer formation, but hardly mitigates the blocking occurrence in membrane pore space. Backwashing by clean water or AnMBR permeates allows to push the flux cross from inside to outside of membrane space, which is conversely to the direction of permeate. By this means, the membrane channel can be flushed to eliminate the foulants blocking membrane pore. Nonetheless, for these two fluidal cleaning strategies, flux rate and frequency of cleaning are crucial factors determining cleaning efficiency.

Physical cleaning by in situ ultrasound treatment can also control membrane fouling. Xu et al. (2010) developed the on-line ultrasonic device for membrane fouling mitigation and optimized the duration of 3 min/h ultrasonic treatment strategy by a power intensity of 0.18 W/cm^2. A further study clarified that ultrasound treatment mainly mitigated membrane fouling caused by the loose cake layer, but hardly eliminated fouling caused by the tight cake layer and gel layer (Xu et al. 2013).

Chemical cleaning is another in situ strategy, which outcompetes physical cleaning for the irreversible fouling caused by pore blocking. Acids, bases, surfactant, oxidants, and antiscalants are commonly used cleaning agents. NaClO is widely used for in situ chemical cleaning. Ramos et al. (2014b) investigated in situ chemical cleaning efficiency regarding different AnMBR configurations. In a submerged

AnMBR, a direct backwashing process with 2000 mg/L NaClO under a backwashing flux of 6 L/m^2 achieved 60% membrane flux recovery in 2 h; while in an external AnMBR, emptying the membrane tank and then backwashing with 1000 mg/L NaClO under a backwashing flux of 10 L/m^2 could recover 80% membrane flux in 3.5 h (Ramos et al. 2014b). Lee et al. (2016) found that chemical cleaning by 600 mg/L NaClO at a flux of 40 LMH recovered the efficiency by 44%. Jiang et al. (2019) delivered that 298 mg/L NaClO backwashing at a flux of 5.2 LMH showed 86% flux recovery. For submerged AnMBR, a crucial issue of in situ chemical cleaning is the inhibition of caustic chemicals to microbial activity (Lee et al. 2019). Thus, a proper chemical concentration is crucial to control membrane fouling without sacrificing microbial activity. An in situ cleaning with 10–20 mmol/L NaOH not only greatly mitigated membrane fouling, but also positively stimulated methanogens activity (Mei et al. 2017). However, a high NaOH concentration would decrease both cleaning efficiencies and microbial activity (Mei et al. 2017). Considering the complexation of foulants composition, cleaning using diverse chemicals is widely applied. Alkaline cleaning was effective to remove organic foulants by hydrolyzing these macro-molecular substances into micro-molecular ones and detached from membrane surface, while citric acid cleaning was effective in removing the inorganic scalants (Martin Vincent et al. 2018). Besides the in situ cleaning strategies mentioned above, some emerging strategies, such as quorum sensing regulation (Liu et al. 2019a; Xu et al. 2022), electrochemical assistance strategy (Hu et al. 2022a), and magnetic dynamics controlling (Bilad et al. 2012), have also been developed.

Although in situ cleaning strategies could mitigate membrane fouling to some extent, ex situ cleaning with higher intensity is also necessary sometimes, which would likely achieve a rapid recovery of membrane flux. Especially for submerged AnMBR, when taking out membrane for ex situ cleaning, there is no need to consider the caustic inhibition of cleaning chemicals on microbial activity. Ex situ cleaning includes physical scouring for cake layer removal, and chemical soaking or backwashing for gel layer control and pore blocking elimination (Shahid et al. 2020). Generally, both membrane fouling rate and cleaning cost determine the frequency of in situ and ex situ cleaning. In situ cleaning frequency always ranges in few weeks and months, while ex situ cleaning is commonly implemented annually.

3 Application of AnMBR for Treating Municipal Wastewater

3.1 Municipal Wastewater

With the introduction of resource recovery rather than pollutant removal, municipal wastewater treatment by AnMBR has received intensive attention (Aslam et al. 2022). Although challenges exist for methane recovery from low-strength municipal wastewater, pilot and full scale AnMBR facilities have been constructed (Kong et al. 2021b).

Through both biodegradation and physical separation, AnMBR can achieve the COD removal percentage above 90% with the permeate effluent COD concentration below 50 mg/L. The conversion proportions of organics in municipal wastewater to methane range between 38 and 97%, depending on the operational conditions such as temperature. In a submerged AnMBR operated at 25 °C, the COD removal efficiency of 89% was achieved when HRT ranging from 6 to 12 h, while shortening HRT to 4 h largely decreased the biogas yield from 0.26 to 0.17 L gas/g COD_{fed} (Ji et al. 2020b). With the HRT of 12 h for treating municipal wastewater, the COD removal efficiencies were about 62% at 19 °C and 75% at 24 °C, respectively (Plevri et al. 2021), confirming the significance of temperature influence on system performance.

Lei et al. (2018) established a correlation between the loading rate and the energy recovery efficiency. The variation of energy recovery efficiency was positively correlated with the sludge loading rate, and the same trend was found between the energy recovery efficiency and the OLR ($\leq$ 2.1 g COD/L/d). The regulation of the system loading rate is feasible for optimizing the energy recovery efficiency. The low organic concentration in municipal wastewater determines the applied OLR generally within 0.2–2 g COD/L/d. Thus, the energy recovered from municipal wastewater is hard to offset the energy consumption for mesophilic or thermophilic operations. This makes ambient temperature operation a common choice for AnMBR treating municipal wastewater, which greatly depresses the methanogenic activity. Also, a low temperature enhances the viscosity of wastewater to accelerate membrane fouling and increases the dissolved methane in the liquid phase (Zhang et al. 2022), which are negative for efficient biogas recovery.

3.2 *Pilot-Scale Applications for Municipal Wastewater Treatment*

McCarty et al. (2011) proposed that AnMBR could be an important technology for sustainable municipal wastewater treatment. Currently, many pilot-scale attempts have demonstrated promising performance of AnMBR regarding treatment efficiency, energy recovery, and operating costs, and proposed the challenges and potential obstacles toward full-scale applications. For pilot-scale AnMBR treating municipal wastewater, key factors influencing system performance include temperature, organic substrate concentration, membrane flux, and membrane fouling.

Martinez-Sosa et al. (2011) reported on a pilot-scale study using external AnMBR to treat municipal wastewater. The working volume was 350 L. A flat sheet membrane module with a total membrane surface of 3.5 m^3 was used. The treatment capacity was approximately 450 L/d. The system was started up under mesophilic condition of 35 °C, which was gradually decreased to 28 and 20 °C during the stable operation phase. A COD removal efficiency of 90% was achieved with a guaranteed effluent COD concentration of $\leq$ 80 mg/L under all conditions. Moreover, this system demonstrated a considerable pathogen removal capacity (5 logs reduction), and the produced effluent can be used for irrigation without any post-treatment. The methane yield was hardly influenced by temperature variation and maintained at 0.2–0.3 L methane/g COD. However, accelerated membrane fouling was observed, potentially attributed to the higher viscosity and soluble COD in the mixed liquor.

Peña et al. (2019) constructed a submerged AnMBR with a 496 L working volume for municipal wastewater treatment. After one-year operation under ambient temperature, temperature was found to be the key parameter influencing system performance. In summer, 90% COD removal efficiency and 0.09–0.14 $Nm^3/kg\ COD_{removed}$ methane yield could be achieved steadily under a high OLR of 5 g COD/L/d and HRT of 8–10 h. However, to maintain similar COD effluent quality (about 15 mg COD/L), the OLR in winter had to be decreased to 1 g COD/L/d, making the energy production rate negligible. In general, the effluent met the Spanish standards for agricultural irrigation, demonstrating the economic benefits of this system.

Lim et al. (2019) comprehensively studied the long-term performance of a pilot-scale AnMBR for municipal wastewater treatment at ambient temperature. When the influent temperature ranged from 12.7 to 31.5 °C (average 18.8 °C), the effluent COD concentration stabled at 58 mg/L, and the average CH_4 yield was 0.14 L/g COD. However, the relatively low temperature resulted in 42% methane dissolved in the effluent, which should be collected by post-treatment. Temperature fluctuations hardly influenced dominant microbial communities in the system, including *Bacteroidetes*, *Proteobacteria*, *Firmicutes*, and *Chloroflexi* at the phylum level. Furthermore, acetoclastic *Methanosaeta* was the dominant archaea. Nonetheless, the occurrence of sulfate reduction in the system illustrated the potential competition between sulfate-reducing bacteria (SRB) and methanogens.

Giménez et al. (2014) established a pilot scale submerged AnMBR with a designed treatment capacity of 28.8 m^3/d and reported the variation of COD/sulfate ratio in municipal wastewater on system performance. The results showed that stable COD removal efficiency of about 87% could be achieved with the COD/sulfate ratio varying from 2 to 12, but this variation greatly influenced methane yield. Increased the sulfate ratio in wastewater stimulated SRB activity for competing electron donors with hydrogenotrophic methanogens, thereby decreasing methane production. Further study exhibited that a seasonal temperature decline did not disturb COD removal efficiency and membrane permeate flux, but mainly decreased methane yield and generated more sludge. Especially, the methane yield greatly decreased from 82.1 L $CH_4/kg\ COD_{removed}$ in summer to 15.4 L $CH_4/kg\ COD_{removed}$ in winter. The lower activities of hydrolytic bacteria and methanogenic archaea, and a larger proportion of dissolved methane concurrently determined the lower methane yield in winter. Robles et al. (2022) showed that the COD/sulfate ratio in wastewater governed the organic stream direction to methanogens and SRB, and the COD mass balance showed that 11.6–43.5% COD flowed to methane, depending on the conditions of temperature, HRT, and COD/sulfate ratio in wastewater. For membrane fouling control, the gas sparging rate (based on the specific gas demand per permeate volume, SGD_P) between 10 and 20 Nm^3 stabled membrane permeate flux between

15 and 25 LMH without requiring chemical cleaning. For more than half of whole operation time, the system could be a net energy producer. The recovery efficiencies of the dissolved methane in the effluent had a great influence on both the net energy production and greenhouse gases (GHG) emission (Jiménez-Benítez et al. 2023).

In a local wastewater treatment plant in Sendai, Japan, a submerged AnMBR was constructed for treating municipal wastewater (Fig. 4). The working volume of the system was 5 m^3. After the system was started up successfully with sugar as the co-substrate, it was then operated at 25 °C with real municipal wastewater as the sole substrate. With HRT ranging from 48 to 6 h over two-hundred days, the system showed stable COD removal with efficiencies over 90%. The methane yield was 0.19–0.22 L/g $COD_{removed}$ (Kong et al. 2021a). *Methanosaeta* was confirmed as the pioneer archaea responsible for methane production, contributing to the stability of system operation together with functional bacterial consortium (Kong et al. 2022). For the suspended solids in wastewater, 57–66% removal was achieved by biodegradation, while the residue existed in the sludge. This resulted in the sludge yield based on mixed liquor suspended solids of 0.19–0.26 g/g $COD_{removed}$ (Kong et al. 2021c). Moreover, chemical cleaning by citric acids and NaClO was applied to keep the stable membrane flux between 2.8 and 17.8 LMH. Theoretical calculation regarding energy balance suggests that although this AnMBR required net energy consumption, which was still much lower than a conventional activated sludge system. Following that, when temperature dropping from 25 to 15 °C, the methane yield decreased from 0.244 to 0.205 L CH_4/g $COD_{removed}$. This was attributed to the increase ratio of dissolved methane from 29 to 43% and reduced methanogenic activity by 25%. Although microbial activity depressed following decreased temperature, physical membrane rejection could still ensure COD removal efficiency of 84.5–90.0%. The stimulated microbial products secretion, increased mix liquor viscosity, and accumulated inorganic foulants concurrently aggravated membrane fouling (Rong et al. 2022a). The energy efficiency of a scaled-up AnMBR with a treatment capacity of 10,000 t/d at both seasonal and annual time-scale was assessed, and underlined that when the dissolved methane was recovered effectively, the annual net electricity demand could be as low as 0.09 kWh/m^3 (Rong et al. 2022b). Meanwhile, effective energy recovery from methane can reduce GHG emission from 0.341 to 0.142 kg CO_2eq/m^3. Consequently, these outcomes shed valuable lights on the future of full-scale AnMBR treating municipal wastewater in temperate climate regions (Rong et al. 2023).

An engineered pilot scale municipal wastewater treatment project with AnMBR as the core unit was constructed in Spernal wastewater treatment plant in UK, and a degassing unit was integrated for the recovery of the dissolved methane (Fig. 5) (Kim et al. 2022). The AnMBR treatment capacity was averaged at 200 m^3/d and maximized to 500 m^3/d. An up-flow anaerobic sludge blanket (UASB) reactor inoculated with mesophilic industrial granular sludge was conducted with an up-flow velocity of 0.8 m/h and an HRT of 8–10 h. The three-polyethylene hollow fiber ultra-filtration membrane reactor had a total membrane area of 1074 m^2 and was sparged with the biogas produced in the UASB reactor. The HRT in the membrane reactor was 1.3 h and the flux was 10 LMH. The membrane contactor unit employing a microporous

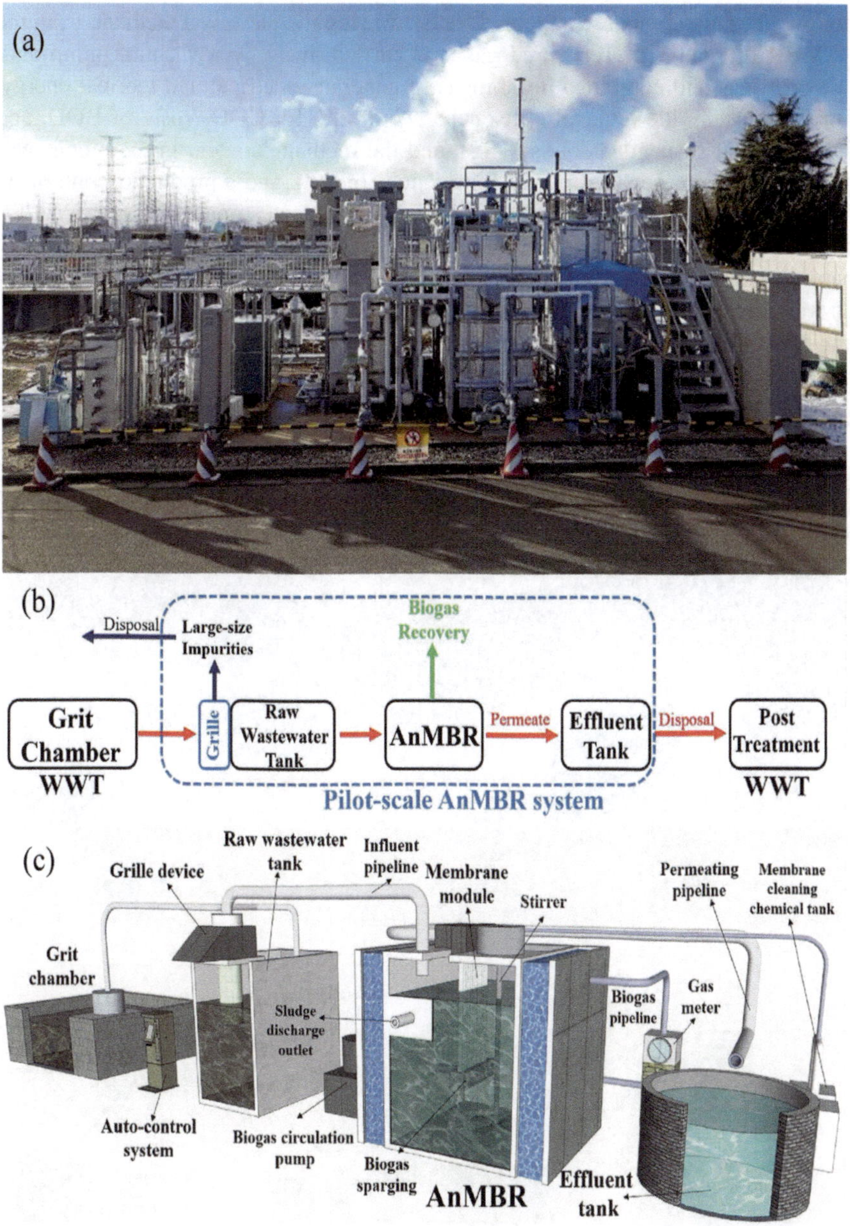

Fig. 4 Photograph and process units of the large pilot-scale AnMBR in Sendai, Japan (Kong et al. 2021b)

hollow fiber membrane was adopted for recovering the dissolved methane from the AnMBR effluent. The recovered gas was carried away by a vacuum pump and/or sweep gas and combined into biogas to generate electrical and thermal energy. For this system, the removal efficiencies were 60% for COD, 76% for BOD, and 71% for total suspended solids (TSS), and the methane content in the biogas was observed to increase steadily, reaching values of 60% and a production rate of 63 L/h. The energy demand for all the process units was estimated at 0.52 kWh/m^3 and the energy production from the biogas was 0.09 kWh/m^3. The overall energy balance was estimated at 0.43 kWh/m^3. As the COD of the influent (average 171 mg/L) was particularly low in this trial, the energy balance was significantly impacted, and energy neutrality was not achieved.

Some emerging AnMBR configuration or operation model have been developed for facilitating system performance. A 0.77 m^3 submerged AnMBR system in

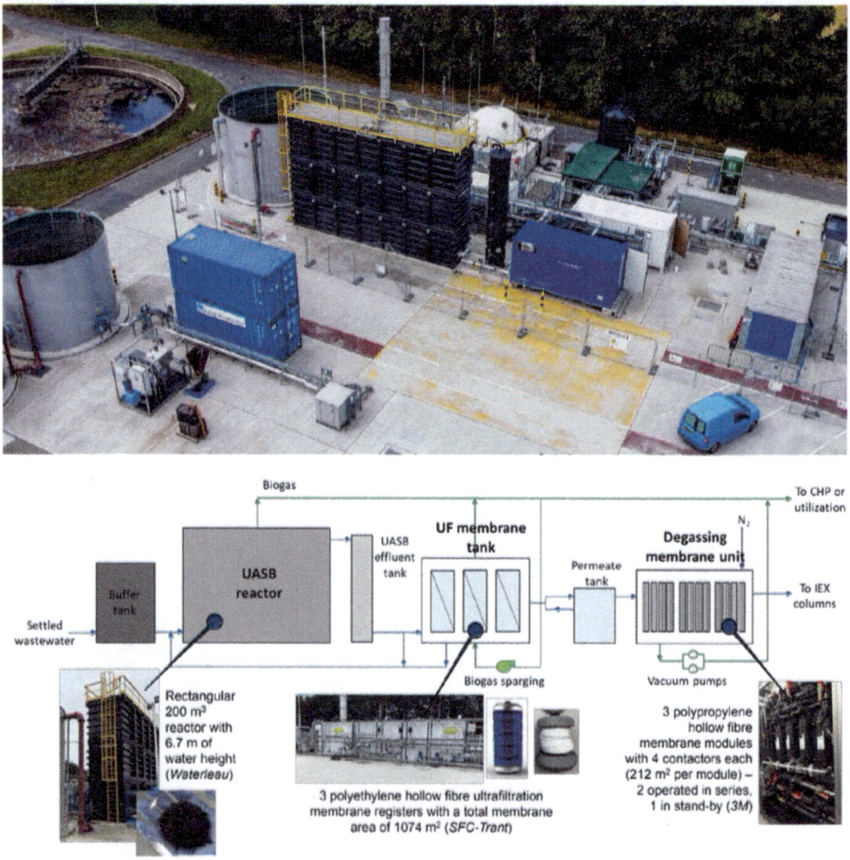

Fig. 5 Areal picture showing the demonstrated AnMBR system including the anaerobic membrane reactor and degassing unit (top) and the schematic representation of the process (bottom) (Kim et al. 2022)

Bucheon (South Korea) municipal sewage treatment plant (Shin et al. 2014) consisted of an anaerobic fluidized bed reactor (AFBR) and submerged fluidized AnMBR. Granular activated carbon (GAC) was added into AFBR and AnMBR. During the 1st year, COD removal efficiency maintained at 81–94%, which was lowest in winter. Nevertheless, in the 2nd year, the COD removal efficiency in winter was promoted to more than 90%, indicating the importance of long-term acclimation for microbes to resist low temperature. Energy balance showed that the gross energy consumption equivalent was 0.23 kWh/m^3 wastewater, which was higher than the energy production of 0.14 kWh/m^3. Meanwhile, the scouring effects of GAC on membrane surface mitigated membrane fouling, and no membrane cleaning was conducted during 485 days' operation. Based on the results, further system modification was developed, consisting of two sequential GAC-assisted AFBRs and a membrane tank (membrane modules of ZeeWeed 500D, SUEZ). The total volume of this system was 4.5 m^3. Bio-carrier enhanced the retaining of microbes for improving COD conversion to methane in the two AFBRs, and the energy recovered from methane exceeded energy consumption of GAC fluidization and gas sparging (Shin et al. 2021). With a GAC-assisted AnMBR treating municipal wastewater at the capacity of 10 m^3/d (Chen et al. 2021) (Fig. 6), the GAC addition in the anaerobic digestion unit was beneficial for maintaining the stable COD removal and CH_4 yield under low temperature condition (10 °C), which could be due to that the introduction of GAC significantly increased the relative abundance of *Methanosaeta* and *Geobacter*, suggesting that direct interspecies electron transfer (DIET) stimulation was a potential mechanism responsible for stable CH_4 production under low temperature conditions.

4 Applications of AnMBR for Treating Industrial Wastewater

4.1 Industrial Bio-Waste Streams

Numerous attempts of AnMBR treating food waste and food (or drinking) processing wastewater have achieved excellent performance regarding energy recovery. The reported OLR ranged from 1.5 to 20 g COD/L/d, with the COD removal efficiency steadily higher than 90%. For treating the food waste with high oil/grease contents, Ramos et al. (2014a) confirmed the advantages of AnMBR by comparing the associated performance with UASB, and at the OLR of 5.1 g COD/L/d, AnMBR achieved 97% COD removal and 100% oil/grease removal. The increase of oil/grease contents mainly caused filtration resistance but hardly declined the critical membrane flux, guaranteeing the stable operation of AnMBR with a wide range oil/grease contents of 4.6–36 g/L (Ramos et al. 2014a). Due to the rapid reaction rate of hydrolysis and acidogenesis, volatile fatty acids (VFA) accumulation is a potential issue of AnMBR treating food waste under high OLR. The accumulated VFA and subsequently deteriorated methane production were likely attributed to the thermodynamic restraints

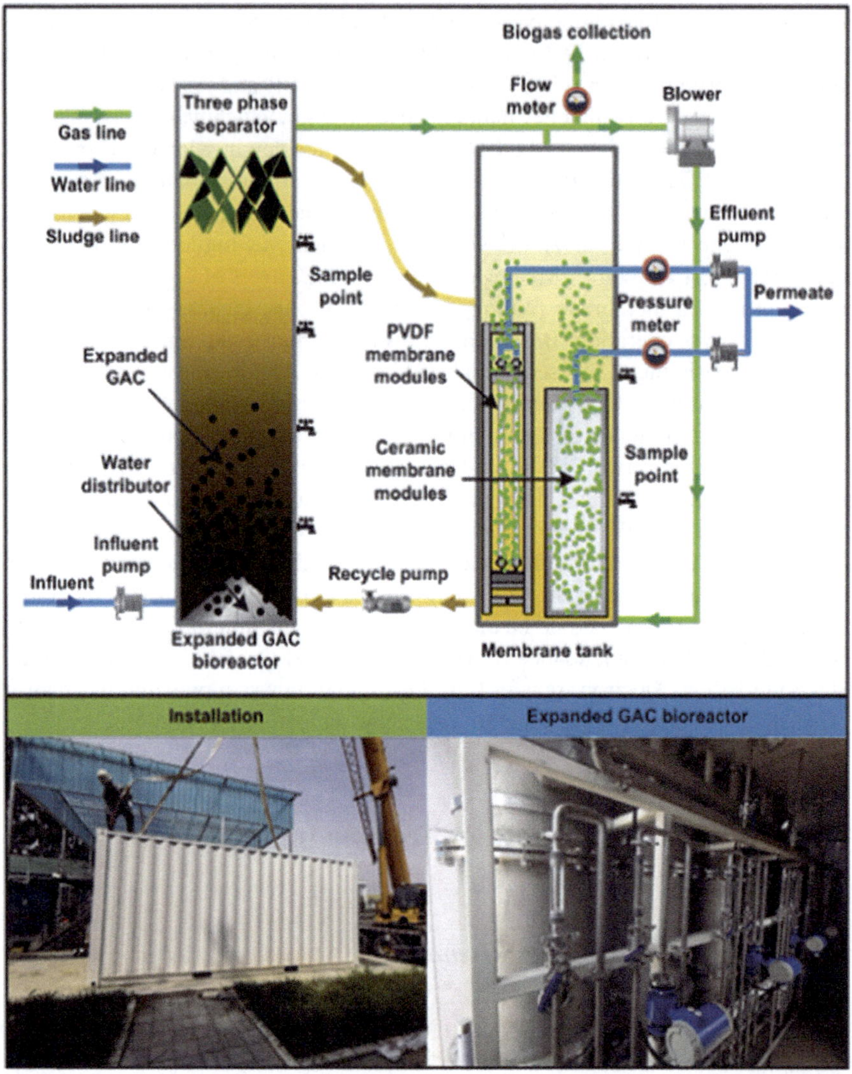

Fig. 6 Large pilot-scale GAC-AnMBR system for municipal wastewater treatment (Chen et al. 2021)

of VFA degradation and the excessive proliferation of VFA-producing bacteria (Wu et al. 2022). For AnMBR operated under OLR from 4.7 to 9.3 g COD/L/d, no VFA accumulation occurred and the adequate substrate for microbes could increase the protease and lipase activities, thereby resulting in a high methane yield (Xiao et al. 2017a). While Qiao et al. (2016) found that when OLR increased to 23.6 g COD/L/d, VFA accumulation and pH decrease occurred, causing the COD removal efficiency to decline to 44.5%. Adding sulfate to stimulate the activity of SRB with propionate as the electron donor is an effective way to recover methanogenesis, which could maintain system stability under this high OLR (Qiao et al. 2016). Also, supplement of NH_4HCO_3 could alleviate pH decrease and recover methanogenic activity (Qiao et al. 2013). Regarding system configuration, AnMBR with two phases of separating hydrolysis-acidogenesis and acetogenesis-methanogenesis could improve both hydrolysis process and accelerate VFA accumulation in first phase, and enrich syntrophic partner to boost methanogenesis in the second phase, showing better performance than the AnMBR with the single-phase mode (Amha et al. 2019).

Livestock wastewater (or bio-waste) derived from husbandry industry contains animal feces, urine, feeding residue and flushing water, thereby possessing not only high contents of suspended organics, ammonia, and phosphorus, but also the emerging contaminants (EC) including antibiotics, heavy metals, and antibiotic resistance gene (ARG) (Lou et al. 2020). These give an opportunity for AnMBR to treat livestock wastewater for energy recovery, nutrients-rich digestate generation, and EC removal. Due to the wide range of COD contents (1–42 g/L) of livestock wastewater, the OLR between 1 and 21 g COD/L/d could achieve acceptable methane yield and maintain stable permeate flux. Tang et al. (2022) applied mesophilic AnMBR for swine wastewater treatment, and at HRT of 15 days, the highest methane yield of 0.24 L/g $COD_{removed}$ was achieved. However, the high suspended solids in livestock wastewater limited the further increase in OLR because the aggravated membrane fouling by cake layer formation reduced the permeate flux significantly. Tang et al. (2022) configured a high-rate AnMBR treating swine wastewater. By shortening the SRT/HRT ratios from 18 to 3.2, stable permeate under an extremely low HRT of 1 day was achieved at the OLR about 17 g COD/L/d. With high-strength swine manure as substrate, Wang et al. (2023a) confirmed that the deteriorated OLR was as high as 21.1 g COD/L/d. The methane yield was not influenced dramatically under this high OLR, but the critical membrane flux could not meet the effluent volume in the low HRT of 5 days because of the rapid membrane fouling rate. Another potential issue of livestock manure stream during anaerobic digestion is the toxicity of ammonia content to microbial activity. Stripping ammonia was a feasible way to remove more than 90% ammonia during AnMBR treating chicken manure (Busato et al. 2020). Moreover, the presence of heavy metals, residue antibiotics and ARG in livestock likely disturbed the methanogenic activity (Yang et al. 2022).

Landfill leachate is generated during landfilling of municipal solid wastes, which contains refractory substances and high concentrations of ammonia. The COD concentration of landfill leachate treated by AnMBR ranged between 1 and 85 g/L. Under mesophilic conditions, the COD removal efficiency ranged between 60 and 98% due to the low biodegradability of landfill leachate. Usually, both long HRT (1.5–11.8 days) and SRT (20–70 days) resulted in stable COD removal efficiency and acceptable methane recovery rate (Issa et al. 2022). Moreover, for the same source leachate, the age greatly determines its biodegradability, and for the fresh leachate, the BOD/COD ratio higher than 0.4 is recommended for AnMBR treatment (Abuabdou et al. 2020). The aging leachate contains mainly recalcitrant organic carbon, thereby declining its methane yield when treated by AnMBR. A high methane yield of 0.34 L/g $COD_{removed}$ was obtained when treating fresh landfill leachate (Xie et al. 2014), while when co-digestion of aged landfill leachate and degradable wastewater, biogas production decreased with increasing ratios of the landfill leachate (Nabi et al. 2022). The high content of ammonia generated from proteaceous substances in landfill leachate, especially the free ammonia, also hinders microbial activity. Furthermore, the sulfate in landfill leachate could cause competition between SRB and methanogens on electron donors. Adding $FeCl_3$ was useful to mitigate these negative influences on methane production and remove sulfate simultaneously.

4.2 Full-Scale Applications to Treat Industrial Wastewater

In January 2006, Kubota Company built an external AnMBR system with a treatment capacity of 15 tons/d for stillage wastewater treatment in Naha, Okinawa (Grant et al. 2008). The system was operated at 55 °C and an OLR of 11 g COD/L/d. The methane content of the biogas was steadily maintained between 63% and 68%, accompanied by a methane yield of 294 mL/g COD(influent) and COD removal efficiency exceeding 95%. To mitigate membrane fouling, a citric acid solution was used for regular membrane cleaning (once every 1–3 months). In March 2007, Denen-Shuzo Company (Japan) built a 20 tons/d AnMBR system for distillery wastewater treatment. By integrating the AnMBR-aerobic MBR processes, this project achieved energy recovery and nitrogen/phosphorus removal from wastewater simultaneously. The advantages for full-scale application of AnMBR are: (1) AnMBR integrated the advantages of biochemical degradation and membrane filtration, guaranteeing low turbidity and COD effluents; (2) stable operation under high OLR (10–15 g COD/L/d) significantly reduces the system footprint; (3) large amounts of suspended solids and lipids are directly fed into AnMBR for digestion, simplifying the treatment process; and (4) no need to consider sludge granulation, which makes system start-up and operation easily (Grant et al. 2008).

In 2008, an ADI® AnMBR established by Evoqua was selected as a technology for upgrading the salad dressings and marinades manufacturing wastewater treatment system of *Ken's Food* in Marlborough (MA, USA) (Allison et al. 2013; Christian et al. 2011). This external AnMBR system consisted of an 8300 m^3 anaerobic digestion unit and four membrane filtration tanks. The designed treatment capacity of this system was 475 m^3/d, and the actual wastewater flow treated was approximately 370 m^3/d. During five years of operation, the average BOD and COD concentrations of raw wastewater were 18.0 and 36.4 g/L, respectively. AnMBR achieved stable BOD and COD removal efficiencies of 99.9% and 99.3%, respectively. Moreover, the fat/oil/grease (FOG) and TSS removal efficiencies were nearly 100% (effluent TSS concentration < 2 mg/L). Besides that, this system produced 5660 m^3 biogas/d, which was partly used for in situ combustion to maintain the operating temperature of the anaerobic digestion system at 35 °C. The residual biogas was used for household heating supply. Circulation of biogas into the membrane filtration unit remarkably mitigated membrane fouling, which maintained its stable performance over a five-year duration. When the TMP reaches 10 kPa, in situ membrane cleaning is necessary, and the period is about once every three months. Overall, this AnMBR system not only achieved the goal of wastewater treatment but also greatly reduced the operating costs (e.g., sludge dewatering).

Another full-scale ADI®AnMBR was in south-central Pennsylvania treating confectionary manufacturing wastewater. The configuration of this system included a completely stirred tank reactor and two membrane tanks. The designed treatment capacity of this AnMBR was 170 m^3/d. The raw wastewater contained 16.6 g/L COD, 8.0 g/L BOD, and 10.0 g/L TSS. For one year operation, the system achieved COD and BOD removal efficiencies higher than 97%, and complete TSS removal. The membrane cleaning period was half a year (Allison et al. 2013).

Veolia also promotes AnMBR development toward full-scale application globally, and has developed the technology named Memthane®. In Alpena (SD, USA), Memthane® solved the issue of low removal efficiency of high-strength oil and grease pollutants in wastewater from the food industry and achieved CHP by integrating biogas desulfurization (sulfothane®). At the Arla Foods dairy processing plant in Aylesbury (UK), a dairy wastewater treatment project achieved a "carbon–neutral" goal with Memthane® (Atkinson 2018). This system was designed to treat 500 m^3 wastewater per day accompanied with 5.5 t COD removal and produced 0.6–0.7 Nm^3 biogas per kg COD removed. At the Mars food processing plant in Veghel (Netherlands), a wastewater treatment system with Memthane® as core technology not only decreased the treatment cost but also achieved a CO_2 equivalent reduction of 150 t/year (https://www.veoliawatertechnologies.com/en/products/memthane). Accordingly, Veolia declared that Memthane® is an ideal solution for high concentrated streams with 15–25 g COD/L, such as dairy whey, the ethanol facilities generated bio-waste, the streams containing FOG such as ice cream, dairy and biodiesel, and the starch slurries.

5 Future Perspectives

5.1 Carrier-Assisted AnMBR

Although AnMBR has been widely used for wastewater treatment, membrane fouling is still a challenge compared with the conventional anaerobic digestion technology. Development of low-cost and effective measurements for in situ controlling membrane fouling is of paramount significance regarding practical applications of AnMBR. With progresses in understanding of anaerobic pollutants removal principles and membrane fouling mechanisms, the addition of functional materials as carriers into AnMBR has been developed as a feasible strategy to cope with the above-mentioned challenges, and both carbon-based and Fe-based materials have been widely investigated (Nabi et al. 2023; Sohn et al. 2021).

The addition of carbon-based materials such as activated carbon and biochar into AnMBR could improve permeate effluent quality, enhance methane yield, and mitigate membrane fouling. First, the porous structure on carbon-based materials could provide attachment sites for microbial growth, protecting them from unfavorable conditions such as ammonia inhibition, antibiotic toxicity, and low temperature. Moreover, some functional organic groups (e.g., phenazine and quinone) and graphite-like micro-structures in these materials could mediate DIET between syntrophic partners, which could not only accelerate the conversion of VFAs to methane, but also drive the degradation of phenolic compounds (Chen et al. 2023a). Biochar could also promote VFA conversion to methane under 10 °C, leading to a 12.5% increase in the methane yield (Lei et al. 2021a). Biochar addition could shorten the start-up of AnMBR and maintain COD removal efficiency higher than 95% at the OLR as high as 21.1 g COD/L/d. Meanwhile, carbon-based materials can shape microbial communities to improve system performance. When treating swine wastewater, Sun et al. (2022) found that the addition of spent coffee ground biochar into an AnMBR could enrich the potential DIETers of *Anaerolinea*, *Methanosarcina*, and *Methanosaeta*, promoting methane yield by 17.4%. Furthermore, biochar could alter the dominant methanogenic pathway from the acetoclastic methanogenesis to the hydrogenotrophic methanogenesis (Wang et al. 2023a). The addition of carbon-based materials into AnMBR treating pharmaceutical wastewater can facilitate the biotransformation of organic halogen and antibiotic pollutants by the adsorption-degradation mechanism. Xiao et al. (2017b) clarified that introducing powder activated carbon (PAC) altered the adsorption surface of pharmaceutical compounds from sludge flocs to PAC, thereby leading to a higher removal efficiency.

The addition of carbon materials can scour membrane surface to alleviate foulants deposition. Carbon-additive materials can adsorb SMP and EPS to alleviate membrane blocking, and may acclimate microorganisms for restraining the cake layer formation and improving the biofouling layer characteristics with higher permeability (Lei et al. 2019, 2021b). For carbon-based materials, GAC could greatly increase EPS contents in the mixed liquor but mitigate membrane fouling by reducing bio-solids deposition process, while PAC achieved the same goal by decreasing the particle size of sludge flocs (Yang et al. 2019a). Chen et al. (2023c) found that after biochar addition, a more homogenous and condensed gel layer was formed by changing the functional groups and protein structures of foulants from lower binding forces ones to higher binding forces ones, contributing to an accumulative permeate flux increase by 2.6 times (Fig. 7). Overall, adding carbon-based materials into AnMBR is considered a sustainable strategy in mitigating biofouling of membrane processes.

Fe-based materials, especially zero-valent iron (ZVI) are also used in AnMBR to promote methane production and mitigate membrane fouling. ZVI with various forms can improve AnMBR performance, while the over dosage of nano-ZVI may generate reactive oxygen species and inhibit methanogenesis (Wu et al. 2023a). In principle, ZVI can provide electron to facilitate the breakage of chemical bonds in various kinds of organic pollutants to drive the biodegradation process, and these extra electrons can be captured by the methanogens with the DIET-based pathway to

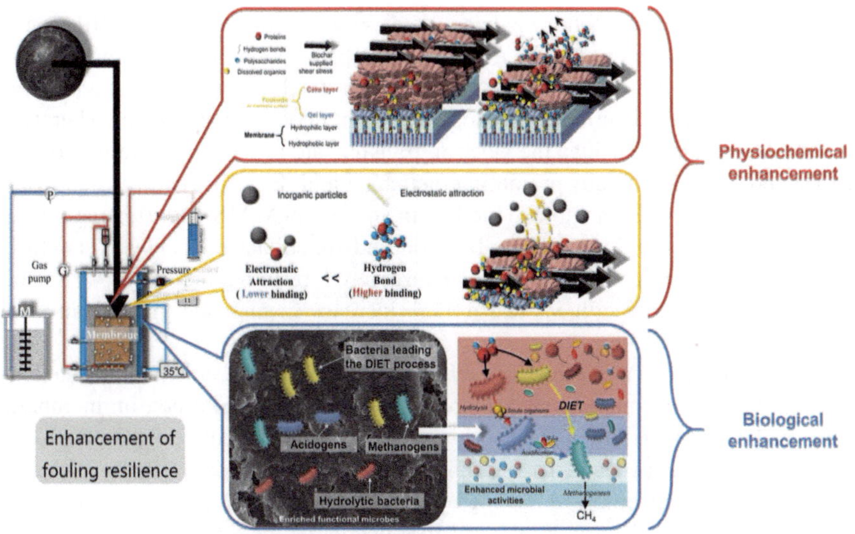

Fig. 7 Physiochemical and biological mechanisms of biochar mitigating membrane fouling (Chen et al. 2023c)

increase methane yield. The flocculant property of ZVI improves sludge flocculation in AnMBR. Zhang et al. (2020) confirmed that the addition of ZVI enlarged the average sludge flocs size by 77.3%, thus mitigating membrane fouling by 20%. Furthermore, the released iron ions from ZVI likely induce complexation or co-precipitation with EPS, inorganic compounds, and other soluble foulants, alleviating membrane fouling by reducing the rate of gel layer formation.

5.2 *Electricity-Assisted AnMBR*

Electricity-assisted AnMBR (electro-AnMBR) is an advanced technology that integrating electrochemical process with AnMBR to enhance the system performances regarding pollutants removal and energy recovery, and to mitigate membrane fouling (Chen et al. 2023b). The configurations of electro-AnMBR varied depending on position nexus of electrodes and AnMBR. For external AnMBR, the pair of electrodes can be placed directly into the anaerobic digestion tank to promote anaerobic digestion performance or placed into membrane tank to control membrane fouling. For submerged AnMBR, both anode and cathode are placed into the system, and either anode or cathode is used by fabricating membrane with conductive material and the other electrode is placed in the suspended sludge system.

Yang et al. (2019b) applied a carbon nanotube-fabricated hollow fiber membrane as cathode in electro-AnMBR and a high COD removal efficiency (> 95%) was achieved, where the cathodic membrane surface enriched more hydrogenotrophic methanogens, inducing better methane yield. When using electro-AnMBR to treat methanolic-containing wastewater, the methane production rate increased from 5.0 to 5.4 L/L/d with the voltage varied from 0 to 0.6 V, while further elevating voltage to 0.8 V negatively disturbed methanogenic activity (Wang et al. 2023b). Therefore, a suitable voltage intensifies the electron transfer rate of electro-active microbe on anode and cathode, thereby promoting methane yield. By constructing a sandwich-type composite anodic membrane module in an electro-AnMBR for treating sewage sludge and food waste, hydrolytic bacteria and functional methanogens could be enriched and the potential DIET could be triggered, resulting in a lower VFA accumulation in permeate effluent and a 12% increase in methane yield (Han et al. 2023).

Electro-AnMBR can mitigate membrane fouling, and the potential mechanisms include the improvement of sludge property with a weak adhesion on membrane surface, the strengthening of the electrostatic repulsive force between membrane and foulants, and the enhancement of fouling layer with a higher permeate flux. Electro-AnMBR mitigates membrane fouling by regulating EPS secretion of microbes, and the introduction of electricity into AnMBR decreases the total concentrations and ratio of protein/polysaccharide (PN/PS) of EPS, thereby resulting in fewer PN-EPS foulants with positive charge deposited on membrane surface. Meanwhile, the lower PN/PS ratio of EPS increases the hydrophilicity of microbes, leading to lower attachment tendency on membrane (Wang et al. 2022a). Cathodic membrane is a promising strategy for controlling membrane biofouling because in principle, the lower voltage potential on membrane surface strengthens the electrostatic repulsive force with microbes and hinders microbial deposition behaviors. Accordingly, in an electro-AnMBR using Mg as anode, accompanied with the decreases of zeta potential, viscosity, and EPS contents of sludge, a more hydrophilic cake layer formed, thereby mitigating membrane fouling by 30%. Consequently, substituting cathodic membrane into regular membrane extended membrane filtration time by 110% (Zhao et al. 2021). Furthermore, the enhanced electrostatic repulsion and gas sparging generated from hydrogen contribute to anti-fouling. Ying et al. (2021) found that −1.0 V potential mitigated membrane fouling by the formed higher porosity of gel layer, −2.0 V potential led to lower TMP rising rate by inactivating adhesive microbe, and −0.5 V potential led to a severer membrane fouling due to formation of a denser microbial cake layer. Compared with cathodic membrane, mitigating membrane fouling by anodic membrane is an emerging research topic. When fabricating membrane surface with positive charge, although a rapid pollutants accumulation occurred in the initial stage because of electrostatic attraction, the electroactive microbes enriched on membrane surface remarkably promoted pollutants degradation (Yu and Zhang 2019). The stable biofilter formed on the membrane surface prevented EPS accumulation, resulting in system's long-term stable operation (Yu and Zhang 2019). Although numerous studies confirmed the benefits of electro-AnMBR for membrane fouling control, the sophisticated mechanisms are still in debate regarding varied substrate type, electrical voltage stress, and system configurations.

5.3 AnMBR for Emerging Contaminants Removal

The feasibility of using AnMBR for removing EC, e.g., PPCPs, antibiotics, virus, steroid hormones and endocrine-disrupting compounds, pesticides, and micro/nanoplastics, has been confirmed. In municipal wastewater, although the EC content is extremely low (at levels of μg/L-ng/L), AnMBR showed good removal efficiency in terms of EC (Zhang et al. 2021). A series of behaviors of EC regulated its transformation synergistically, including sorption on sludge, biodegradation by microbes, and deposition in membrane pore and the associated fouling layer.

For the sorption process, the characteristics of hydrophobicity, electrostatic forces, and hydrogen bond strength of EC determine the interaction between EC and sludge. For example, for the EC containing positively charged groups, it can be adsorbed onto the negatively charged surface of biomass by electrostatic interaction force, which can be characterized by the index of acid dissociation constant. Moreover, for some hydrophobic EC, a temporary transformation from liquid phase to the hydrophobic phase of biomass, e.g., the lipid fraction, might occur, resulting in an enhanced biodegradation by microbes.

For the degradable EC, the biodegradation pathway includes mainly co-metabolism, because the low EC concentration cannot support microbial growth. Co-metabolism with other carbon sources contributes greatly to the removal of EC. In this process, EC are converted into more bio-transformable intermediates by specific bacteria and oxidoreductase enzymes.

The function of membrane separation to retard EC is contributed to both membrane itself and the fouling layer. For membrane itself, the retarding capacity is determined by pore size, hydrophobicity, and electrostatic properties of membrane materials. Membrane fouling in MBRs has been shown to have a positive correlation with EC retention (Sawaya and Harb 2021; Sanguanpak et al. 2019). The type of EC compound significantly influences its retention in membrane biofilms, with hydrophobic EC exhibiting higher affinity for fouling layers, particularly due to interactions with EPS. Although pore blockage by EPS and SMP can aid in EC rejection, it can also lead to increase TMP. Research on EC retention in AnMBRs revealed that the majority of EC removed in the fouling layer were hydrophobic, including compounds like 17α-estradiol, bisphenol A, and triclosan, with biofilm retention levels ranging from 10 to 80 mg/m^2 of membrane surface area. The contribution of membrane retention to ARG removal was confirmed by Zarei-Baygi et al. (2019). A further study of this research group proposed the opposite impacts of AnMBR on intracellular and extracellular ARG (iARG and eARG) removal, where membrane with high fouling level effectively decrease the eARG in the effluent, but had little impacts on retarding the iARG level (Zarei-Baygi et al. 2020). Cheng and Hong (2017) explored the degree of membrane fouling on antibiotic resistance bacteria (ARB) and ARG retarding. For ARB, membranes with a minimal foulant layer could remove more than 5 log units, while the efficiency declined as increasing TMPs. For ARG, the fouling layer contributed mostly to its removal. As the TMP increase and the formation of fouling layer, the ARG showed a tendency from suspended sludge to fouling layer (Cheng and Hong 2017).

6 Take-Home Messages

(1) AnMBR emerges as a promising option in the wastewater treatment industry, since it can not only effectively treat both high-strength and municipal wastewaters, but also demonstrate outstanding performance in reducing carbon emissions and increasing resource recovery.
(2) Municipal wastewater treatment with AnMBR encounters challenges such as low organic content and operating at low temperatures, hindering its widespread application. Optional strategies for future development include but not limits in integrating wastewater concentrating as a pretreatment unit and dissolved methane recovery as post-treatment with the AnMBR core unit.
(3) In industrial wastewater treatment, there is a need for technological innovation to enhance the treatment efficiency of AnMBR. This includes optimizing the anaerobic digestion unit configuration to improve methane recovery efficiency and developing new membrane materials to resist membrane fouling.
(4) The removal of EC by AnMBR is a growing research focus due to the global concerns of environmental health. The next generation of AnMBR with targets of resource recovery from bulk organics and environmental risk abatement of EC is encouraged.

References

Abuabdou SM, Ahmad W, Aun NC, Bashir MJ (2020) A review of anaerobic membrane bioreactors (AnMBR) for the treatment of highly contaminated landfill leachate and biogas production: effectiveness, limitations and future perspectives. J Clean Prod 255:120215

Allison M, Grant S, Christian S, Wilson D (2013) Full-scale operating experience with USA-based ADI-AnMBR systems for food wastes. Proc Water Environ Fed 10:5255–5270

Amha YM, Corbett M, Smith AL (2019) Two-phase improves performance of anaerobic membrane bioreactor treatment of food waste at high organic loading rates. Environ Sci Technol 53(16):9572–9983

An Z, Zhu J, Zhang M, Zhou Y, Su X, Lin H, Sun F (2023) Anaerobic membrane bioreactor for the treatment of high-strength waste/wastewater: a critical review and update. Chem Eng J 470:144322

Aslam A, Khan SJ, Shahzad HM (2022) Anaerobic membrane bioreactors (AnMBRs) for municipal wastewater treatment-potential benefits, constraints, and future perspectives: an updated review. Sci Total Environ 802:149612

Atkinson S (2018) Veolia helps UK dairy reduce its impact on the environment. Membr Technol 1:8

Bilad MR, Mezohegyi G, Declerck P, Vankelecom IF (2012) Novel magnetically induced membrane vibration (MMV) for fouling control in membrane bioreactors. Water Res 46(1):63–72

Bolton G, LaCasse D, Kuriyel R (2006) Combined models of membrane fouling: development and application to microfiltration and ultrafiltration of biological fluids. J Membr Sci 277(1–2):75–84
Busato CJ, Da Ros C, Pellay R, Barbierato P, Pavan P (2020) Anaerobic membrane reactor: biomethane from chicken manure and high-quality effluent. Renew Energy 145:1647–1657
Charfi A, Amar NB, Harmand J (2012) Analysis of fouling mechanisms in anaerobic membrane bioreactors. Water Res 46(8):2637–2650
Chen C, Sun M, Liu Z, Zhang J, Xiao K, Zhang X, Song G, Chang J, Liu G, Wang H, Huang X (2021) Robustness of granular activated carbon-synergized anaerobic membrane bioreactor for pilot-scale application over a wide seasonal temperature change. Water Res 189:116552
Chen L, Fang W, Liang J, Nabi M, Cai Y, Wang Q, Zhang P, Zhang G (2023a) Biochar application in anaerobic digestion: performances, mechanisms, environmental assessment and circular economy. Resour Conserv Recycl 188:106720
Chen M, Li Y, Sun X, Dai R, Zheng J, Wang X, Wang Z (2023b) Recent advances in electrochemical processes integrated with anaerobic membrane bioreactor in wastewater treatment. Chem Eng J 468:143822
Chen Y, Xie H, Wang Y, Cao W, Zhang Y (2023c) Multi-routes and mechanisms of biochar in enhancing the fouling resilience in an anaerobic membrane bioreactor during municipal wastewater treatment. Chem Eng J 474:145759
Cheng H, Hong PY (2017) Removal of antibiotic-resistant bacteria and antibiotic resistance genes affected by varying degrees of fouling on anaerobic microfiltration membranes. Environ Sci Technol 51(21):12200–12209
Christian S, Grant S, McCarthy P, Wilson D, Mills D (2011) The first two years of full-scale anaerobic membrane bioreactor (AnMBR) operation treating high-strength industrial wastewater. Water Pract Tech 6(2):wpt2011032
De Vela RJ (2021) A review of the factors affecting the performance of anaerobic membrane bioreactor and strategies to control membrane fouling. Rev Environ Sci Bio/Tech 20(3):607–644
Diaz-Elsayed N, Rezaei N, Guo T, Mohebbi S, Zhang Q (2019) Wastewater-based resource recovery technologies across scale: a review. Resour Conserv Recycl 145:94–112
Ghyoot WR, Verstraete WH (1997) Coupling membrane filtration to anaerobic primary sludge digestion. Environ Technol 18(6):569–580
Giménez JB, Martí N, Robles A, Ferrer J, Seco A (2014) Anaerobic treatment of urban wastewater in membrane bioreactors: evaluation of seasonal temperature variations. Water Sci Technol 69(7):1581–1588
Grant S, Page I, Moro M, Yamamoto T (2008) Full-scale applications of the anaerobic membrane bioreactor process for treatment of stillage from alcohol production in Japan. Proc Water Environ Fed 7:7556–7570
Grethlein HE (1978) Anaerobic digestion and membrane separation of domestic wastewater. J (Water Pollut Control Fed) 50:754–763
Han Y, Cai T, Yin J, Li W, Li S, Qiu B, Lu X, Zhou Y, Zhen G (2023) Impact of sandwich-type composite anodic membrane on membrane fouling and methane recovery from sewage sludge and food waste via electrochemical anaerobic membrane bioreactor. Biores Technol 382:129222
Hu Y, Cheng H, Ji J, Li YY (2020) A review of anaerobic membrane bioreactors for municipal wastewater treatment with a focus on multicomponent biogas and membrane fouling control. Environ Sci: Water Res Technol 6(10):2641–2663
Hu D, Liu L, Liu W, Yu L, Dong J, Han F, Wang H, Chen Z, Ge H, Jiang B, Wang X (2022a) Improvement of sludge characteristics and mitigation of membrane fouling in the treatment of pesticide wastewater by electrochemical anaerobic membrane bioreactor. Water Res 213:118153
Hu Y, Cai X, Du R, Yang Y, Rong C, Qin Y, Li YY (2022b) A review on anaerobic membrane bioreactors for enhanced valorization of urban organic wastes: achievements, limitations, energy balance and future perspectives. Sci Total Environ 820:153284
Issa L, El Kik O, El-Fadel M (2022) AnMBR technology for landfill leachate treatment: a framework towards improved performance. Rev Environ Sci Bio/Tech 21(2):517–538

Jeong Y, Kim Y, Jin Y, Hong S, Park C (2018) Comparison of filtration and treatment performance between polymeric and ceramic membranes in anaerobic membrane bioreactor treatment of domestic wastewater. Sep Purif Technol 199:182–188

Ji J, Kakade A, Yu Z, Khan A, Liu P, Li X (2020a) Anaerobic membrane bioreactors for treatment of emerging contaminants: a review. J Environ Manage 270:110913

Ji J, Sakuma S, Ni J, Chen Y, Hu Y, Ohtsu A, Chen R, Cheng H, Qin Y, Hojo T, Kubota K (2020b) Application of two anaerobic membrane bioreactors with different pore size membranes for municipal wastewater treatment. Sci Total Environ 745:140903

Jiang CK, Tang X, Tan H, Feng F, Xu ZM, Mahmood Q, Zeng W, Min XB, Tang CJ (2019) Effect of scrubbing by NaClO backwashing on membrane fouling in anammox MBR. Sci Total Environ 670:149–157

Jiang M, Huang J, Li P, Ataa B, Gu J, Wu Z, Qiao W (2023) Optimization of membrane filtration and cleaning strategy in a high solid thermophilic AnMBR treating food waste. Chemosphere 342:140151

Jiménez-Benítez A, Ruiz-Martinez A, Robles Á, Serralta J, Ribes J, Rogalla F, Seco A, Ferrer J (2023) A semi-industrial AnMBR plant for urban wastewater treatment at ambient temperature: analysis of the filtration process, energy balance and quantification of GHG emissions. J Environ Chem Eng 11(2):109454

Kim J, Hofman J, Kenyeres I, Poor-Pocsi E, Kleybocker A, Heinze J, Kraus F, Soares A, Paissoni E, Vale P, Palmer M, Bloemendal M, Beernink S, Ros S, Hartog N, Frijns J (2022) New approaches and best practices for closing the energy cycle in the water sector. https://nextgenwater.eu

Kong Z, Li L, Wu J, Rong C, Wang T, Chen R, Sano D, Li YY (2022) Unveiling the characterization and development of prokaryotic community during the start-up and long-term operation of a pilot-scale anaerobic membrane bioreactor for the treatment of real municipal wastewater. Sci Total Environ 813:152643

Kong Z, Li L, Wu J, Wang T, Rong C, Luo Z, Pan Y, Li D, Li Y, Huang Y, Li YY (2021a) Evaluation of bio-energy recovery from the anaerobic treatment of municipal wastewater by a pilot-scale submerged anaerobic membrane bioreactor (AnMBR) at ambient temperature. Biores Technol 339:125551

Kong Z, Wu J, Rong C, Wang T, Li L, Luo Z, Ji J, Hanaoka T, Sakemi S, Ito M, Kobayashi S (2021b) Large pilot-scale submerged anaerobic membrane bioreactor for the treatment of municipal wastewater and biogas production at 25 °C. Biores Technol 319:124123

Kong Z, Wu J, Rong C, Wang T, Li L, Luo Z, Ji J, Hanaoka T, Sakemi S, Ito M, Kobayashi S (2021c) Sludge yield and degradation of suspended solids by a large pilot-scale anaerobic membrane bioreactor for the treatment of real municipal wastewater at 25 °C. Sci Total Environ 759:143526

Lee EJ, An AK, Hadi P, Yan DY (2016) Characterizing flat sheet membrane resistance fraction of chemically enhanced backflush. Chem Eng J 284:61–67

Lee E, Rout PR, Shin C, Bae J (2019) Effects of sodium hypochlorite concentration on the methanogenic activity in an anaerobic fluidized membrane bioreactor. Sci Total Environ 678:85–93

Lei Z, Yang S, Li YY, Wen W, Wang XC, Chen R (2018) Application of anaerobic membrane bioreactors to municipal wastewater treatment at ambient temperature: a review of achievements, challenges, and perspectives. Biores Technol 267:756–768

Lei Z, Yang S, Li X, Wen W, Huang X, Yang Y, Wang X, Li YY, Sano D, Chen R (2019) Revisiting the effects of powdered activated carbon on membrane fouling mitigation in an anaerobic membrane bioreactor by evaluating long-term impacts on the surface layer. Water Res 167:115137

Lei Z, Ma Y, Wang J, Wang XC, Li Q, Chen R (2021a) Biochar addition supports high digestion performance and low membrane fouling rate in an anaerobic membrane bioreactor under low temperatures. Biores Technol 330:124966

Lei Z, Wang J, Leng L, Yang S, Dzakpasu M, Li Q, Li YY, Wang XC, Chen R (2021b) New insight into the membrane fouling of anaerobic membrane bioreactors treating sewage: physicochemical and biological characterization of cake and gel layers. J Membr Sci 632:119383

Lim K, Evans PJ, Parameswaran P (2019) Long-term performance of a pilot-scale gas-sparged anaerobic membrane bioreactor under ambient temperatures for holistic wastewater treatment. Environ Sci Technol 53(13):7347–7354

Liu J, Eng CY, Ho JS, Chong TH, Wang L, Zhang P, Zhou Y (2019a) Quorum quenching in anaerobic membrane bioreactor for fouling control. Water Res 156:159–167

Liu Z, Zhu X, Liang P, Zhang X, Kimura K, Huang X (2019b) Distinction between polymeric and ceramic membrane in AnMBR treating municipal wastewater: in terms of irremovable fouling. J Membr Sci 588:117229

Lou EG, Harb M, Smith AL, Stadler LB (2020) Livestock manure improved antibiotic resistance gene removal during co-treatment of domestic wastewater in an anaerobic membrane bioreactor. Environ Sci: Water Res Technol 6(10):2832–2842

Lutze R, Engelhart M (2021) Effects of sludge characteristics on the critical flux of an AnMBR for sludge treatment. Chem Ing Tec 93(9):1375–1382

Mahmoud I, Liao B (2017) Effects of sludge concentration and biogas sparging rate on critical flux in a submerged anaerobic membrane bioreactor. J Water Process Eng 20:51–60

Mahmood Z, Cheng H, Tian M (2022) A critical review on advanced anaerobic membrane bioreactors (AnMBRs) for wastewater treatment: advanced membrane materials and energy demand. Environ Sci: Water Res Technol 8(10):2126–2144

Martin Vincent N, Tong J, Yu D, Zhang J, Wei Y (2018) Membrane fouling characteristics of a side-stream tubular anaerobic membrane bioreactor (AnMBR) treating domestic wastewater. Processes 6(5):50

Martinez-Sosa D, Helmreich B, Netter T, Paris S, Bischof F, Horn H (2011) Anaerobic submerged membrane bioreactor (AnSMBR) for municipal wastewater treatment under mesophilic and psychrophilic temperature conditions. Biores Technol 102(22):10377–10385

McCarty PL, Bae J, Kim J (2011) Domestic wastewater treatment as a net energy producer–can this be achieved? Environ Sci Technol 45(17):7100–7106

Mei X, Quek PJ, Wang Z, Ng HY (2017) Alkali-assisted membrane cleaning for fouling control of anaerobic ceramic membrane bioreactor. Biores Technol 240:25–32

Meng F, Zhang S, Oh Y, Zhou Z, Shin HS, Chae SR (2017) Fouling in membrane bioreactors: an updated review. Water Res 114:151–180

Nabi M, Gao D, Liang H, Cheng L, Yang W, Li Y (2022) Landfill leachate treatment by graphite engineered anaerobic membrane bioreactor: performance enhancement and membrane fouling mitigation. Environ Res 214:114010

Nabi M, Liang H, Zhou Q, Cao J, Gao D (2023) In-situ membrane fouling control and performance improvement by adding materials in anaerobic membrane bioreactor: a review. Sci Total Environ 865:161262

Ni BJ, Rittmann BE, Yu HQ (2011) Soluble microbial products and their implications in mixed culture biotechnology. Trends Biotechnol 29(9):454–463

Odriozola M, Lousada-Ferreira M, Spanjers H, van Lier JB (2021) Effect of sludge characteristics on optimal required dosage of flux enhancer in anaerobic membrane bioreactors. J Membr Sci 619:118776

Peña M, do Nascimento T, Gouveia J, Escudero J, Gómez A, Letona A, Arrieta J, Fdz-Polanco F (2019) Anaerobic submerged membrane bioreactor (AnSMBR) treating municipal wastewater at ambient temperature: operation and potential use for agricultural irrigation. Bioresource Technol 282:285–293

Plevri A, Mamais D, Noutsopoulos C (2021) Anaerobic MBR technology for treating municipal wastewater at ambient temperatures. Chemosphere 275:129961

Qiao W, Takayanagi K, Niu Q, Shofie M, Li YY (2013) Long-term stability of thermophilic co-digestion submerged anaerobic membrane reactor encountering high organic loading rate, persistent propionate and detectable hydrogen in biogas. Biores Technol 149:92–102

Qiao W, Takayanagi K, Li Q, Shofie M, Gao F, Dong R, Li YY (2016) Thermodynamically enhancing propionic acid degradation by using sulfate as an external electron acceptor in a thermophilic anaerobic membrane reactor. Water Res 106:320–329

Ramos C, García A, Diez V (2014a) Performance of an AnMBR pilot plant treating high-strength lipid wastewater: biological and filtration processes. Water Res 67:203–215
Ramos C, Zecchino F, Ezquerra D, Diez V (2014b) Chemical cleaning of membranes from an anaerobic membrane bioreactor treating food industry wastewater. J Membr Sci 458:179–188
Robles Á, Jiménez-Benítez A, Giménez JB, Durán F, Ribes J, Serralta J, Ferrer J, Rogalla F, Seco A (2022) A semi-industrial scale AnMBR for municipal wastewater treatment at ambient temperature: performance of the biological process. Water Res 215:118249
Rong C, Wang T, Luo Z, Hu Y, Kong Z, Qin Y, Hanaoka T, Ito M, Kobayashi M, Li YY (2022a) Pilot plant demonstration of temperature impacts on the methanogenic performance and membrane fouling control of the anaerobic membrane bioreactor in treating real municipal wastewater. Biores Technol 354:127167
Rong C, Wang T, Luo Z, Hu Y, Kong Z, Qin Y, Li YY (2022b) Seasonal and annual energy efficiency of mainstream anaerobic membrane bioreactor (AnMBR) in temperate climates: assessment in onsite pilot plant and estimation in scaled-up plant. Biores Technol 360:127542
Rong C, Wang T, Luo Z, Li YY (2023) Achieving low-carbon municipal wastewater treatment by anaerobic membrane bioreactor at seasonal temperatures: a pilot scale investigation on reducing sludge yield and greenhouse gas emissions. Chem Eng J 463:142415
Sanguanpak S, Chiemchaisri W, Chiemchaisri C (2019) Membrane fouling and micro-pollutant removal of membrane bioreactor treating landfill leachate. Rev Environ Sci Bio/Tech 18(4):715–740
Sawaya CB, Harb M (2021) Considering the prospect of utilizing anaerobic membrane biofouling layers advantageously for the removal of emerging contaminants. Front Chem Eng 3:642280
Shahid MK, Kashif A, Rout PR, Aslam M, Fuwad A, Choi Y, Park JH, Kumar G (2020) A brief review of anaerobic membrane bioreactors emphasizing recent advancements, fouling issues and future perspectives. J Environ Manage 270:110909
Shin C, McCarty PL, Kim J, Bae J (2014) Pilot-scale temperate-climate treatment of domestic wastewater with a staged anaerobic fluidized membrane bioreactor (SAF-MBR). Biores Technol 159:95–103
Shin C, Tilmans SH, Chen F, McCarty PL, Criddle CS (2021) Temperate climate energy-positive anaerobic secondary treatment of domestic wastewater at pilot-scale. Water Res 204:117598
Sohn W, Guo W, Ngo HH, Deng L, Cheng D, Zhang X (2021) A review on membrane fouling control in anaerobic membrane bioreactors by adding performance enhancers. J Water Process Eng 40:101867
Sun C, Du Q, Zhang X, Wang Z, Zheng J, Wu Q, Li Z, Long T, Guo W, Ngo HH (2022) Role of spent coffee ground biochar in an anaerobic membrane bioreactor for treating synthetic swine wastewater. J Water Proc Eng 49:102981
Tang J, Pu Y, Zeng T, Hu Y, Huang J, Pan S, Wang XC, Li Y, Abomohra AE (2022) Enhanced methane production coupled with livestock wastewater treatment using anaerobic membrane bioreactor: performance and membrane filtration properties. Biores Technol 345:126470
Wang G, Li Q, Gao X, Wang XC (2018a) Synergetic promotion of syntrophic methane production from anaerobic digestion of complex organic wastes by biochar: performance and associated mechanisms. Biores Technol 250:812–820
Wang KM, Cingolani D, Eusebi AL, Soares A, Jefferson B, McAdam EJ (2018b) Identification of gas sparging regimes for granular anaerobic membrane bioreactor to enable energy neutral municipal wastewater treatment. J Membr Sci 555:125–133
Wang L, Wu Y, Fu Y, Deng L, Wang Y, Ren Y, Zhang H (2022a) Low electric field assisted surface conductive membrane in AnMBR: strengthening effect and fouling behavior. Chem Eng J 431:133185
Wang T, Jin Z, Yang Y, Ma J, Aghbashlo M, Zhang H, Sun S, Tabatabaei M, Pan J (2022b) In-depth insights into the temporal-based fouling mechanism and its exploration in anaerobic membrane bioreactors: a review. J Clean Prod 375:134110

Wang G, Liu G, Yao G, Fu P, Sun C, Li Y, Li Q, Li YY, Chen R (2023a) Biochar-assisted anaerobic membrane bioreactor towards high-efficient energy recovery from swine wastewater: performances and the potential mechanisms. Biores Technol 369:128480

Wang Y, Gao Y, Lu X, Gadow SI, Zhuo G, Hu W, Song Y, Zhen G (2023b) Bioelectrochemical anaerobic membrane bioreactor enables high methane production from methanolic wastewater: roles of microbial ecology and microstructural integrity of anaerobic biomass. Chemosphere 339:139676

Wu D, Li L, Zhen F, Liu H, Xiao F, Sun Y, Peng X, Li Y, Wang X (2022) Thermodynamics of volatile fatty acid degradation during anaerobic digestion under organic overload stress: the potential to better identify process stability. Water Res 214:118187

Wu C, Qi C, Zhai G, Han S, Song X (2023a) Zero-valent iron to advance anaerobic membrane bioreactors for wastewater treatment and reuse: a critical review. Environ Technol Innov 32:103394

Wu Z, Qiao W, Dong R (2023b) Elucidating methanogenesis and membrane filtration performance of a ceramic AnMBR treating municipal waste leachate. J Water Process Eng 53:103905

Xiao X, Shi W, Huang Z, Ruan W, Miao H, Ren H, Zhao M (2017a) Process stability and microbial response of anaerobic membrane bioreactor treating high-strength kitchen waste slurry under different organic loading rates. Int Biodeterior Biodegradation 121:35–43

Xiao Y, Yaohari H, De Araujo C, Sze CC, Stuckey DC (2017b) Removal of selected pharmaceuticals in an anaerobic membrane bioreactor (AnMBR) with/without powdered activated carbon (PAC). Chem Eng J 321:335–345

Xie Z, Wang Z, Wang Q, Zhu C, Wu Z (2014) An anaerobic dynamic membrane bioreactor (AnDMBR) for landfill leachate treatment: performance and microbial community identification. Biores Technol 161:29–39

Xin Y, Bligh MW, Kinsela AS, Wang Y, Waite TD (2015) Calcium-mediated polysaccharide gel formation and breakage: impact on membrane foulant hydraulic properties. J Membr Sci 475:395–405

Xiong Y, Harb M, Hong PY (2016) Characterization of biofoulants illustrates different membrane fouling mechanisms for aerobic and anaerobic membrane bioreactors. Sep Purif Technol 157:192–202

Xu M, Wen X, Huang X, Li Y (2010) Membrane fouling control in an anaerobic membrane bioreactor coupled with online ultrasound equipment for digestion of waste activated sludge. Sep Sci Technol 45(7):941–947

Xu M, Wen X, Huang X, Yu Z, Zhu M (2013) Mechanisms of membrane fouling controlled by online ultrasound in an anaerobic membrane bioreactor for digestion of waste activated sludge. J Membr Sci 445:119–126

Xu B, Cho QA, Ng TC, Huang S, Ng HY (2022) Enriched autoinducer-2 (AI-2)-based quorum quenching consortium in a ceramic anaerobic membrane bioreactor (AnMBR) for biofouling retardation. Water Res 214:118203

Yang S, Zhang Q, Lei Z, Wen W, Huang X, Chen R (2019a) Comparing powdered and granular activated carbon addition on membrane fouling control through evaluating the impacts on mixed liquor and cake layer properties in anaerobic membrane bioreactors. Biores Technol 294:122137

Yang Y, Qiao S, Jin R, Zhou J, Quan X (2019b) Novel anaerobic electrochemical membrane bioreactor with a CNTs hollow fiber membrane cathode to mitigate membrane fouling and enhance energy recovery. Environ Sci Technol 53(2):1014–1021

Yang L, Si B, Tan X, Xu J, Xu W, Zhou L, Chen J, Zhang Y, Zhou X (2022) Valorization of livestock manure for bioenergy production: a perspective on the fates and conversion of antibiotics. Resour Conserv Recycl 183:106352

Ying XB, Huang JJ, Shen DS, Feng HJ, Jia YF, Guo QQ (2021) Fouling behaviors are different at various negative potentials in electrochemical anaerobic membrane bioreactors with conductive ceramic membranes. Sci Total Environ 761:143199

Yu Q, Zhang Y (2019) Fouling-resistant biofilter of an anaerobic electrochemical membrane reactor. Nat Commun 10(1):4860

Yue X, Koh YK, Ng HY (2015) Treatment of domestic wastewater with an anaerobic ceramic membrane bioreactor (AnCMBR). Water Sci Technol 72(12):2301–2307
Zarei-Baygi A, Harb M, Wang P, Stadler LB, Smith AL (2019) Evaluating antibiotic resistance gene correlations with antibiotic exposure conditions in anaerobic membrane bioreactors. Environ Sci Technol 53(7):3599–3609
Zarei-Baygi A, Wang P, Harb M, Stadler LB, Smith AL (2020) Membrane fouling inversely impacts intracellular and extracellular antibiotic resistance gene abundances in the effluent of an anaerobic membrane bioreactor. Environ Sci Technol 54(19):12742–12751
Zhang S, Zhao Y, Yang K, Liu W, Xu Y, Liang P, Zhang X, Huang X (2020) Versatile zero valent iron applied in anaerobic membrane reactor for treating municipal wastewater: performances and mechanisms. Chem Eng J 382:123000
Zhang S, Lei Z, Dzakpasu M, Li Q, Li YY, Chen R (2021) Removal of trace organic contaminants in municipal wastewater by anaerobic membrane bioreactor: efficiencies, fates and impact factors. J Water Process Eng 40:101953
Zhang X, Gu J, Meng S, Liu Y (2022) Dissolved methane in anaerobic effluent: emission or recovery? Front Environ Sci Eng 16(4):54
Zhao K, Su F, Gu K, Qi J, Liu R, Hu C (2021) Antifouling potential and microbial characterization of an electrochemical anaerobic membrane bioreactor utilizing membrane cathode and iron anode. Biores Technol 334:125230
Zhen G, Pan Y, Lu X, Li YY, Zhang Z, Niu C, Kumar G, Kobayashi T, Zhao Y, Xu K (2019) Anaerobic membrane bioreactor towards biowaste biorefinery and chemical energy harvest: recent progress, membrane fouling and future perspectives. Renew Sustain Energy Rev 115:109392
Zhou Z, Tao Y, Zhang S, Xiao Y, Meng F, Stuckey DC (2019) Size-dependent microbial diversity of sub-visible particles in a submerged anaerobic membrane bioreactor (SAnMBR): implications for membrane fouling. Water Res 159:20–29

Anaerobic Treatment of Dye Wastewater

Xingchen Huang, Kun Dai, and Fang Zhang

Abstract The printing and dyeing industry generates a significant volume of wastewater, which poses a considerable challenge for treatment due to the presence of various toxic compounds, dyes, and other recalcitrant substances. Anaerobic degradation offers an economically feasible, environmentally friendly, and innovative solution for treating wastewater that contains azo dyes, which enables decolorization and safe discharge or reuse of water. The biological decolorization mechanisms of extracellular electron transport and Ni-Fe hydrogenase exist in two genera of *Shewanella* and *Caldicellulosiruptor*, respectively. The commonly used anaerobic reactors for decolorization include the upflow anaerobic sludge blanket, expanded granular sludge bed, anaerobic baffled reactor, and anaerobic membrane bioreactor. Recently, novel thermophilic bacteria such as *Caldicellulosiruptor*, *Caldanaerobacter*, *Caldibacillus*, and *Aeribacillus* have been utilized to develop a process for high temperature decolorization. To refine the treatment processes of dye wastewater, novel strategies, including machine learning, advanced multi-omics technologies, robust monitoring and control systems, and by-product identification, can be employed in future.

Keywords Anaerobic biological technologies · Decolorization mechanism · Dye wastewater · Thermophilic decolorization

X. Huang · K. Dai · F. Zhang (✉)
Fujian Provincial Key Laboratory of Soil Environmental Health and Regulation, College of Resources and Environment, Fujian Agriculture and Forestry University, Fuzhou 350002, Fujian, China
e-mail: zhfang@mail.ustc.edu.cn

X. Huang
e-mail: xchuang@fafu.edu.cn

K. Dai
e-mail: daikun@mail.ustc.edu.cn

G. Wu (ed.), *Anaerobic Digestion*, Green Energy and Technology,
https://doi.org/10.1007/978-3-031-69378-6_15

1 Introduction

The textile industry is the chemical processing of textile materials, such as fiber, yarn, thread, and fabric, including four processing steps of pretreatment, dyeing, printing, and finishing (Clark 2023; Rathi and Kumar 2022; Yang et al. 2018b). Since the discovery of the first synthetic dye (i.e., mauveine) in 1856, more than 10,000 different types of textile dyes have been commercialized worldwide (Lin et al. 2023; Wang et al. 2022). Synthetic dyes are a significant group of persistent compounds widely used in the textile industry's dyeing process. Immense amounts of processing water and chemicals are consumed in the textile industry. For instance, approximately 10–15% of the dyes are released into the effluent during the dyeing process (Bai et al. 2022; Holkar et al. 2016; Tawfik et al. 2021). It was estimated that 21–377 m^3 of water is consumed for one ton of textile product and chemical consumption normally varies from 10 to over 100% of the weight of the cloth (Asghar et al. 2015; Bai et al. 2022). Thus, around 280,000 tons of textile dyes are discharged every year into water sinks through textile effluents (Asghar et al. 2015; Wang et al. 2022).

Various techniques have been employed to enhance the dye properties, which can result in improved delivery to fabrics, resistance to fading, and a wider range of shades (Kabra et al. 2013; Rathi and Kumar 2022). However, these additional properties make them highly resistant to environmental degradation. Moreover, textile dyes, like many other industrial pollutants, possess high toxicity and potential carcinogenicity, resulting in severe environmental damage and causing numerous new health issues for animals and humans (Al-Amrani et al. 2022; Islam et al. 2023; Josephy and Allen-Vercoe 2023). Concurrently, printing and dyeing wastewater contains a variety of heavy metals, such as Zn, Cd, Pb, Hg, Cr, As, Fe, Co, Al, and Cu (Berkessa et al. 2020). These heavy metals can cause a range of health issues, including skeletal diseases, various cancers, kidney failure, brain damage, gastrointestinal diseases, memory issues, and neurotoxicity (Khan et al. 2022; Mishra et al. 2019; Sekomo et al. 2012). Thus, the generation of a large amount of textile printing and dyeing wastewater can lead to severe environmental pollution without adequate treatment (Wang et al. 2015).

Chemical, physical, and biological techniques are used to treat printing and dyeing effluents (Lin et al. 2023; Mate and Pathade 2012; Zhang et al. 2017). Due to their high removal efficiency and ease of operation, physical adsorption and chemical oxidation processes are commonly used (Liu et al. 2015, 2017; Teng et al. 2018). However, these processes are typically energy-intensive and costly, and require additional steps for regenerating adsorbents and/or decomposing by-products (Yang et al. 2018a). Anaerobic degradation provides an economically feasible, environmentally friendly, and innovative method for treating wastewater containing azo dyes, enabling decolorization and safe water discharge or reuse (He et al. 2020; Xu et al. 2018b).

Table 1 Characteristics and hazards of printing and dyeing wastewater

Parameters	Range	Hazardous and other possible effects
Temperature	45–80 °C	High temperatures may increase the solubility of organic matter and reduce the growth and activity of microorganisms, consequently impacting the efficiency of biological treatment methods (Chen et al. 2018; Shehab et al. 2017; Song et al. 2023; Tang et al. 2018; Varma et al. 2021)
pH	3–12	Wastewater with acidic or alkaline pH levels significantly affects the environment by harming aquatic organisms and disrupting the ecological equilibrium of water ecosystems (Kishor et al. 2021; Mantzavinos and Psillakis 2004)
BOD (mg/L)	100–6000	High BOD and COD concentrations may decrease dissolved oxygen levels in water and hinder normal respiration in organisms, significantly reducing the survival and reproduction of aquatic life (Can et al. 2006; Cerqueira et al. 2009; Gosavi and Sharma 2014; Herrera et al. 2023)
COD (mg/L)	500–12,000	
Salinity (mg/L)	1000–5000	Elevated salinity raises wastewater conductivity, impacting the efficiency of biological treatment (Alvares et al. 2001)
Color (Pt-Co)	50–2500	Impairs water clarity and aesthetics, consequently influencing downstream treatment processes (Cerqueira et al. 2009; Jia et al. 2024)
Metal ions, such as iron, copper, zinc, and chromium (mg/L)	10–50	Induce severe harm to the nervous system and multiple organs, resulting in enduring health risks like cancer (Al-Tohamy et al. 2022; Gosavi and Sharma 2014; Park et al. 1991; Sun et al. 2021)

2 Typical Properties of Dye Wastewater

2.1 Properties of Dye Wastewater

As shown in Table 1, the textile effluents are characterized by high temperature, high salinity, high chemical oxygen demand (COD), high biological oxygen demand (BOD), extreme pH (acidic or alkaline) and color, and the existence of heavy metals (Asghar et al. 2015; Chen et al. 2021; Islam et al. 2023; Kabra et al. 2013). These toxic components are revealed to have carcinogenic and mutagenic effects, posing a potential threat to all life forms.

2.2 Typical Dye Wastewater Containing Azo Dyes

Nowadays, the industry utilizes over 2000 different types of azo dyes, mainly found in textiles, leather, and cosmetics (Wang et al. 2020). As shown in Table 2, the azo dyes are an important class of synthetic dyes, which are usually used in orange (such as

Acid Orange 52), red (such as Disperse Red 1), and yellow (such as Disperse Yellow 3) (dos Santos et al. 2007, 2018). Azo dyes contain aromatic structures substituted by one or more azo bonds (Ismail et al. 2020; Oke and Mohan 2022; Yadav et al. 2022), accounting for 60–70% of synthetic dyes in the textile and dye products industry (Castro et al. 2017; Lin et al. 2023; Mahmood et al. 2016).

Table 2 Summary of well-used azo dyes and their molecular structure

Azo dye	Chemical name	Molecular formula	Molecular structure
Single azo chemicals	Acid Orange 52	$C_{14}H_{14}N_3NaO_3S$	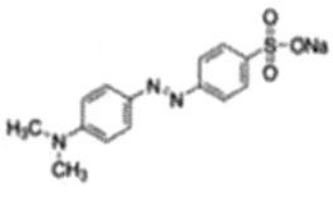
	Disperse Red 1	$C_{16}H_{18}N_4O_3$	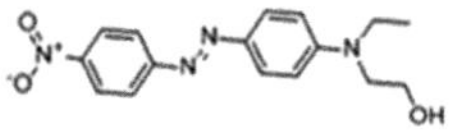
	Disperse Yellow 3	$C_{15}H_{15}N_3O_2$	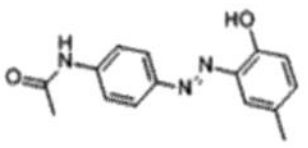
Double azo chemicals	Direct Scarlet 4BS	$C_{35}H_{25}N_7Na_2O_{10}S_2$	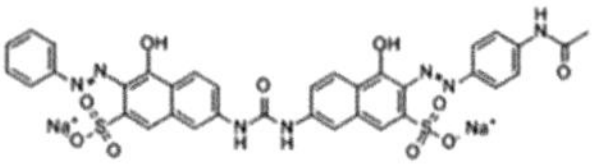
	Direct Blue 14	$C_{34}H_{24}N_6Na_4O_{14}S_4$	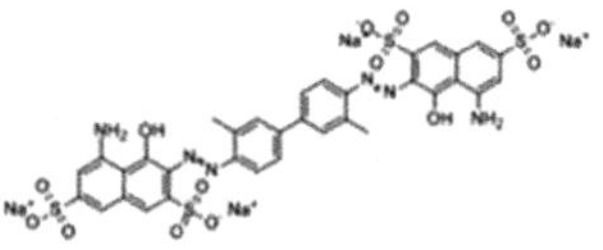
Multi-azo chemicals	Direct Brown 31	$C_{46}H_{30}N_{10}Na_4O_{13}S_3$	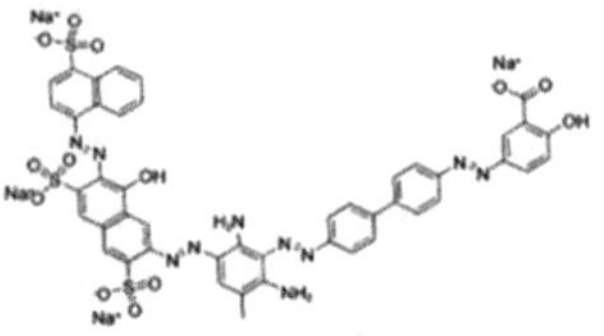

3 Typical Decolorization Mechanisms During Dye Wastewater Treatment

3.1 Decolorization Through Extracellular Electron Transfer

The –N = N– bonds in azo dyes are known as chromophores, which are responsible for the dye color and account for the dye degradation resistance (Castro et al. 2017). Several strains have been reported to possess the ability to decolorize azo dyes. *Shewanella* strains are highly regarded for their exceptional decolorization performance through extracellular electron transfer (EET) (Cai et al. 2012; Liu et al. 2016, 2018). For the genus *Shewanella,* several components within the electron transfer pathway are crucial for its decolorization capacity (Cai et al. 2012; Liu et al. 2016; Xiao et al. 2012). The intracellular tetraheme c-type cytochrome CymA accepts electrons from the quinone pool and subsequently delivers them to various respiratory pathways, including Mtr, Dms, NarfA/B, and NrfA pathways (Gao et al. 2009; Schwalb et al. 2003). In *Shewanella oneidensis* MR-1, the Mtr respiratory pathway, which consists of MtrA, MtrB, MtrC, and OmcA, has been identified to participate in the decolorization of diverse azo dyes (Cai et al. 2012; Xiao et al. 2012).

As shown in Fig. 1, Liu et al. (2020) proposed a molecular mechanism model of cyclic AMP receptor protein (Crp) regulating the biological decolorization of Acid Yellow 36 (AY). Initially, Crp forms a complex with cAMP, directly activating the transcription of cymA. The construction of deletion mutants MtrC, MtrA, and MtrB in the frame was used to determine whether the Mtr respiratory pathway in *S. putrefaciens* CN32 was involved in AY decolorization. It was found that the decolorization efficiencies of MtrC, MtrA, and MtrB decreased by 10.6%, 9.8%, and 14.5% at 4 h, respectively, indicating that in *S. putrefaciens* CN32, the respiratory pathway of Mtr only partially participated in the electron transfer required in the decolorization process of AY. Subsequently, CymA transfers some intracellular electrons through MtrC, MtrA, and MtrB to the extracellular dye receptor for reduction, but the main electron transfer pathway from CymA to dye molecules remains unclear. Simultaneously, flavin can act as an electron shuttle to switch back and forth between oxidation state and reduction state and accelerate the electron transfer between cells and dye molecules. Under the action of electron reduction, dye molecules are decomposed into colorless degradation products. Gu et al. (2021) proposed an indirect pathway that the electrons generated from organic carbon metabolism would transfer through c-type cytochrome in *Geobacter* and *Staphylococcus* to functional bacteria responsible for azo dye degradation, such as *Bacillus* and *Desulfovibrio.* This electron transfer between organic carbon metabolism bacteria and azo dye degrading bacteria can also rely on some electron carriers, such as conductive Fe_3O_4 (Gu et al. 2021; Wang et al. 2018).

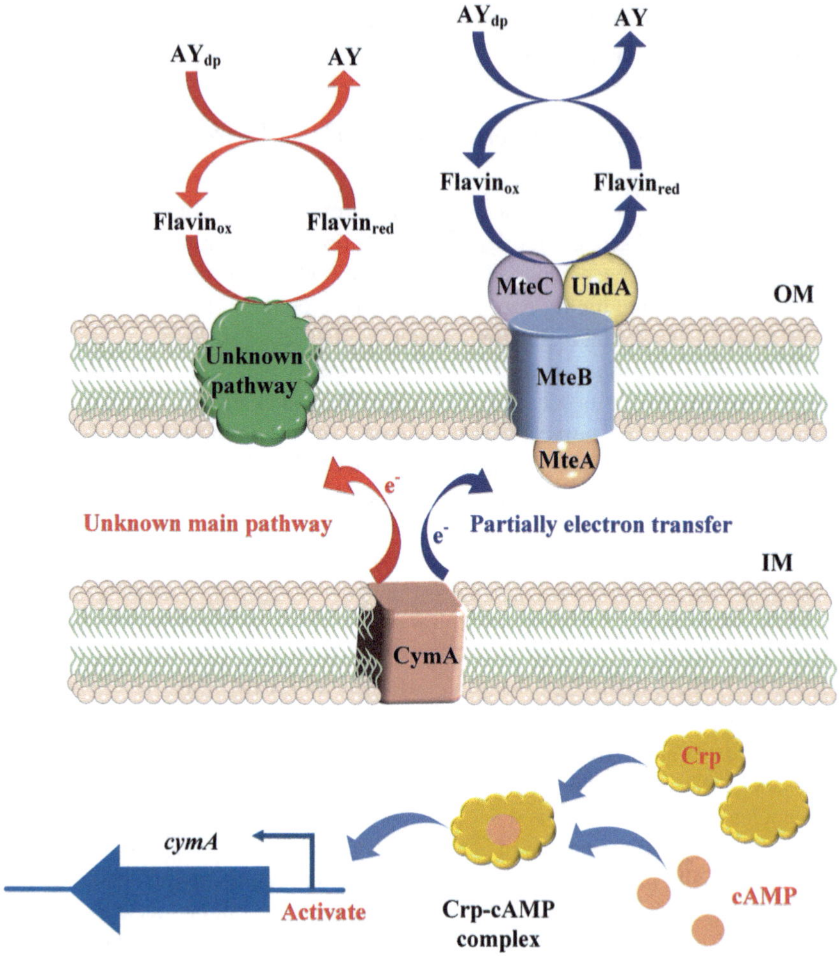

Fig. 1 Proposed mechanism of AY decolorization in *S. putrefaciens* CN32 via Crp regulation (Liu et al. 2020)

3.2 Decolorization Through Ni-Fe Hydrogenase

Ni-Fe hydrogenase is involved in the uptake and release of H_2 (Meshulam-Simon et al. 2007). When dissolved hydrogen is reduced by Ni-Fe hydrogenase, the generated electrons can be transferred to other electron acceptors, such as heavy metals. For example, the extracellular iron reduction of neutral red by *Escherichia coli* can be conducted with the help of H_2-mediated hydrogenase (McKinlay and Zeikus 2004). During fermentation, H_2 is commonly produced as an intermediate product and acts as a potent reducing agent, enabling the reduction of heavy metals by Ni-Fe hydrogenase. Consequently, it could participate in the decolorization of azo dyes through a similar pathway. *Caldicellulosiruptor saccharolyticus*, a bacterium thriving in extremely thermophilic conditions, has been isolated from hot springs, with an optimal growth temperature of 70 °C. *C. saccharolyticus* can grow using (poly) sugar, with a high H_2 yield up to 3.5 mol H_2/mol glucose (Willquist and Van Niel 2012). Particularly, it hosts Ni-Fe hydrogenase bound to the cell membrane (Van de Werken et al. 2008). Consequently, *C. saccharolyticus* has the capability to decolorize azo dyes due to these attributes.

As shown in Fig. 2, Shen et al. (2015) proposed the mechanism of decolorization of Methyl Orange azo bond catalyzed by Ni-Fe hydrogenase in *C. saccharolyticus*. Dissolved hydrogen generates protons and electrons under the action of Ni-Fe hydrogenase, and the electrons are transferred to methyl orange to break up the azo bond to complete decolorization. The degradation products include N', N-dimethyl-p-phenylenediamine (DPD) and 4-aminobenzenesulfonic acid (4-ABA), indicating that dissolved hydrogen serves as an electron donor for the decolorization of Methyl Orange. When the hydrogen pressure decreases to below 0.3 atm, ferredoxins can continue to reduce H^+ to H_2. If the reducing power during glucose fermentation is limited and part of the reducing power is utilized for the degradation of Methyl Orange, other reducing products in the *C. saccharolyticus* system, such as ethanol and lactic acid, will decrease.

4 Technologies for Anaerobic Decolorization of Dye Wastewater

The anaerobic processes have been demonstrated to effectively treat high-concentration textile wastewater, exhibiting good performance in pollutant removal and low energy demand (Amaral et al. 2014; Li et al. 2018). For instance, it has been demonstrated that the addition of nutrient-rich domestic wastewater can effectively enhance the anaerobic degradation of textile wastewater, resulting in a 70% reduction in wastewater color and a 72% reduction in COD (Bidu et al. 2023; Osama et al. 2020). High-rate anaerobic reactors, such as upflow anaerobic sludge blanket (UASB), expanded granular sludge bed (EGSB), anaerobic baffled reactor (ABR), and anaerobic membrane bioreactor (AnMBR), are the well-used anaerobic reactors for anaerobic decolorization (Fig. 3).

4.1 Dye Wastewater Treatment Using Typical Anaerobic Technologies

Figure 3a illustrates the schematic diagram of the UASB reactor. The UASB reactor is a popular anaerobic system widely used for treating various industrial wastewater due to its low investment, high efficiency, stable operation, and strong adaptability to

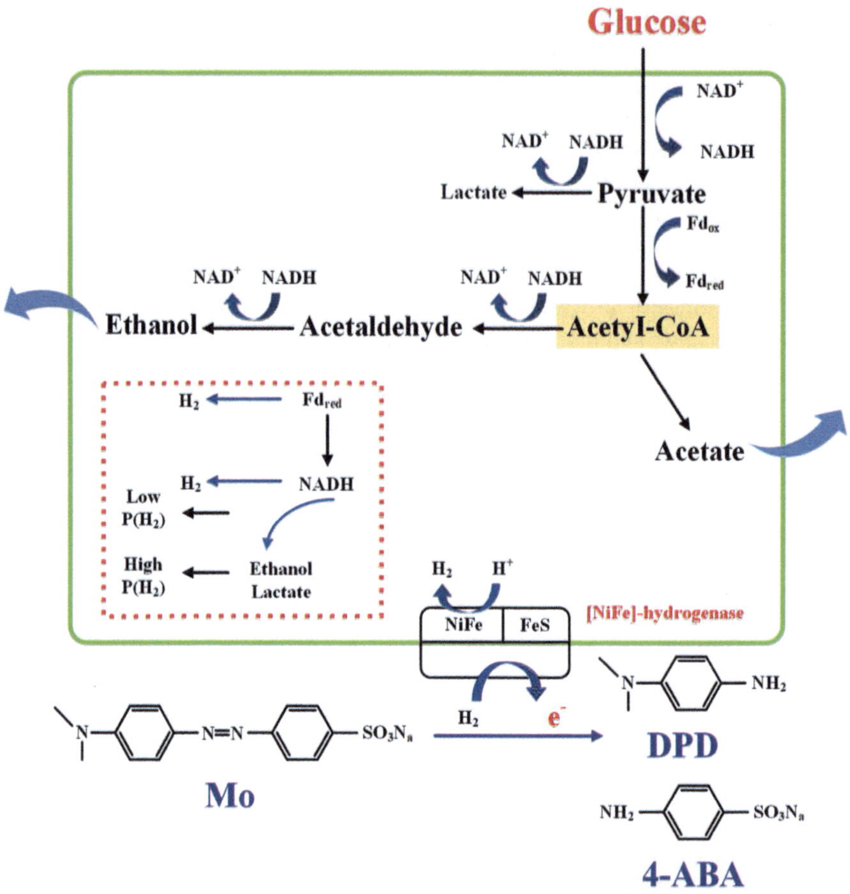

Fig. 2 Proposed mechanism of Methyl Orange degradation during glucose fermentation by *C. saccharolyticus* (Shen et al. 2015)

different types of wastewaters (Qiu et al. 2013). The UASB reactor has been proven effective in treating textile dyeing wastewater with high color removal efficiency due to its robust resistance to toxic compounds in the medium, such as dyes, additives, and aromatic amines (Senthilkumar et al. 2011; Wijetunga et al. 2010). This could be attributed to the presence of reductases under anaerobic conditions, leading to the degradation of chromophoric groups such as azo bonds. For instance, Somasiri et al. (2008) confirmed that despite the high influent COD concentration of up to 6000 mg/L, the decolorization efficiency of industrial textile wastewater using the UASB reactor exceeded 90%. Additionally, the operational parameters of anaerobic reactors, such as co-metabolism substrates and hydraulic retention time (HRT), along with water quality, significantly affect their performance. For example, Wijetunga et al. (2010) indicated that with an HRT of 24 h and an organic loading rate (OLR) of 5.6 kg COD/m^3/d, the percentage of optimal COD removal could reach 53.1%. Their

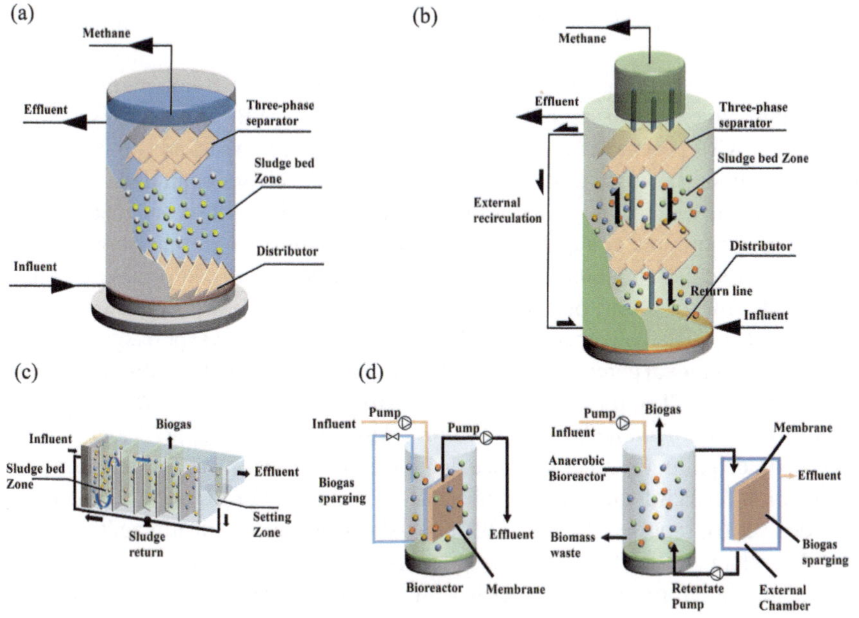

Fig. 3 Schematic diagrams of high-rate anaerobic reactors: **a** UASB, **b** EGSB, **c** ABR, and **d** AnMBR

results also indicated that further degradation of these refractory components could be achieved by extending the HRT. In recent years, several combined processes have been developed based on the UASB reactor and applied in the treatment of textile dyeing and printing wastewater (Yang et al. 2018b). Among these, the UASB-aerobic combined technology is considered one of the most economically effective techniques.

The EGSB reactor was developed as a representative of the third generation of anaerobic bioreactors to improve color removal. The EGSB reactor is based on the UASB reactor (Fig. 3b). The EGSB reactor is equipped with an external circulation system to improve mass transfer and enhance the contact between wastewater and biomass, in contrast to the UASB reactor (Xu et al. 2018a). This design aims to increase efficiency in the treatment process. Consequently, the EGSB reactor has several advantages, such as high OLRs, a small footprint, increased methane production, and seismic resistance (Xu et al. 2018a). Currently, the application of EGSB reactors is primarily concentrated on treating pharmaceutical wastewater, starch, and sugar wastewater, as well as alcohol wastewater (Xu et al. 2018b). EGSB reactors can also be considered as a promising technology for the effective treatment of wastewater containing recalcitrant substances, including dyes (Cruz-Salomón et al. 2019). For example, Inaba et al. (2022) evaluated the decolorization of 11 dyes in EGSB reactors and observed that the biological decolorization of over 94% was achieved for Reactive Red 21, 23, and 180, and Reactive Yellow 15, 17, and 23 after a 7-day incubation.

As a high-rate reactor, the ABR typically consists of 3–8 compartments, and each compartment can be regarded as an independent UASB reactor. The schematic diagram of ABR is depicted in Fig. 3c. In ABR, the liquid flows up and down between the compartments, and this special design facilitates the phase separation of microorganisms within a single unit, such as the separation of acidogenesis and methanogenesis. Furthermore, the configuration of ABR is straightforward, requiring no agitation devices, support materials, or liquid-gas separation systems. Several studies have investigated the feasibility of ABR for treating textile dyeing and printing wastewater. Ozdemir et al. (2013) obtained that ABR achieved a COD removal percentage of 98% and a sulfate removal percentage of 93% for synthetic dyeing and printing wastewater with a 48-h HRT. The results indicated that ABR provided a favorable environment for various microbial biochemical reactions. Although anaerobic processes may not fully degrade pollutants, they enhance the biodegradability of wastewater, thus facilitating their further degradation in subsequent processes. For instance, Reynaud and Buckley (2016) reported ABR's COD removal performance is insufficient, resulting in ABR effluent failing to meet wastewater discharge standards and requiring further treatment. Therefore, combining ABR with other aerobic systems may offer a viable solution to this issue.

AnMBR is an advanced wastewater treatment technology that combines anaerobic biotechnology with membrane separation technology. AnMBRs can be categorized into external and submerged types based on the configuration of membrane components, as illustrated in Fig. 3d. In recent years, submerged AnMBRs have attracted much attention due to their low energy consumption, small footprint, compact structure, and low operating costs, demonstrating excellent performance in treating textile dyeing wastewater. Spagni et al. (2012) demonstrated that the use of submerged AnMBR for treating textile dyeing wastewater could effectively remove COD and achieve a decolorization percentage of over 99%. Additionally, MBR has the advantages of short HRT and high OLR. For instance, Yilmaz et al. (2024) recently demonstrated the removal percentage of COD in a real textile wastewater reached above 70% even at a sludge retention time of 0.5 d and HRT of 1.2 h in an aerobic MBR system. However, the high salinity may reduce the performance of decolorization. For example, Yurtsever et al. (2016) showed that increasing salinity (0–1000 mg/L) appreciably decreased biomass concentrations even in both sequential anaerobic and aerobic MBRs. It is also worth noting that addressing the severe membrane fouling is crucial for the practical application of these MBR technologies.

4.2 Decolorization by Enriching the Thermophilic Bacteria

Printing and dyeing wastewater is mainly generated during the dyeing and washing processes, with discharge temperatures reaching 45–80 °C, which severely inhibits the activity of common microorganisms, hindering the efficient treatment of printing and dyeing wastewater (Yaseen and Scholz 2019; Zhang et al. 2019). However, utilizing thermophilic bacteria to treat high-temperature printing and dyeing wastewater offers advantages such as low treatment costs, no need to reduce wastewater temperature, rapid degradation rates of thermophilic bacteria at high temperatures, and simultaneous eradication of pathogens in wastewater. Moreover, the thermostable enzymes produced by thermophilic bacteria exhibit strong adaptability, enabling efficient degradation of toxic pollutants. For example, Shen et al. (2015) reported that *Caldicellulosiruptor saccharolyticus* could degrade Methyl Orange using glucose as an electron donor at temperatures below 70 °C. Zhang et al. (2019) enriched an extreme thermophilic bacterial community capable of degrading azo dye Acid Orange 7 at 70 °C, mainly composed of *Caldanaerobacter* (64.0%) and *Pseudomonas* (25.4%). The genus *Caldanaerobacter* can decouple the decolorization and sulfate reduction that no sulfide was detected during the Acid Orange 7 decolorization process in the presence of sulfate (Yan et al. 2021). Interestingly, a thermophilic bacterial consortium mainly composed of *Caldibacillus* (70.90%) and *Aeribacillus* (17.63%) was enriched, which could decolorize Brilliant Crocein GR under dye concentrations (100–400 mg/L), high temperatures (50–75 °C), wide pH values (6~8), and salinities (1–5%) (Tian et al. 2024). The multi-omics technologies may also provide a powerful tool to reveal the bacteria or enzymes with high efficiency of decolorization (Dwivedi et al. 2023; Rodríguez et al. 2015). Thus, decolorization by thermophilic bacteria provides a new paradigm to promote azo dye decolorization and minimize sulfate reduction, which deserves further investigation in the future.

5 Future Perspectives

Although anaerobic biological technology has been widely applied in the treatment of textile dyeing and printing wastewater, combining it with other wastewater treatment methods (such as aerobic biological treatment and chemical treatment) can further improve the performance of decolorization and COD removal. Achieving a stable system operation may require advanced online monitoring technologies and optimization of operational parameters and nutrient conditions. Novel methods such as machine learning have the potential to significantly reduce reactor startup times and improve treatment efficiency. These methods should be considered in future studies.

Secondly, a better understanding of microbial community composition and functional dynamics will inform the development of improved processes and innovative bioreactor designs for dye wastewater treatment. The application of advanced multi-omics technologies offers a promising avenue for the identification of functional microorganisms, with the potential to enhance the efficiency of decolorization.

Thirdly, the toxicity of by-products generated during the process of dye wastewater treatment warrants further investigation. Anaerobic biological technology has been shown to produce intermediates or end products during wastewater treatment, and their impacts on the environment and human health require further assessment and control. Consequently, the generation mechanisms, transformation pathways, and potential environmental impacts of by-products can assist in the refinement of wastewater treatment processes and the minimization of adverse effects.

Finally, from the perspective of a circular economy, highly saline textile effluent from exhausted dye baths represents a potential resource, with the residual dyes and inorganic salts able to be reused. Highly selective membranes, including ultrafiltration and nanofiltration membranes, enable high dye retention and low salt rejection. It is imperative that pilot and full-scale demonstrations using highly selective membranes will be conducted to ascertain their suitability for applications in textile wastewater treatment.

6 Take-Home Messages

(1) Printing and dyeing industry produces a large amount of wastewater, which presents a significant challenge for treatment due to the presence of numerous toxic compounds, dyes, and other recalcitrant substances.
(2) Typical biological decolorization mechanisms include extracellular electron transport and Ni-Fe hydrogenase.
(3) Not only conventional anaerobic digestion technologies such as UASB, EGSB, ABR, and AnMBR, but also high temperature decolorization technologies, can be applied for the treatment of dye wastewater.
(4) Novel strategies, including robust monitoring and control systems, machine learning, advanced multi-omics technologies, by-product identification, and highly selective recovery of dyes and salts, can assist in the refinement of processes for the treatment of dye wastewater.

References

Al-Amrani WA, Hanafiah MA, Mohammed AH (2022) A comprehensive review of anionic azo dyes adsorption on surface-functionalised silicas. Environ Sci Pollut Res 29(51):76565–76610

Al-Tohamy R, Ali SS, Li F, Okasha KM, Mahmoud YA, Elsamahy T, Jiao H, Fu Y, Sun J (2022) A critical review on the treatment of dye-containing wastewater: ecotoxicological and health concerns of textile dyes and possible remediation approaches for environmental safety. Ecotoxicol Environ Saf 231:113160

Alvares AB, Diaper C, Parsons SA (2001) Partial oxidation by ozone to remove recalcitrance from wastewaters—a review. Environ Technol 22(4):409–427

Amaral FM, Kato MT, Florêncio L, Gavazza S (2014) Color, organic matter and sulfate removal from textile effluents by anaerobic and aerobic processes. Biores Technol 163:364–369

Asghar A, Raman AA, Daud WM (2015) Advanced oxidation processes for in-situ production of hydrogen peroxide/hydroxyl radical for textile wastewater treatment: a review. J Clean Prod 87:826–838

Bai Y, Wang X, Zhang F, Zeng RJ (2022) Acid Orange 7 degradation using methane as the sole carbon source and electron donor. Front Environ Sci Eng 16(3):34

Berkessa YW, Yan B, Li T, Jegatheesan V, Zhang Y (2020) Treatment of anthraquinone dye textile wastewater using anaerobic dynamic membrane bioreactor: performance and microbial dynamics. Chemosphere 238:124539

Bidu JM, Njau KN, Rwiza M, Van der Bruggen B (2023) Textile wastewater treatment in anaerobic reactor: influence of domestic wastewater as co-substrate in color and COD removal. S Afr J Chem Eng 43(1):112–121

Cai PJ, Xiao X, He YR, Li WW, Chu J, Wu C, He MX, Zhang Z, Sheng GP, Lam MH, Xu F (2012) Anaerobic biodecolorization mechanism of methyl orange by *Shewanella oneidensis* MR-1. Appl Microbiol Biotechnol 93:1769–1776

Can OT, Kobya M, Demirbas E, Bayramoglu M (2006) Treatment of the textile wastewater by combined electrocoagulation. Chemosphere 62(2):181–187

Castro FD, Bassin JP, Dezotti M (2017) Treatment of a simulated textile wastewater containing the Reactive Orange 16 azo dye by a combination of ozonation and moving-bed biofilm reactor: evaluating the performance, toxicity, and oxidation by-products. Environ Sci Pollut Res 24(7):6307–6316

Cerqueira A, Russo C, Marques MR (2009) Electroflocculation for textile wastewater treatment. Braz J Chem Eng 26(4):659–668

Chen Y, Feng L, Li H, Wang Y, Chen G, Zhang Q (2018) Biodegradation and detoxification of Direct Black G textile dye by a newly isolated thermophilic microflora. Biores Technol 250:650–657

Chen H, Yu X, Wang X, He Y, Zhang C, Xue G, Liu Z, Lao H, Song H, Chen W, Qian Y (2021) Dyeing and finishing wastewater treatment in China: state of the art and perspective. J Clean Prod 326:129353

Clark JH (2023) Textile waste–an opportunity as well as a threat. Green Carbon 1(2):146–149

Cruz-Salomón A, Ríos-Valdovinos E, Pola-Albores F, Lagunas-Rivera S, Meza-Gordillo R, Ruíz-Valdiviezo VM, Cruz-Salomón KC (2019) Expanded granular sludge bed bioreactor in wastewater treatment. Global J Environ Sci Manag 5(1):119–138

Dos Santos AB, Cervantes FJ, Van Lier JB (2007) Review paper on current technologies for decolourisation of textile wastewaters: perspectives for anaerobic biotechnology. Biores Technol 98(12):2369–2385

Dos Santos RF, Ramlow H, Dolzan N, Machado RA, de Aguiar CR, Marangoni C (2018) Influence of different textile fibers on characterization of dyeing wastewater and final effluent. Environ Monit Assess 190:693

Dwivedi S, Gupta S, Tanveer A, Anand G, Yadav S, Yadav D (2023) Mining of novel microbial enzymes using metagenomics approach for efficient bioremediation: an overview. In: Kumar V, Bilal M, Ferreira LFR, Iqbal HMN (eds) Genomics approach to bioremediation: principles, tools, and emerging technologies. John Wiley & Sons Inc., pp 183–198

Gao H, Yang ZK, Barua S, Reed SB, Romine MF, Nealson KH, Fredrickson JK, Tiedje JM, Zhou J (2009) Reduction of nitrate in *Shewanella oneidensis* depends on atypical NAP and NRF systems with NapB as a preferred electron transport protein from CymA to NapA. ISME J 3(8):966–976

Gosavi VD, Sharma S (2014) A general review on various treatment methods for textile wastewater. J Environ Sci, Comp Sci Eng Tech 3(1):29–39

Gu M, Yin Q, Wu G (2021) Metagenomic analysis of facilitation mechanism for azo dye reactive red 2 degradation with the dosage of ferroferric oxide. J Water Process Eng 41:102010

He X, Qi Z, Gao J, Huang K, Li M, Springael D, Zhang XX (2020) Nonylphenol ethoxylates biodegradation increases estrogenicity of textile wastewater in biological treatment systems. Water Res 184:116137

Herrera AS, Esparza MD, Arias MP (2023) Melanin: the philosophical stone of molecular biology and its implications on dissolved oxygen levels in water and fertility of agricultural soil. Int J Agricult Environ Res 9(3):265–288

Holkar CR, Jadhav AJ, Pinjari DV, Mahamuni NM, Pandit AB (2016) A critical review on textile wastewater treatments: possible approaches. J Environ Manage 182:351–366

Inaba T, Yamaguchi M, Taniguchi A, Sato Y, Aoyagi T, Hori T, Inoue H, Fujita M, Iwata M, Iwata Y, Habe H (2022) Evaluation of dye decolorization using anaerobic granular sludge from an expanded granular sludge bed based on spectrometric and microbiome analyses. J Gen Appl Microbiol 68(5):242–247

Islam T, Repon MR, Islam T, Sarwar Z, Rahman MM (2023) Impact of textile dyes on health and ecosystem: a review of structure, causes, and potential solutions. Environ Sci Pollut Res 30(4):9207–9242

Ismail S, Nasr M, Abdelrazek E, Awad HM, Zhaof S, Meng F, Tawfik A (2020) Techno-economic feasibility of energy-saving self-aerated sponge tower combined with up-flow anaerobic sludge blanket reactor for treatment of hazardous landfill leachate. J Water Process Eng 37:101415

Jia J, Wu D, Yu J, Gao T, Guo L, Li F (2024) Upgraded β-cyclodextrin-based broad-spectrum adsorbents with enhanced antibacterial property for high-efficient dyeing wastewater remediation. J Hazard Mater 461:132610

Josephy PD, Allen-Vercoe E (2023) Reductive metabolism of azo dyes and drugs: toxicological implications. Food Chem Toxicol 178:113932

Kabra AN, Khandare RV, Govindwar SP (2013) Development of a bioreactor for remediation of textile effluent and dye mixture: a plant–bacterial synergistic strategy. Water Res 47(3):1035–1048

Khan AA, Gul J, Naqvi SR, Ali I, Farooq W, Liaqat R, AlMohamadi H, Štěpanec L, Juchelková D (2022) Recent progress in microalgae-derived biochar for the treatment of textile industry wastewater. Chemosphere 306:135565

Kishor R, Purchase D, Saratale GD, Saratale RG, Ferreira LF, Bilal M, Chandra R, Bharagava RN (2021) Ecotoxicological and health concerns of persistent coloring pollutants of textile industry wastewater and treatment approaches for environmental safety. J Environ Chem Eng 9(2):105012

Li F, Xia Q, Gao Y, Cheng Q, Ding L, Yang B, Tian Q, Ma C, Sand W, Liu Y (2018) Anaerobic biodegradation and decolorization of a refractory acid dye by a forward osmosis membrane bioreactor. Environ Sci: Water Res Technol 4(2):272–280

Lin J, Ye W, Xie M, Seo DH, Luo J, Wan Y, Van der Bruggen B (2023) Environmental impacts and remediation of dye-containing wastewater. Nat Rev Earth Environ 4(11):785–803

Liu Y, Liu H, Zhou Z, Wang T, Ong CN, Vecitis CD (2015) Degradation of the common aqueous antibiotic tetracycline using a carbon nanotube electrochemical filter. Environ Sci Technol 49(13):7974–9780

Liu W, Liu L, Liu C, Hao Y, Yang H, Yuan B, Jiang J (2016) Methylene blue enhances the anaerobic decolorization and detoxication of azo dye by *Shewanella onediensis* MR-1. Biochem Eng J 110:115–124

Liu Y, Zheng Y, Du B, Nasaruddin RR, Chen T, Xie J (2017) Golden carbon nanotube membrane for continuous flow catalysis. Ind Eng Chem Res 56(11):2999–3007

Liu W, You Y, Sun D, Wang S, Zhu J, Liu C (2018) Decolorization and detoxification of water-insoluble Sudan dye by *Shewanella putrefaciens* CN32 co-cultured with *Bacillus circulans BWL1061*. Ecotoxicol Environ Saf 166:11–17

Liu W, Chen Y, Zhou X, Liu J, Zhu J, Wang S, Liu C, Sun D (2020) The cyclic AMP receptor protein, Crp, is required for the decolorization of acid yellow 36 in *Shewanella putrefaciens* CN32. Front Microbiol 11:596372

Mahmood S, Khalid A, Arshad M, Mahmood T, Crowley DE (2016) Detoxification of azo dyes by bacterial oxidoreductase enzymes. Crit Rev Biotechnol 36(4):639–651

Mantzavinos D, Psillakis E (2004) Enhancement of biodegradability of industrial wastewaters by chemical oxidation pre-treatment. J Chem Tech Biotech: Int Res Process, Environ Clean Tech 79(5):431–454

Mate MS, Pathade G (2012) Biodegradation of C.I. Reactive Red 195 by *Enterococcus faecalis* strain YZ66. World J Microbiol Biotech 28(3):815–826

McKinlay JB, Zeikus JG (2004) Extracellular iron reduction is mediated in part by neutral red and hydrogenase in *Escherichia coli*. Appl Environ Microbiol 70(6):3467–3474

Meshulam-Simon G, Behrens S, Choo AD, Spormann AM (2007) Hydrogen metabolism in *Shewanella oneidensis* MR-1. Appl Environ Microbiol 73(4):1153–1165

Mishra S, Bharagava RN, More N, Yadav A, Zainith S, Mani S, Chowdhary P (2019) Heavy metal contamination: an alarming threat to environment and human health. In: Sobti R, Arora N, Kothari R (eds) Environmental biotechnology: for sustainable future. Springer, Singapore, pp 103–125

Oke N, Mohan S (2022) Development of nanoporous textile sludge based adsorbent for the dye removal from industrial textile effluent. J Hazard Mater 422:126864

Osama R, Awad HM, Ibrahim MG, Tawfik A (2020) Mechanistic and economic assessment of polyester wastewater treatment via baffled duckweed pond. J Water Process Eng 35:101179

Ozdemir S, Cirik K, Akman D, Sahinkaya E, Cinar O (2013) Treatment of azo dye-containing synthetic textile dye effluent using sulfidogenic anaerobic baffled reactor. Biores Technol 146:135–143

Park HS, Lee MK, Kim BO, Lee KJ, Roh JH, Moon YH, Hong CS (1991) Clinical and immunologic evaluations of reactive dye-exposed workers. J Allergy Clin Immunol 87(3):639–649

Qiu G, Song YH, Zeng P, Duan L, Xiao S (2013) Characterization of bacterial communities in hybrid upflow anaerobic sludge blanket (UASB)–membrane bioreactor (MBR) process for berberine antibiotic wastewater treatment. Biores Technol 142:52–62

Rathi BS, Kumar SP (2022) Sustainable approach on the biodegradation of azo dyes: a short review. Current Opinion Green Sustain Chem 33:100578

Reynaud N, Buckley CA (2016) The anaerobic baffled reactor (ABR) treating communal wastewater under mesophilic conditions: a review. Water Sci Technol 73(3):463–478

Rodríguez E, García-Encina PA, Stams AJ, Maphosa F, Sousa DZ (2015) Meta-omics approaches to understand and improve wastewater treatment systems. Rev Environ Sci Bio/Tech 14:385–406

Schwalb C, Chapman SK, Reid GA (2003) The tetraheme cytochrome CymA is required for anaerobic respiration with dimethyl sulfoxide and nitrite in *Shewanella oneidensis*. Biochemistry 42(31):9491–9497

Sekomo CB, Rousseau DP, Saleh SA, Lens PN (2012) Heavy metal removal in duckweed and algae ponds as a polishing step for textile wastewater treatment. Ecol Eng 44:102–110

Senthilkumar M, Gnanapragasam G, Arutchelvan V, Nagarajan S (2011) Treatment of textile dyeing wastewater using two-phase pilot plant UASB reactor with sago wastewater as co-substrate. Chem Eng J 166(1):10–14

Shehab NA, Ortiz-Medina JF, Katuri KP, Hari AR, Amy G, Logan BE, Saikaly PE (2017) Enrichment of extremophilic exoelectrogens in microbial electrolysis cells using Red Sea brine pools as inocula. Biores Technol 239:82–86

Shen N, Huo YC, Chen JJ, Zhang F, Zheng H, Zeng RJ (2015) Decolorization by *Caldicellulosiruptor saccharolyticus* with dissolved hydrogen under extreme thermophilic conditions. Chem Eng J 262:847–853

Somasiri W, Li XF, Ruan WQ, Jian C (2008) Evaluation of the efficacy of upflow anaerobic sludge blanket reactor in removal of colour and reduction of COD in real textile wastewater. Biores Technol 99(9):3692–3699

Song Y, Wang L, Qiang X, Gu W, Ma Z, Wang G (2023) An overview of biological mechanisms and strategies for treating wastewater from printing and dyeing processes. J Water Process Eng 55:104242

Spagni A, Casu S, Grilli S (2012) Decolourisation of textile wastewater in a submerged anaerobic membrane bioreactor. Biores Technol 117:180–185

Sun Y, Gu Y, Zha Q (2021) A novel surface imprinted resin for the selective removal of metal-complexed dyes from aqueous solution in batch experiments: ACB GGN as a representative contaminant. Chemosphere 280:130611

Tang J, Wang Y, Yang G, Luo H, Zhuang L, Yu Z, Zhou S (2018) Complete genome sequence of the dissimilatory azo reducing thermophilic bacterium *Novibacillus thermophiles* SG-1. J Biotechnol 284:6–10

Tawfik A, Ali M, Danial A, Zhao S, Meng F, Nasr M (2021) 2-biofuels (H_2 and CH_4) production from anaerobic digestion of biscuits wastewater: experimental study and techno-economic analysis. J Water Process Eng 39:101736

Teng J, Shen L, He Y, Liao BQ, Wu G, Lin H (2018) Novel insights into membrane fouling in a membrane bioreactor: elucidating interfacial interactions with real membrane surface. Chemosphere 210:769–778

Tian F, Guo G, Fu W, Li S, Ding K, Yang F, Liang C (2024) Decolorization and detoxification of Brilliant Crocein GR by a newly enriched thermophilic consortium. J Environ Manage 356:120623

Van de Werken HJ, Verhaart MR, VanFossen AL, Willquist K, Lewis DL, Nichols JD, Goorissen HP, Mongodin EF, Nelson KE, van Niel EW, Stams AJ (2008) Hydrogenomics of the extremely thermophilic bacterium *Caldicellulosiruptor saccharolyticus*. Appl Environ Microbiol 74(21):6720–6729

Varma M, Gupta AK, Ghosal PS, Majumder A (2021) A review on performance of constructed wetlands in tropical and cold climate: insights of mechanism, role of influencing factors, and system modification in low temperature. Sci Total Environ 755:142540

Wang J, Yan J, Xu W (2015) Treatment of dyeing wastewater by MIC anaerobic reactor. Biochem Eng J 101:179–184

Wang Z, Yin Q, Gu M, He K, Wu G (2018) Enhanced azo dye Reactive Red 2 degradation in anaerobic reactors by dosing conductive material of ferroferric oxide. J Hazard Mater 357:226–234

Wang Y, Wang H, Wang X, Xiao Y, Zhou Y, Su X, Cai J, Sun F (2020) Resuscitation, isolation and immobilization of bacterial species for efficient textile wastewater treatment: a critical review and update. Sci Total Environ 730:139034

Wang X, Jiang J, Gao W (2022) Reviewing textile wastewater produced by industries: characteristics, environmental impacts, and treatment strategies. Water Sci Technol 85(7):2076–2096

Wijetunga S, Li XF, Jian C (2010) Effect of organic load on decolourization of textile wastewater containing acid dyes in upflow anaerobic sludge blanket reactor. J Hazard Mater 177(1–3):792–798

Willquist K, Van Niel EW (2012) Growth and hydrogen production characteristics of *Caldicellulosiruptor saccharolyticus* on chemically defined minimal media. Int J Hydrogen Energy 37(6):4925–4929

Xiao X, Xu CC, Wu YM, Cai PJ, Li WW, Du DL, Yu HQ (2012) Biodecolorization of Naphthol Green B dye by *Shewanella oneidensis* MR-1 under anaerobic conditions. Biores Technol 110:86–90

Xu H, Liu Y, Gao Y, Li F, Yang B, Wang M, Ma C, Tian Q, Song X, Sand W (2018a) Granulation process in an expanded granular sludge blanket (EGSB) reactor for domestic sewage treatment: impact of extracellular polymeric substances compositions and evolution of microbial population. Biores Technol 269:153–161

Xu H, Yang B, Liu Y, Li F, Shen C, Ma C, Tian Q, Song X, Sand W (2018b) Recent advances in anaerobic biological processes for textile printing and dyeing wastewater treatment: a mini-review. World J Microbiol Biotechnol 34(11):165

Yadav A, Kumar P, Rawat D, Garg S, Mukherjee P, Farooqi F, Roy A, Sundaram S, Sharma RS, Mishra V (2022) Microbial fuel cells for mineralization and decolorization of azo dyes: recent advances in design and materials. Sci Total Environ 826:154038

Yan Y, Geng ZQ, Dai K, Guo X, Zeng RJ, Zhang F (2021) Decoupling mechanism of Acid Orange 7 decolorization and sulfate reduction by a *Caldanaerobacter* dominated extreme-thermophilic consortium. J Hazard Mater 419:126498

Yang B, Xu H, Wang J, Yan D, Zhong Q, Yu H (2018a) Performance evaluation of anaerobic baffled reactor (ABR) for treating alkali-decrement wastewater of polyester fabrics at incremental organic loading rates. Water Sci Technol 77(10):2445–2453

Yang B, Xu H, Yang S, Bi S, Li F, Shen C, Ma C, Tian Q, Liu J, Song X, Sand W (2018b) Treatment of industrial dyeing wastewater with a pilot-scale strengthened circulation anaerobic reactor. Biores Technol 264:154–162
Yaseen DA, Scholz M (2019) Textile dye wastewater characteristics and constituents of synthetic effluents: a critical review. Int J Environ Sci Technol 16:1193–1226
Yilmaz T, Demir EK, Aşık G, Başaran ST, Cokgor E, Sözen S, Sahinkaya E (2024) Performance of a high-rate membrane bioreactor for energy-efficient treatment of textile wastewater. J Environ Manage 358:120845
Yurtsever A, Calimlioglu B, Görür M, Çınar Ö, Sahinkaya E (2016) Effect of NaCl concentration on the performance of sequential anaerobic and aerobic membrane bioreactors treating textile wastewater. Chem Eng J 287:456–465
Zhang M, Lin H, Shen L, Liao BQ, Wu X, Li R (2017) Effect of calcium ions on fouling properties of alginate solution and its mechanisms. J Membr Sci 525:320–329
Zhang F, Guo X, Qian DK, Sun T, Zhang W, Dai K, Zeng RJ (2019) Decolorization of acid orange 7 by extreme-thermophilic mixed culture. Biores Technol 291:121875

Enhancing Anaerobic Digestion with the Addition of Conductive Materials

Haowen Zhang, Haoyong Li, and Yan Dang

Abstract The application of anaerobic digestion technology in organic waste treatment has been increasingly emphasized, but its practical application is limited due to its susceptibility to factors such as microbial inhibition and increased organic loading. Conductive materials can be used as effective additives to enhance microbial activity, increase the efficiency of electron transfer, ensure the stability of the anaerobic digestion system, and improve the efficiency of anaerobic digestion. However, the use of conductive materials is limited by high operating cost, increased process complexity, and poor material stability. These limitations can be effectively addressed by modifying the surface of conductive materials, optimizing the preparation process and material structure, and exploring cost-effective raw materials. The influencing factors and application constraints of anaerobic digestion, the mechanism of conductive materials to enhance anaerobic digestion, the limitations and challenges of conductive materials in the application of anaerobic digestion, and the proposed feasible solutions were comprehensively summarized. Conductive materials can improve the stability and processing efficiency of the anaerobic digestion system and promote the development of anaerobic digestion technology, which will bring new opportunities in the field of organic waste treatment and provide sustainable solutions for environmental protection and resource recovery.

Keywords Anaerobic digestion · Conductive materials · Organic waste

H. Zhang · H. Li · Y. Dang (✉)
College of Environmental Science and Engineering, Beijing Forestry University, Beijing 100083, China
e-mail: yandang@bjfu.edu.cn

H. Zhang
e-mail: zhanghw1998@bjfu.edu.cn

H. Li
e-mail: haoyongli@bjfu.edu.cn

G. Wu (ed.), *Anaerobic Digestion*, Green Energy and Technology,
https://doi.org/10.1007/978-3-031-69378-6_16

1 Introduction

Anaerobic digestion technology has gained popularity as an eco-friendly and efficient waste treatment method. The biogas produced during anaerobic digestion can be used as renewable energy to achieve waste resource utilization and energy recovery (Meng et al. 2022; Yadav et al. 2022). However, conventional anaerobic digestion technology still faces challenges in treating complex organic wastes and improving gas production efficiency. These challenges include low gas production rates, long reaction durations, low mass transfer efficiency, and process instability, which limit its effectiveness in practical applications (Nabi et al. 2022; Yang et al. 2023). To address these limitations, anaerobic digestion systems have incorporated conductive materials in recent years to enhance performance and stability.

The use of conductive materials presents new opportunities for the development of anaerobic digestion technology. Conductive materials possess excellent electrical conductivity and biocompatibility, promoting microbial attachment and electron transfer (Chen et al. 2022; Fagbohungbe et al. 2016). Conductive materials can also optimize substrate utilization pathways, balance pH, regulate oxidation-reduction potential (ORP), improve fermentation efficiency, and enhance system stability (Sarker and Nikhil 2023). Additionally, conductive materials increase the rate and volume of gas production, thus improving the economic efficiency and environmental friendliness of the anaerobic digestion system (Liu et al. 2021; Wen et al. 2022).

Nevertheless, the conductive material-enhanced anaerobic digestion technology still faces some challenges. Conductive materials increase operating costs, complicate processes, and are susceptible to corrosion damage (Abdolahi et al. 2014; Jadhav et al. 2022). Therefore, it is important to carefully select suitable conductive materials, optimize their structural design, choose appropriate dosages, modify them to enhance corrosion resistance, and improve biocompatibility (Garcia-Cabezon et al. 2020). To tackle these challenges, it is necessary to investigate the interaction between conductive materials and microorganisms in-depth, optimize the surface properties of conductive materials, select appropriate raw materials, and explore the application of new conductive materials.

2 Types and Characteristics of Conductive Materials

The materials used in the anaerobic digestion process include carbon-based materials such as carbon nanotubes (CNT) and carbon fibers, metal-based materials such as nano-magnetite, conductive polymers, and other conductive materials (Fig. 1). These materials possess excellent electron conduction properties, which can accelerate the metabolic reaction inside microorganisms. At the same time, some have a large specific surface area, which promotes the attachment and growth of microorganisms, enhancing their metabolic efficiency. Additionally, these materials exhibit excellent

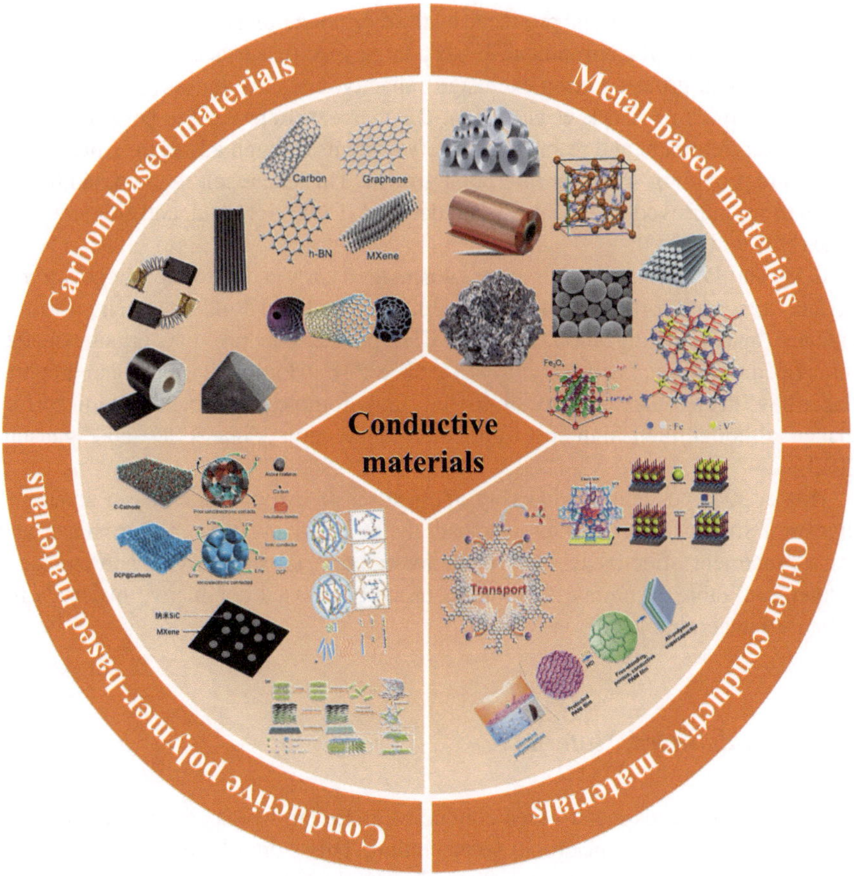

Fig. 1 Types of conductive materials suitable for anaerobic digestion

chemical and structural stability, maintaining long-term electrical conductivity in harsh environments.

2.1 *Carbon-Based Materials*

Carbon materials are utilized in various wastewater treatment and resource recovery technologies, such as biochar, granular activated carbon (GAC), powdered activated carbon (PAC), carbon cloth, carbon mats, carbon fibers, graphene, graphene oxide (GO), reduced graphene oxide (rGO), graphite, and CNT.

Biochar is a solid material produced by the thermochemical conversion of biomass in an oxygen-limited environment. It has a large specific surface area, porous structure, functional groups, cation exchange capacity, and excellent electrical conductivity (2–5 μS/cm) (Baek et al. 2018). Biochar can improve the performance of wastewater treatment and resource recovery systems through adsorption, pH adjustment, and providing space for microbial growth. This can result in increased biogas production, improved chemical oxygen demand (COD) removal, and shortened lag time (Nie et al. 2023).

GAC and PAC are common forms of activated carbon due to their large specific surface area and porosity (Chen et al. 2023a). Modified PAC has been shown to enhance anaerobic digestion, increase CH_4 production, promote organic matter removal, and shorten lag time (Orrantia et al. 2024). Additionally, CNT and graphite flakes have been found to reduce biotoxicity and promote anaerobic digestion performance, although their high cost limits their application (Ziganshina and Ziganshin 2023).

However, large-scale engineering application of carbon materials in anaerobic digestion still faces challenges, particularly in balancing cost, performance, and waste treatment effectiveness. Further investigation is necessary to address these challenges. Table 1 summarizes studies on carbon-based conductive materials for enhanced anaerobic digestion.

2.2 Metal-Based Materials

Magnetite is a porous metal-based conductive material, which can adsorb and store hydrogen produced by anaerobic fermentation. This stored hydrogen can be used by hydrogenotrophic methanogens, which reduces the hydrogen partial pressure. Additionally, these materials can absorb and remove inhibitory substances, such as heavy metals, thereby improving the overall efficiency of anaerobic digestion systems. Iron-based nanoparticles, such as magnetite and hematite, can enhance the anaerobic digestion of leachate from municipal solid waste.

Introduction of metal-based materials promotes direct interspecies electron transfer (DIET) by encouraging symbiosis between bacteria and archaea. For example, the addition of magnetite to upflow anaerobic sludge blanket (UASB) reactors treating leachate enhanced the symbiotic relationship between *Geobacter* and *Methanosaeta* (Lei et al. 2016; Wang et al. 2019a). This results in a significant decrease in volatile fatty acids (VFAs) concentrations, indicating improved VFA biodegradation compared to the control system.

The addition of zero-valent iron (ZVI) to anaerobic digesters has been shown to alleviate inhibition caused by ammonium and VFAs, and can also enhance the reactor's buffering capacity and CH_4 production rate (Antwi et al. 2020). ZVI has the potential to improve propionate degradation and boost methanogenic activity (Puyol et al. 2018). The inclusion of iron and its oxides enhanced the growth of Fe^{3+} reducing microorganisms that could utilize various organic substrates, thereby

Table 1 Carbon-based conductive materials for enhanced anaerobic digestion

Conductive material	Dosage	Enhanced performance	Evidence on direct interspecies electron transfer (DIET)	References
Activated carbon	2–12 g/L	CH_4 yield: 22.3% increase	Microbial community shift	Sun et al. (2020)
Biochar	20 g/L	CH_4 yield/ production rate: increased	Na	Rasapoor et al. (2020)
	2.5 g/L	CH_4 yield: 24.9% increase	Na	Indren et al. (2020)
	15 g/L	Lag time: significantly shorten CH_4 production rate: increased	Microbial community shift	Wang et al. (2020b)
	2–15 g/L	CH_4 production rate: 17.9% increase	Microbial community shift/ syntrophic degradation of butyrate to acetate at high H_2	Wang et al. (2018b)
	5 g/L	CH_4 yield: 18% increase	Electroactive microbial enrichment/ΔG characteristics of methanogenesis	Lim et al. (2020)
	10 g/L	CH_4 production rate: increased Lag time: shortened	Microbial community shift	Li et al. (2018b)
	4 g/L	Lag time: 28.6% reduction	Microbial community shift/ granulation enhanced/ conductivity enhancement	Wang et al. (2018a)
Carbon cloth	3 pieces per reactor	CH_4 production rate: 222.3% increase	Increase of electron transfer rate	Li et al. (2018a)
	10 pieces per reactor	Organic loading rate (OLR): 34.2% increase	Microbial community shift/ Conductivity enhancement	Lei et al. (2016)
	Na	COD removal efficiency: 98% at influent COD of 10,000 mg/L	Microbial community shift	Jia et al. (2020)

(continued)

Table 1 (continued)

Conductive material	Dosage	Enhanced performance	Evidence on direct interspecies electron transfer (DIET)	References
Carbon felt	Na	CH_4 yield: 12.9% increase Sludge: 17.2% reduction	Microbial community shift	Zhao et al. (2016)
	30 g/L	Propionate accumulation at high OLR: suppression	Microbial community shift	Xu et al. (2016)
Carbon fiber	Na	CH_4 production rate: increase	Microbial community shift/ DIET genes/ proteins characteristic/ Electrocatalytic features of biomass	Lin et al. (2020)
	Na	CH_4 yield: 240% increase CH_4 production rate: 670% increase	Microbial community shift/ conductivity enhanced/ electrocatalytic activity increased	Barua et al. (2018)
CNT	5 g/L	Butyrate oxidation and CH_4 production: enhanced	Formation of CNT-microbials aggregates	Zhang et al. (2018)
	5 g/L	CH_4 production rate (pure culture): 1700% increase VFA conversion efficiency (coculture): 150% increase	Na	Salvador et al. (2017a)
GAC	15 g/L	CH_4 yield: 45% increase	Enhancement of VFA degradation/ biofilm formation	Johnravindar et al. (2020)
	25 g/L	Lag phase: reduced, CH_4 yield (35°C): 200% increase	Microbial community shift	Ryue et al. (2019)
	25 g/L	COD removal rate: 13% increase CH_4 yield: 131 mg COD/d increase per reactor	Microbial community shift	Zhang et al. (2020a)

(continued)

Table 1 (continued)

Conductive material	Dosage	Enhanced performance	Evidence on direct interspecies electron transfer (DIET)	References
	10 g/L	OLR variation: enhanced	Microbial community shift	Ferrero et al. (2020)
	25 g/L	Organic matter removal: 1375 mg COD/L increase	Microbial community shift	Guo et al. (2020)
Graphene	1 g/L	CH_4 production rate: 25–26.4% increase	Microbial community shift	Lin et al. (2018a)
	1 g/L	CH_4 yield: 25% increase CH_4 production rate: 19.5% increase	Microbial community shift/ electron transfer rate (ETR) enhancement	Lin et al. (2017)
	1 g/L	CH_4 production rate: 28% increase	Microbial community shift/ ETR enhancement	Lin et al. (2018b)
PAC	3 g/batch	CH_4 production rate and degradation of recalcitrant substrate: promoted	Microbial community shift/ change of FTIR characteristic	Li et al. (2021)
	15 g/L	Anaerobic digestion performance: enhanced (brewery spent yeast)	NA	Valero et al. (2020)

NA: not available; AC: activated carbon; CNT: carbon nanotube; GAC: granular-activated carbon; PAC: powdered-activated carbon

aiding in the breakdown of complex polymers (Baek et al. 2015). Table 2 summarizes studies on metal-based conductive materials to enhance anaerobic digestion.

2.3 Conductive Polymer-Based Materials

Organic polymers can be classified into two categories based on their electrical conductivity: insulators, represented by conventional rubbers whose electrical conductivity is usually below 10^{-10} S/cm, and conductive polymers, which are typically synthesized through oxidative polymerization. Until the preparation of iodine-doped polyacetylene (PA), organic polymers were considered insulating materials. This breakthrough showed that the conductivity of organic polymers can be similar to metals, challenging the traditional concept that organic polymers are not conductive

Table 2 Metal-based conductive materials for enhanced anaerobic digestion

Conductive material	Reactor type	Effect performance	Enriched bacteria	Dominant Archaea	References
Magnetite	AnSBR	CH_4 yield: 50% increase	*Desulfovibrionales*	*Methanosarcina*	Wang et al. (2020a)
	UASB	CH_4 yield: 36–58% increase	*Geobacter*	*Methanosarcina*	Zhuang et al. (2018)
	UASB	CH_4 production: 4082 ± 334 mL	*Pseudomonas*	*Methanosarcina*	Chen et al. (2020)
	AnSBR	CH_4 yield: 44% increase	*Thauera, Dechloromonas, Tepidoanaerobacter* and *Coprothermobacter*	*Methanothrix*	Jing et al. (2017)
	AnSBR	CH_4 yield: 16% increase	*Pelobacter*	*Methanosarcina*	Im et al. (2019)
	EGSB	Organic loading: 20 kg COD / $m^3 \cdot d$	*Geobacter*	*Methanothrix*	Wang et al. (2019a)
	UASB	CH_4 content: 32% increase	*Anaerolineaceae* and *Eubacteriaceae*	*Methanosarcina and Methanothrix*	Lei et al. (2018)
	AnSBR	CH_4 yield: 78.3% increase	*Proteiniclasticum* and *Prolixibacter*	*Methanosarcina*	Yin et al. (2018)
	AnSBR	CH_4 yield: 12.2% increase	*Proteiniclasticum*	*Methanosarcina*	Yin et al. (2017)
ZVI	UASB	CH_4 content: 36% increase	*Chloroflexi*	*Euryarchaeota*	Antwi et al. (2020)
	AnSBR	CH_4 yield: 19.7% increase	Na	Na	Liang et al. (2017)
Nano Fe_3O_4	CSTR	Biogas production: 24.44% increase	Na	*Methanosarcina*	Xiang et al. (2019)
	CSTR	CH_4 yield: 39.4% increase	Na	Na	Noonari et al. (2019)
Fe_2O_3	EGSB	12.6% more mL/g VS CH_4 than the control reactor	Na	*Methanothrix*	Ambuchi et al. (2017)
		The accumulative CH_4 production: improved by a maximum of 11.06%	*Ruminococcaceae*	*Methanoculleus*	Lu et al. (2019)

(continued)

Table 2 (continued)

Conductive material	Reactor type	Effect performance	Enriched bacteria	Dominant Archaea	References
Goethite	Na	Biogas and CH_4 production: increased by 16.1% and 24.1%	*Pseudomonas putida*	*Methanosarcina mazei*	Tan et al. (2015)
	CSTR	The methane yield reached 263 ± 5 mL/g	*Clostridium*	*Methanosaeta*	Yue et al. (2015)

Na: not available, AnMBR: anaerobic membrane bioreactor, UASB: upflow anaerobic sludge blanket, EGSB: expanded granular sludge bed, CSTR: continuous stirred tank reactor

(Shirakawa et al. 1977). Through continuous research and effective doping strategies, conductive organic polymers such as polypyrrole (PPy), polyaniline (PANI), and polythiophene (PT) have been synthesized (Stejskal 2020).

2.4 *Other Conductive Materials*

New conductive materials are being explored, such as DNA-fullerene complexes and DNA-peptide nanowires with electrical conductivity 1000 times that of DNA itself (Jiménez-Monroy et al. 2017). DNA-related molecules exhibit good conductivity and potential for extracellular electron transfer (EET). Furthermore, the binding of phenazine to DNA improves retention in biological membranes and facilitates efficient EET. DNA-based composites utilize DNA as the foundational structure of the conductive material, with nanowires serving as the primary electron carriers alongside other binding compounds. These materials offer potential advantages over traditional conductive materials, such as better biocompatibility and the ability to serve as templates for the creation of other nanomaterials (Cheong et al. 2019; Thivya and Wilson 2019). However, demonstrating electron transfer function between microorganisms via these DNA materials under anaerobic conditions remains challenging. Therefore, further research is necessary to gain a deeper understanding of how DNA-based materials facilitate DIET.

3 Mechanisms of Anaerobic Digestion Enhanced by Conductive Materials

Conductive materials can enhance the efficiency of anaerobic digestion through diverse mechanisms (Fig. 2). Materials with a large specific surface area provide active sites for microbial attachment and immobilization. Additionally, conductive materials can accelerate the degradation of organic waste by promoting electron transfer between microorganisms, improving conversion efficiency. At the same time, conductive materials can adjust the ORP within the system, providing a more suitable growth environment for microorganisms and enhancing their ability to degrade organic matter under anaerobic conditions. The conductive material also stabilizes the pH of the digestion solution, reducing the impact of fluctuations in acidity and alkalinity on the stability of the system. The conductive material may stimulate the secretion of extracellular polymeric substances (EPS) and improve cellular metabolism. Conductive materials provide an effective solution for improving biogas production and optimizing waste treatment systems.

3.1 Increasing the Efficiency of Electron Transfer

Electron transfer in DIET differs from interspecies hydrogen/formate transfer. In DIET, electrons are directly transferred from the donor to the acceptor through electrical contact, rather than using an intermediate electron carrier. The use of conductive materials can increase the efficiency of electron transfer between microorganisms, thereby enhancing anaerobic digestion. The way conductive materials facilitating electron transfer is distinct (Fig. 3). The discovery of the DIET phenomenon and its application in anaerobic digestion reactors occurred recently, but its crucial role

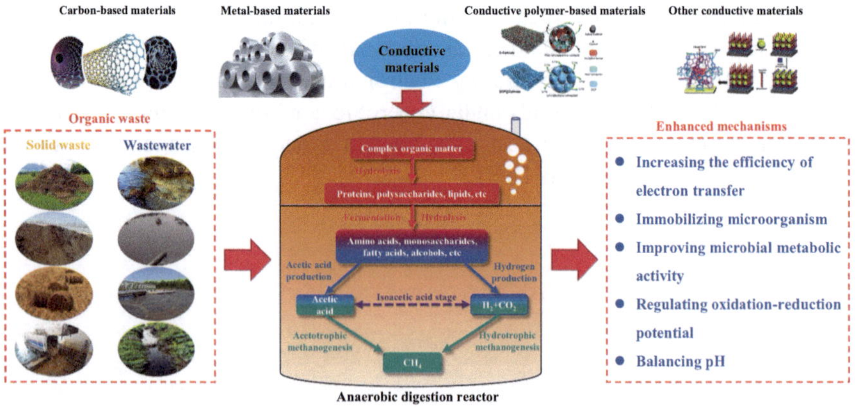

Fig. 2 Mechanism of anaerobic digestion of organic waste enhanced by conductive materials

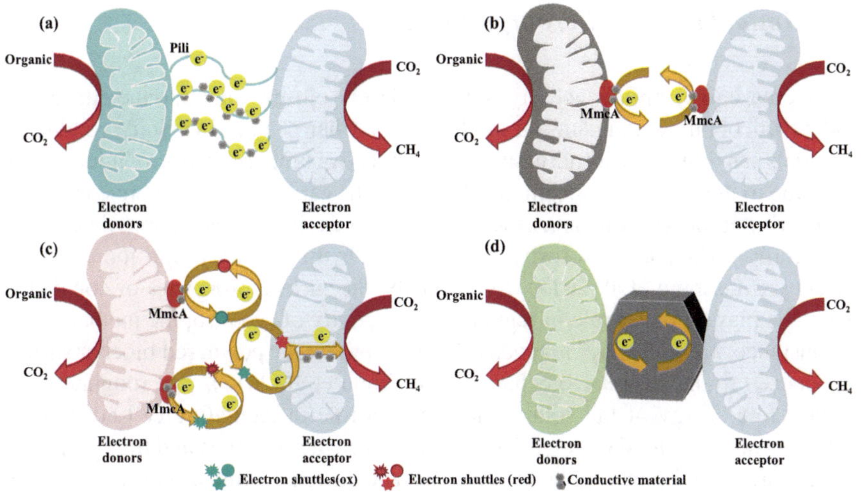

Fig. 3 Conducting material-enhanced pili (**a**), c-type cytochromes (**b**), electron shuttles (**c**), and serve as mediator (**d**) in diverse electron transfer pathways

has been demonstrated in enhancing pollutant removal (Lovley 2011; Rotaru et al. 2014). Enhancing DIET with conductive materials is generally believed to promote the growth of methanogens, thereby improving the CH_4 production of anaerobic digestion. However, current evidence is primarily indirect. For instance, the presence of a conductive material can facilitate the degradation of certain bio-refractory organic compounds (e.g., propionate), suggesting that electrons can be transferred between microbial species via the DIET pathway, independent of H_2 diffusion in the system (Guo et al. 2021; Zhang et al. 2020b).

The addition of conductive materials can promote mutualistic metabolism between various microorganisms, leading to the enrichment of electroactive microorganisms for long-distance electron transfer. Genera with the DIET potential were identified in magnetite dosing systems, such as *Syntrophomonas* and *Methanosarcina*, *Thauera* and *Methanobacterium*, and *Chloroflexi* and *Methanosarcina* (Wang et al. 2018c; Yang et al. 2016). The dosage of biochar has been shown to enhance interspecies electron transfer between *Geobacter* and *Methanothrix*, *Anaerolineaceae* and *Methanothrix*, as well as *Syntrophomonas*/*Clostridiaceae* and *Methanothrix*/*Methanosarcina* (Luo et al. 2015; Wang et al. 2018a, b). Inoculum sources, substrates, material doses, reactor characteristics, and operating conditions shape diverse microorganisms and affect their dynamics, while different microorganisms may perform a similar DIET function during anaerobic digestion.

3.2 *Immobilizing Microorganisms*

The increase in symbiotic activity is believed to be closely linked to the attachment and growth of microorganisms on solid surfaces (Chen et al. 2014). Some conductive materials can provide adsorption sites for the effective immobilization of microorganisms due to their inherent specificity (Fig. 4a). The capacity of conductive materials to adsorb microorganisms is a crucial factor influencing this adhesion, determined by specific surface area, porosity, particle size, hydrophobicity, and ion exchange capacity (Masebinu et al. 2019). Additionally, the surface properties of conductive materials play a pivotal role in regulating the activity of acetotrophic methanogens. For instance, *Methanosarcina* tends to thrive more on the powdered biochar surface as opposed to the granular biochar surface (Zhou et al. 2017). The roughness and surface free energy of the solid surface play a role in regulating cell attachment, with microorganisms showing a preference for rougher surfaces and higher specific surface areas, and materials with lower surface energy facilitate biofilm formation (Ammar et al. 2015; Habouzit et al. 2011).

Biochar serves as an effective tool for enhancing anaerobic digestion by selectively enriching functional microorganisms (He et al. 2020). Its ability to form biofilms on

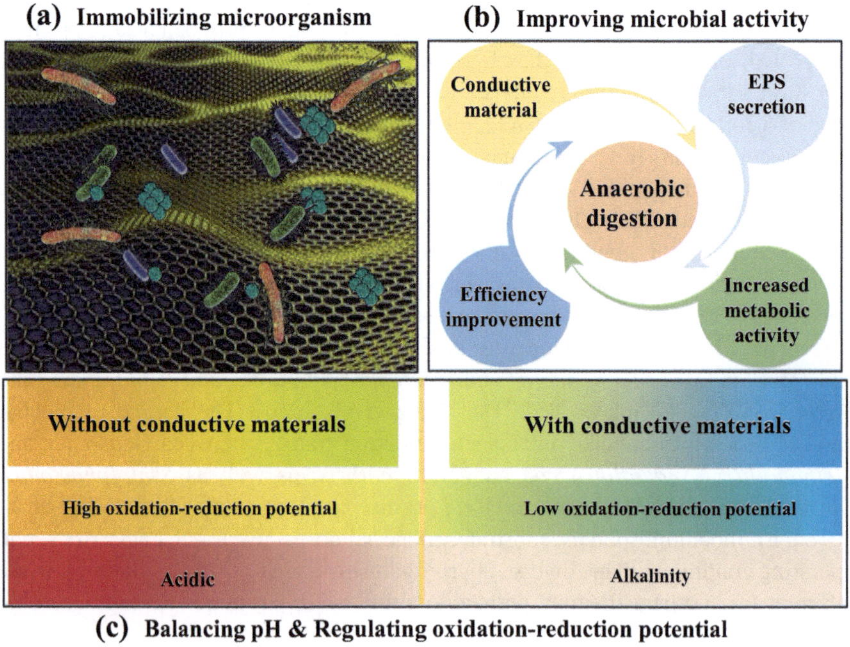

Fig. 4 Enhanced anaerobic digestion with the dosage of conductive materials by: **a** immobilizing microorganism, **b** improving microbial metabolic activity, **c** balancing pH and regulating of oxidation–reduction potential

both the surface and internal microtubules allows for the immobilization of microorganisms. Furthermore, porous biochar has been found to absorb ammonia, reducing its inhibitory effect on microorganisms and immobilizing methanogens (Luo et al. 2015). Similarly, carbon fibers act as a supportive medium in anaerobic fermentation systems, facilitating biofilm formation and increasing biomass retention, ultimately promoting the efficiency of anaerobic digestion reactors. This is evidenced by the higher abundance of *Methanosaeta* and *Methanosarcina* in bioreactors fed with carbon fibers, indicating enhanced methanogenic capacity due to microorganism immobilization (Wang et al. 2019b). In another study, the addition of GAC immobilized with riboflavin promoted the growth of *Methanothrix* and bacteria in the anaerobic digestion system, thus enhancing pollutant treatment efficiency (Huang et al. 2022).

3.3 *Improving Microbial Metabolic Activity*

Microbial metabolic activity is associated with effective degradation of organic wastes, accelerated decomposition of wastes, and increased biogas production and digestion efficiency. As shown in Fig. 4b, conductive materials play a crucial role in stimulating microbial secretion of EPS and enhancing microbial metabolic activity (Yang et al. 2024). These polymers can serve as an electron transfer carrier, promoting electron transfer within microorganisms, accelerating metabolic reactions, and improving the metabolic rate and product generation efficiency of microorganisms (Paritosh et al. 2020). EPS have an adsorption capacity and can adsorb substrates and nutrients, providing the necessary energy and nutrients for microorganisms to carry out metabolic reactions (Ye et al. 2018). This increases the efficiency of substrate utilization and promotes metabolic activity.

The binding mechanism of conductive materials, such as nano-zero-valent iron (nZVI), nano-magnetite, iron oxide nanoparticles, and silver nanoparticles, with microorganisms opens new research and application possibilities in the field of biotechnology. These materials can facilitate electron transfer between microbial cells, extend the electron transfer chain, and accelerate the efficiency of metabolic processes and biological reactions (Li et al. 2019, 2022). These materials can enhance the metabolic activity of microorganisms by stimulating the production of EPS when acting together.

3.4 *Regulating Oxidation–Reduction Potential (ORP)*

ORP has been identified as a critical factor in enhancing the hydrolysis of biodegradable components in biomass or wastewater. The incorporation of conductive materials serves to regulate the ORP (Fig. 4c). Numerous studies investigated ORP dynamics during CH_4 production and found that adding conductive agents can change ORP from −200 to −400 mV (Hirano et al. 2013). The enhancement of CH_4 production by biochar mainly depends on its redox properties and charge/discharge capacity (Yuan et al. 2018). Salvador et al. (2017b) discovered that high concentrations of CNT, specifically at 5 g/L, resulted in a reduction of ORP within the anaerobic system, favoring the growth of methanogens.

Zero-valent metals (ZVMs) and metal catalysts have emerged as a promising option due to their accessibility and cost-effectiveness. ZVMs have a significant effect on ORP of anaerobic digestion systems. For example, ZVI, with a standard reduction potential of −440 mV, reduced ORP of the anaerobic digestion system and created a favorable environment for the efficient enrichment of anaerobic bacteria (Jiang et al. 2011).

3.5 Balancing pH

Anaerobic digestion reactors with acidified substrates typically exhibit reduced stability, inefficient VFA conversion, and reduced CH_4 production (Wang et al. 2017). Conductive materials can also balance pH (Fig. 4c). Maintaining a stable pH is critical to the success of the anaerobic digestion process (Shen et al. 2015). Biochar, with its acid buffering properties, can play an important role in anaerobic digestion. For instance, biochar derived from vermicompost can effectively regulate the pH of the reactor in the range of 5.0 to 9.0, while the reactor without biochar showed pH fluctuations from 4.0 to 3.0 (Wang et al. 2017). There was a direct relationship between the pH values of biochar, indicating its rectifying effect under acidic or alkaline conditions (Komnitsas et al. 2015). In addition, the carbon cloth had the ability to mitigate acidification when the pH was adjusted from 5.0 to 7.0 (Zhao et al. 2017).

4 Application of Conductive Materials in Anaerobic Digestion

4.1 Application

The purification of biogas has been a focus due to its composition, comprising CH_4 (50–75%), CO_2 (25–50%), N_2 (0–10%), and hydrogen sulfide (H_2S; 0–3%). The presence of CO_2 poses a challenge to the efficiency of biogas, prompting increased interest in biogas upgrading to biomethane for use as a vehicle fuel and in national grids (Ryckebosch et al. 2011). H_2S, known for its malodorous and corrosive nature, can lead to environmental and economic concerns through the emission of SO_2 during combustion, which corrodes metallic pipelines, engines, and combustors (Barik et al. 2013). Consequently, purification techniques have been developed to selectively absorb CO_2 and H_2S, thereby enhancing the potential utilization of biogas and minimizing the associated hazards.

Given that the CO_2 content in biogas can exceed 60%, efforts have been made to selectively absorb and sequester CO_2. For instance, the incorporation of SAPO-34 zeolite into polyvinylidene fluoride membranes has shown promise due to its high affinity for CO_2, attributed to its chabazite (CHA) structure with a framework pore size of 0.38 nm, making it suitable for absorbing small gas molecules within a biogas mixture. The addition of 10 wt% SAPO-34 zeolite has been found to enhance the CO_2 flux of PVDF membranes by 60% through increased membrane porosity and improved affinity toward CO_2. Additionally, the inclusion of 5wt% polybenzimidazole (PBI) was observed to enhance membrane stability in corrosive solutions, albeit resulting in a lower CO_2 flux (Ahmad et al. 2016). These advancements aim to effectively mitigate the challenges posed by CO_2 and H_2S while elevating the potential applications of biogas.

4.2 Limitations and Challenges

Conductive materials have several advantages in improving anaerobic digestion by facilitating electron transfer between microorganisms, improving waste decomposition efficiency, and increasing biogas production. However, the cost of conductive material dosage, the long-term effectiveness of conductive materials in practical applications, and the increased complexity of the anaerobic digestion process may limit their practical application.

Research on the enhancement of anaerobic digestion by conductive materials has made great progress, but the cost issue is an important factor to be considered for practical application. Some efficient conductive materials, such as CNT and conductive polymers, are more expensive, which will directly increase the construction and operation costs of anaerobic digestion systems (Sánchez 2022). Especially, in full-scale projects, large amounts of conductive materials are required, which further increases the investment cost. Second, the maintenance cost of conductive materials is also a consideration. Some conductive materials may need to be replaced or maintained on a regular basis, which increases operating and labor costs. In addition, integrating conductive materials into existing anaerobic digestion systems may require additional engineering and technical support to ensure their effectiveness and safety, which also increases costs.

The stability of conductive materials is a critical factor in the enhanced anaerobic digestion process and a major contributor to its limitations. Destabilization of conductive materials due to external environmental influences can have a direct impact on their long-term effectiveness and cost-efficiency in anaerobic digestion systems. Under the influence of pressure, temperature, pH, and other factors, conductive materials may be denatured or degraded, resulting in a decrease in their conductivity or even rendering them ineffective. This lack of stability can limit the ability of conductive materials to last in anaerobic digestion systems, increasing maintenance costs and system risks (Wang et al. 2023b).

Additionally, microbial metabolites can also affect the stability of conductive materials. Anaerobic digestion systems produce various metabolites through microbial metabolic activity. Some of these metabolites, such as sulfides, acids, and other substances, can react chemically with conductive materials, leading to structural damage or performance degradation (Chen et al. 2023b). This, in turn, affects the application of the material in the system and its interaction with microbial communities. Therefore, it is important to ensure the stability of conductive materials. Furthermore, certain conductive materials may corrode or change in nature, which can inhibit the growth of microorganisms or prevent the formation of a stable symbiotic relationship with it, thus affecting the overall performance of the anaerobic digestion system (He et al. 2021). This is particularly important to consider and manage in full-scale anaerobic digestion projects with long operating cycles, where the stability of conductive materials is crucial.

The use of conductive materials can enhance microbial metabolic activity and improve the efficiency of organic waste treatment in anaerobic digestion. However, it may also increase process complexity. The use of conductive materials must be carefully considered with regards to the application method, amount, and location to enhance system efficiency while controlling costs. Customized selection of conductive materials is necessary, requiring careful system design and control. The placement of conductive materials must be determined rationally to ensure effective interaction with the microbial community. This increases the technical difficulty and management cost of system operation.

Additionally, the introduction of conductive materials may complicate the regulation of system operating parameters. The presence of conductive materials can alter the microenvironment within the system and impact the growth and metabolic processes of microorganisms (Wang et al. 2023c). To achieve optimal treatment results, it may be necessary to adjust the system's parameters, such as temperature, pH, and ORP, more frequently and precisely. Additionally, the introduction of conductive materials may increase the complexity of system monitoring and control. To monitor and evaluate the impact of conductive materials in real-time, it is essential to establish monitoring indicators, detection methods, and a comprehensive data analysis system. Additionally, for the interaction between conductive materials and microorganisms, a more sophisticated intelligent control system may be necessary to automate and optimize system operations.

4.3 Proposed Solutions

4.3.1 Optimizing Preparation Processes and Selecting Cost-Effective Raw Materials

Optimizing the preparation process can improve the structure and properties of conductive materials, enhancing their stability and durability. Suitable synthesis methods and control conditions can achieve homogeneity and purity, reducing defects and contamination on the material surface. This improves contact efficiency with microorganisms, facilitates the electron transfer process, and enhances the efficiency and stability of the anaerobic digestion system. Finding cost-effective raw materials for conductive materials can reduce system construction and operating costs and improve economic sustainability. Affordable and readily available raw materials can lower production costs of conductive materials, promoting their widespread use in anaerobic digestion applications. Additionally, the selection of raw materials must consider their renewability and environmental impact to ensure the sustainability of the system. For example, researchers have used rapid microwave heating (tens of seconds) to synthesize inexpensive amorphous materials with excellent cycling and electrical conductivity properties using natural pyroxene ores and charcoal as feedstocks (Wang et al. 2023a).

Additionally, the performance and applicability of conductive materials can be improved to meet the needs of different anaerobic digestion systems by optimizing the preparation process and selecting cost-effective raw materials. Feng et al. (2024) successfully prepared highly conductive poly(3,4-ethylenedioxythiophene):poly(4-styrenesulfonate) aerogels by ice template preassembly and a specially designed thawing process in H_2SO_4 solution, which exhibited excellent conductivity, low density, improved mechanical properties, and competitive electromagnetic interference (EMI) shielding efficiency, as well as good performance in supercapacitors. The properties and structures of the conductive materials can be adjusted for different application scenarios and treatment objects to better adapt to different microbial communities and reaction conditions, achieving more efficient waste degradation and biogas generation. Optimizing the preparation process and selecting cost-effective raw materials can effectively address the limitations and challenges of applying conductive materials in anaerobic digestion. This improves the performance and economic benefits of the system and promotes the widespread application and dissemination of this technology in waste treatment.

4.3.2 Optimized Dosing Method and Surface Modification

By optimizing the dosing method of conductive materials, the contact efficiency with microorganisms can be improved and the electron transfer between microorganisms can be promoted, thus accelerating the process of waste decomposition and biogas production. By selecting the appropriate dosing method and evenly dispersing the conductive material in the reactor or mixing it with the waste before adding it to the reactor, the contact area between the conductive material and the microorganisms can be maximized. This will improve electron transfer efficiency, enhance system stability, and improve treatment effectiveness.

The surface modification of the conductive material can improve its compatibility and affinity with microorganisms, reduce the impact on the microbial community structure, and improve its applicability and stability in anaerobic digestion systems. Through surface modification, functionalization, or coating, the surface properties and chemical activity of conductive materials can be regulated to better adapt to the growth environment of microorganisms and reduce the possible adverse effects on microorganisms. This helps to maintain the stability and efficient operation of the system. For example, Tan et al. (2024) used 3-mercaptopropyltrimethoxysilane (3-MT) to modify viscose nonwovens, resulting in 3-MT@Cu@Viscose with very low surface and volume resistivity (346.6 and 333.2 mΩ·m), and corrosion rate reduced by 58%.

Surface modifications for optimizing conductive materials can also improve their utilization, reduce operating costs, and decrease the demand for raw materials. By rationally designing dosing methods and modification programs, conductive material loss and waste can be reduced, extending their service life and reducing maintenance costs and energy consumption. This improves the economic sustainability of the system.

5 Future Perspectives

Conductive materials have the potential to efficiently reinforce anaerobic digestion processes. They can improve microbial activity and electron transfer efficiency, regulate pH and ORP, and immobilize active microorganisms. This significantly enhances the treatment efficiency and stability of anaerobic digestion systems. Studies have shown that the introduction of conductive materials can effectively promote microbial attachment and growth, improve the efficiency of waste degradation, and promote resource recovery in the anaerobic digestion process. Additionally, conductive materials can optimize the efficiency of substrate utilization, thereby improving the gas production rate. However, the application process still faces some challenges, such as bioattachment on the surface of conductive materials and the stability of such materials. Addressing these issues requires further in-depth research and innovation.

Enhanced anaerobic digestion by dosing conductive materials is a topic of extensive research. Future studies should focus on the interaction mechanism between conductive materials and microorganisms to optimize the design and application of conductive materials for more efficient waste degradation and resource recovery. Secondly, the development of new conductive materials should be explored to enhance their conductivity and corrosion resistance. This can be achieved by combining nanomaterials and functionalized surface technologies to design conductive materials with excellent electrical conductivity, biocompatibility, and resistance to acid and alkali corrosion, which will help to address the current technical challenge of conductive materials that are unable to operate stably over long periods of time. Finally, the applicability of conductive materials in various environmental conditions should be examined. Through performance optimization and application engineering of conductive materials, the stability and economics of the anaerobic digestion system can be further improved, promoting its widespread use in waste treatment. Continuous research and innovation are expected to lead to breakthroughs and progress in the field of waste treatment through conductive material-enhanced anaerobic digestion technology, contributing to sustainable development.

6 Take-Home Messages

(1) Conductive materials present new opportunities for the development of anaerobic digestion technology, which can accelerate electron transfer, increase the rate and volume of gas production, and improve the economic efficiency and environmental friendliness of the anaerobic digestion system.
(2) Typical conductive materials include carbon-based materials such as carbon nanotubes and carbon fibers, metal-based materials such as nano-magnetite, conductive polymers, and other conductive materials.
(3) The addition of conductive materials enriches electroactive microorganisms, increases microbial interactions, and provides a novel strategy for long-distance electron transfer between microorganisms.
(4) The development of conductive material-enhanced anaerobic digestion can be further promoted by selecting cost-effective raw materials, optimizing the preparation process, exploring reasonable dosing methods, and modifying material surfaces.

References

Abdolahi A, Hamzah E, Ibrahim Z, Hashim S (2014) Application of environmentally-friendly coatings toward inhibiting the microbially influenced corrosion (MIC) of steel: a review. Polym Rev 54(4):702–745

Ahmad N, Leo C, Junaidi M, Ahmad A (2016) PVDF/PBI membrane incorporated with SAPO-34 zeolite for membrane gas absorption. J Taiwan Inst Chem Eng 63:143–150

Ambuchi JJ, Zhang Z, Shan L, Liang D, Zhang P, Feng Y (2017) Response of anaerobic granular sludge to iron oxide nanoparticles and multi-wall carbon nanotubes during beet sugar industrial wastewater treatment. Water Res 117:87–94

Ammar Y, Swailes D, Bridgens B, Chen J (2015) Influence of surface roughness on the initial formation of biofilm. Surf Coat Technol 284:410–416

Antwi P, Zhang D, Luo W, Kabutey FT, Li J, Su H, Wu M, Liu Z (2020) Response of hydrolysis, methanogenesis, and microbial community structure to iron dose during anaerobic digestion of food waste leachate. Biomass Conv Biorefinery 12:5959–5973

Baek G, Kim J, Cho K, Bae H, Lee C (2015) The biostimulation of anaerobic digestion with (semi) conductive ferric oxides: their potential for enhanced biomethanation. Appl Microbiol Biotechnol 99:10355–10366

Baek G, Kim J, Kim J, Lee C (2018) Role and potential of direct interspecies electron transfer in anaerobic digestion. Energies 11(1):107

Barik D, Sah S, Murugan S (2013) Biogas production and storage for fueling internal combustion engines. Int J Emerging Tech Adv Eng 3(3):193–202

Barua S, Zakaria BS, Dhar BR (2018) Enhanced methanogenic co-degradation of propionate and butyrate by anaerobic microbiome enriched on conductive carbon fibers. Biores Technol 266:259–266

Chen L, Fang W, Chang J, Liang J, Zhang P, Zhang G (2022) Improvement of direct interspecies electron transfer via adding conductive materials in anaerobic digestion: mechanisms, performances, and challenges. Front Microbiol 13:860749

Chen L, Zhang R, Li Y, Zhang Y, Fang W, Zhang P, Zhang G (2023a) Internal driving mechanisms of microbial community and metabolism for granular activated carbon enhancing anaerobic digestion of high-strength sulfate organic wastewater. Chem Eng J 476:146648

Chen L, Zhang Y, Liang J, Li Y, Zhang J, Fang W, Zhang P, Zhang G, Ngo HH (2023b) Improvement of anaerobic digestion containing sulfur with conductive materials: focusing on recent advances and internal biological mechanisms. Chem Eng J 472:144867

Chen Q, Liu CQ, Liu XY, Sun DZ, Li PS, Qiu B, Dang Y, Karpinski NA, Smith JA, Holmes DE (2020) Magnetite enhances anaerobic digestion of high salinity organic wastewater. Environ Res 189:109884

Chen S, Rotaru AE, Liu F, Philips J, Woodard TL, Nevin KP, Lovley DR (2014) Carbon cloth stimulates direct interspecies electron transfer in syntrophic co-cultures. Biores Technol 173:82–86

Cheong LZ, Zhao W, Song S, Shen C (2019) Lab on a tip: applications of functional atomic force microscopy for the study of electrical properties in biology. Acta Biomater 99:33–52

Fagbohungbe MO, Herbert BM, Hurst L, Li H, Usmani SQ, Semple KT (2016) Impact of biochar on the anaerobic digestion of citrus peel waste. Biores Technol 216:142–149

Feng J, Zhuang Z, Zhou Y, Li C (2024) Highly conductive PEDOT: PSS aerogels. Adv Func Mat, 2315188

Ferrero P, Izquierdo M, Álvarez-Hornos FJ, Penya-Roja JM, Martínez-Soria V (2020) Influence of activated carbon on performance and microbial communities in the treatment of solvent pollutant mixtures in a continuous stirred tank reactor. Environ Sci: Water Res Technol 6(5):1445–1455

Garcia-Cabezon C, Salvo-Comino C, Garcia-Hernandez C, Rodriguez-Mendez ML, Martin-Pedrosa F (2020) Nanocomposites of conductive polymers and nanoparticles deposited on porous material as a strategy to improve its corrosion resistance. Surf Coat Technol 403:126395

Guo B, Zhang Y, Yu N, Liu Y (2021) Impacts of conductive materials on microbial community during syntrophic propionate oxidization for biomethane recovery. Water Environ Res 93(1):84–93

Guo B, Zhang Y, Zhang L, Zhou Y, Liu Y (2020) RNA-based spatial community analysis revealed intra-reactor variation and expanded collection of direct interspecies electron transfer microorganisms in anaerobic digestion. Biores Technol 298:122534

Habouzit F, Gévaudan G, Hamelin J, Steyer JP, Bernet N (2011) Influence of support material properties on the potential selection of Archaea during initial adhesion of a methanogenic consortium. Biores Technol 102(5):4054–4060

He P, Zhang H, Duan H, Shao L, Lü F (2020) Continuity of biochar-associated biofilm in anaerobic digestion. Chem Eng J 390:124605

He Z, Yang C, Tang C, Liu W, Zhou A, Ren Y, Wang A (2021) Response of anaerobic digestion of waste activated sludge to residual ferric ions. Biores Technol 322:124536

Hirano S, Matsumoto N, Morita M, Sasaki K, Ohmura N (2013) Electrochemical control of redox potential affects methanogenesis of the hydrogenotrophic methanogen *Methanothermobacter thermautotrophicus*. Lett Appl Microbiol 56(5):315–321

Huang Y, Cai B, Dong H, Li H, Yuan J, Xu H, Wu H, Xu Z, Sun D, Dang Y, Holmes DE (2022) Enhancing anaerobic digestion of food waste with granular activated carbon immobilized with riboflavin. Sci Total Environ 851:158172

Im S, Yun Y, Song Y, Kim DH (2019) Enhanced anaerobic digestion of glycerol by promoting DIET reaction. Biochem Eng J 142:18–26

Indren M, Birzer CH, Kidd SP, Medwell PR (2020) Effect of total solids content on anaerobic digestion of poultry litter with biochar. J Environ Manage 255:109744

Jadhav P, Nasrullah M, Zularisam AW, Bhuyar P, Krishnan S, Mishra P (2022) Direct interspecies electron transfer performance through nanoparticles (NPs) for biogas production in the anaerobic digestion process. Int J Environ Sci Technol 19(10):10427–10439

Jia R, Sun D, Dang Y, Meier D, Holmes DE, Smith JA (2020) Carbon cloth enhances treatment of high-strength brewery wastewater in anaerobic dynamic membrane bioreactors. Biores Technol 298:122547

Jiang Z, Lv L, Zhang W, Du Q, Pan B, Yang L, Zhang Q (2011) Nitrate reduction using nanosized zero-valent iron supported by polystyrene resins: role of surface functional groups. Water Res 45(6):2191–2198

Jiménez-Monroy KL, Renaud N, Drijkoningen J, Cortens D, Schouteden K, Van Haesendonck C, Guedens WJ, Manca JV, Siebbeles LD, Grozema FC (2017) High electronic conductance through double-helix DNA molecules with fullerene anchoring groups. J Phys Chem A 121(6):1182–1188

Jing Y, Wan J, Angelidaki I, Zhang S, Luo G (2017) ITRAQ quantitative proteomic analysis reveals the pathways for methanation of propionate facilitated by magnetite. Water Res 108:212–221

Johnravindar D, Liang B, Fu R, Luo G, Meruvu H, Yang S, Yuan B, Fei Q (2020) Supplementing granular activated carbon for enhanced methane production in anaerobic co-digestion of post-consumer substrates. Biomass Bioenerg 136:105543

Komnitsas K, Zaharaki D, Pyliotis I, Vamvuka D, Bartzas G (2015) Assessment of pistachio shell biochar quality and its potential for adsorption of heavy metals. Waste Biomass Valorizat 6:805–816

Lei Y, Sun D, Dang Y, Chen H, Zhao Z, Zhang Y, Holmes DE (2016) Stimulation of methanogenesis in anaerobic digesters treating leachate from a municipal solid waste incineration plant with carbon cloth. Biores Technol 222:270–276

Lei Y, Wei L, Liu T, Xiao Y, Dang Y, Sun D, Holmes DE (2018) Magnetite enhances anaerobic digestion and methanogenesis of fresh leachate from a municipal solid waste incineration plant. Chem Eng J 348:992–999

Li J, Xiao L, Zheng S, Zhang Y, Luo M, Tong C, Xu H, Tan Y, Liu J, Wang O, Liu F (2018a) A new insight into the strategy for methane production affected by conductive carbon cloth in wetland soil: beneficial to acetoclastic methanogenesis instead of CO_2 reduction. Sci Total Environ 643:1024–1030

Li Q, Xu M, Wang G, Chen R, Qiao W, Wang X (2018b) Biochar assisted thermophilic co-digestion of food waste and waste activated sludge under high feedstock to seed sludge ratio in batch experiment. Biores Technol 249:1009–1016

Li S, Cao Y, Zhao Z, Zhang Y (2019) Regulating secretion of extracellular polymeric substances through dosing magnetite and zerovalent iron nanoparticles to affect anaerobic digestion mode. ACS Sustain Chem Eng 7(10):9655–9662

Li X, Li Q, He J, Zhang Y, Zhu N (2021) Application of activated carbon to enhance biogas production rate of *Flammulina velutipes* residues with composting pretreatment. Waste Biomass Valori 12(2):823–831

Li X, Li Q, He J, Zhang Y, Zhu N (2022) Application of activated carbon to enhance biogas production rate of *Flammulina velutipes* residues with composting pretreatment. Waste Biomass Valorizat 12:823–831

Liang Y, Li X, Zhang J, Zhang L, Cheng B (2017) Effect of microscale ZVI/magnetite on methane production and bioavailability of heavy metals during anaerobic digestion of diluted pig manure. Environ Sci Pollut Res 24(13):12328–12337

Lim E, Tian H, Chen Y, Ni K, Zhang J, Tong Y (2020) Methanogenic pathway and microbial succession during start-up and stabilization of thermophilic food waste anaerobic digestion with biochar. Biores Technol 314:123751

Lin L, Chowdhury B, Zakaria BS, Dhar BR (2020) Temperature-dependent (20–55°C) electrocatalytic characteristics during ethanol/propionate degradation by methanogenic communities grown on conductive carbon fibers. Chem Eng J 391:123566

Lin R, Cheng J, Zhang J, Zhou J, Cen K, Murphy JD (2017) Boosting biomethane yield and production rate with graphene: the potential of direct interspecies electron transfer in anaerobic digestion. Biores Technol 239:345–352

Lin R, Cheng J, Ding L, Murphy JD (2018a) Improved efficiency of anaerobic digestion through direct interspecies electron transfer at mesophilic and thermophilic temperature ranges. Chem Eng J 350:681–691

Lin R, Deng C, Cheng J, Xia A, Lens PNL, Jackson SA, Dobson ADW, Murphy JD (2018b) Graphene facilitates biomethane production from protein-derived glycine in anaerobic digestion. iScience 10:158–170

Liu Y, Li X, Wu S, Tan Z, Yang C (2021) Enhancing anaerobic digestion process with addition of conductive materials. Chemosphere 278:130449

Lovley DR (2011) Reach out and touch someone: potential impact of DIET (direct interspecies energy transfer) on anaerobic biogeochemistry, bioremediation, and bioenergy. Rev Environ Sci Bio/Tech 10:101–105

Lu T, Zhang J, Wei Y, Shen P (2019) Effects of ferric oxide on the microbial community and functioning during anaerobic digestion of swine manure. Biores Technol 287:121393

Luo C, Lü F, Shao L, He P (2015) Application of eco-compatible biochar in anaerobic digestion to relieve acid stress and promote the selective colonization of functional microbes. Water Res 68:710–718

Masebinu S, Akinlabi E, Muzenda E, Aboyade A (2019) A review of biochar properties and their roles in mitigating challenges with anaerobic digestion. Renew Sustain Energy Rev 103:291–307

Meng Q, Liu H, Zhang H, Xu S, Lichtfouse E, Yun Y (2022) Anaerobic digestion and recycling of kitchen waste: a review. Environ Chem Lett 20(3):1745–1762

Nabi M, Liang H, Cheng L, Yang W, Gao D (2022) A comprehensive review on the use of conductive materials to improve anaerobic digestion: focusing on landfill leachate treatment. J Environ Manage 309:114540

Nie W, He S, Lin Y, Cheng J, Yang C (2023) Functional biochar in enhanced anaerobic digestion: synthesis, performances, and mechanisms. Sci Total Environ 906:167681

Noonari AA, Mahar RB, Sahito AR, Brohi KM (2019) Anaerobic co-digestion of canola straw and banana plant wastes with buffalo dung: effect of Fe_3O_4 nanoparticles on methane yield. Renewable Energy 133:1046–1054

Orrantia M, Armenta M, Alvarez LH, Burboa-Charis VA, Meza-Escalante ER, Olivas A, Arroyo E, Maytorena V (2024) Enhanced methane production via anaerobic digestion assisted with Fe_3O_4 nanoparticles supported on microporous granular activated carbon. Fuel 360:130517

Paritosh K, Yadav M, Chawade A, Sahoo D, Kesharwani N, Pareek N, Vivekanand V (2020) Additives as a support structure for specific biochemical activity boosts in anaerobic digestion: a review. Front Energy Res 8:88

Puyol D, Flores-Alsina X, Segura Y, Molina R, Padrino B, Fierro JLG, Gernaey KV, Melero JA, Martínez F (2018) Exploring the effects of ZVI addition on resource recovery in the anaerobic digestion process. Chem Eng J 335:703–711

Rasapoor M, Young B, Asadov A, Brar R, Sarmah AK, Zhuang WQ, Baroutian S (2020) Effects of biochar and activated carbon on biogas generation: a thermogravimetric and chemical analysis approach. Energy Convers Manage 203:112221

Rotaru AE, Shrestha PM, Liu F, Shrestha M, Shrestha D, Embree M, Zengler K, Wardman C, Nevin KP, Lovley DR (2014) A new model for electron flow during anaerobic digestion: direct interspecies electron transfer to *Methanosaeta* for the reduction of carbon dioxide to methane. Energy Environ Sci 7(1):408–415

Ryckebosch E, Drouillon M, Vervaeren H (2011) Techniques for transformation of biogas to biomethane. Biomass Bioenerg 35(5):1633–1645

Ryue J, Lin L, Liu Y, Lu W, McCartney D, Dhar BR (2019) Comparative effects of GAC addition on methane productivity and microbial community in mesophilic and thermophilic anaerobic digestion of food waste. Biochem Eng J 146:79–87

Salvador AF, Martins G, Melle-Franco M, Serpa R, Stams AJM, Cavaleiro AJ, Pereira MA, Alves MM (2017a) Carbon nanotubes accelerate methane production in pure cultures of methanogens and in a syntrophic coculture. Environ Microbiol 19(7):2727–2739

Sánchez A (2022) Biogas improvement as renewable energy through conversion into methanol: a perspective of new catalysts based on nanomaterials and metal organic frameworks. Front Nanotech 4:1012384

Sarker A, Nikhil GN (2023) Bioprospecting syntrophic microbial-material interactions during anaerobic digestion: a review on applications of biocatalysis and challenges. Biomass Conv Biorefinery. https://doi.org/10.1007/s13399-023-04737-1

Shen Y, Linville JL, Urgun-Demirtas M, Schoene RP, Snyder SW (2015) Producing pipeline-quality biomethane via anaerobic digestion of sludge amended with corn stover biochar with in-situ CO_2 removal. Appl Energy 158:300–309

Shirakawa H, Louis EJ, MacDiarmid AG, Chiang CK, Heeger AJ (1977) Synthesis of electrically conducting organic polymers: halogen derivatives of polyacetylene, (CH)x. J Chem Soc, Chem Commun 16:578–580

Stejskal J (2020) Interaction of conducting polymers, polyaniline and polypyrrole, with organic dyes: polymer morphology control, dye adsorption and photocatalytic decomposition. Chem Pap 74(1):1–54

Sun W, Fu S, Zhu R, Wang Z, Zou H, Zheng Y (2020) Improved anaerobic digestion efficiency of high-solid sewage sludge by enhanced direct interspecies electron transfer with activated carbon mediator. Biores Technol 313:123648

Tan J, Wang J, Xue J, Liu S, Peng SC, Ma D, Chen TH, Yue Z (2015) Methane production and microbial community analysis in the goethite facilitated anaerobic reactors using algal biomass. Fuel 145:196–201

Tan X, Jiang Y, Puchalski M, Peng Q, Hu S, Xiong W, Saskova J, Wiener J, Venkataraman M, Militky J (2024) The multifunctional flexible conductive viscose fabric prepared by thiol modification followed by copper plating. Cellulose 31:3169–3184

Thivya P, Wilson RRJ (2019) Environmental pollutants simultaneous determination: DNA catalyst mediated polyaniline biocomposite nanostructures. Biocatal Agric Biotechnol 21:101352

Valero D, Alzate-Gaviria L, Montes JA, Rico C (2020) Influence of a conductive material and different anaerobic inocula on biochemical methane potential of substrates from alcoholic beverage production. Waste Biomass Valorizat 11(11):5957–5964

Wang C, Liu Y, Gao X, Chen H, Xu X, Zhu L (2018a) Role of biochar in the granulation of anaerobic sludge and improvement of electron transfer characteristics. Biores Technol 268:28–35

Wang C, Wang C, Jin L, Lu D, Chen H, Zhu W, Xu X, Zhu L (2019a) Response of syntrophic aggregates to the magnetite loss in continuous anaerobic bioreactor. Water Res 164:114925

Wang C, Wang C, Liu J, Xu Q, Han Z, Xu X, Zhu L (2020a) Tolerance of aceticlastic methanogenesis enhanced by magnetite under the condition of ammonia stress. ACS Sustain Chem Eng 8(3):1417–1426

Wang D, Ai J, Shen F, Yang G, Zhang Y, Deng S, Zhang J, Zeng Y, Song C (2017) Improving anaerobic digestion of easy-acidification substrates by promoting buffering capacity using biochar derived from vermicompost. Biores Technol 227:286–296

Wang G, Gao X, Li Q, Zhao H, Liu Y, Wang XC, Chen R (2020b) Redox-based electron exchange capacity of biowaste-derived biochar accelerates syntrophic phenol oxidation for methanogenesis via direct interspecies electron transfer. J Hazard Mater 390:121726

Wang G, Li Q, Gao X, Wang XC (2018b) Synergetic promotion of syntrophic methane production from anaerobic digestion of complex organic wastes by biochar: performance and associated mechanisms. Biores Technol 250:812–820

Wang S, Hao Y, Zhang K, Liu H, Liao L (2023a) Amorphous C/SbS x composites from natural stibnite as low cost and high performance lithium/sodium-ion battery anodes. J Mat Chem A 11(45):24959–24969

Wang T, Zhang D, Dai L, Dong B, Dai X (2018c) Magnetite triggering enhanced direct interspecies electron transfer: a scavenger for the blockage of electron transfer in anaerobic digestion of high-solids sewage sludge. Environ Sci Technol 52(12):7160–7169

Wang Y, He J, Xu B, Osman NA, Algadi H, Abo-Dief HM, Lu N, Wasnik P, Sridhar D, Alanazi AK (2023b) Iron-material-facilitated methane production from anaerobic wastewater treatment. Emerging Mat Res 12(3):231–240

Wang Z, Chun-Hai W, Yu H, Qu F, Rong H, He J, Liu G, Huang X, Ngo HH (2023c) Preparation and mechanism of carbon felt supported iron trioxide and zero-valent iron for enhancing anaerobic digestion performance. Chem Eng J 468:143565

Wang Z, Chen C, Liu H, Hrynshpan D, Savitskaya T, Chen J, Chen J (2019b) Effects of carbon nanotube on denitrification performance of *Alcaligenes sp.* TB: promotion of electron generation, transportation and consumption. Ecotoxicol Environ Safety 183:109507

Wen L, Huang L, Wang Y, Yuan Y, Zhou L (2022) Facet-engineered hematite boosts microbial electrogenesis by synergy of promoting electroactive biofilm formation and extracellular electron transfer. Sci Total Environ 819:153154

Xiang Y, Yang Z, Zhang Y, Xu R, Zheng Y, Hu J, Li X, Jia M, Xiong W, Cao J (2019) Influence of nanoscale zero-valent iron and magnetite nanoparticles on anaerobic digestion performance and macrolide, aminoglycoside, β-lactam resistance genes reduction. Biores Technol 294:122139

Xu H, Wang C, Yan K, Wu J, Zuo J, Wang K (2016) Anaerobic granule-based biofilms formation reduces propionate accumulation under high H_2 partial pressure using conductive carbon felt particles. Biores Technol 216:677–683

Yadav M, Joshi C, Paritosh K, Thakur J, Pareek N, Masakapalli SK, Vivekanand V (2022) Reprint of organic waste conversion through anaerobic digestion: a critical insight into the metabolic pathways and microbial interactions. Metab Eng 71:62–76

Yang J, Sim YB, Joo HH, Mašek O, Kim SH (2024) Synergetic effects of conductive materials and bacterial population in inoculum on mixed-culture biohydrogen production. Int J Hydrogen Energy 53:1293–1302

Yang S, Li Z, Wang X, Cheng S, Zhu L (2023) Research progress of enhanced methane production by anaerobic digestion of blackwater. J Water Proc Eng 53:103719

Yang Z, Guo R, Shi X, Wang C, Wang L, Dai M (2016) Magnetite nanoparticles enable a rapid conversion of volatile fatty acids to methane. RSC Adv 6(31):25662–25668

Ye J, Hu A, Ren G, Chen M, Tang J, Zhang P, Zhou S, He Z (2018) Enhancing sludge methanogenesis with improved redox activity of extracellular polymeric substances by hematite in red mud. Water Res 134:54–62

Yin Q, He K, Liu A, Wu G (2017) Enhanced system performance by dosing ferroferric oxide during the anaerobic treatment of tryptone-based high-strength wastewater. Appl Microbiol Biotechnol 101(9):3929–3939

Yin Q, Yang S, Wang Z, Xing L, Wu G (2018) Clarifying electron transfer and metagenomic analysis of microbial community in the methane production process with the addition of ferroferric oxide. Chem Eng J 333:216–225

Yuan H, Ding L, Zama EF, Liu P, Hozzein WN, Zhu Y (2018) Biochar modulates methanogenesis through electron syntrophy of microorganisms with ethanol as a substrate. Environ Sci Technol 52(21):12198–12207

Yue Z, Ma D, Wang J, Tan J, Peng S, Chen T (2015) Goethite promoted anaerobic digestion of algal biomass in continuous stirring-tank reactors. Fuel 159:883–886

Zhang W, Zhang J, Lu Y (2018) Stimulation of carbon nanomaterials on syntrophic oxidation of butyrate in sediment enrichments and a defined coculture. Sci Rep 8(1):12185

Zhang Y, Guo B, Zhang L, Liu Y (2020a) Key syntrophic partnerships identified in a granular activated carbon amended UASB treating municipal sewage under low temperature conditions. Biores Technol 312:123556

Zhang Y, Zhang L, Guo B, Zhou Y, Gao M, Sharaf A, Liu Y (2020b) Granular activated carbon stimulated microbial physiological changes for enhanced anaerobic digestion of municipal sewage. Chem Eng J 400:125838

Zhao Z, Zhang Y, Li Y, Dang Y, Zhu T, Quan X (2017) Potentially shifting from interspecies hydrogen transfer to direct interspecies electron transfer for syntrophic metabolism to resist acidic impact with conductive carbon cloth. Chem Eng J 313:10–18

Zhao Z, Zhang Y, Quan X, Zhao H (2016) Evaluation on direct interspecies electron transfer in anaerobic sludge digestion of microbial electrolysis cell. Biores Technol 200:235–244

Zhou G, Yang X, Marshall CW, Li H, Zheng B, Yan Y, Su J, Zhu Y (2017) Biochar addition increases the rates of dissimilatory iron reduction and methanogenesis in ferrihydrite enrichments. Front Microbiol 8:256906

Zhuang L, Ma J, Yu Z, Wang Y, Tang J (2018) Magnetite accelerates syntrophic acetate oxidation in methanogenic systems with high ammonia concentrations. Microb Biotechnol 11(4):710–720

Ziganshina EE, Ziganshin AM (2023) Magnetite nanoparticles and carbon nanotubes for improving the operation of mesophilic anaerobic digesters. Microorganisms 11(4):938

Bioelectrochemical System-Integrated Anaerobic Digestion

Xinbo Zhang, Qing Du, and Huizhong Wang

Abstract The widespread application of anaerobic digestion technologies faces significant challenges, including low processing efficiency, poor stability, and low methane yield. The bioelectrochemical system, such as microbial fuel cell and microbial electrolysis cell, is proposed to integrate with anaerobic digestion for performance enhancement by provision of exogenous electrons. Bioelectrochemical system-anaerobic digestion (BES-AD) enhances the electron transfer activity between microorganisms and their resistance to toxic and hazardous substances by means of electrical currents. By optimizing key factors, such as reactor configurations, electrode materials, pH, organic loading rate, and temperature, the BES-AD performance can be improved. BES-AD has shown excellent performance in the removal of highly concentrated organic compounds and refractory substances in wastewater. BES-AD has great potential to break through bottlenecks of anaerobic digestion, and further application of the coupled system is conducive to the development of green, low-carbon, and circular economy.

Keywords Anaerobic digestion · Bioelectrochemical system · Electrochemically active bacteria · Microbial electrolysis cell · Microbial fuel cell

X. Zhang (✉) · Q. Du · H. Wang
Joint Research Centre for Protective Infrastructure Technology and Environmental Green Bioprocess, School of Environmental and Municipal Engineering, Tianjin Chengjian University, Tianjin 300384, China
e-mail: xinbozhang@tcu.edu.cn

Q. Du
e-mail: duqing1@tcu.edu.cn

H. Wang
e-mail: whuizhong@tcu.edu.cn

G. Wu (ed.), *Anaerobic Digestion*, Green Energy and Technology,
https://doi.org/10.1007/978-3-031-69378-6_17

1 Introduction

Anaerobic digestion is a wastewater treatment and resource recovery technology in which anaerobic microorganisms convert biodegradable organic matter into renewable and clean energy sources such as CH_4 and H_2. However, the anaerobic digestion process suffers from long start-up period, stringent requirements for operating conditions, and incomplete removal of refractory pollutants (Modestra et al. 2020). Bioelectrochemical system (BES) is a promising technology for treating wastewater containing highly concentrated organic compounds and refractory substances, in which electrode reaction is driven by electrochemically active bacteria (EAB) to generate electricity, volatile fatty acids (VFAs), and biofuels (e.g., H_2 and CH_4) through the degradation of organic substances. The recalcitrant organics, such as nitrobenzene, azo dye, and chlorophenol, can be transformed to their corresponding low toxicity or easily degraded reductive products on the cathode by EAB (such as *Geobacter, Bacteroides, Shewanella, Pseudomonas,* and *Clostridium*) (Park et al. 2020).

The coupled BES and anaerobic digestion (BES-AD) systems might overcome the limitations of anaerobic digestion and improve the efficiency of wastewater treatment and energy recovery. However, the enhancement of anaerobic digestion performance by BES is multifaceted. BES can use electrical currents to activate microbial activity and change the microbial community structure, especially improving the degradation ability of microbial flora on refractory substances. For instance, some dominant genera in BES-AD systems, such as *Pseudomonas*, *Methylobacillus*, *Arcobacter*, and *Desulfovibrio*, can metabolize refractory aromatic compounds and toxic pollutants efficiently (Jiang et al. 2016). In addition, by adjusting the operating parameters of the BES-AD system and designing optimized electrodes, the growth and activity of methanogenic archaea can be promoted, thus increasing methane production. Overall, BES-AD enhances the treatment performance of anaerobic reactors through the provision of exogenous electrons to enhance microbial metabolic activity and promote the degradation of refractory pollutants.

2 Principles of BES

BES uses microorganisms as the driving force to generate electricity or hydrogen, and effectively degrades a wide range of substrates, such as carbohydrates, proteins, etc. Typically, BES consists of an anode and a cathode chamber, connected by an external circuit. EAB are the catalysts at the anode to oxidize organic matter, and the produced electrons are transferred to the cathode through the external circuit. BES is distinguished from traditional fuel cells by their low energy consumption, longevity, and absence of secondary pollution. Recently, a variety of derivative technologies have emerged, including microbial fuel cell (MFC), microbial electrolysis cell (MEC), microbial desalination cell (MDC), microbial electrosynthesis (MES), and microbial electroremediating cell (MERC), etc (Goswami and Mishra 2018).

2.1 *Principles of MFC*

MFC uses EAB as catalysts to oxidize organic substrates and harnesses the potential difference between various redox reactions to convert bioenergy into electrical energy. EAB can produce electron-rich metabolites, maintain redox gradients, generate redox mediators, and transfer electrons to the electrode either directly or via soluble electron transfer mediators, ultimately generating an electric current. MFC typically consists of an anaerobic anode chamber and an aerobic cathode chamber, separated by a proton exchange membrane (PEM) (Fig. 1). PEM selectively inhibits oxygen diffusion to the anode chamber while allowing protons to migrate to the cathode.

The fundamental working principle of MFC is akin to fuel cells: the organic substrate (such as glucose) is oxidized by electrogenic microbes at the anode into CO_2, producing protons (H^+) and electrons (e^-). Electrons are transported to the anode through mediators or across cell membranes and migrated to the cathode via an external circuit. Metabolic pathways of the anode electrogenic microbes, primarily anaerobes, significantly influence the flux of protons and electrons and are thus crucial to the generation of electricity. Meanwhile, protons diffuse through the electrolyte to the cathode, where they combine with oxygen to produce clean water, thus achieving charge and mass balances within the system and using the electricity generated to power external devices. The cathode of MFC does not require microbes and can function both in the presence or absence of oxygen. Multiple factors affect the performance of MFC, including reactor design, electrode materials, microbial species, cathode electron acceptors, PEM, substrates, pH, and temperature. Optimization of these parameters can substantially enhance MFC's performance (Gul et al. 2021).

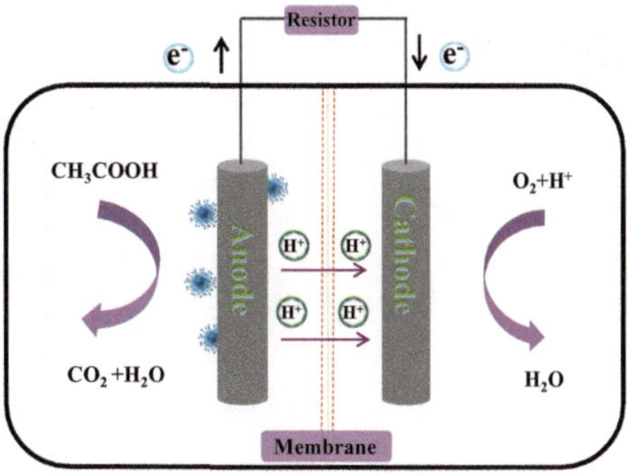

Fig. 1 Working principle of MFC

2.2 Principles of MEC

MEC is a modified MFC technology (Fig. 2) and was developed to drive a redox reaction by applying voltage to a system, which cannot occur under natural conditions (Arun et al. 2024). When the applied potential is greater than the reaction potential of cytochromes and electron mediators, an oxidation reaction can occur on the anode, allowing microorganisms to respire at the anode and driving a reduction reaction at the cathode. When a biocathode is used, the electron donor requirements of the cathodic microorganisms can be supplied by an applied voltage, and a variety of chemical products (H_2 and CH_4) with high economic value can be synthesized. In MEC, electroactive microorganisms (EAM; mainly *Pseudomonas* sp., *Flavobacterium* sp., *Ochrobactrum* sp., etc.) on the anode act as catalysts to enhance the oxidation of organic compounds (proteins, sodium acetate, alcohols, etc.) and the generation of CO_2, H^+, and e^- (Rozendal et al. 2006). Moreover, driven by an external electrical potential, anodic oxidation reactions can improve the treatment efficiency of organic pollutants, as well as antibiotics and pharmaceuticals (Croese et al. 2014). For example, sulfate can be utilized by microorganisms at the anode as the electron acceptor (Yuan et al. 2022), and 2,4-dichloronitrobenzene (DClNB) can be dechlorinated at the anode to produce nitrobenzene (NB) (Chen et al. 2017).

Biocathodes are mainly responsible for the conversion of anodic metabolites such as VFAs and ethanol into clean energy such as H_2 and CH_4. The dominant species include *Methanobacterium*, *Methanosarcina*, *Methanosaeta*, etc. (Gharbi et al. 2022). *Methanobacterium*, as a hydrogenotrophic methanogen, can use H_2/formic acid to reduce CO_2 for methane production. *Methanosarcina* is a multipotent methanogen that can produce methane using methanol, acetic acid, and H_2 as the substrate, respectively (Guo et al. 2015), and it can also participate in the direct interspecies electron transfer (DIET) pathway to accept electrons for the reduction of CO_2 to produce methane. While *Methanothrix* can only utilize acetic acid for methane production (Wang et al. 2020b) and can maintain the acid–base balance in MEC. These methanogens can utilize the electrons or hydrogen produced by the anode more efficiently when driven by an external potential to facilitate methane

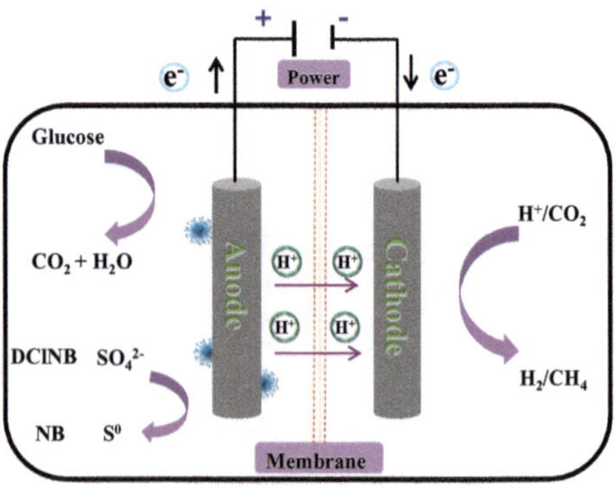

Fig. 2 Working principle of MEC

production, through overcoming thermodynamic limitations and the bottleneck of the fermentation barrier (Sun et al. 2022). Ion-selective membranes (anion-exchange and cation-exchange membranes) are commonly used to prevent the reaction between oxygen and the generated hydrogen, regulatorily controlling the MEC end products.

3 Types and Configurations of BES-AD Systems

3.1 Types and Configurations of MFC-AD Systems

3.1.1 MFC-CSTR

The continuous stirred tank reactor (CSTR) is an anaerobic treatment technology that maintains fermentation substrates and microorganisms in a fully mixed state. However, traditional CSTRs face challenges such as low methane production capacity, high variability, and reactor inhomogeneity. Recently, a novel energy conversion technology that combines MFC with CSTR has been proposed to achieve microbial cultivation, electricity generation, and wastewater treatment. The coupled MFC-CSTR system offers continuous operation, efficient pollutant removal, and low operational costs, showing great prospects for efficient wastewater treatment and energy recovery.

The coupled MFC-CSTR system comprises an anode chamber, a cathode chamber, and an external resistor (Fig. 3a). The two electrode chambers are separated by a PEM. Substrates in the anolyte are directly converted into protons, electrons, and metabolic products by microbes. The electrons travel from the anode through the external circuit to the cathode of the cell, while protons diffuse from the anode chamber to the cathode chamber through the PEM. In the cathode chamber, electrons passed through the external resistor are transmitted to the cathode surface, where the oxidized substances (oxidants) combine with the protons and electrons transferred from the anode to undergo a reduction reaction. Kuntke et al. (2014) combined CSTR with dual-chamber MFC for treating molasses wastewater in the anode chamber and simulated electroplating wastewater (mainly consisting of $AgNO_3$) in the cathode chamber. With graphite brush anodes and cathodes placed in two CSTRs and separated by a perfluorinated PEM, a stable chemical oxygen demand (COD) removal percentage of 81% and a silver ion removal percentage of about 90% was obtained under steady state conditions (Kuntke et al. 2014).

To enhance system performance, Zhao et al. (2010) utilized a modified CSTR (up-flow type) to construct a coupled MFC-CSTR system to evaluate the continuous voltage output with cow dung as the substrate during the hydrolysis-acidification process (Fig. 3b). The modified CSTR serves as the anode chamber, equipped with dual anodes which were made of carbon brush. Catholyte is a potassium permanganate buffer solution. The two chambers are separated by a PEM and the electrodes are connected by copper wires. The system maintained a stable electricity generation

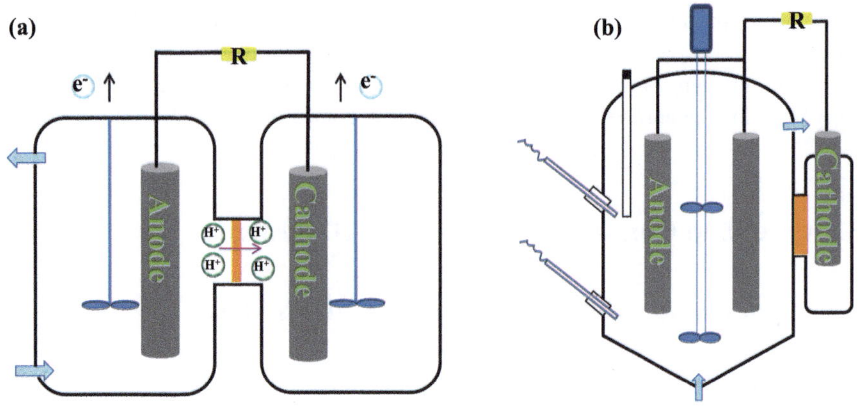

Fig. 3 Schematics diagrams of MFC-CSTR systems. **a** conventional MFC-CSTR system; **b** an MFC-CSTR system equipped with dual anodes

of around 470 mV during steady operation, with the main components of VFAs in the anode chamber being acetic, propionic, and butyric acids.

3.1.2 MFC-AnMBR

Although anaerobic membrane bioreactor (AnMBR) possesses high biomass concentration, excellent effluent quality, and independent hydraulic and sludge retention times, the unavoidable issue of membrane fouling significantly hampers its widespread application. The introduction of a low-intensity electric field positively affects microbial activity and reduces membrane fouling by altering the properties of the sludge mixture. Consequently, a coupled MFC and AnMBR (MFC-AnMBR) system was designed to alleviate the high energy consumption and membrane fouling of AnMBR, achieving efficient treatment of high COD loading wastewater and enhancing energy conversion efficiency and recovery.

An MFC-AnMBR system typically comprises an immersed AnMBR and an MFC (Fig. 4a, b). In this setup, AnMBR serves as the anode chamber where specific microbes (electron producers) metabolize organic substances into electrons, protons, and carbon dioxide. Electrons flow to the cathode chamber through an external circuit, generating a current. Depending on its design, the cathode chamber is divided into a cavity cathode and an air cathode, where electrons, protons, and nearby oxygen (or other acceptors) are reacted to produce water or other chemicals. Liu et al. (2022) developed a coupled MFC-AnMBR system (Fig. 4c) using a polypyrrole/anthraquinone-2,6-disulfonate/polytetrafluoroethylene (PPy/AQDS/PTFE) conductive membrane as both the MFC's anode and AnMBR's filtration membrane, with a stainless steel mesh bearing two layers of carbon felt as the cathode. In the presence of electrons, high molecular weight polysaccharides in the synthetic wastewater can be transformed into low molecular weight-hydroxylated compounds, which are more readily degraded and utilized by microbes, thus preventing membrane fouling (Liu et al. 2022). Hao et al. (2022) devised a novel MFC-AnMBR system (Fig. 4d) by integrating AnMBR and single-chamber air cathode MFC within the same reactor chamber to enhance methane production while reducing membrane fouling. With AnMBR acting as the MFC's anode chamber and carbon felt as the anode electrode material, the MFC-AnMBR system provided a stable microbial electric field, significantly improved methane production, shortened the start-up period, and mitigated membrane fouling (Hao et al. 2022).

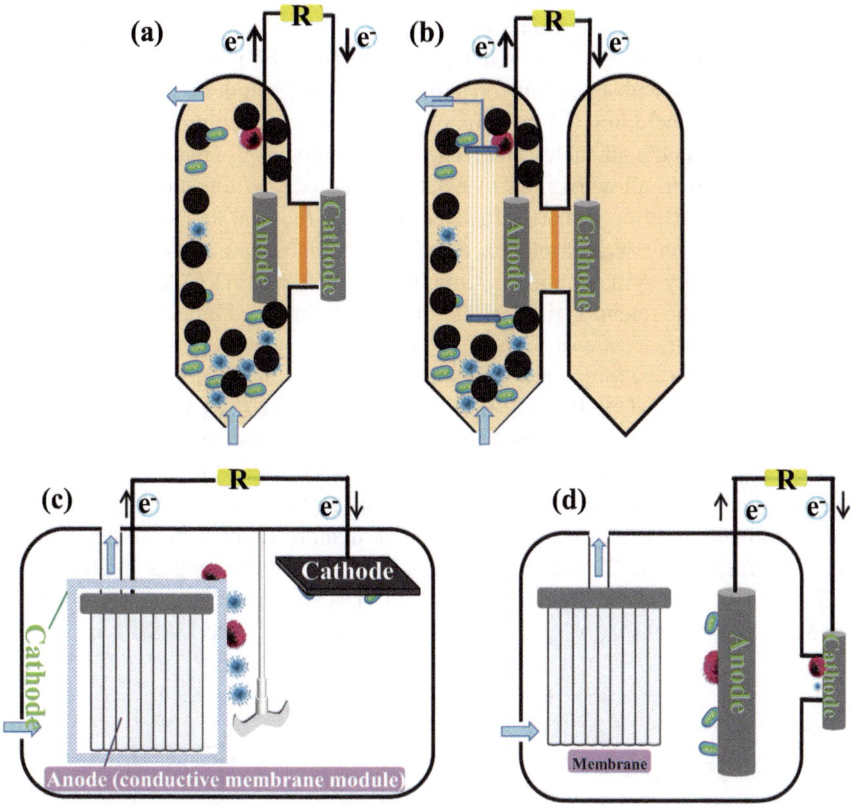

Fig. 4 Schematic diagrams of MFC-AnMBR systems. **a** conventional single-chamber MFC-AnMBR system; **b** conventional double-chamber MFC-AnMBR system; **c** MFC-AnMBR system using a conductive membrane as both the MFC's anode and AnMBR's filtration membrane; **d** single-chamber air cathode MFC-AnMBR

3.1.3 MFC-UASB

The up-flow anaerobic sludge blanket (UASB) reactor is suitable for treating high-concentration organic wastewater, offering advantages such as high volumetric loading rates and simple operational requirements. However, it has limitations of relatively high costs and energy consumption. In recent years, a novel coupling technique that integrates MFC with UASB has been introduced. This coupled system, which can efficiently process high influent COD load and recover energy, has been proven to be an easily operable configuration (Zhang et al. 2012).

The coupled MFC-UASB system can be implemented in either a parallel (Fig. 5a) or serial (Fig. 5b) configurations. Zhang et al. (2012) constructed a serial MFC-UASB reactor for treating sulfate-rich wastewater. The MFC reactor adopted a single-chamber structure, with a graphite rod anode and a carbon cloth air cathode, featuring a catalyst layer (including 0.5 mg/cm^2 of Pt) on the water-facing side, and a PTFE diffusion layer on the air-facing side. The entire cathode was encased in thick resin glass with holes to allow oxygen to reach the cathode. Within the entire coupled system, the UASB unit primarily focused on COD removal and sulfate reduction, while the MFC unit was employed to oxidize sulfides into elemental sulfur and generate electricity. A maximum power output was achieved at a COD/sulfate ratio of 2.3 and a hydraulic retention time (HRT) of 54.3 h, while the highest sulfate removal rate at a COD/sulfate ratio of 3.7 and an HRT of 55.6 h. Furthermore, Zhang et al. (2009) integrated the aforementioned coupled system with a biological aerated filter (BAF) to construct an UASB-MFC-BAF system. Both the UASB reactor and the BAF unit were made of polymethyl methacrylate, cylindrical in shape, and a rectangular air cathode, single-chamber, and membrane-less MFC unit were constructed using flat organic glass. This integrated system was applied for simultaneous bioelectricity generation and molasses wastewater treatment, demonstrating removal percentages of 53.2% for COD, 52.7% for sulfate, and 41.1% for color (Zhang et al. 2009).

Jia et al. (2016) constructed a parallel MFC-UASB coupling system, employing the MFC as a biosensor for real-time online monitoring of the UASB reactor. The

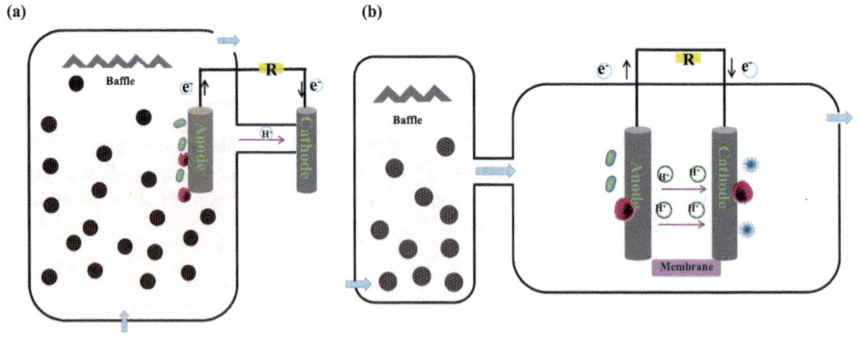

Fig. 5 Schematic diagrams of MFC-UASB systems. **a** conventional parallel MFC-UASB system; **b** conventional serial MFC-UASB system

MFC was a single-chamber structure without a PEM, with the UASB reactor serving as the anode chamber and carbon felt as the anode electrode submerged in the UASB reactor. The carbon cloth acted as a cathode, with air continuously sparged inside the cathode chamber and functioned as the final electron acceptor. The circuit between the anode electrode and the air cathode was connected by copper wire. The findings revealed that although only 0.94% of the influent COD was converted into electrical energy for biosensing, this integrated system exhibited significant potential for real-time monitoring without additional COD consumption (Jia et al. 2016).

3.1.4 MFC-AFB

The anaerobic fluidized bed (AFB) reactor improves anaerobic digestion efficiency by filling tiny carriers, allowing anaerobic microorganisms to grow and form a biofilm on the surface of the carriers, and increasing contact between the microorganisms and the wastewater through the fluidization of the carriers. However, the conventional AFB reactors have low stability at high organic loading rates (OLRs), and low stability with toxic and hazardous substances, which limits their application (Asensio et al. 2021). The combined MFC-AFB system has the advantages of compact structure, high resistance to loading impact, fast biochemical reaction rate, high pollutant removal rate, and so on (Song et al. 2023).

The typical MFC-AFB system consists of an anode chamber, a cathode chamber, and an external resistor. Generally, the AFB reactor serves as the anode chamber (Fig. 6a, b, c), where copper wires are used to connect the cathode and anode electrodes. Substrate flows into the anode chamber from the bottom of the reactor, and then flows upward to drive the carriers. Microorganisms attached to the anode and carriers degrade the pollutants while generating electrons and protons. In the coupled system, the fluidization of the carriers reduce the internal electrochemical resistance of the coupled system, while the carriers collide frequently with the anode. This results in the rapid transfer of electrons to the anode and the accelerated transfer of protons to the cathode, accelerating the biochemical reaction rate and enhancing the removal of pollutants. Huang et al. (2011) achieved a high output voltage of 611.0 mV, a high-power density of 124.0 mW/m^2, and a COD removal percentage of 80–90% by using a double-chamber MFC-AFB coupled system to treat real distillery wastewater, thus verifying the feasibility of efficient wastewater treatment and power generation by MFC-AFB. Niu et al. (2020) utilized MFC-AFB filled with macroporous adsorbent resin particles to treat isopropanol wastewater, and a maximum power density of 135.73 $\pm$ 0.17 mW/m^2 and a COD removal percentage of 68.21 $\pm$ 0.24% under the optimal condition were obtained.

Modified MFC-AFB systems have been proposed to enhance performance of anaerobic digestion through extended contact time of wastewater with microorganisms. Song et al. (2023) constructed a novel MFC-pulsed AFB (MFC-PAFB) system to enhance the electrochemical performance of the conventional MFC-AFB, which combined an MFC with a pulsed intermittent liquid-solid fluidized bed as the anode chamber. In this system, the liquid reciprocates upward and downward in the fluidized

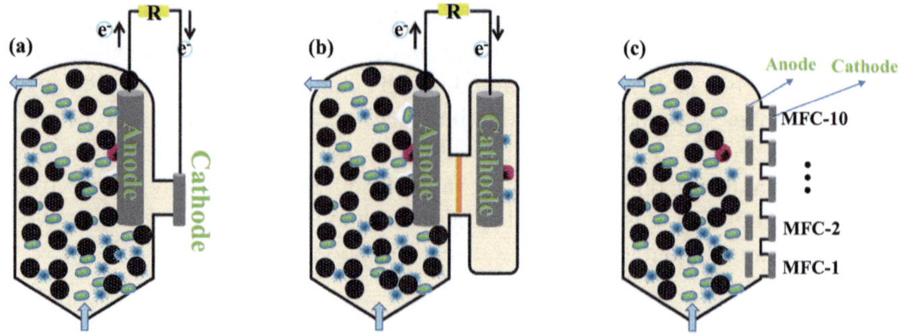

Fig. 6 Schematic diagrams of MFC-AFB systems. **a** single-chamber MFC-AFB system; **b** double-chamber MFC-AFB system; and **c** stacked MFC-AFB consisting of 10 individual air cathode MFCs

bed in the form of a sinusoidal curve. The amplitude and period of the liquid flow rate were controlled by motorized valves. The relative velocities between the liquid and solid phases were increased by pulsed intermittent liquid flow, which effectively improved the mass transfer between microorganisms and wastewater. Compared with the conventional MFC-AFB, MFC-PAFB has the advantages of better power generation and shorter wastewater treatment time. The power generation of MFC-PAFB increased by about 0.5%–8.1% and the wastewater treatment time was shortened by about 1.7–2.1 times. A stacked MFC-AFB, consisting of 10 individual air cathode MFCs (Fig. 6c), generated open-circuit voltages reaching peaks of 617.6 mV and 5.3 V in parallel and serial modes, respectively. Additionally, the highest power density and COD removal percentage of 3440.1 mW/m^2 and 96.9% were achieved in the serial mode (Chen et al. 2024).

3.1.5 MFC-ABR

Anaerobic baffled reactor (ABR) is closer to plug flow by using vertical baffles to divide the reactor into multiple reaction compartments in series. Each compartment can cultivate a microbial community acclimated to the incoming substrate, resulting in the separation of the acidogenesis and methanogenesis phases of the anaerobic reaction along the flow path. The coupled MFC-ABR system has been proposed to enhance the activity of methanogens by avoiding the accumulation of VFAs that can lead to the imbalance of the system.

The coupled MFC-ABR system includes multiple anode chambers and a shared air cathode (Fig. 7a). The overall structure is similar to multiple single-chamber air cathode MFCs connected sequentially. This coupled system can achieve high power generation efficiency and current output, while improving the system's resistance to shock loading. Liu et al. (2021) constructed a novel MFC-ABR system with a rectangular carbon blanket as the cathode to treat dye wastewater, which could achieve higher dye removal and methane production, with the obtained maximum

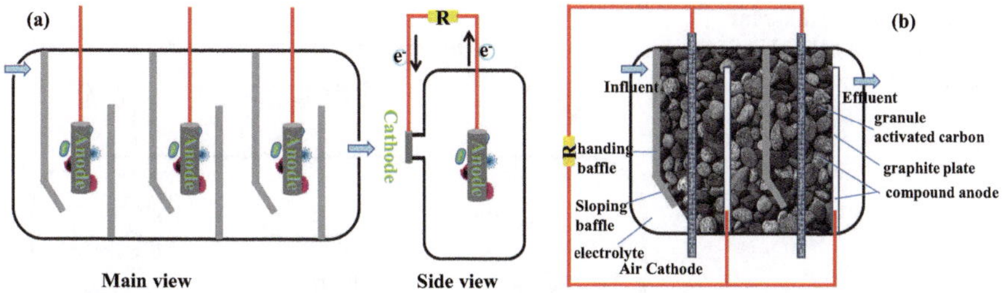

Fig. 7 Schematic diagrams of MFC-AFB systems. **a** conventional MFC-ABR system; **b** graphite particle-filled MFC-ABR system

power density of up to 1,226.4 mW/m^3 and dye removal percentage of 91.7%. Liu et al. (2015) invented a promising MFC-ABR system for removing sulfide and organic matter, 70.0% and 54.6% removal percentages of sulfide and COD could be achieved, while the maximum voltage of 2340 mV in series and the maximum current of 3.29 mA/m^2 in parallel were obtained, respectively. Feng et al. (2010) designed a graphite granule-filled MFC-ABR system to treat high-loaded wastewater generated from an actual corn stover fermentation process, and a power density of 10.7 W/m^2 and a COD removal percentage of 89.1% were obtained. The designed MFC-AFB had two identical compartments (Fig. 7b), each of which was regarded as a baffled unit constructed by an air–cathode, a vertical hanging baffle plate, a graphite plate and graphite granule bed.

3.2 *Types and Configurations of MEC-AD Systems*

3.2.1 MEC-UASB

The coupled MEC-UASB system is mainly submerged (Fig. 8). The cathode and anode materials in the coupled MEC-UASB system are usually carbon rods and felts, which are directly inserted into the UASB reactor. The anode and cathode are connected to the direct-current power source through wires. The electrodes are mounted above the bottom of the reactor, where full contact with the sludge layer is possible. Zhu et al. (2022) found that MEC-UASB not only promoted electromethanogenesis but also provided high-quality biogas. More importantly, the coupled MEC-UASB system also exhibited better process stability compared to conventional anaerobic digesters, mitigating the accumulation of short-chain VFAs at higher OLRs and the inhibition of a wide range of toxic and refractory compounds.

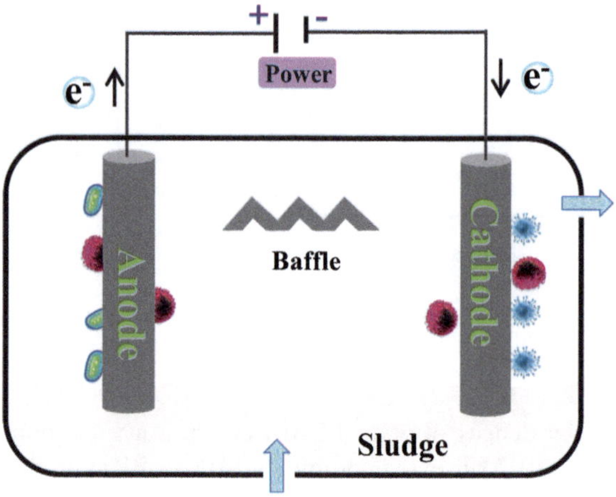

Fig. 8 Schematic diagram of an immersed MEC-UASB system

3.2.2 MEC-AnMBR

A coupled system of MEC and AnMBR (MEC-AnMBR) overcomes the problems of sensitive biological communities, low degradation rate of toxic pollutants, and membrane fouling that exists in traditional AnMBR, and makes up for the low methane recovery rate of the MEC system, making it an emerging green technology (Ding et al. 2018). Direct immersion of the cathode and anode of MEC into the AnMBR is the main coupling method. However, there are various electrode materials used in MEC, which directly affect the installation position of the electrode and the membrane module. When both the cathode and anode of MEC are made of carbon rods or carbon felts, the cathode and anode are usually arranged on both sides of the membrane module (Fig. 9a). When the cathode is titanium mesh, the cathode can be used to wrap the membrane module (Fig. 9b). The titanium mesh is rolled into a cylinder, and then the membrane module is pressed into the cylinder of titanium mesh, subsequently the titanium mesh and membrane module are tightened using titanium wire. Carbon felt can be used for the anode, and the titanium mesh is connected to the external power source via titanium wire. Submerged coupled system facilitates the microbial activity of the AnMBR using the high current density in the MEC and increases the sludge surface charge. Therefore, it can reduce the stability and densification of the sludge cake layer formed on the membrane surface and effectively alleviates membrane pore clogging (Sun et al. 2017).

In the coupled system, *Desulfuromonas* and *Geobacter* spp. are the main dominant species, and both can transfer electrons directly to methanogens via DIET (Zhao et al. 2015). In addition, *Methanosarcina* and *Methanothrix* are common methanogenic genera in coupled systems, and the consumption of acetate via DIET is the main pathway for methanogenesis (Barua et al. 2019). The presence of these EAM greatly enhanced the treatment performance of MEC-AnMBR. For example, Kik et al. (2021)

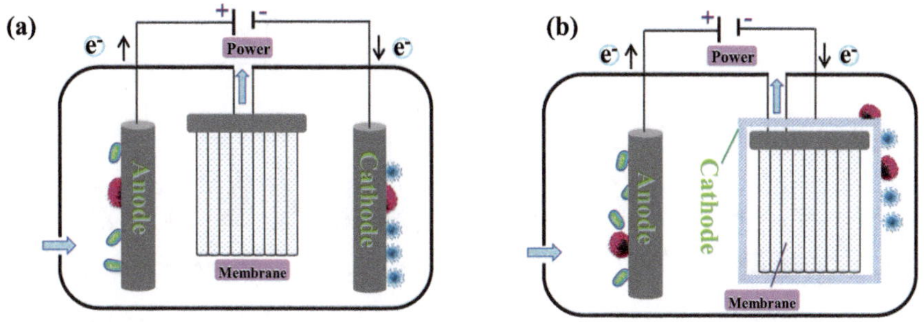

Fig. 9 Schematic diagrams of MEC-AnMBR systems. **a** MEC-AnMBR system with carbon rods as the cathode; **b** MEC-AnMBR system with titanium mesh as the cathode

used MEC-AnMBR to shorten the start-up time of AnMBR by 25 days, meanwhile, the average methane yield increased by 56.0%, and the membrane fouling cycle was extended by 6.5 times. More importantly, MEC-AnMBR had a desirable degradation effect of antibiotics, such as sulfonamides, nitroimidazoles, β-lactams, and chloramphenicol (Guo et al. 2020).

3.2.3 MEC-ABR

A coupled MEC and ABR (MEC-ABR) system has been used for the treatment of industrial wastewater. The coupled MEC-ABR system can be assembled in series (Fig. 10a) or parallel (Fig. 10b). In the serial MEC-ABR system, the microorganisms in the ABR do not have direct contact with the electrodes (carbon rods or felts) in the MEC. The organic matter in the wastewater first passes through the acid-producing zone of the ABR, where it undergoes hydrolysis and fermentation to produce various VFAs and alcohols, and then enters the methanogenic zone where the produced products are converted by methanogens to CH_4. The refractory organics and toxic substances enter the MEC and are degraded by EAM. In the parallel MEC-ABR system, the cathode and anode (granular graphite) of the MEC are mounted into four compartments divided by three overflow plate panels (Fig. 10b). The graphite is tightly packed in a rectangle made of titanium to act as a collector, and a titanium rod is welded to the electrode for conductivity and external circuit connections (Liu et al. 2017). This configuration allows the MEC-ABR system to have an anaerobic granular sludge bed at the bottom and a bioelectrolysis zone at the top.

The dominant species in the acid-producing zone of the coupled MEC-ABR system are *Syntrophobacter* and *Thermodesulfovibrio*, whose presence favoring the oxidation of propionate and butyrate (Cui et al. 2021). While the genera enriched in the methanogenic zone are *Methanobacterium* and *Methanolinea*, and both are hydrogenotrophic methanogens. These electrochemical microorganisms are extremely active, resisting toxic substances and facilitating the effective removal of pollutants (Zhang et al. 2016). For example, Cui et al. (2014) found that MEC-ABR

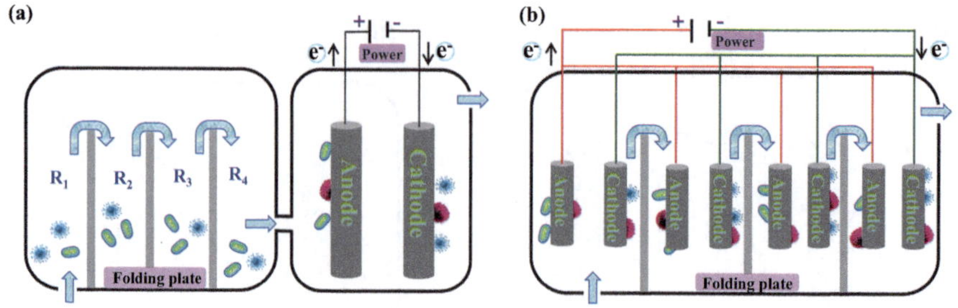

Fig. 10 Schematic diagrams of MEC-ABR systems. **a** conventional serial MEC-ABR system; **b** conventional parallel MEC-ABR system

had a powerful decolorization efficiency for dye wastewater, with a 9.3% increase in the decolorization efficiency of alizarin yellow compared to conventional ABR and a significant increase in the rate of acetate consumption at 0.5 V.

3.2.4 MEC-TPAD

The two-phase anaerobic digestion (TPAD) process separates the acidification and methanogenesis phases in two serial reactors, so that hydrolytic/fermentative bacteria and methanogens can grow under their optimal environmental conditions. However, the traditional TPAD has the problem of insufficient volumetric loading rate when treating high-concentration organic wastewater, resulting in incomplete hydrolysis of organic matter. Moreover, toxic and hazardous substances can destroy microorganisms in both acid and methanogenic phases, substantially reducing system stability. A coupled MEC and TPAD (MEC-TPAD) system can solve the problem of sensitive microbial communities and improve the treatment efficiency of complex industrial wastewater.

The coupled MEC-TPAD system is dominated by parallel type (Fig. 11). Electrodes (carbon brushes or rods) for MEC are built into the two-phase anaerobic reactor, and the coupled system is separated into acid-producing and methanogenic chambers using PEMs. The acid-producing chamber is used to house the acid-producing sludge and anode, and the methanogenic chamber is used to house the methanogenic sludge and cathode, which receives the effluent from the acid-producing chamber. The wall between the two chambers is uniformly arranged with circular holes to facilitate the transfer of metabolites. Such a structure allows the use of electric current to cultivate more EAM for the purpose of catalyzing both anodic and cathodic reactions (Yuan et al. 2022).

The dominant phyla on the anode biofilm are Firmicutes and Proteobacteria, which are electroactive hydrolytic acidifying bacteria. Proteobacteria are responsible for the oxidation of sulfide to S^0, *Desulfomicrobium* can reduce SO_4^{2-} in an acid-producing environment, and Firmicutes can utilize VFAs and carbohydrates

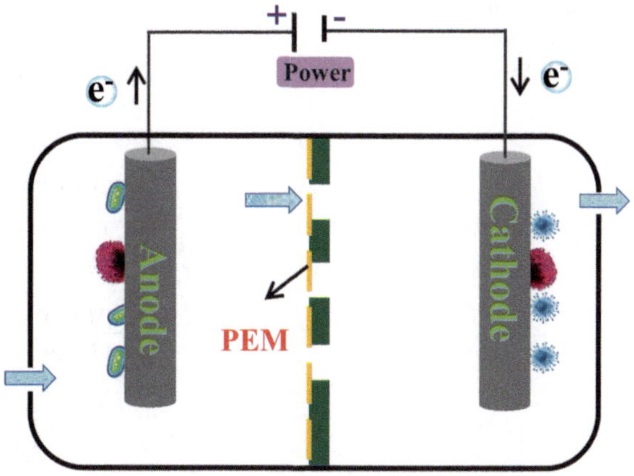

Fig. 11 Schematic diagram of the MEC-TPAD system

and their electrohydrogenation products (Baek et al. 2021). In contrast, *Methanothrix* and *Methanobacterium* are the main methanogenic genera enriched in cathodic biofilms, and both utilize electrons by interspecies hydrogen transfer. Driven by electric potential, these electroactive methanogens can efficiently utilize electrons and H_2 to convert CO_2 to CH_4. In addition, *Geobacter* showed high abundance in both anode and cathode biofilms of the MEC-TPAD reactor, and it was able to transfer electrons on the electrode surfaces, which served to enhance interspecies electron transfer (Yuan et al. 2022). Thus, MEC-TPAD can remove toxic inhibitors and products (sulfides, ammonia, and harmful metal ions) and then enhance CH_4 production in the subsequent methanogenic phase.

4 Key Factors for Design and Operation of BES-AD Systems

Factors affecting the performance of BES-AD include substrate type, OLR, pH, temperature, HRT, electrode material, and so on (Fig. 12). For example, different substrate types acclimate different types of dominant microbial communities, and then directly determine the performance of anaerobic systems. While the OLR restricts the efficiency of power generation, pollutant removal, and biomass production, thereby shaping the composition of microbial communities.

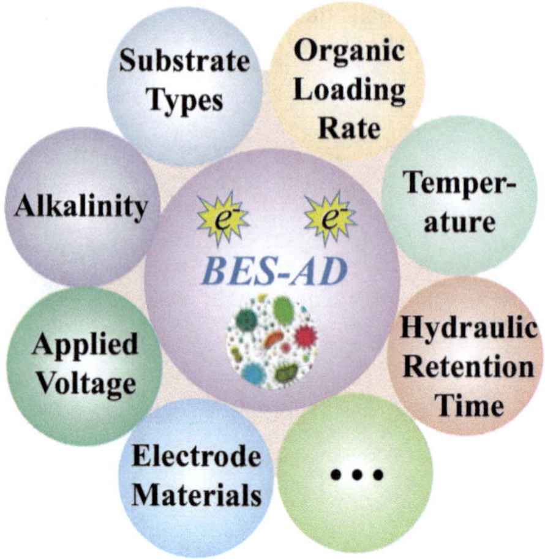

Fig. 12 Key factors for design and operation of BES-AD systems

4.1 Key Factors for Design and Operation of MFC-AD Systems

4.1.1 Substrate Types

The substrate, as a carbon source for microorganisms, plays a crucial role in the microbial community in the coupled MFC-AD system. In particular, the biodegradability of the substrate directly determines the efficiency of biomass energy conversion into electricity and heat in the system, and thus the substrate type and concentration are the most critical factors affecting the operation of MFC-AD systems (Bazina et al. 2023). Substrate types can be simple compounds or complex mixtures of organic and inorganic compounds. These substrates come from a variety of sources and can be broadly categorized into three groups based on source and nature: easily degradable wastewater, high-strength wastewater, and toxic and hazardous wastewater.

Easily degradable wastewater is defined as wastewater with a simple and easily degradable organic composition or low concentration of organic matter that promotes more efficient electricity and methane production, including domestic wastewater and municipal wastewater. Although such wastewater has a low concentration of total organic carbon, it usually requires pretreatment with a filtration unit due to the presence of suspended solids. Such wastewater can often be easily treated by MFC-AD and energy recovery can be realized. For example, Ren et al. (2014) utilized the anaerobic fluidized bed membrane bioreactor (MFC-AFMBR) system to treat domestic wastewater with a total chemical oxygen demand (TCOD) concentration of 210.0 ± 11.0 mg/L, and at an HRT of 9 h, the effluent TCOD concentration was 16.0 ± 3.0 mg/L (92.5% removal percentage) and total suspended solids were almost completely removed from 45.0 ± 10.0 mg/L to less than 1.0 mg/L.

High-strength wastewater contains many different types of macromolecular organic matter (cellulose, fats, proteins, etc.) and has a high organic matter concentration, which is commonly found in the production of several agricultural products, food processing farms, dairies, and confectionery factories. However, due to their complex composition and structure, these wastes must be sufficiently hydrolyzed before they can be utilized by electrophiles and methanogens, and hydrolysis often becomes the rate-limiting step. To solve the problem, Liu et al. (2020) achieved a maximum COD removal of about 90.0% when treating brewery wastewater using MFC-UASB, with a MFC voltage range of 0.3–0.4 V.

Toxic and hazardous wastewater contains many substances that inhibit the activity of microorganisms, such as heavy metals, pharmaceutical and personal care products, sulfides, aromatic compounds, and phenols. This type of wastewater is commonly found in a variety of industrial wastewater, landfill leachate, antibiotic wastewater, and petrochemical wastewater. For example, Niu et al. (2020) utilized MFC-AFB filled with macroporous adsorbent resin particles to treat isopropanol wastewater, and a maximum power density of 135.73 ± 0.17 mW/m^2 and a COD removal percentage of $68.21 \pm 0.24\%$ were obtained under the optimal condition.

4.1.2 OLR

OLR is an important factor affecting the performance of the MFC-AD system. Too high or too low OLR can inhibit the growth of microorganisms, and affect the COD removal, biogas production and microbial community structure. The OLR determines the specific flow of organic substrates into the reactor and the microbial populations required to degrade the contaminants, so optimizing the OLR for different types of wastewater is critical to improve the operation of the MFC-AD system.

Compared with other low-concentration wastewater, high-concentration livestock and poultry breeding wastewater can provide sufficient carbon source for the long-term stable operation of MFC-AD systems, which is conducive to the generation of electricity. Hao et al. (2022) used the ScMFC-AnMBR system to treat synthetic swine wastewater with glucose as the carbon source, and when OLR increased from 0.4 kg COD/(m^3·d) to 0.8 kg COD/(m^3·d), the power density, current density and coulombic efficiency decreased, the average removal percentage of COD increased by nearly 50.0% with the highest COD removal percentage of 99.3%, and the methane yield increased from 67.8 mL/g COD to 85.9 mL/g COD.

The MFC-AD system is mainly used in the treatment of industrial organic wastewater, such as alcohol wastewater, papermaking wastewater, textile wastewater, and so on. Zhang et al. (2009) used an UASB-MFC-BAF system to treat high-concentration molasses wastewater, and under different OLRs of 6.4 kg COD/(m^3·d), 25.5 kg COD/(m^3·d), and 127.5 kg COD/(m^3·d), the voltage output ranges were 200–300, 300–350, and 350–500 mV, and the COD removal percentages were 30.0%, 60.0%, and 53.2%, respectively.

4.1.3 Temperature

Temperature can affect metabolic activity of microbial communities, thereby affecting the efficiency and stability of anaerobic digestion. Within an appropriate range, increased temperatures can significantly enhance anaerobic digestion performance. Most laboratory MFC-AD coupled systems are operating at around 30 °C. Khandaker et al. (2021) treated high organic load wastewater from the vegetable oil industry with a two-chamber anaerobic MFC at two different temperatures (namely 25 °C and 35 °C), and COD removal percentages at 35 °C ranged from 60 to 90% whereas at 25 °C, the percentages were between 40 and 80%. The increase in temperature can enhance the activity of mesophilic microbes in the anode chamber, thus improving the COD removal efficiency. At lower temperatures, the mesophilic bacterial community undergoes an adaptation process, during which their activity sharply decreases. The electricity generation of MFCs improves with increasing temperatures. At 35 °C, the COD removal efficiency is highest (80–90%), with a maximum voltage of 5839 mV (Shen et al. 2018).

Compared to standalone anaerobic digestion systems, coupled MFC-AD systems can operate over a broader temperature range, increasing the possibility of continuous operation under adverse climatic conditions, albeit with some performance losses. Thus, when selecting the operating temperature, it needs to balance the efficiency and biological tolerance to maintain stable operation of the MFC-AD system.

4.1.4 HRT

HRT not only has an important impact on the removal of organic matter, gas production, and microbial community composition, but also has an important relationship with the size of the reactor volume. A long HRT increases the reactor volume and the capital cost, leading to the depletion of carbon-containing substances in the microbial cell. However, if the HRT is too short, the microorganisms will not be able to fully degrade the organic matter, thereby reducing the COD removal efficiency and affecting power production. HRT may lead to changes in microbial communities, which will directly affect wastewater treatment performance.

The HRT commonly used in the coupled MFC-AD system is 2.0–40.0 h. A long HRT results in a long residence time for the organics in the anode chamber and provides a higher contact time for microorganisms to oxidize the organics and transfer electrons, resulting in higher COD removal efficiency and higher energy output. Huang et al. (2011) treated the wastewater of the alcohol plant with a fixed HRT of 15.3 h in an MFC-AFB system, and the maximum COD removal efficiency was 90.0%. In particular, Zhang et al. (2012) found that both sulfate and COD could be effectively removed at an HRT of 40.0 h in the MFC-UASB system; when the HRT was longer, the sulfides were over-oxidized to sulfate in the MFC unit, affecting energy recovery. However, higher HRT can still increase power generation (Zhang et al. 2012).

Table 1 Performance of the MEC-AD systems under different conditions

Influencing factors		Operating performance	References
Electrode material	Platinum electrode	Hydrogen recovery rate of 72.2%	Selembo et al. (2009)
	Biocathode	Hydrogen recovery rate of 45.4%	Selembo et al. (2009)
	Platinum electrode	Hydrogen recovery rate of 72.2%	Selembo et al. (2009)
Voltage applied	1.0 V	COD removal percentage of 83.6%	Flores-Rodriguez et al. (2019)
	2.0 V	COD removal percentage of 96.5%	Arvin et al. (2019)
Influent C/N	Influent COD was 16.7 g/L	COD removal rate of 36.2 ± 24.4%	Zhang et al. (2013)
	Influent COD was 8.5 g/L	COD removal percentage of 20.0%	Cerrillo et al. (2016)
Alkalinity	pH = 5.0	Caproic acid yield was 7.2 g COD/L	Ren et al. (2023)
	pH = 9.0	Methane yield of 224 mL CH_4/g VS	Yuan et al. (2016)

4.2 *Key Factors for Design and Operation of MEC-AD Systems*

The MEC-AD system has superior treatment capability for all kinds of wastewater, but factors such as electrode materials, magnitude of applied voltage, etc., affect the economy and conductivity of the coupled system. The characteristics of wastewater (including influent carbon to nitrogen (C/N) ratio and pH, etc.) also cause certain impact on the coupled system, which may directly lead to the decrease of its treatment efficiency and the collapse of the biodegradation system. Table 1 gives performance of the coupled system under different conditions.

4.2.1 Electrode Materials

Carbon-based, metallic, and composite materials are the most widely used electrode materials in the MEC-AD system. Carbon-based materials, such as carbon cloth, carbon paper, and graphite, have been widely used in most MEC-AD systems due to their good stability in microbial inoculum, high electrical conductivity, and low toxicity. The anaerobic digestion performance of MEC-AD varies considerably when different carbon-based electrodes are used. Yang et al. (2023) examined the effect of graphite rod, carbon felt, and carbon fiber brush on methane yield of MEC-AD systems and found that carbon felt had a better net energy output than graphite rod and carbon fiber brush. Because *Methanobacterium*, electroactive and known as hydrogenotrophic methanogen, has a high abundance with carbon felt electrode, facilitating the interspecies electron transfer in the BES system. In contrast, the larger specific surface area and higher electrical conductivity of graphite felt and graphite fiber brushes contribute to the formation of dense biofilms on the electrode surface, substantially increasing the methane yield of the MEC system (Choi et al. 2017). Zhen et al. (2016) used graphite felt-modified carbon rods as the cathode, and the methane yield of the system reached 80.9 mL/L after 24 h, where the graphite felt enhanced the electrocatalytic activity of microorganisms and reduced the cathodic overpotential required for methane production.

Metallic materials, such as stainless steel, nickel and copper, have the advantages of high electrical conductivity and mechanical strength. Stainless steel materials show good durability and low cost compared to graphite felts. Bo et al. (2014) found that using stainless steel as a cathode can result in a methane volume fraction of up to 98.1%, allowing it to be used directly as a biofuel in place of fossil fuels. In addition, Wang et al. (2019) compared the suitability of stainless steel mesh, nickel mesh, and copper mesh as cathodes in MEC-AD, and showed that the cathode methane production was ranked in ascending order as nickel mesh, copper mesh, and stainless steel mesh, and nickel mesh had a higher hydrogen production activity than stainless steel mesh and copper mesh, leading to the difference in the methanogenic efficacy of cathodes of the three metals. Hu et al. (2009) found that the hydrogen yield of nickel cathodes was 17.0% higher than that of stainless steel cathodes.

Synthetic electrodes using metal-modified carbon-based materials can enhance the methane production of MEC-AD at lower energy inputs compared to conventional pure carbon-based and pure metal electrodes. For example, Liu et al. (2016) used platinum (Pt)-modified carbon cloth as a cathode to enhance the methane production of the MEC-AD system, and the methane production rate of the Pt-carbon cloth cathode reached 138 mL/(L·d) at an applied voltage of 0.8 V, which was three times higher than that of the conventional anaerobic digestion system (Liu et al. 2016). Nickel nanoparticle-modified electrodes replacing noble metal electrodes can further improve the electrochemical performance and specific surface area of the electrodes. Kim et al. (2017) loaded nickel nanoparticles on the surface of granular activated carbon and found that nickel nanoparticles could increase the methane yield of MEC-AD by 25.6% through improving the electrical conductivity and DIET of granular-activated carbon. The loaded metal increased the current response and decreased the charge transfer impedance and cathodic overpotential of the system. However, this modified composite leads to an increase in the cost of electrode fabrication, so a balance of electrode material availability and efficiency needs to be considered when selecting electrode materials.

4.2.2 Applied Voltage

Voltage has a significant impact on system performance and the operating cost of the MEC-AD system. Boosting the voltage within a certain range (0 to 2.0 V) will improve the organic matter removal in MEC-AD. In a MEC-UASB system with acetate as the substrate, a sharp decrease in soluble COD concentration was observed over eight days when the voltage increased from 0 to 1.0 V, and a maximum COD removal percentage of 83.6% was obtained at 1.0 V (Flores-Rodriguez et al. 2019). Similarly, COD removal gradually rose from 66.7 to 96.5% when the external voltage increased from 0 to 2.0 V in ABR-MEC (Arvin et al. 2019).

The applied voltage can affect methane production. When the cathode potential increased from −0.6 to 1.0 V, methane yields gradually increased, and the highest methane yields of 408.3 mL/g COD was obtained at an applied voltage of 1.0 V (Choi et al. 2017). When the applied voltage of MEC-UASB for treating dog food waste raised from 0.6 V to 1.0 V, both methane yields and methane content increased with the maximum content of 72% (Wang et al. 2021).

The external voltage also affects the rate of membrane fouling. As the voltage of the MEC-AnMBR system increased from 0 to 1.0 V, the membrane fouling cycle can be extended from 60 to 98 h (Ding et al. 2018). Even the sludge viscosity decreased from 4.3 to 3.3 mPa·s when the voltage increased from 0 to 1.0 V (Anjum et al. 2021). The increase in voltage would decrease the ratio of extracellular polyprotein to extracellular polysaccharide in the sludge and raise the absolute value of the ζ-potential of the sludge particles, causing a gradually slowed down in membrane fouling of the MEC-AnMBR system.

The enhancement of MEC-AD performance by applied voltage is attributed to the optimization of microbial community structure. During anaerobic digestion of protein wastewater, the abundance of fermentative bacteria at the anode increased by 46.7% with increasing voltages, whereas the abundance of methanogens at the cathode increased from 10.4 to 84.3% (Zhao et al. 2021). Moreover, from a thermodynamic point of view, the applied external voltage led to a decrease in the cathode potential, which favored the enhancement of methane production through the enrichment of interspecies electron transfer mediated by hydrogenotrophic methanogens (De Vrieze et al. 2018).

Applied voltage affects cell metabolism and cell rupture of microorganisms. Ding et al. (2016) used adenosine triphosphate (ATP) and lactate dehydrogenase (LDH) as indicators of cell metabolism and cell rupture, respectively, to investigate the optimal voltage of the MEC-AD system, where the activity of LDH increased by 14.0% and 10.0% in anode and by 35.0% and 41.0% in cathode under 1.0 and 2.0 V. While the activity of ATP decreased sharply at both voltages, 27.0% and 55.0% at the anodic voltage and 46.0% and 66.0% at the cathodic voltage, respectively. When the voltage is greater than 1.0 V, the more severe the cell rupture, the lower the growth rate and the lower the metabolic activity as the potential increases. Therefore, when choosing the applied voltage, the treatment efficiency and the tolerance threshold should be considered to maintain the efficient operation of the MEC-AD system.

4.2.3 OLR

A high OLR may deteriorate system performance. Huang et al. (2021) found that the hydrolysis kinetics of MEC-AD deteriorated when the OLR increased from 0.7 to 3.0 g COD/(L·d), and the methane percentage in biogas decreased from 42.4 to 35.2%. Similarly, when using MEC-AnMBR for blackwater treatment, a high OLR limits the hydrolysis capacity of the coupled system. The performance of both the MEC-AD system and the control system decreased when the OLR was increased from 3.0 to 4.5 g/(L·d) (Huang et al. 2023).

Industrial wastewater has the characteristics of a wide range of organic matter concentrations, and usually contains many refractory and toxic compounds that further reduce the biodegradability of organic matter. Yu et al. (2017) evaluated the performance of a MEC-AD system for treating soybean cooking oil refinery wastewater with low, middle, middle high, and high OLRs, and the highest methane yields of 45.4 ± 1.1 L/kg COD was obtained at the middle high OLR, which was much higher than 8.40 ± 0.6 L/kg COD at the low OLR and 7.8 ± 0.26 L/kg COD at the high OLR. This may be attributed to the accumulation of inhibitory substances such as long-chain fatty acids, polyphenols, and pectins, thus inhibiting the anaerobic digestion of edible oil processing wastewater. At a high OLR, hydrolysis/acetogenesis and methanogenesis may be imbalanced, which could lead to the rapid accumulation of VFAs in the MEC-AD system and a significant reduction in methane production (Kanellos et al. 2024).

4.2.4 Alkalinity

pH 5.0–6.5 favors generating VFAs in MEC-AD systems. Ren et al. (2023) constructed a MEC-AD system to convert lactic and acetic acid to caproic acid under the electric field of 0.05 V. Under the condition of pH 5.0, *Ruminococcaceae_bacterium* and *Candidatus_Pseudoramibacter_fermentans*, having the function of synthesizing caproic acid, were enriched in this system, and the caproate yield was 3.3 g/L. The optimal pH of *Ruminococcaceae_bacterium* is 5.0–6.5 (Chivero et al. 2022). While the optimal pH of *Candidatus_Pseudoramibacter_fermentans* is 5.5 (Matthew et al. 2020). In addition, Ma et al. (2022) used MEC-AD to treat Chinese liquor wastewater producing caproic acid with sludge as the inoculum at cathodic potential of −0.8 V and found that caproate concentration was 1.9 g/L in MEC-AD operating at pH 6.5. *Clostridium* and *Caproiciproducens*, which can synthesize caproic acid, were enriched by 38.8% in this MEC-AD system (Ma et al. 2022). The optimal pH for *Clostridium* and *Caproiciproducens* was 6.0–6.5 (Candry et al. 2020). As a result, MEC-AD system operating at pH 5.0–6.5 favors the acid-producing bacteria enrichment to enhance the acid production of MEC-AD systems.

When MEC-AD operates at pH 6.5–7.0, the production of hydrogen favors. Yu et al. (2021) used MEC-AD to treat sludge lysate at an external voltage of 1.0 V and found that the hydrogen production rate was 0.2 L/d when operated at around pH of 7.0. Moreover, Shao et al. (2019) used MEC-AD to treat sodium acetate at an external voltage of 1.0 V and showed that the hydrogen production was 0.1 L/d operating at pH 7.0. *Desulfuromonas* with the ability to produce hydrogen, was enriched with the relative abundance of 82.0% in the system. *Desulfuromonas* has optimal pH of 6.5–7.0. Hence, MEC-AD operating at pH 6.5–7.0 favors the hydrogen-producing bacteria to enhance the hydrogen production of MEC-AD systems.

Methane production is favored when MEC-AD operates at pH 7.0–7.5. Xu et al. (2016) operated MEC-AD to treat waste-activated sludge at an external voltage of 0.6 V and found that the methane production was 224 mL/g VS operated at pH 7.0. Meanwhile, the relative abundance of *Methanothrix* in MEC-AD was 53.5%. Also, *Methanothrix* has an optimal pH of 7.0–7.6. Wang et al. (2021) used MEC-AD to treat waste-activated sludge at an external voltage of 0.8 V and found the methane production was 248 mL/g VSS operating at pH 7.0. The relative abundance of *Methanobacterium* in the system reached 61.0%. The optimal pH of *Methanobacterium* was 6.9–7.6. Consequently, MEC-AD operating at pH 7.0–7.5 favors the enrichment of methanogenic archaea that enhances the capacity of MEC-AD for methane production.

5 Application of BES-AD

Even though application of anaerobic digestion to degrade organic pollutants and produce biogas has achieved great success, the treatment of highly concentrated organic wastewater and wastewater containing refractory substances still has problems such as instability, low substrate removal percentage, and low biogas yield. BES-AD enhances the electron transfer activity between microorganisms and their resistance to toxic and hazardous substances by means of electrical currents, and it may provide an ideal solution to achieve high-quality effluent, with less energy requirement (than single anaerobic digestion systems).

MFC-AD systems can efficiently treat the highly concentrated organic wastewater and recover energy under appropriate system operational settings. Wang et al. (2015) developed a continuous stirred microbial electrochemical reactor (CSMER) for the treatment of cellulosic ethanol wastewater, and COD removal and energy recovery efficiency were 87.1% and 32.1%, which were 1.6 and 2.5 times higher than that of a CSTR. Larger amount of deltaproteobacteria (5.3%) and hydrogenotrophic methanogens (47%) could account for the better performance of CSMER, since syntrophic associations among them provided more degradation pathways compared to the CSTR (Wang et al. 2015). Huang et al. (2011) designed an MFC-AFB system to treat alcohol distillery wastewater and generate electricity simultaneously. The COD removal efficiency was from 80–90% in stabilization period, and a maximum power density of 124.03 mW/m^2 was achieved (Huang et al. 2011). Tan et al. (2017) developed an MFC-AnMBR system, in which the MFC acted as a pretreatment unit prior to the AnMBR, for the treatment of palm oil mill wastewater. MFC-AnMBR system increased COD removal from 77.13 $\pm$ 1.07% to 97.07 $\pm$ 0.70%, and the time required for MFC-AnMBR system to reach the threshold (1 kPa) was relatively longer compared to AnMBRs (Tan et al. 2017).

MFC-AD systems can effectively increase the resistance of the reactor and improve the performance of the reactor for the treatment of toxic and refractory organic wastewater. Niu et al. (2020) used an MFC-AFB system with macroporous adsorbent resin as a multifunctional bio-carrier to treat isopropanol wastewater. The maximum power density is 135.73 ± 0.17 mW/m^2 and COD removal is $68.21 \pm 0.24\%$ (Niu et al. 2020). Liu et al. (2021) constructed an MFC-ABR system to treat simulated printing and dyeing wastewater, and a dye removal percentage of 91.7% and a maximum power density of 1226.4 mW/m^3 were obtained. Besides, methane production and microbial species in the ABR-MFC system were higher than those in the single ABR (Liu et al. 2021). Wang et al. (2020a) used an MFC-ABR system to treat pharmaceutical wastewater, and around $35.2 \pm 12.8\%$ of COD reduction and $67.2 \pm 21.4\%$ of color removal efficiency were observed during about six months' operation, which was $16.7 \pm 13.2\%$ and $59.4 \pm 27.9\%$ higher than those of the conventional ABR.

MECs have been used to enhance denitrification, dechlorination, and reduction of oxides, especially NB, chloronitrobenzene (ClNB), and p-chloronitrobenzene (p-ClNB), due to the input energy being much lower than typical electrochemical process (Chen et al. 2017). When treating wastewater containing NB (Zhang et al. 2015), the MEC-UASB removed 98.1% of NB at 400 mg/L, while traditional UASB removed 80.7%. MEC-UASB produced more methane (0.79 L/d) than UASB (0.60 L/d). *Sphingomonas* and *Pseudomonas* accounted for 22% and 10% in MEC-UASB, respectively, and 8% and 0% in UASB. These two microorganisms enriched by MEC-UASB may enhance the energy recovery of highly concentrated NB wastewater. In addition, Xu et al. (2016) established a 4.8 L MEC-UASB system equipped with a graphite felt anode and a stainless steel mesh as the cathode. In the MEC-UASB system, p-ClNB and p-chloroaniline removal efficiency reached $99.63 \pm 0.37\%$ and $40.39 \pm 9.26\%$, respectively, which was significantly improved in comparison with the control UASB system (Xu et al. 2016). Applied voltage would significantly influence the succession of microbial community and promote oriented enrichment of the functional bacteria, which could be the underlying reasons for the improved performance (Xu et al. 2016). MEC-AD system had a promising application to remove the recalcitrant pollutants effectively.

6 Future Perspectives

Since Prof. Willy Verstraete (Pham et al. 2006) proposed the concept of electrochemical enhanced anaerobic digestion, there have been lots of studies to verify the feasibility of the coupled technology. Compared with traditional anaerobic digestion, anaerobic digestion systems by applying external voltage can significantly increase methane production (Malaeb et al. 2013), removal of refractory substances (Park et al. 2020), and microbial stability (Wang et al. 2022). However, previous studies have been mainly focusing on the treatment ability of waste/wastewater containing toxic and refractory compounds, and further studies are required to better understand the underlying mechanisms responsible for the better performances of MEC-AD systems. Especially, functional microbes, microbial interactions, metabolic pathways, and functional gene expression responsible for the degradation of these compounds should be investigated.

The extracellular electron transfer (EET) rate and metabolic pathway of the electroactive biofilms on the surface of BES-AD electrodes are the key factors determining product conversion and pollutant removal. An ideal microbial catalyst needs to not only quickly absorb extracellular electrode electrons and couple them with intracellular redox metabolism, but also possess metabolic pathways for degrading refractory organic matter. However, naturally EEA with high EET abilities (such as *Geobacter* spp. and *Shewanella* spp.) often lack metabolic pathways for reducing and synthesizing high value-added products or oxidizing refractory substances, and most industrial bacteria generally do not have the EET ability. Therefore, appropriate strategies are required to integrate these two abilities within one system. Exogenous EET pathways can be transported into industrial bacteria through gene editing, but current research mainly focuses on *Escherichia coli* and is not yet universal. The metabolic pathways of industrial bacteria can be genetically modified to produce more valuable products. However, excessive genetic and metabolic modifications often have adverse effects on the growth rate, resistance, and other aspects of the strains. Finally, a limited mixed microbial community can be constructed to improve the speed and controllability of electron and material flow through different ecological niches and microbial interactions.

The pilot- and full-scale application of the BES-AD system is a great challenge. BES-AD might significantly accelerate conversion rates in larger reactors, while a large reactor has difficulty in maintaining a biofilm on the electrodes. The optimization of electrode materials is another important factor for the scale-up of MEC-AD systems. The electrode material with high catalytic performance and low cost would be the primary consideration. Furthermore, the conductivity, biocompatibility, and electron transfer kinetics of electrodes can be improved through surface modification. The design of innovative reactor configurations and examination of optimal operational conditions through intensive research should be conducted for treating different types of wastewaters to further expand the application of BES-AD.

7 Take-Home Messages

(1) BES-AD improves the resource recovery efficiency and system stability of conventional anaerobic digestion, contributing to the strategic goal of green, low-carbon, and recycling development.
(2) Diverse types of BES-AD configurations can be selected for treating different types of wastewater with the consideration of energy recovery or system monitoring.

(3) Factors such as temperature, electrode materials, applied voltage, HRT, and OLR significantly influence operational performance and economic viability of BES-AD.
(4) BES-AD expands the practical application of anaerobic digestion in treating diverse types of wastewater, due to its ability to effectively degrade refractory substances and high adaptability to complex wastewater environments.
(5) Continuous parameter tuning and process optimization from laboratory scale to practical application are prerequisites for the widespread use of BES-AD.

References

Anjum F, Khan IM, Kim J, Aslam M, Blandin G, Heran M, Lesage G (2021) Trends and progress in AnMBR for domestic wastewater treatment and their impacts on process efficiency and membrane fouling. Environ Technol Innov 21:101204

Arun J, SundarRajan P, Grace Pavithra K, Priyadharsini P, Shyam S, Goutham R, Le Hoang Q, Pugazhendhi A (2024) New insights into microbial electrolysis cells (MEC) and microbial fuel cells (MFC) for simultaneous wastewater treatment and green fuel (hydrogen) generation. Fuel 355:129530

Arvin A, Hosseini M, Amin M, Darzi G, Ghasemi Y (2019) A comparative study of the anaerobic baffled reactor and an integrated anaerobic baffled reactor and microbial electrolysis cell for treatment of petrochemical wastewater. Biochem Eng J 144:157–165

Asensio Y, Llorente M, Fernández P, Tejedor-Sanz S, Ortiz JM, Ciriza JF, Monsalvo V, Rogalla F, Esteve-Núñez A (2021) Upgrading fluidized bed bioelectrochemical reactors for treating brewery wastewater by using a fluid-like electrode. Chem Eng J 406:127103

Baek G, Shi L, Rossi R, Logan BE (2021) The effect of high applied voltages on bioanodes of microbial electrolysis cells in the presence of chlorides. Chem Eng J 405:126742

Barua S, Zakaria BS, Lin L, Dhar BR (2019) Shaping microbial communities with conductive carbon fibers to enhance methane productivity and kinetics. Bioresource Tech Rep 5:20–27

Bazina N, Ahmed TG, Almdaaf M, Jibia S, Sarker M (2023) Power generation from wastewater using microbial fuel cells: a review. J Biotechnol 374:17–30

Bo T, Zhu X, Zhang L, Tao Y, He X, Li D, Yan Z (2014) A new upgraded biogas production process: coupling microbial electrolysis cell and anaerobic digestion in single-chamber, barrel-shape stainless steel reactor. Electrochem Commun 45:67–70

Candry P, Radić L, Favere J, Carvajal-Arroyo JM, Rabaey K, Ganigué R (2020) Mildly acidic pH selects for chain elongation to caproic acid over alternative pathways during lactic acid fermentation. Water Res 186:116396

Cerrillo M, Oliveras J, Viñas M, Bonmatí A (2016) Comparative assessment of raw and digested pig slurry treatment in bioelectrochemical systems. Bioelectrochemistry 110:69–78

Chen H, Gao X, Wang C, Shao J, Xu X, Zhu L (2017) Efficient 2,4-dichloronitrobenzene removal in the coupled BES-UASB reactor: effect of external voltage mode. Biores Technol 241:879–886

Chen H, Jiang H, Song Y, Wang M, Shi R, Fu Y, Zhao C (2024) Impact of liquid velocity and stacking modes on the performance of anaerobic fluidized bed microbial fuel cell. Powder Technol 433:119264

Chivero ET, Sil S, Kumar M, Buch S (2022) Substance use, microbiome and psychiatric disorders. Pharmacol Biochem Behav 219:173432

Choi K-S, Kondaveeti S, Min B (2017) Bioelectrochemical methane (CH_4) production in anaerobic digestion at different supplemental voltages. Biores Technol 245:826–832

Croese E, Jeremiasse AW, Marshall IPG, Spormann AM, Euverink G-JW, Geelhoed JS, Stams AJM, Plugge CM (2014) Influence of setup and carbon source on the bacterial community of biocathodes in microbial electrolysis cells. Enzyme Microb Technol 61–62:67–75

Cui M, Liu W, Tang Z, Cui D (2021) Recent advancements in azo dye decolorization in bio-electrochemical systems (BESs): insights into decolorization mechanism and practical application. Water Res 203:117512

Cui D, Guo Y-Q, Lee H-S, Wu W-M, Liang B, Wang A-J, Cheng H-Y (2014) Enhanced decolorization of azo dye in a small pilot-scale anaerobic baffled reactor coupled with biocatalyzed electrolysis system (ABR–BES): a design suitable for scaling-up. Biores Technol 163:254–261

De Vrieze J, Arends JBA, Verbeeck K, Gildemyn S, Rabaey K (2018) Interfacing anaerobic digestion with (bio)electrochemical systems: potentials and challenges. Water Res 146:244–255

Ding A, Fan Q, Cheng R, Sun G, Zhang M, Wu D (2018) Impacts of applied voltage on microbial electrolysis cell-anaerobic membrane bioreactor (MEC-AnMBR) and its membrane fouling mitigation mechanism. Chem Eng J 333:630–635

Ding A, Yang Y, Sun G, Wu D (2016) Impact of applied voltage on methane generation and microbial activities in an anaerobic microbial electrolysis cell (MEC). Chem Eng J 283:260–265

Feng Y, Lee H, Wang X, Liu Y, He W (2010) Continuous electricity generation by a graphite granule baffled air–cathode microbial fuel cell. Biores Technol 101:632–638

Flores-Rodriguez C, Nagendranatha Reddy C, Min B (2019) Enhanced methane production from acetate intermediate by bioelectrochemical anaerobic digestion at optimal applied voltages. Biomass Bioenerg 127:105261

Gharbi R, Gomez Vidales A, Omanovic S, Tartakovsky B (2022) Mathematical model of a microbial electrosynthesis cell for the conversion of carbon dioxide into methane and acetate. J CO_2 Utili 59:101956

Goswami R, Mishra VK (2018) A review of design, operational conditions and applications of microbial fuel cells. Biofuels 9:203–220

Gul H, Raza W, Lee J, Azam M, Ashraf M, Kim KH (2021) Progress in microbial fuel cell technology for wastewater treatment and energy harvesting. Chemosphere 281:130828

Guo J, Peng Y, Ni BJ, Han X, Fan L, Yuan Z (2015) Dissecting microbial community structure and methane-producing pathways of a full-scale anaerobic reactor digesting activated sludge from wastewater treatment by metagenomic sequencing. Microb Cell Fact 14:33

Guo W, Cheng D, Ngo HH, Chang SW, Nguyen DD, Nguyen DP, Bui XT (2020) 9 - Anaerobic membrane bioreactors for antibiotic wastewater treatment. In: Ngo HH, Guo W, Ng HY, Mannina G, Pandey A (eds) Current developments in biotechnology and bioengineering. Elsevier, pp 219–239

Hao Y, Zhang X, Du Q, Wang H, Ngo HH, Guo W, Zhang Y, Long T, Qi L (2022) A new integrated single-chamber air-cathode microbial fuel cell—anaerobic membrane bioreactor system for improving methane production and membrane fouling mitigation. J Membr Sci 655:120591

Hu H, Fan Y, Liu H (2009) Hydrogen production in single-chamber tubular microbial electrolysis cells using non-precious-metal catalysts. Int J Hydrogen Energy 34:8535–8542

Huang J, Yang P, Guo Y, Zhang K (2011) Electricity generation during wastewater treatment: an approach using an AFB-MFC for alcohol distillery wastewater. Desalination 276:373–378

Huang Q, Liu Y, Dhar BR (2021) Pushing the organic loading rate in electrochemically assisted anaerobic digestion of blackwater at ambient temperature: insights into microbial community dynamics. Sci Total Environ 781:146694

Huang Q, Liu Y, Dhar BR (2023) Boosting resilience of microbial electrolysis cell-assisted anaerobic digestion of blackwater with granular activated carbon amendment. Biores Technol 381:129136

Jia H, Yang G, Wang J, Ngo HH, Guo W, Zhang H, Zhang X (2016) Performance of a microbial fuel cell-based biosensor for online monitoring in an integrated system combining microbial fuel cell and upflow anaerobic sludge bed reactor. Biores Technol 218:286–293

Jiang X, Shen J, Han Y, Lou S, Han W, Sun X, Li J, Mu Y, Wang L (2016) Efficient nitro reduction and dechlorination of 2,4-dinitrochlorobenzene through the integration of bioelectrochemical system into upflow anaerobic sludge blanket: a comprehensive study. Water Res 88:257–265

Kanellos G, Tremouli A, Arvanitakis G, Lyberatos G (2024) Boosting methane production and raw waste activated sludge treatment in a microbial electrolysis cell-anaerobic digestion (MEC-AD) system: the effect of organic loading rate. Bioelectrochemistry 155:108555

Khandaker S, Das S, Hossainm T (2021) Sustainable approach for wastewater treatment using microbial fuel cells and green energy generation—a comprehensive review. J Mol Liq 344:117795

Kik O, Issa L, Katuri P (2021) Coupling anaerobic fluidized membrane bioreactors with microbial electrolysis cells towards improved wastewater reuse and energy recovery. J Environ Chem Eng 9(5):105974

Kim K-R, Kang J, Chae K-J (2017) Improvement in methanogenesis by incorporating transition metal nanoparticles and granular activated carbon composites in microbial electrolysis cells. Int J Hydrogen Energy 42(45):27623–27629

Kuntke P, Sleutels THJA, Saakes M, Buisman CJN (2014) Hydrogen production and ammonium recovery from urine by a microbial electrolysis cell. Int J Hydrogen Energy 39:4771–4778

Liu H, Leng F, Guan Y, Yao Y, Li Y, Xu S (2017) Simultaneous pollutant removal and electricity generation in a combined ABR-MFC-MEC system treating fecal wastewater. Water Air Soil Pollut 228:179

Liu H, Zhang B, Liu Y, Wang Z, Hao L (2015) Continuous bioelectricity generation with simultaneous sulfide and organics removals in an anaerobic baffled stacking microbial fuel cell. Int J Hydrogen Energy 40:8128–8136

Liu N, Yun Y, Hu L, Xin L, Han M, Zhang P (2021) Study on Start-up membraneless anaerobic baffled reactor coupled with microbial fuel cell for dye wastewater treatment. ACS Omega 6:23515–23527

Liu W, Yang G, Jia H, Wang J (2020) A novel UASB-MFC dual sensors system for wastewater treatment: on-line sensor recovery and electrode cleaning in the long-term operation. Chemosphere 246:125751

Liu WZ, Cai WW, Guo ZC, Wang L, Yang CX, Varrone C, Wang AJ (2016) Microbial electrolysis contribution to anaerobic digestion of waste activated sludge, leading to accelerated methane production. Renew Energy 91:334–339

Liu Y, Cao X, Zhang J, Fang Z, Zhang H, Gao X, Li X (2022) The use of a self-generated current in a coupled MFC-AnMBR system to alleviate membrane fouling. Chem Eng J 442:136090

Ma H, Wu W, Yu Z, Zhao J, Fu P, Xia C, Lam SS, Wang Q, Gao M (2022) Medium-chain fatty acid production from Chinese liquor brewing yellow water by electro-fermentation: division of fermentation process and segmented electrical stimulation. Biores Technol 360:127510

Malaeb L, Katuri KP, Logan BE, Maab H, Nunes SP, Saikaly PE (2013) A hybrid microbial fuel cell membrane bioreactor with a conductive ultrafiltration membrane biocathode for wastewater treatment. Environ Sci Technol 47(20):11821–11828

Modestra JA, Reddy CN, Krishna KV, Min B, Mohan SV (2020) Regulated surface potential impacts bioelectrogenic activity, interfacial electron transfer and microbial dynamics in microbial fuel cell. Renew Energy 149:424–434

Niu Y, Liu X, Chang G (2020) Treatment of isopropanol wastewater in an anaerobic fluidized bed microbial fuel cell filled with macroporous adsorptive resin as multifunctional biocarrier. Sci Total Environ 719:137495

Park J-G, Jiang D, Lee B, Jun H-B (2020) Towards the practical application of bioelectrochemical anaerobic digestion (BEAD): insights into electrode materials, reactor configurations, and process designs. Water Res 184:116214

Pham TH, Rabaey K, Aelterman P, Clauwaert P, De Schamphelaire L, Boon N, Verstraete W (2006) Microbial fuel cells in relation to conventional anaerobic digestion technology. Eng Life Sci 6(3):285–292

Ren L, Ahn Y, Logan BE (2014) A two-stage microbial fuel cell and anaerobic fluidized bed membrane bioreactor (MFC-AFMBR) system for effective domestic wastewater treatment. Environ Sci Technol 48:4199–4206

Ren W, Wu Q, Deng L, Wang H, Zhang Y, Guo W (2023) Insights into chain elongation mechanisms of weak electric-field-stimulated continuous caproate biosynthesis: key enzymes, specific species functions, and microbial collaboration. ACS ES&T Engineering 3:1649–1660

Rozendal RA, Hamelers HVM, Euverink GJW, Metz SJ, Buisman CJN (2006) Principle and perspectives of hydrogen production through biocatalyzed electrolysis. Int J Hydrogen Energy 31:1632–1640

Matthew JS, Myers Kevin S, Donohue Timothy J, Noguera Daniel R (2020) Medium-chain fatty acid synthesis by "*Candidatus Weimeria bifida" gen. nov.,* sp. nov., and "*Candidatus Pseudoramibacter fermentans*" sp. nov. Appl Environ Microbiol 86:e02242-19

Selembo PA, Merrill MD, Logan BE (2009) The use of stainless steel and nickel alloys as low-cost cathodes in microbial electrolysis cells. J Power Sources 190:271–278

Shao Q, Li J, Yang S, Sun H (2019) Effects of different substrates on microbial electrolysis cell (MEC) anodic membrane: biodiversity and hydrogen production performance. Water Sci Technol 79:1123–1133

Shen R, Jiang Y, Ge Z, Lu J, Zhang Y, Liu Z, Ren ZJ (2018) Microbial electrolysis treatment of post-hydrothermal liquefaction wastewater with hydrogen generation. Appl Energy 212:509–515

Song Y, Wang X, Liu Y, Chen H, Zhao C, Liu L, Jiang H, Zhu L, Ai T (2023) Power generation characteristics of pulsed anaerobic fluidized bed microbial fuel cell. Powder Technol 416:118215

Sun J, Hu C, Tong T, Zhao K, Qu J, Liu H, Elimelech M (2017) Performance and mechanisms of ultrafiltration membrane fouling mitigation by coupling coagulation and applied electric field in a novel electrocoagulation membrane reactor. Environ Sci Technol 51:8544–8551

Sun Z-F, Zhao L, Wu K-K, Wang Z-H, Wu J-t, Chen C, Yang S-S, Wang A-J, Ren N-Q (2022) Overview of recent progress in exogenous hydrogen supply biogas upgrading and future perspective. Sci Total Environ 848:157824

Tan SP, Kong HF, Bashir MJ, Lo PK, Ho CD, Ng CA (2017) Treatment of palm oil mill effluent using combination system of microbial fuel cell and anaerobic membrane bioreactor. Biores Technol 245:916–924

Wang AJ, Wang HC, Cheng HY, Liang B, Liu WZ, Han JL, Zhang B, Wang SS (2020a) Electrochemistry-stimulated environmental bioremediation: development of applicable modular electrode and system scale-up. Environ Sci Ecotech 3:100050

Wang G, Gao X, Li Q, Zhao H, Liu Y, Wang XC, Chen R (2020b) Redox-based electron exchange capacity of biowaste-derived biochar accelerates syntrophic phenol oxidation for methanogenesis via direct interspecies electron transfer. J Hazard Mater 390:121726

Wang H, Liu Y, Du H, Zhu J, Peng L, Yang C, Luo F (2021) Exploring the effect of voltage on biogas production performance and the methanogenic pathway of microbial electrosynthesis. Biochem Eng J 171:108028

Wang H, Qu Y, Li D, Zhou X, Feng Y (2015) Evaluation of an integrated continuous stirred microbial electrochemical reactor: wastewater treatment, energy recovery and microbial community. Biores Technol 195:89–95

Wang L, He Z, Guo Z, Sangeetha T, Yang C, Gao L, Wang A, Liu W (2019) Microbial community development on different cathode metals in a bioelectrolysis enhanced methane production system. J Power Sources 444:227306

Wang XT, Zhang YF, Wang B, Wang S, Xing X, Xu XJ, Liu WZ, Ren NQ, Lee DJ, Chen C (2022) Enhancement of methane production from waste activated sludge using hybrid microbial electrolysis cells-anaerobic digestion (MEC-AD) process–a review. Biores Technol 346:126641

Xu X, Shao J, Li M, Gao K, Jin J, Zhu L (2016) Reductive transformation of p-chloronitrobenzene in the upflow anaerobic sludge blanket reactor coupled with microbial electrolysis cell: performance and microbial community. Biores Technol 218:1037–1045

Yang L, Chen L, Chen K, Zhu H (2023) Improved net energy recovery in a sludge anaerobic digestion process by coupling an electrochemical system: electrode material and its impact on suspended microbial community. Environ Sci Pollut Res 30:99473–99483

Yu N, Xing D, Li W, Yang Y, Li Z, Li Y, Ren N-Q (2017) Electricity and methane production from soybean edible oil refinery wastewater using microbial electrochemical systems. Int J Hydrogen Energy 42:96–102

Yu Z, Liu W, Shi Y, Wang B, Huang C, Liu C, Wang A (2021) Microbial electrolysis enhanced bioconversion of waste sludge lysate for hydrogen production compared with anaerobic digestion. Sci Total Environ 767:144344

Yuan H, Chen Y, Dai X, Zhu N (2016) Kinetics and microbial community analysis of sludge anaerobic digestion based on Micro-direct current treatment under different initial pH values. Energy 116:677–686

Yuan Y, Zhang L, Chen T, Huang Y, Qian X, He J, Li Z, Ding C, Wang A (2022) Simultaneous recovery of bio-sulfur and bio-methane from sulfate-rich wastewater by a bioelectrocatalysis coupled two-phase anaerobic reactor. Biores Technol 363:127883

Zhang B, Zhang J, Yang Q, Feng C, Zhu Y, Ye Z, Ni J (2012) Investigation and optimization of the novel UASB–MFC integrated system for sulfate removal and bioelectricity generation using the response surface methodology (RSM). Biores Technol 124:1–7

Zhang B, Zhao H, Zhou S, Shi C, Wang C, Ni J (2009) A novel UASB–MFC–BAF integrated system for high strength molasses wastewater treatment and bioelectricity generation. Biores Technol 100:5687–5693

Zhang F, Ge Z, Grimaud J, Hurst J, He Z (2013) Long-term performance of liter-scale microbial fuel cells treating primary effluent installed in a municipal wastewater treatment facility. Environ Sci Technol 47:4941–4948

Zhang J, Zhang Y, Quan X (2015) Bio-electrochemical enhancement of anaerobic reduction of nitrobenzene and its effects on microbial community. Biochem Eng J 94:85–91

Zhang J, Zhu G, Pan X, Jha A, Liu L, Huang X (2016) Effects of hydraulic retention time and influent alkalinity on the performance of bio-electrochemical system assisted anaerobic baffled reactor. Desalin Water Treat 57(53):25399–25410

Zhao G, Wei L, Ma F, Chua H, Wang Z (2010) Electricity generation from cattle dung using microbial fuel cell in acidification process of two-phase anaerobic digestion system. Adv Mat Res 113–116:286–290

Zhao L, Wang X-T, Chen K-Y, Wang Z-H, Xu X-J, Zhou X, Xing D-F, Ren N-Q, Lee D-J, Chen C (2021) The underlying mechanism of enhanced methane production using microbial electrolysis cell assisted anaerobic digestion (MEC-AD) of proteins. Water Res 201:117325

Zhao Z, Zhang Y, Wang L, Quan X (2015) Potential for direct interspecies electron transfer in an electric-anaerobic system to increase methane production from sludge digestion. Sci Rep 5:11094

Zhen G, Lu X, Kobayashi T, Kumar G, Xu K (2016) Promoted electromethanosynthesis in a two-chamber microbial electrolysis cells (MECs) containing a hybrid biocathode covered with graphite felt (GF). Chem Eng J 284:1146–1155

Efficient Anaerobic Co-digestion of Sewage Sludge and Food Waste Incorporating Qualitative and Phase Separation

Caiping Sun, Kai Feng, and Huan Li

Abstract As a byproduct of wastewater treatment plants, sewage sludge contains various types of pollutants, including antibiotics, heavy metals, parasitic eggs, and viruses. Anaerobic digestion is the primary method for sludge stabilization, widely used for resource recovery from and detoxification of sewage sludge. Conventional anaerobic digestion of sludge faces technical challenges such as low carbon-to-nitrogen ratio and low organic matter content. Co-digestion sewage sludge with food waste can effectively address these issues and improve system performance. Under high-solid conditions, anaerobic co-digestion of sewage sludge and food waste is influenced by many factors, and system performance can be enhanced through various pretreatment approaches. A novel anaerobic treatment process has been proposed, in which hydrothermally pretreated sludge and fermented food waste undergo synergistic anaerobic digestion. This process effectively addresses issues such as the slow hydrolysis rate of sewage sludge and the high content of recalcitrant organic matter in the co-digestion system. Additionally, the process provides insights into the transformation of humic acids during hydrothermal treatment of sludge and the fermentation patterns of food waste, providing a sound technology for the design of high solids co-digestion systems for treating sewage sludge and food waste.

Keywords Co-digestion · Food waste · Qualitative separation · Sewage sludge

1 Introduction

Sewage sludge is a secondary pollutant generated from various wastewater treatment processes in municipal wastewater treatment plants. Sewage sludge primarily consists of microbial biomass, with partially recalcitrant organic matter as well as

C. Sun · K. Feng · H. Li (✉)
Tsinghua Shenzhen International Graduate School, Tsinghua University, Shenzhen 518055, China
e-mail: li.huan@sz.tsinghua.edu.cn

C. Sun
e-mail: scp23@mails.tsinghua.edu.cn

G. Wu (ed.), *Anaerobic Digestion*, Green Energy and Technology,
https://doi.org/10.1007/978-3-031-69378-6_18

toxic and harmful substances such as heavy metals, pathogens, and parasite eggs. Without proper treatment, sewage sludge can adversely affect the environment and public health. Typical sludge treatment and disposal technologies include sanitary landfilling, incineration, aerobic composting, and anaerobic digestion. Among these, anaerobic digestion is extensively employed for resource recovery and harmless treatment of sludge organic matter. Based on the total solids (TS) content of sludge within anaerobic digestion reactors, conventional anaerobic digestion (CAD, TS below 6%) and high-solid anaerobic digestion (HSAD, TS above 6%) are classified. CAD requires long sludge retention time (SRT) and large footprint of digestion facilities, and possesses low gas production efficiency. Conversely, HSAD effectively increases the organic matter content per unit volume within digestion tanks, thus enhancing volumetric gas production rates and reducing facility footprint, suitable for centralized treatment and disposal of sludge.

HSAD faces some technical challenges when treating sewage sludge. Firstly, a low sludge carbon-to-nitrogen (C/N) ratio of 6–9 exists in sludge, which is much lower than the optimal C/N for anaerobic digestion of 15–25. Low C/N ratios can lead to nutrient imbalance for anaerobic microorganisms, potentially resulting in ammonia nitrogen accumulation in anaerobic reactors and the inhibition of microbial activity (Zhang et al. 2015). Secondly, sewage sludge may have low organic matter content, even below 50%. The low organic matter content in sludge usually induces low biogas production. Co-digestion with other organic waste is an essential means for improving anaerobic digestion performance by increasing the organic matter content. Food waste, as a significant component of urban domestic waste, has a high organic matter content, making it a suitable co-digestion substrate with sewage sludge. Co-digestion sewage sludge with food waste can improve the nutrient balance for anaerobic microorganisms and prevent the accumulation of intermediate products such as volatile fatty acids (VFAs) and ammonia nitrogen (Dai et al. 2013), thereby improving gas production in mesophilic or thermophilic anaerobic digestion systems (Kim et al. 2003). Therefore, co-digestion in HSAD is a promising technology for the management of diverse types of wastes.

2 Key Factors Affecting Anaerobic Co-digestion of Sewage Sludge and Food Waste

2.1 Substrate Ratio

Appropriate co-digestion substrate ratios can effectively enhance gas production, surpassing the sum of methane production from individual substrates. The synergistic effect between substrates occurs during co-digestion, mutually promoting the conversion of organic matter to methane (Dennehy et al. 2016; Wickham et al. 2016; Xie et al. 2017; Zhao et al. 2014). Kim et al. (2003) found that during mesophilic and thermophilic anaerobic co-digestion of sewage sludge and kitchen waste, the optimal ratios of kitchen waste were 39.3% and 50.1%, respectively. Prabhu and Mutnuri (2016) noted that at the SRT of 21 days, anaerobic co-digestion of kitchen waste and sewage sludge at a 1:2 ratio yielded the maximum methane production of 823 mL/g VS_{added} (volume of gas produced per gram of added volatile solids), with an average methane content of 60%. Kuo-Dahab et al. (2014) confirmed that when the addition of kitchen waste accounted for 50% of the total feeding solids mass, methane production, organic matter degradation rate, and system stability were all improved. Dai et al. (2013) indicated that when anaerobic co-digestion of kitchen waste and sewage sludge, methane production and organic matter degradation rate increased linearly with the proportion of kitchen waste, primarily influenced by the C/N ratio of the sludge and kitchen waste. The optimal methane production was achieved when food waste accounted for 11% of the mass fraction, with a substrate C/N ratio of 15, while methane production decreased with further increasing the food waste content (Siddiqui et al. 2011).

2.2 *Salt*

High-salt content causes inhibition on anaerobic co-digestion of sewage sludge and food waste. Generally, NaCl is one of the main salt components in food waste, with mass fractions ranging from 0.3% to 1.4% (Li et al. 2016). In areas such as South Korea, due to the eating of high-salt foods of kimchi, fermented bean paste, and chili paste, the NaCl mass fraction in food waste can reach from 1.8% to 3.0% (Lim et al. 2012; Oh et al. 2008). For mesophilic anaerobic microorganisms, low concentrations of NaCl serve as a nutrient (McCarty and McKinney 1961) and can promote methane production by activating nicotinamide adenine dinucleotide for generating more adenosine triphosphate (Dimroth and Thomer 1989). Moreover, moderate NaCl concentrations can counteract the toxic effect of ammonia nitrogen on anaerobic microorganisms (Kugelman and McCarty 1965). Conversely, high NaCl concentrations cause microbial cell dehydration due to osmotic stress (Liao and Li 2015), with the cations in salt being the primary contributors to toxicity (McCarty and McKinney 1961). During anaerobic digestion of food waste, the accumulation of propionic acid is a significant cause of excessive acidification in anaerobic systems (Zhou et al. 2014), and Na^+ is more toxic to propionate utilizing microorganisms than acetate utilizing ones (Soto et al. 1993). Therefore, excessively high concentrations of Na^+ may lead to propionic acid accumulation, thereby affecting the stability of the anaerobic digestion system.

HSAD is generally more stable than CAD because it can acclimate microorganisms for better withstanding harsh conditions such as high ammonia nitrogen or VFA concentrations (Liao and Li 2015). When the Na^+ concentration reached to 3.4 g/L (corresponding to the NaCl concentration of 8.6 g/L), no inhibitory effect of Na^+ was observed in high solids anaerobic co-digestion of food waste and sewage sludge (Dai et al. 2013). All these suggest that a balanced nutrient of the mixed food waste and sewage sludge, combined with the inherent tolerance of acclimated microorganisms to high-salt environments, results in good digestion performance under high NaCl concentrations.

2.3 Grease

The content of grease in food waste varies greatly, which can have a significant impact on anaerobic digestion. During anaerobic digestion, the degradation rate of grease is lower than proteins and carbohydrates, although grease has a high methane production potential (Scherer et al. 2000). In addition, grease may cause prolonged lag phases during anaerobic digestion (Cirne et al. 2007). The hydrophobicity of grease prevents effective contact between encapsulated anaerobic microorganisms and soluble substances, thereby interrupting methanogenic conversion of VFAs into methane and leading to the accumulation of VFAs. After degradation, the produced long-chain fatty acids (LCFAs) from grease can encapsulate methanogens, bind to methanogen cell walls, and reduce methanogen activity (Hanaki et al. 1981; Silvestre et al. 2014). Increasing the grease content (in terms of chemical oxygen demand, COD) of lipid-rich waste from 18% to 31%–47% can cause severe inhibition of anaerobic digestion (Cirne et al. 2007). Furthermore, high-grease waste may not only inhibit microorganisms but also hinder substrate mass transfer, causing sludge flotation, and leading to reactor foaming and pump blockages (Long et al. 2012).

2.4 Temperature

Microbial populations have their optimal temperature ranges for growth, such as 37–39 °C for *Methanobrevibacter spp.* and 65–70 °C for thermophilic methanogens. Within their respective optimal temperature ranges, microbial growth rates reach their maximum, corresponding to peak gas production. However, enzymes involved in microbial metabolism are sensitive to temperature changes. Inappropriate temperature can affect enzyme activity and further causes the inhibition of microbial growth and organic matter degradation. Particularly, enzymes may undergo irreversible changes and become completely deactivated at a high temperature. Temperature also influences physical and chemical parameters and biochemical reactions, including the solubility of gas compounds, viscosity of liquids and solids, pH, Gibbs free energy, etc. Lower temperatures result in higher viscosity of liquids or solids. The solubility of gas compounds such as CH_4, H_2S, and CO_2 increases with decreasing temperature. Additionally, with increasing temperature, the Gibbs free energy of some essential metabolic reactions will decrease, and the reactions will become more favorable.

Methanogens usually have strict temperature requirements for growth. Therefore, short-term fluctuations in reactor temperature can have a significant impact on the fermentation process. It is generally believed that daily temperature fluctuations in anaerobic digestion systems should not exceed 2–3 °C. Thermophilic methanogens are more sensitive to temperature fluctuations, and a temperature fluctuation of 3 °C during high-temperature fermentation may cause changes in the anaerobic fermentation process. Furthermore, the sensitivity of anaerobic digestion to temperature increases with increasing organic load rates. Currently, in anaerobic digestion processes, mesophilic processes operating at 30–40 °C are more common, with optimal temperatures between 35 and 40 °C, while thermophilic processes are typically operated between 50 and 60 °C, with optimal temperatures between 50 and 55 °C, because different temperature ranges can acclimate different types of dominant microbial populations. However, when designing co-digestion processes, energy efficiency should also be considered for thermophilic or mesophilic anaerobic digestion systems.

2.5 pH

pH determines the stability of anaerobic digestion processes and has a significant impact on anaerobic co-digestion processes. For acidogenic fermentation, acid-producing bacteria can tolerate a wide range pH, generally from 4.0 to 8.0. The influence of pH on acid production mainly manifests in two aspects. On the one hand, pH affects the activity of hydrolytic enzymes during acid production, with a low pH typically inhibiting the hydrolysis of macromolecular organic matter. On the other hand, pH affects the formation of metabolic products, with low pH promoting the formation of free carboxylic acids which can easily pass through the cell membrane and inhibit microbial metabolism. After entering the cell, H^+ will be released to lower the intracellular pH and causes the inhibition of metabolic processes (Elbeshbishy et al. 2017). The methane production stage requires a relatively stable pH range, and the appropriate pH range for single-phase anaerobic reactors is generally 6.8–8.0. Methanogenic archaea are sensitive to changes in pH, with the optimal pH range being 6.8–7.2. pH values above 8.0 or below 6.0 will inhibit their growth, metabolism, and reproduction, thereby adversely affecting the entire anaerobic digestion process. The performance of anaerobic co-digestion systems largely depends on their buffering capacity, which can be achieved by adjusting the ratio of sewage sludge to food waste.

2.6 Rheological Behavior of Sewage Sludge with High Solids Content

The rheological behavior of the low concentration sludge was akin to water. However, under high solids conditions, these characteristics significantly diverged. Commonly, the rheological properties of sludge at different concentrations are analyzed using upward and downward flow curves. If the measured viscosity remains constant, the fluid is classified as a Newtonian fluid. If the upward and downward curves overlap, the fluid is identified as a pseudoplastic fluid. If the downward curve is below the upward curve, forming a hysteresis loop, the fluid exhibits time-dependent viscosity and is classified as a thixotropic fluid. Conversely, if the downward curve is above the upward curve, the fluid is characterized as a rheopectic (anti-thixotropic) fluid.

With increasing solid contents, sludge transitions from behaving nearly liked a Newtonian fluid to a non-thixotropic pseudoplastic fluid. Beyond 8%–12% solid contents, it became a thixotropic pseudoplastic fluid, while above 15%, it exhibited viscoelastic properties (Cheng and Li 2015). The viscosity escalated with higher solid contents (Fig. 1), following an exponential relationship:

$$\mu = e^{(a \cdot \text{TS}+b)} \cdot \gamma^{(c \cdot \text{TS}+d)} \tag{1}$$

where a, b, c, and d are constants, e is the Euler number, μ represents viscosity (mPa·s), TS is the total solids concentration of sludge (%), and γ denotes the shear rate (s^{-1}).

Generally, viscosity decreases with increasing temperatures. At solid contents of 1%–3%, the impact of temperature on viscosity was minimal. For solid contents of 6%, 8%, 10%, and 12%, viscosity diminishes by 36.54%, 49.55%, 53.97%, and

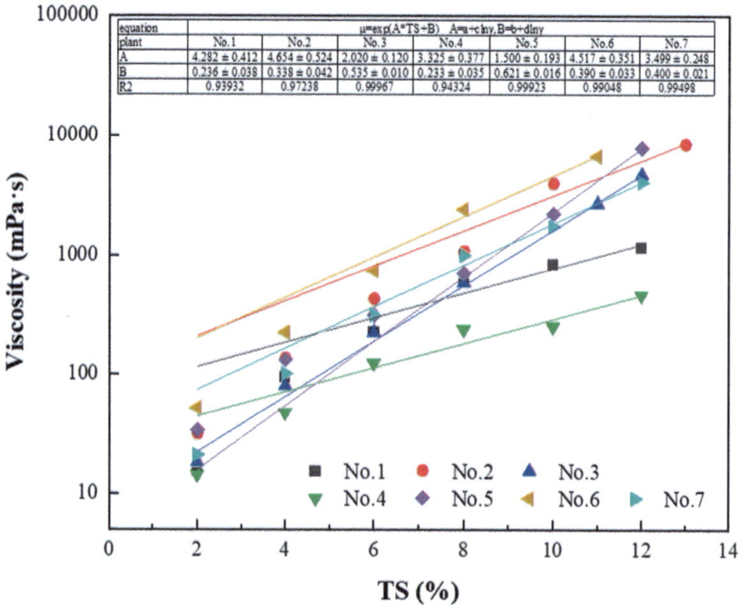

Fig. 1 Effect of sludge TS on viscosities at shear rate 15 s^{-1} (sludge samples were derived from seven wastewater treatment plants)

65.44%, respectively, when temperature was increased from 9 to 55 °C. This reduction was attributed to the disintegration of sludge particle structures and the deformation and hydrolysis of organic polymers at elevated temperatures, suggesting the possible benefit of thermal pretreatment before processing high solids content sludge. The viscosity was also influenced by the composition of organic matter, the content of extracellular polymers, and the residual flocculants. Notably, a sludge with 12% solid content experienced a 25-fold increase in viscosity when its organic content rose from 28% to 56%.

Changes in the rheological properties of sludge significantly affect its energy transfer capabilities. Changes in thermal conductivity of sewage sludge are shown in Fig. 2, and their thermal diffusivities are shown in Fig. 3. Sludge with less than 12%–15% solid content, resembling a suspension, can approximate its thermal conductivity to the value of water (with a mere 2%–3% relative standard deviation). However, at solid contents exceeding 12%–15%, the thermal constants of sludge notably decreased as approaching a viscoelastic state. Therefore, for anaerobic digestion systems with solid contents below 12%–15%, thermal constants of water could be applied. In contrast, high solids content sludge required the measured values.

The sludge concentration markedly impacts its liquid-phase diffusion coefficient. Increasing solid contents leads to a significant reduction in the water content, particularly free water within the sludge flocs. This, coupled with the capillary structure of high solids content sludge, affects internal porosity and curvature and elongates

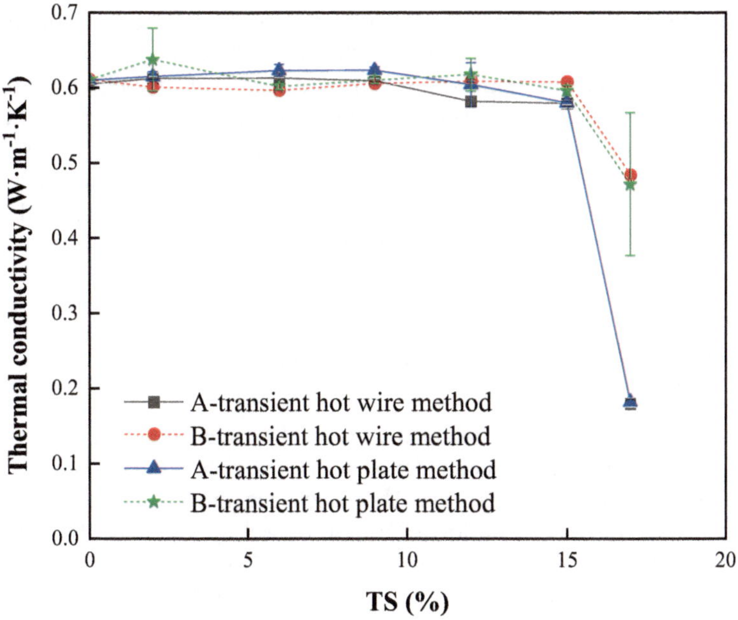

Fig. 2 Measurement of sludges (A and B) thermal conductivity using transient hot-wire method and transient hot-plate method

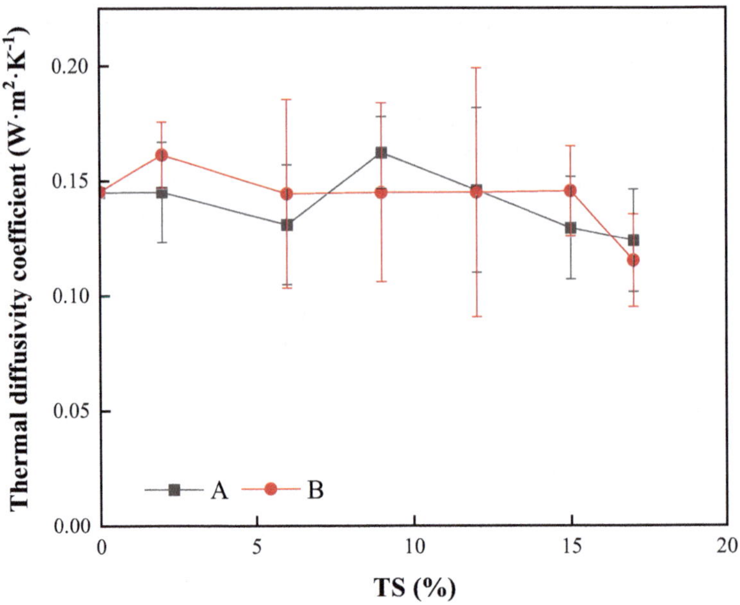

Fig. 3 Thermal diffusivity coefficients of two types of sludges (A and B) using the laser flash method

the solute diffusion path, thereby decelerating the diffusion process and diminishing diffusion capacity. The membrane pool method (with KI as the calibration solution) can be used to measure the liquid-phase diffusion coefficient of sludge with different solid contents. In dehydrated and digested sludge with a 15% solid content, the diffusion coefficient declined by 8–90 times compared to that in water (Fig. 4; Zhang et al. 2016). The decrease implied slower material transfer during the biochemical transformation of sludge, leading to potential uneven distribution, accumulation of intermediates, and inhibition of biochemical reactions.

2.7 Accumulation of Intermediate Products

Sewage sludge with high organic load rates exhibits relatively poor mass transfer, consequently leading to the accumulation of intermediate products such as VFAs, total ammonia nitrogen (TAN), and H_2, which can impact anaerobic digestion. Sewage sludge typically has a low C/N ratio (generally less than 10) and a relatively high protein content. After microbial degradation of proteins, the organic nitrogen is converted to ammonia nitrogen. If the accumulation of ammonia nitrogen exceeds a specific concentration, the activity of functional microbial communities can be inhibited (Heinrichs et al. 1990). Particularly, when the organic matter content of sewage sludge is high, uneven mass transfer occurs during anaerobic digestion,

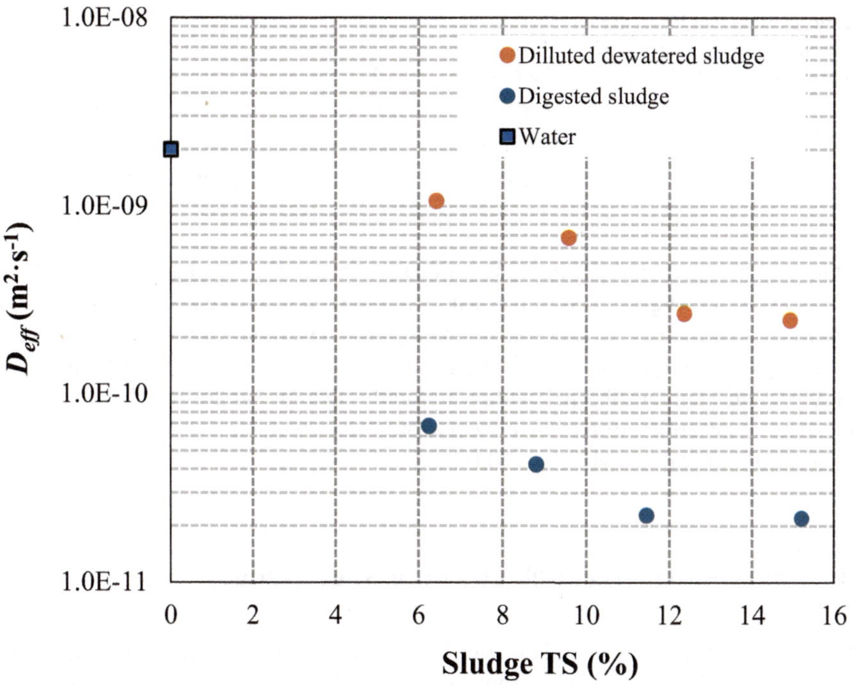

Fig. 4 Variation of iodine ion diffusion coefficient with sludge concentrations

with ammonia nitrogen emerging as the primary factor inhibiting anaerobic digestion. Duan et al. (2012) determined the ranges of TAN and free ammonia nitrogen (FAN) concentrations causing mild, moderate, and severe inhibition in mesophilic HSAD systems (Table 1). Vanvelsen (1979) found that methanogens adapted to high TAN concentrations (2420 mg/L) can immediately metabolize and produce CH_4 upon entering an environment with a TAN concentration of 3000 mg/L without a lag phase, which could be attributed to microbial adaptability.

Microorganisms in HSAD systems also exhibit robust tolerance to VFAs. Although the concentrations of acetate and total acid in HSAD systems are higher than the inhibition threshold of CAD systems, no evident inhibition has been observed in HSAD systems (Karim et al. 2005; Weiland 2006). Duan et al. (2012) summarized the inhibitory concentration of VFAs in HSAD systems (Table 2). In reality, VFAs,

Table 1 Range of ammonia nitrogen inhibiting anaerobic digestion of high solids content sludge

Inhibition level	FAN (mg/L)	TAN (mg/L)
None	< 250	< 2000
Mild	250–400	2000–4000
Moderate	400–600	3000–4000
Severe	600–800	3000–4000

Table 2 Range of VFAs that inhibit anaerobic digestion of high solids content sludge

Inhibition level	VFAs (mg/L)	VFAs/TAK	Stability level
None	< 400	0.02–0.04	Stable
Mild	400–800	0.03–0.12	Stable
Moderate	1000–3000	0.05–0.20	Inhibited
Severe	3000–4500	0.19–0.26	Collapsed

ammonia nitrogen, and pH interact with each other. VFAs can cause a decrease in pH of anaerobic digestion systems, while ammonia nitrogen leads to an increase in pH values. The carbon dioxide-bicarbonate pair in anaerobic digestion has a specific buffering capacity (Murto et al. 2004), and the higher total alkalinity (TAK) in HSAD systems can maintain the pH within a slightly alkaline range, which may be the reason that anaerobic digestion systems could remain stable even under high VFA concentrations (Sosnowski et al. 2008).

3 Pretreatment Technologies of Sewage Sludge

In sewage sludge, most organic matter is encapsulated within microbial cells by the cell wall, which undergoes slow rupture and dissolution during anaerobic digestion. It is generally considered that the hydrolysis-acidogenesis process is the rate-limiting step during anaerobic digestion, resulting in prolonged retention times, larger reactor volumes, and lower organic matter degradation rates, particularly affecting anaerobic digestion of high solids content sludge. To improve organic matter utilization in sludge, pretreatment of the sludge is required to enhance its anaerobic digestion.

3.1 Thermal Pretreatment Technology

Thermal treatment (e.g., 60–270 °C) of sewage sludge promotes solid dissolution and floc fragmentation, as well as organic matter hydrolysis. At temperatures between 45 and 65 °C, cell membrane is ruptured and rRNA is degraded; at temperatures between 65 and 90 °C, cell walls disrupt; and at temperatures between 70 and 95 °C, protein denaturation occurs (McCarty and McKinney 1961). Thermal pretreatment also causes the dissolution of extracellular polymeric substances (EPS) from sludge flocs into the liquid phase. Therefore, thermal treatment of high solids content sludge not only accelerates anaerobic digestion rates but also enhances the degradation of organic matter in sewage sludge.

Within a specific temperature range, a high-temperature results in more effective thermal hydrolysis. Usually, when thermal treatment temperatures exceed 100 °C, the effect of temperature is more pronounced than reaction duration (Valo et al. 2004). However, when temperatures exceed 180 °C, intermediate products of hydrolysis (aldehyde groups of reducing sugars and amino groups of amino acids) undergo rapid browning reactions, known as the Maillard reaction, producing recalcitrant substances that reduce the biodegradability of organic matter in thermally treated sludge, thereby weakening subsequent anaerobic digestion. In addition, the properties (such as organic matter composition and concentration) of the sludge have a significant effect on the efficiency of thermal treatment and anaerobic digestion (Gavala et al. 2003). Appels et al. (2010) found that pretreatment of sludge at 80 and 90 °C significantly increased anaerobic digestion gas production, while treatment at 70 °C did not markedly increase gas production. In addition, the drawback of high-temperature thermal pretreatment is high energy consumption and complex requirement for equipment and operation, as water has a high heat capacity, and significant energy is required to heat sludge to the desired temperature.

High solids content sludge exhibits high viscosity, poor flowability, and difficulty mixing. Theoretically, thermal treatment can alter physical properties of high solids content sludge. After thermal treatment, part of solid materials dissolves and the solids content decreases. Moreover, due to disintegration and fragmentation of sludge floc structures, EPS is hydrolyzed and the particle size of sludge changes. Therefore, sludge viscosity decreases after thermal treatment, improving mixing efficiency and reducing mixing energy consumption in anaerobic digestion systems. When the pretreatment temperature increases from 20 to 150 °C, sludge viscosity decreases; while when the pretreatment temperature exceeds 150 °C, sludge viscosity remains almost unchanged (Bougrier et al. 2008).

3.2 Other Pretreatment Technologies

Various chemical and physical pretreatment methods, such as alkali pretreatment, acid-alkali combined pretreatment, ozone pretreatment, and ultrasound and microwave pretreatment methods, can also effectively enhance the co-digestion performance of sewage sludge and food waste. These methods show promising in improving the utilization efficiency of sludge organic matter in anaerobic digestion. However, these methods also increase processing costs due to higher energy consumption and additional chemical additives.

Li et al. (2012) found that the optimal NaOH concentration for sludge alkali pretreatment was 0.1 mol/L, achieving a methane yield of 0.65 L/g VSS and an 8% increase in organic matter degradation rate compared to the control group. Ozone pretreatment is also one of the pretreatment methods, and ozone pretreatment was found to double the gas production during anaerobic digestion of sewage sludge (Bougrier et al. 2006; Weemaes et al. 2000). However, the high cost and difficulty in controlling dosage of ozone have led to its relatively fewer application. Apul and Sanin (2010) utilized an ultrasonic device (24 kHz, 400 W) to treat sludge and found a 50-fold increase in soluble chemical oxygen demand (SCOD) content, indicating that ultrasound facilitates the transformation of sludge organic matter from particulate to soluble form. Similarly, Tiehm et al. (2001) found that ultrasound treatment effectively shortened the time required for subsequent anaerobic digestion while increasing the volatile solids removal rate, demonstrating that microwave treatment promoted the hydrolysis-acidogenesis process of organic matter and enhanced gas production efficiency.

4 Technology for Anaerobic Co-digestion of Sewage Sludge and Food Waste

4.1 The Proposed Novel Treatment Process

Although co-digesting sewage sludge and food waste can adjust the C/N ratio of the materials and improve digestion efficiency, it still fails to address several key issues. These include the low utilization rate of organic matter in food waste, slow sludge hydrolysis, interference from refractory organic substances such as humic acids, and the instability of high-loaded anaerobic digestion. These challenges hinder the widespread adoption and implementation of anaerobic digestion, thereby restricting the resource recovery process of organic solid waste. Besides sewage sludge and food waste, other urban organic wastes, such as dead poultry and livestock, as well as manure residues, also require proper disposal. However, constructing separate treatment facilities for different types of waste presents issues such as large total land area requirements and high costs for residual waste liquid treatment. In tackling these issues, a novel co-digestion methodology was developed, merging hydrolysis with a two-phase anaerobic digestion process, as shown in Fig. 5.

The utilization of high-temperature hydrothermal treatment is key to disintegrating sludge particles and solubilizing the organic solid phase for decomposition. The resultant hydrothermal filtrate, rich in easily degradable small molecular organic acids and separated from solids, is a prime candidate for methane generation. This process effectively removes the rate-limiting step of sludge hydrolysis in anaerobic digestion. Concurrently, the dewaterability of sludge is considerably enhanced, and the moisture content of the resulting cake is reduced to 50%.

For food waste, the material undergoes a purification process to remove oils and debris, followed by controlled heterotypic lactic acid fermentation. This converts the organic matter into a liquefied state, predominantly yielding fermentation products like lactic and acetic acids. Co-digesting these products with the hydrothermal filtrate could avoid the issue of rapid acidification that often plagues standalone food waste digestion, preventing system failure. In this approach, refractory substances such as humic acids from the sludge predominantly stay within the solid-phase post-hydrothermal treatment and solid-liquid separation. This leaves the liquid-phase humic substances with markedly improved molecular weight and biodegradability, thus mitigating their inhibitory impact on anaerobic digestion. When the

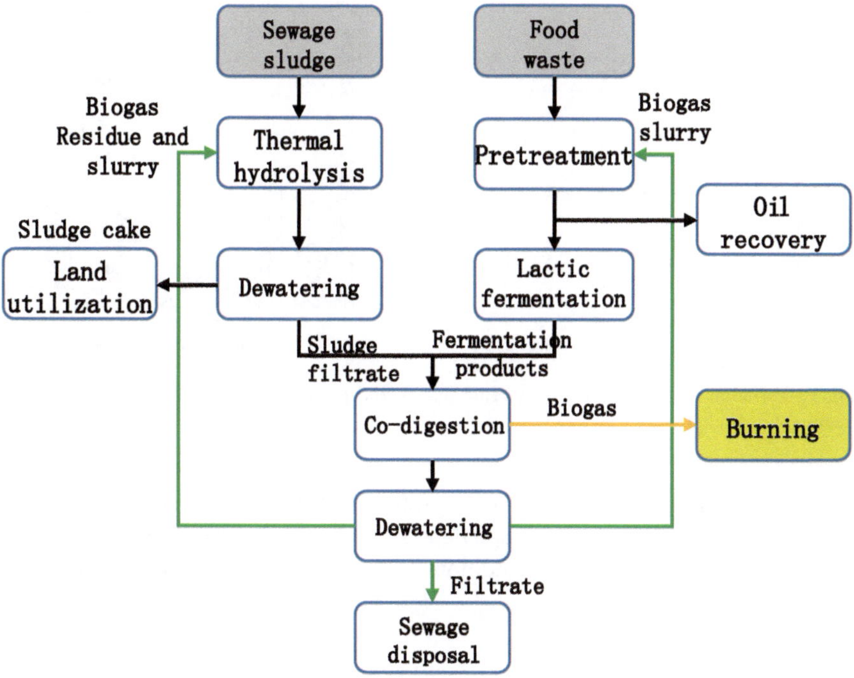

Fig. 5 Efficient anaerobic co-digestion of municipal sludge and food waste through qualitative and phase separation

hydrothermal filtrate blends with the fermentation liquid from food waste, the anaerobic digestion process predominantly involves soluble, easily digestible small molecular organics. This mix, characterized by a low solid content, ensures high efficiency in heat and mass transfer within the system. Additionally, the balanced C/N ratio and trace elements in the digestion substrate contribute to the high processing efficiency and stable operation of the anaerobic digestion system.

4.2 Hydrothermal Treatment of Sludge

4.2.1 Overall Performance

The purpose of hydrothermal treatment is to promote the solubilization of organic matter and improve dewatering performance, achieving the separation of easily degradable organic matter and inert components. Hydrothermal treatment at 160–180 °C could cause a large amount of organic matter to dissolve, with the SCOD being more than twice that of the 100–120 °C hydrothermal treatment (Fig. 6). After hydrothermal treatment, the dewatering performance of the sludge was well, and the

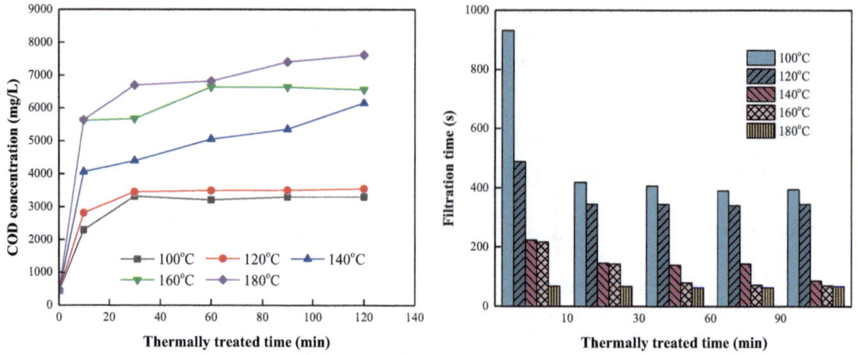

Fig. 6 Variation of the leaching of organic matter and the rate of filtration after hydrothermal treatment of sludge

dewatering performance at 180 °C was significantly better than at 160 °C. The impact of hydrothermal treatment on sludge dewatering is mainly concentrated within the first 30 min, so the appropriate conditions for hydrothermal treatment of sludge are 180 °C for 30 min.

4.2.2 Humic Acids' Changes in the Sludge

Humic acids, mainly comprising fulvic acid and humic acid, accounted for about 6–20% of the organic components of sludge. They are difficult to biodegrade and can affect the degradation of other organics during anaerobic digestion, which was one of the reasons for the low degradation rate of organic matter in anaerobic digestion. Before hydrothermal treatment, most of the humic acids in the sludge were in the solid phase, while after hydrothermal treatment at 180 °C for 30 min, the total amount of humic acids did not change significantly. However, some of the solid-phase humic acids dissolved, with 32% of the humic acids eventually distributed in the liquid phase and 68% remaining in the solid phase (Table 3). At the same time, 35% of the humic acid in the sludge was converted to fulvic acid, indicating that some humic acids decomposed during high-temperature hydrothermal treatment, and fulvic acid was more biodegradable than humic acid. Carbon and oxygen were the main elements in sludge humic acids (Table 4). The oxygen-to-carbon (O/C) ratio decreased after high-temperature hydrothermal treatment, indicating a reduction in oxygen-containing functional groups, and a decrease in the hydrogen-to-carbon (H/C) ratio indicated an increase in the aromaticity and unsaturation of humic acid and fulvic acid.

Table 3 Concentration of humic acid (HA) and fulvic acid (FA) in sludge before and after hydrothermal treatment

Items (mg/L)	HA in raw sludge	FA in raw sludge	HA in hydrothermal sludge	FA in hydrothermal sludge
Solid phase	2382.21 ± 14.95	361.93 ± 5.27	1089.60 ± 22.80	711.83 ± 18.80
Liquid phase	11.64 ± 1.10	1.16 ± 0.20	543.60 ± 49.40	416.90 ± 18.10
Total concentration	2393.85 ± 16.05	363.09 ± 5.47	1633.20 ± 72.20	1128.73 ± 36.90

Table 4 Elemental analysis of humic acid (HA) and fulvic acid (FA) before and after hydrothermal treatment

	Content/%						Atomic ratio		
	C	O	H	N	S	Ash	H/C	O/C	C/N
HA in raw sludge	54.20 ± 0.10	29.39 ± 0.74	8.01 ± 0.02	6.66 ± 0.02	1.22 ± 0.05	0.52 ± 0.02	1.77	0.41	9.49
HA in hydrothermal sludge	58.46 ± 0.12	26.33 ± 0.47	7.30 ± 0.21	6.04 ± 0.02	1.37 ± 0.02	0.50 ± 0.01	1.50	0.34	11.29
FA in raw sludge	44.66 ± 0.19	38.29 ± 0.85	7.22 ± 0.25	4.26 ± 0.04	1.07 ± 0.03	4.50 ± 0.05	1.94	0.64	12.23
FA in hydrothermal sludge	46.87 ± 0.20	37.71 ± 0.53	5.31 ± 0.23	4.73 ± 0.10	1.09 ± 0.03	4.30 ± 0.04	1.36	0.60	11.56

4.2.3 Filtrate Compositions

After hydrothermal treatment, there was little residue of proteins, lipids, and carbohydrates in the filtrate (Table 5), indicating that most of these organics dissolved and degraded into small molecule substances (such as VFAs); the dissolved humic acids had a small molecular weight, mainly below 3 kDa, consisting mostly of fulvic acid, which was more degradable. Therefore, after hydrothermal treatment and solid-liquid separation, the sludge not only yielded small molecular organic substances that were easily converted to methane but also removed most of the humic acids that inhibited hydrolysis and methane production. However, the C/N ratio of the hydrothermal filtrate was still low, so co-digestion with food waste fermentation liquid is still necessary.

Table 5 Properties of the leachate from high solids sludge after high-temperature hydrothermal treatment

Items	Concentration (mg/L)	Items	Concentration (mg/L)
pH	7.32	Protein	123.2 ± 4.3
Total organic carbon (TOC)	2265.2 ± 302.0	Lipid	400.0 ± 20.0
Soluble chemical oxygen demand (SCOD)	7790.0 ± 50.0	Carbohydrate	16.2 ± 0.5
Total nitrogen (TN)	603.3 ± 139.0	Nucleic acid	56.9 ± 4.7
Total ammonia nitrogen (TAN)	600.6 ± 21.1	PO_4^{3-}	3.0 ± 0.8
Total phosphorus (TP)	3.7 ± 0.5		

4.3 Fermentation of Food Waste

The anaerobic digestion of food waste is prone to acidification and system instability. By enriching *Lactobacillus* under mixed bacterial inoculation conditions, high-load lactic acid fermentation of food waste was achieved. At the pH of 3.2–4.2, the lactic acid content exceeded 80%, with food waste primarily converted into lactic acid by *Lactobacillus*, indicating a homolactic fermentation (Figs. 7 and 8; Feng et al. 2018, 2020). At a pH of 4.5, the acetic acid content was second only to lactic acid, and *Bifidobacterium* became the second dominant genus after *Lactobacillus*, capable of converting glucose into lactic and acetic acids, representing a heterolactic fermentation. Between pH 4.7 and 5.0, the system shifted to butyric acid fermentation, with butyric and acetic acids as the main products. At pH 4.7, *Bifidobacterium* and *Megasphaera* dominated, while at pH 5.0, the dominant genera included *Lactobacillus*, *Bifidobacterium*, and *Olsenella* related to lactic acid production, and *Caproiciproducens*, *Solobacterium*, and *Eubacteria nodatum* related to hydrogen production. Under a pH of 6.0, butyric, valeric, acetic, and propionic acids were the main products and distributed more evenly, indicating mixed acid fermentation, with *Prevotella* and *Megasphaera* as the dominant genera.

4.4 Anaerobic Co-digestion Performance of Hydrothermal Filtrate and Fermentation Effluent

During co-digestion, the mix of food waste fermentation liquor and raw sludge did not consistently exhibit synergistic effects across different ratios (Fig. 9; Feng et al. 2023). However, the co-digestion of hydrothermally treated sludge and hydrothermal filtrate with food waste lactic acid fermentation liquor consistently yielded methane volumes surpassing theoretical expectations (Fig. 10; Feng et al. 2023). Notably, the

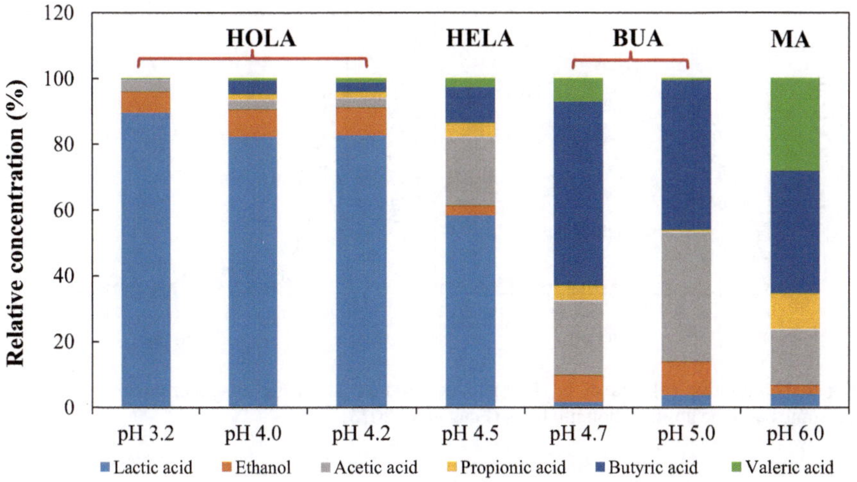

Fig. 7 Distribution of fermentation products for different fermentation types (HOLA: homolactic acid fermentation; HELA: heterolactic acid fermentation; BUA: butyric acid fermentation; MA: mixed acid fermentation) corresponding to different pH values (Feng et al. 2020)

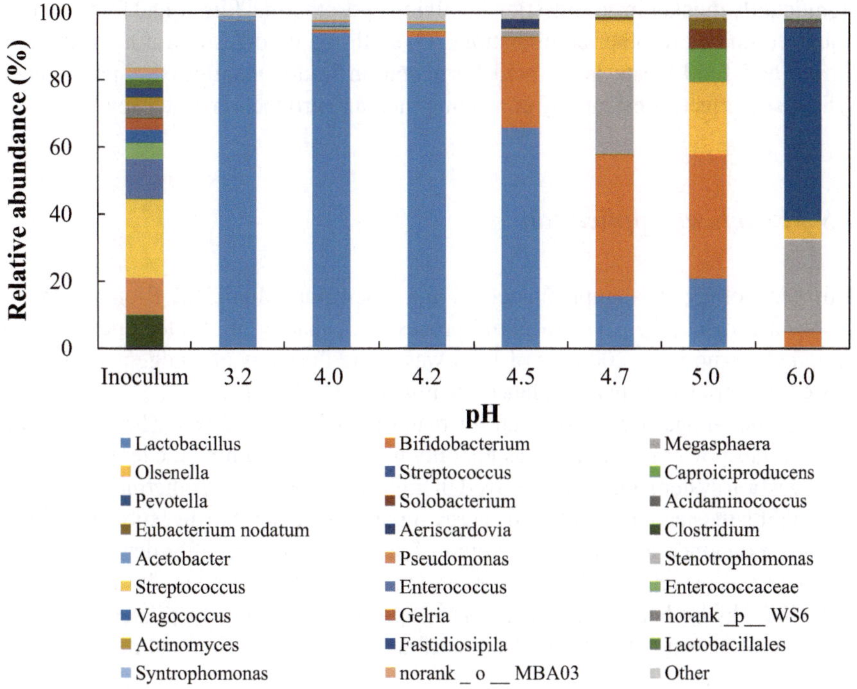

Fig. 8 Taxonomic classification of main bacterial genera (with relative abundance higher than 1% in the inoculum) during anaerobic fermentation of food waste under different pHs (Feng et al. 2018)

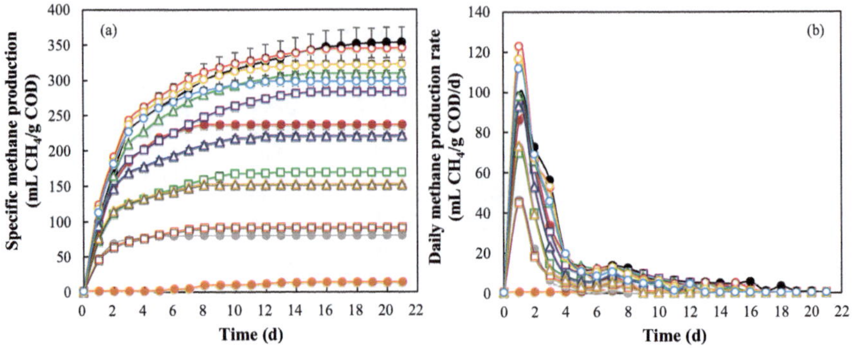

Fig. 9 Cumulative (**a**) and daily (**b**) methane production during all experiments for the sole substrates and the co-digestion mixtures (Feng et al. 2023)

ratio between the measured methane production and the theoretical methane production decreased from 10.8% to 3.7% as the proportion of hydrothermally treated sludge increased. Conversely, an increase in the hydrothermal filtrate ratio led to an increase in the above ratio from 5.2% to 11.7%. Furthermore, the methane production rate was significantly quicker in co-digestion involving hydrothermal filtrate and fermentation liquor compared to other combinations. Overall, the co-digestion of hydrothermal filtrate and food waste lactic acid fermentation liquor manifested a pronounced synergistic effect, substantially enhancing methane production efficiency.

4.5 Practical Application

This technology was implemented in the Shenzhen Municipal Biomass Waste Disposal Project (Fig. 11). An initial processing capacity of 500 tons per day was designed, comprising 200 tons of food waste and 300 tons of sludge per day. The process material and energy balance are illustrated in Fig. 12.

The system had been optimized for material and energy flows. The excess heat from biogas power generation was not only used for maintaining the temperature of the digestion system but also employed in high-temperature hydrothermal treatment, eliminating the need for additional energy input. Post-anaerobic digestion, the digestate still contained a significant amount of solids. Therefore, after dewatering, the digestate was recycled back into the hydrothermal treatment system for a secondary, intensified hydrothermal treatment. This approach not only facilitated enhanced recovery of organic matter but also contributed to the reduction and detoxification of the digestate.

The digestate from the secondary hydrothermal process, along with the hydrothermally treated sludge cake, was eventually transported for landfill or incineration. The filtrate was directed to the wastewater treatment system for final management. This

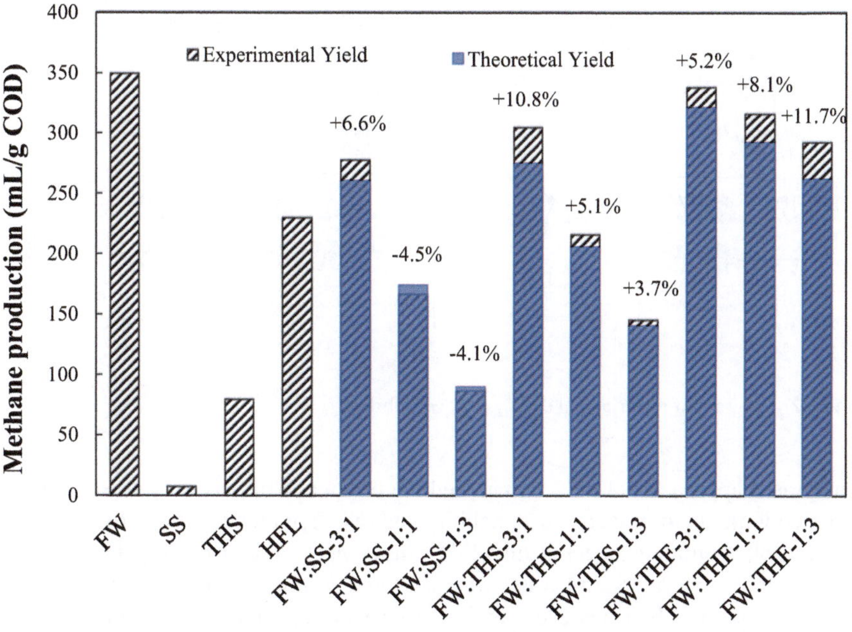

Fig. 10 Theoretical and experimental methane yields (Feng et al. 2023)

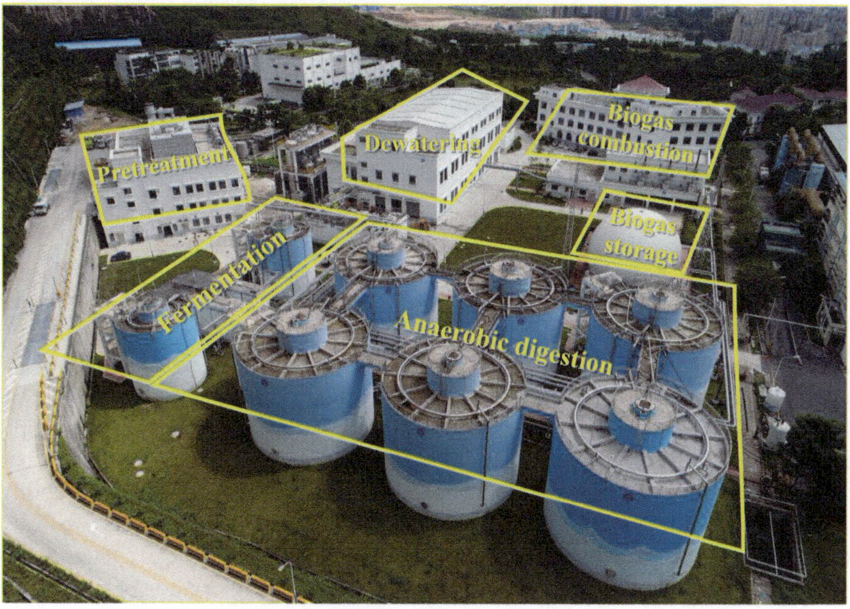

Fig. 11 Aerial view of the Shenzhen municipal biomass waste disposal project

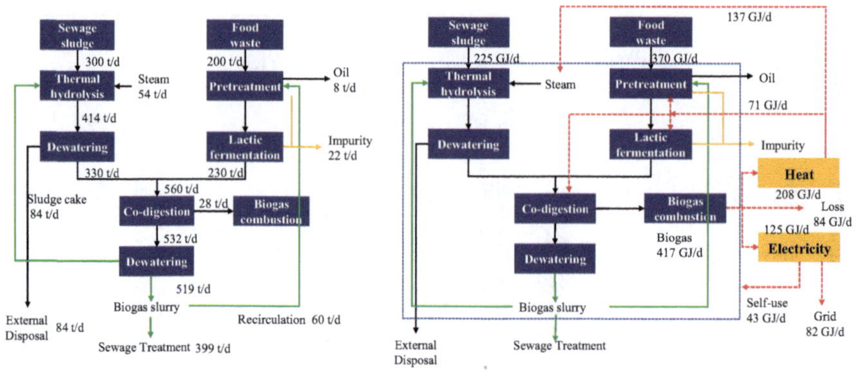

Fig. 12 Diagram of material and energy flow equilibrium

integrated approach not only maximized the efficiency of waste treatment but also aligned with environmental sustainability and resource recovery objectives.

Since its commencement in July 2018, this project has consistently demonstrated exceptional performance in waste treatment and resource recovery. The anaerobic digestion system had achieved an impressive organic matter conversion rate of approximately 80%, substantially outperforming the average rates of 35%–40% for conventional sludge anaerobic digestion and even surpassing the typical 65% efficiency for food waste digestion.

The project annually recycled over 150,000 tons of organic waste, converting it into valuable resources. This included the production of over 3,000 tons of industrial oils and fats. In terms of energy generation, the project produced more than 8 million kilowatt-hours of electricity for the grid. This operational efficiency and resource conversion have translated into substantial annual revenues.

5 Future Perspectives

Co-digestion technology can significantly enhance anaerobic digestion efficiency and save land, making it highly promotable in many urban areas with substantial development potential. Co-digestion facilities can become regional clean energy supply centers. However, co-digestion technology also faces several challenges. Sludge and food waste can be co-digested in various proportions, offering broad adaptability, but it is essential to consider the differences in the characteristics of food waste and sludge across different regions, particularly when some sewage sludge contains toxic elements. The co-digestion process and the storage and transportation of sludge may lead to methane leakage, which can significantly impact the environmental benefits of the system and thus require attention. The co-digestion of different materials will affect the properties of the digestate and digested slurry, necessitating an evaluation of the environmental risks associated with their subsequent treatment. In large cities, due to the lack of agricultural land for disposal, the digestate from co-digestion needs to be treated, which is costly. Therefore, it is crucial to find economic and green treatment or utilization technologies.

6 Take-Home Messages

(1) Anaerobic co-digestion of food waste with sewage sludge can effectively improve the organic content and C/N ratio of substrates, thereby enhancing biogas production and organic matter utilization.
(2) Under high solids conditions, the co-digestion of sewage sludge and food waste is influenced by various factors, including substrate addition ratio, salinity, grease content, temperature, pH, sludge rheological properties, and the accumulation of intermediate products.
(3) Various pretreatment methods, such as thermal, alkaline, ozone, and ultrasonic treatments, can effectively improve anaerobic co-digestion of sewage sludge and food waste.
(4) A novel case study of anaerobic co-digestion of sewage sludge and food waste has been proposed, which employs a "qualitative separation" method. This method effectively avoids the damage caused by recalcitrant organics, fully exploits the high efficiency of the liquid-phase anaerobic system, shortens the digestion cycle, and significantly improves the conversion rate of organic matter in the sludge.

References

Appels L, Degreve J, Van der Bruggen B, Van Impe J, Dewil R (2010) Influence of low temperature thermal pre-treatment on sludge solubilisation, heavy metal release and anaerobic digestion. Biores Technol 101(15):5743–5748

Apul OG, Sanin FD (2010) Ultrasonic pretreatment and subsequent anaerobic digestion under different operational conditions. Biores Technol 101(23):8984–8992

Bougrier C, Albasi C, Delgenes JP, Carrere H (2006) Effect of ultrasonic, thermal and ozone pre-treatments on waste activated sludge solubilisation and anaerobic biodegradability. Chem Eng Proc-Process Intensif 45(8):711–718

Bougrier C, Delgenes JP, Carrere H (2008) Effects of thermal treatments on five different waste activated sludge samples solubilisation, physical properties and anaerobic digestion. Chem Eng J 139(2):236–244

Cheng YC, Li H (2015) Rheological behavior of sewage sludge with high solid content. Water Sci Technol 71(11):1686–1693

Cirne DG, Paloumet X, Bjornsson L, Alves MM, Mattiasson B (2007) Anaerobic digestion of lipid-rich waste: effects of lipid concentration. Renew Energy 32(6):965–975

Dai X, Duan N, Dong B, Dai L (2013) High-solids anaerobic co-digestion of sewage sludge and food waste in comparison with mono digestions: stability and performance. Waste Manage 33(2):308–316

Dennehy C, Lawlor PG, Croize T, Jiang Y, Morrison L, Gardiner GE, Zhan X (2016) Synergism and effect of high initial volatile fatty acid concentrations during food waste and pig manure anaerobic co-digestion. Waste Manage 56:173–180

Dimroth P, Thomer A (1989) A primary respiratory Na^+ pump of an anaerobic bacterium: the Na^+-dependent NADH: quinone oxidoreductase of klebsiella pneumoniae. Arch Microbiol 151(5):439–444

Duan N, Dong B, Wu B, Dai X (2012) High-solid anaerobic digestion of sewage sludge under mesophilic conditions: feasibility study. Biores Technol 104:150–156

Elbeshbishy E, Dhar BR, Nakhla G, Lee HS (2017) A critical review on inhibition of dark biohydrogen fermentation. Renew Sustain Energy Rev 79:656–668

Feng K, Li H, Zheng CZ (2018) Shifting product spectrum by pH adjustment during long-term continuous anaerobic fermentation of food waste. Biores Technol 270:180–188

Feng K, Li H, Deng Z, Wang Q, Zhang YY, Zheng CZ (2020) Effect of pre-fermentation types on the potential of methane production and energy recovery from food waste. Renewable Energy 146:1588–1595

Feng K, Yang LX, Chen L, Jing Q, Li H, Li DB, Liu JY, Cai JB, Deng Z (2023) Enhanced methane production from the co-digestion of food waste and thermally hydrolyzed sludge filtrate. J Environ Manage 342:118169

Gavala HN, Yenal U, Skiadas IV, Westermann P, Ahring BK (2003) Mesophilic and thermophilic anaerobic digestion of primary and secondary sludge. Effect of pre-treatment at elevated temperature. Water Res 37(19):4561–4572

Hanaki K, Nagase M, Matsuo T (1981) Mechanism of inhibition caused by long-chain fatty-acids in anaerobic-digestion process. Biotechnol Bioeng 23(7):1591–1610

Heinrichs DM, Poggivaraldo HM, Oleszkiewicz JA (1990) Effects of ammonia on anaerobic-digestion of simple organic substrates. J Environ Eng-ASCE 116(4):698–710

Karim K, Hoffmann R, Klasson T, Al-Dahhan MH (2005) Anaerobic digestion of animal waste: waste strength versus impact of mixing. Biores Technol 96(16):1771–1781

Kim HW, Han SK, Shin HS (2003) The optimisation of food waste addition as a co-substrate in anaerobic digestion of sewage sludge. Waste Manage Res 21(6):515–526

Kugelman IJ, McCarty PL (1965) Cation toxicity and stimulation in anaerobic waste treatment. J (Water Pollution Cont Federat) 37(1):97–116

Kuo-Dahab WC, Amirhor P, Zona M, Duest D, Park C (2014) Investigating anaerobic co-digestion of sewage sludge and food waste using a bench-scale pilot study. Proc Water Environ Fed 9:1–23

Li H, Li C, Liu W, Zou S (2012) Optimized alkaline pretreatment of sludge before anaerobic digestion. Biores Technol 123:189–194

Li Y, Jin Y, Li J, Chen Y, Gong Y, Li Y, Zhang J (2016) Current situation and development of kitchen waste treatment in China. Procedia Environ Sci 31:40–49

Liao X, Li H (2015) Biogas production from low-organic-content sludge using a high-solids anaerobic digester with improved agitation. Appl Energy 148:252–259

Lim BS, Kim BC, Chung I (2012) Anaerobic treatment of food waste leachate for biogas production using a novel digestion system. Environ Eng Res 17(1):41–46

Long JH, Aziz TN, de los Reyes FL, III, Ducoste JJ (2012) Anaerobic co-digestion of fat, oil, and grease (FOG): a review of gas production and process limitations. Proc Safety Environ Protect 90(3):231–245

McCarty PL, McKinney RE (1961) Salt toxicity in anaerobic digestion. J Water Pollut Cont Federation 33(4):399–415

Murto M, Björnsson L, Mattiasson B (2004) Impact of food industrial waste on anaerobic co-digestion of sewage sludge and pig manure. J Environ Manage 70(2):101–107

Oh G, Zhang L, Jahng D (2008) Osmoprotectants enhance methane production from the anaerobic digestion of food wastes containing a high content of salt. J Chem Techn Biotech: Int Res Process, Environ Clean Tech 83(9):1204–1210

Prabhu MS, Mutnuri S (2016) Anaerobic co-digestion of sewage sludge and food waste. Waste Manage Res 34(4):307–315

Scherer P, Vollmer GR, Fakhouri T, Martensen S (2000) Development of a methanogenic process to degrade exhaustively the organic fraction of municipal "grey waste" under thermophilic and hyperthermophilic conditions. Water Sci Technol 41(3):83–91
Siddiqui Z, Horan NJ, Anaman K (2011) Optimisation of C:N ratio for co-digested processed industrial food waste and sewage sludge using the BMP test. Int J Chem React Eng 9:Article S4
Silvestre G, Illa J, Fernandez B, Bonmati A (2014) Thermophilic anaerobic co-digestion of sewage sludge with grease waste: effect of long chain fatty acids in the methane yield and its dewatering properties. Appl Energy 117:87–94
Sosnowski P, Klepacz-Smolka A, Kaczorek K, Ledakowicz S (2008) Kinetic investigations of methane co-fermentation of sewage sludge and organic fraction of municipal solid wastes. Biores Technol 99(13):5731–5737
Soto M, Mendez R, Lema JM (1993) Sodium inhibition and sulfate reduction in the anaerobic treatment of mussel processing wastewaters. J Chem Technol Biotechnol 58(1):1–7
Tiehm A, Nickel K, Zellhorn M, Neis U (2001) Ultrasonic waste activated sludge disintegration for improving anaerobic stabilization. Water Res 35(8):2003–2009
Valo A, Carrère H, Delgenès JP (2004) Thermal, chemical and thermo-chemical pre-treatment of waste activated sludge for anaerobic digestion. J Chem Technol Biotechnol 79(11):1197–1203
Vanvelsen AFM (1979) Adaptation of methanogenic sludge to high ammonia-nitrogen concentrations. Water Res 13(10):995–999
Weemaes M, Grootaerd H, Simoens F, Verstraete W (2000) Anaerobic digestion of ozonized biosolids. Water Res 34(8):2330–2336
Weiland P (2006) Biomass digestion in agriculture: a successful pathway for the energy production and waste treatment in Germany. Eng Life Sci 6(3):302–309
Wickham R, Galway B, Bustamante H, Nghiem LD (2016) Biomethane potential evaluation of co-digestion of sewage sludge and organic wastes. Int Biodeterior Biodegradation 113:3–8
Xie S, Wickham R, Nghiem LD (2017) Synergistic effect from anaerobic co-digestion of sewage sludge and organic wastes. Int Biodeterior Biodegradation 116:191–197
Zhang Y, Li H, Liu C, Cheng Y (2015) Influencing mechanism of high solid concentration on anaerobic mono-digestion of sewage sludge without agitation. Front Environ Sci Eng 9(6):1108–1116
Zhang Y, Li H, Cheng Y, Liu C (2016) Influence of solids concentration on diffusion behavior in sewage sludge and its digestate. Chem Eng Sci 152:674–677
Zhao M, Wang Y, Zhang C, Li S, Huang Z, Ruan W (2014) Synergistic and pretreatment effect on anaerobic co-digestion from rice straw and municipal sewage sludge. BioResources 9(4):5871–5882
Zhou Q, Shen F, Yuan H, Zou D, Liu Y, Zhu B, Jaffu M, Chufo A, Li X (2014) Minimizing asynchronism to improve the performances of anaerobic co-digestion of food waste and corn stover. Biores Technol 166:31–36

Printed by Printforce, the Netherlands